Advanced Microscopy

This book covers the principle, structure, enhancement of sensitivity and resolution power of photothermal and Raman microscopies. It includes real-world applications to biological and medical targets.

Advanced Microscopy: Photo-Thermal and Induced-Raman Microscopy introduces clear descriptions of various Raman processes such as spontaneous, stimulates, coherent anti-Stokes Raman (CARS), Raman loss and Stokes Raman (gain). It covers pump-probe microscopies using actinic (pump) laser and sensing (probe) laser resulting in improvement due to intrinsic nonlinearity, which provides an advantage in the imaging of nonfluorescent targets. The author also provides solutions to noise and sensitivity problems which are two of the most important concerns in the microscopy applications. Finally, the book also draws direct comparisons of the advantages and drawbacks of Raman microscopes in comparison with photothermal microscopes.

The book will be useful to researchers and non-specialists in biomedical fields using optics and electronics relevant to (optical) microscopes. It will also be a helpful resource to graduate students in the fields of biology and medical research who are using photothermal microscopes in their research.

Advanced Microscopy

Photo-Thermal and Induced-Raman Microscopy

Takayoshi Kobayashi

CRC Press
Taylor & Francis Group
Boca Raton London New York

CRC Press is an imprint of the
Taylor & Francis Group, an **informa** business

Designed cover image: Takayoshi Kobayashi

First edition published 2025
by CRC Press
2385 NW Executive Center Drive, Suite 320, Boca Raton FL 33431

and by CRC Press
4 Park Square, Milton Park, Abingdon, Oxon, OX14 4RN

CRC Press is an imprint of Taylor & Francis Group, LLC

ISBN: 978-1-032-75885-5 (hbk)
ISBN: 978-1-032-75886-2 (pbk)
ISBN: 978-1-003-47603-0 (ebk)

DOI: 10.1201/9781003476030

Typeset in Times
by Deanta Global Publishing Services, Chennai, India

Contents

Preface

The contents of this book are composed of research for more than five years in the University of Electro-Communications with many collaborators in the Center for Neurosciences and Biomedical Engineering, the University of Electro-Communications, National Chiao-Tung Brain Science, Advanced Ultrafast Laser Center, of University of National Yng-Ming University, Tokyo University of Science. They have conducted photothermal microscopy and Raman microscopy, both of which are classified in pump-probe microscopy. The target materials discussed are semiconductor quantum dots, gold nanoparticles, endogenous brain from mouse, mouse brain tissues, malignant melanoma and benign melanoma samples from mouse skin, ovary and intestine from rabbit. I hope this book is useful for those young people who are interested in this subject area. I paid special attention to providing basic information explaining technical terms relevant to physiology and pathology in the contents.

The contents include principles of optical microscopes, especially pump-probe photothermal microscopes and Raman microscopes in terms of spatial resolution, sensitivity, image acquisition speed, parallel data acquisition by multi-color imaging, and the theory and methodology of image data analysis. The contents of this book were partially published by the following publishing companies and associations. The author would like to express his thanks to them all for their support at the time of their publications. They include the following:

Optics Express, Optica, the *Journal of Optical Society of America,* the *Journal of Biomedical Optics* (Optica Publishing Group)
Optics Communications and *Journal of Optics* (Elsevier)
Optical Review (Springer)
Advances by American Institute of Physics (AIP)
Science Reports (Nature Portfolio)
Journal of Biomedical Optics [SPIE (Society of Photo-Optical Instrumentation Engineers)]
Journal of Physics D [Institute of Physics (IOP)]
Journal of Raman Spectroscopy (John Wiley & Sons)

For all of the above I am sharing copyrights with the publication organizations and/or publication companies, further request is not needed from them.

Special thanks to my wife Yoshino for her continuing support to my long research life.

About the Author

Takayoshi Kobayashi was born in Niigata Prefecture in 1944. He is a professor emeritus at the University of Tokyo, from which he previously graduated with Bachelor's, Master's, and Doctorate degrees. He joined the Institute of Physical and Chemical Research (Riken), and between 1977 and 1979, he was a temporary member of the Technical Staff at Bell Laboratories. In 1980 he joined the Department of Physics at the University of Tokyo as an associate professor and was promoted to full professor in 1994. In March 2006 he retired from the university and moved to the Department of Applied Physics and Chemistry at the University of Electro-Communications, in Tokyo. He was appointed Chair Professor and the Director of the National Chiao-tung University in 2006, which became the National Yang-Ming Chiao-Tung University in 2021. His research interests include quantum electronics, laser physics, femtosecond spectroscopy, ultrafast nonlinear optics, quantum optics, chemical physics, quantum information science and technology.

1 Introduction

We are describing very briefly several types of microscopes on principles, their specific features, and applications in this Introduction.

The optical microscope, also referred to as a light microscope, is a type of microscope that commonly uses visible light and a system of lenses to generate magnified images of small objects. Optical microscopes are the oldest design of microscope and were possibly invented in their present compound form in the 17th century. Basic optical microscopes can be very simple, although many complex designs aim to improve resolution and sample contrast.

I OPTICAL MICROSCOPE IN GENERAL

1. Stereo microscope: A microscope used to observe objects as they are at low magnifications of about 2 to 80 times. This microscope is also called a binocular stereomicroscope. Binoculars or field glasses are two refracting telescopes mounted side-by-side and aligned to point in the same direction, allowing the viewer to use both eyes (binocular vision) when viewing distant objects. In addition to similar observations, it is possible also to observe the subject 360 degrees from the side without touching it.
2. Conventional (most commonly used) optical microscope, also referred to as a light microscope, is a type of microscope that commonly uses visible light and a system of lenses to generate magnified images of small objects. Optical microscopes are the oldest design of microscope and were possibly invented in their present compound form in the 17th century. Basic optical microscopes can be very simple, although many complex designs aim to improve resolution and sample contrast.

The object is placed on a stage and may be directly viewed through one or two eyepieces on the microscope. In high-power microscopes, both eyepieces typically show the same image, but with a stereo microscope, slightly different images are used to create a 3-D effect. A camera is typically used to capture the image (micrograph).

3. Polarized light microscope: irradiates a subject with polarized light and observes the subject's polarization characteristics as changes in brightness or color. Ideal for observing minerals and crystals. Polarized light observation is possible just by attaching an adapter to the lens.
4. Biological microscope: Observe objects that transmit light, such as thin sections of living tissues, cells, and bacteria. There is a light underneath the subject and it is observed through the light. Easily switches between transmission and epi-illumination.

DOI: 10.1201/9781003476030-1

5. Metallurgical microscope: A microscope for magnifying observation of metal structures using bright field illumination at medium to high magnifications of about 100 to 1000 times. Similar observations can be made by simply switching the lens illumination.

II CLASSIFICATION OF MICROSCOPE BY OPTICAL CONFIGURATION

1. Simple microscope

A simple microscope uses a lens or set of lenses to enlarge an object through angular magnification alone, giving the viewer an erect enlarged virtual image. Use of a single convex lens or groups of lenses are found in simple magnification devices such as the magnifying glass, loupes, and eyepieces for telescopes and microscopes.

A range of objective lenses with different magnification are usually provided mounted on a turret, allowing them to be rotated into place and providing an ability to zoom-in. The maximum magnification power of optical microscopes is typically limited to around 1000x because of the limited resolving power of visible light. While larger magnifications are possible, no additional details of the object are resolved.

Alternatives to optical microscopy which do not use visible light include scanning electron microscopy, transmission electron microscopy, and scanning-probe microscopy and as a result, can achieve much greater magnifications.

2. Compound microscope

A compound microscope uses a lens close to the object being viewed to collect light (called the objective lens), which focuses a real image of the object inside the microscope (image 1). That image is then magnified by a second lens or group of lenses (called the eyepiece) that gives the viewer an enlarged inverted virtual image of the object (image 2). The use of a compound objective/eyepiece combination allows for much higher magnification. Common compound microscopes often feature exchangeable objective lenses, allowing the user to quickly adjust the magnification. A compound microscope also enables more advanced illumination setups, such as phase contrast.

Other microscope variants

There are many variants of the compound optical microscope design for specialized purposes. Some of these are physical design differences allowing specialization for certain purposes:

3. Stereo microscope, a low-powered microscope which provides a stereoscopic view of the sample, commonly used for dissection.

4. Comparison microscope has two separate light paths allowing direct comparison of two samples via one image in each eye.
5. Inverted microscope, for studying samples from below; useful for cell cultures in liquid or for metallography.
6. Fiber optic connector inspection microscope, designed for connector endface inspection.
7. Traveling microscope for measuring length along the light propagation direction with a resolution typically in the order of 0.01mm.

Other microscope variants are designed for different illumination techniques as follows:

8. Petrographic microscope, whose design usually includes a polarizing filter, rotating stage, and gypsum plate to facilitate the study of minerals or other crystalline materials whose optical properties can vary with orientation.
9. Polarizing microscope, similar to the petrographic microscope.
10. Phase-contrast microscope, which applies the phase-contrast illumination method.
11. Epifluorescence microscope, designed for analysis of samples that include fluorophores.
12. Confocal microscope, a widely used variant of epifluorescent illumination that uses a scanning laser to illuminate a sample for fluorescence.
13. Two-photon microscope, used to image fluorescence deeper in scattering media and reduce photobleaching, especially in living samples.
14. Ultramicroscope, an adapted light microscope that uses light scattering to allow viewing of tiny particles whose diameter is below or near the wavelength of visible light (around 500 nanometers); mostly obsolete since the advent of electron microscopes
15. Tip-enhanced Raman microscope, is a variant of optical microscope based on tip-enhanced Raman spectroscopy (TERS), without traditional wavelength-based resolution limits. This microscope is primarily realized on the scanning-probe microscope platforms using all optical tools.
16. Differential Interference Microscope: The difference in the refractive index of substances inside cells is used for magnification observation of living cells without staining. Similar observations can be made simply by attaching a differential interference adapter to the lens.

II CLASSIFICATION OF OPTICAL MICROSCOPE BY IMAGE INFORMATION CARRYING MEDIA AND TARGET OBJECTS

1. Digital microscope

A digital microscope is a microscope equipped with a digital camera allowing observation of a sample via a computer. Microscopes can also be partly or wholly computer-controlled with various levels of automation. Digital microscopy allows greater analysis of a microscope image, for example, measurements of distances and areas and quantitation of a fluorescent or histological stain.

Digital microscopy with very low light levels to avoid damage to vulnerable biological samples is available using sensitive photon-counting digital cameras. It has been demonstrated that a light source providing pairs of entangled photons may minimize the risk of damage to most of light-sensitive samples. In this application of ghost imaging to photon-sparse microscopy, the sample is illuminated with infrared photons, each spatially correlated with an entangled partner in the visible band for efficient imaging by a photon-counting camera.

2. Low-powered digital microscopes, USB microscopes, are also commercially available. These are essentially webcams with a high-powered macro lens and generally do not use transillumination. The camera is attached directly to a computer's USB port to show the images directly on the monitor. They offer modest magnifications (up to about 200×) without the need to use eyepieces and at a very low cost. High-power illumination is usually provided by an LED source or sources adjacent to the camera lens.
3. Polarized light microscope: irradiates a subject with polarized light and observes the subject's polarization characteristics as changes in brightness or color. Ideal for observing minerals and crystals. Polarized light observation is possible just by attaching an adapter to the lens.
4. A miniature USB microscope.

Low-powered digital microscopes, USB microscopes, are also commercially available. These are essentially webcams with a high-powered macro lens and generally do not use transillumination. The camera is attached directly to a computer's USB port to show the images directly on the monitor. They offer modest magnifications (up to about 200×) without the need to use eyepieces and at a very low cost. High-power illumination is usually provided by an LED source or sources adjacent to the camera lens.

Digital microscope is a microscope equipped with a digital camera allowing observation of a sample via a computer. Microscopes can also be partly or wholly computer-controlled with various levels of automation. Digital microscopy allows greater analysis of a microscope image, for example, measurements of distances and areas and quantitation of a fluorescent or histological stain.

Digital microscopy with very low light levels to avoid damage to vulnerable biological samples is available using sensitive photon-counting digital cameras. It has been demonstrated that a light source providing pairs of entangled photons may minimize the risk of damage to the most light-sensitive samples. In this application of ghost imaging to photon-sparse microscopy, the sample is illuminated with infrared photons, each spatially correlated with an entangled partner in the visible band for efficient imaging by a photon-counting camera.

5. Configurations of a camera and object

5.1 Viewing: The object is placed on a stage and may be directly viewed through one or two eyepieces on the microscope. In high-power microscopes, both eyepieces typically show the same image, but with a stereo microscope, slightly different images are used to create a 3-D effect. A camera is typically used to capture the image (micrograph).

5.2 Lighting: The sample can be lit in a variety of ways. Transparent objects can be lit from below and solid objects can be lit with light coming through (bright field) or around (dark field) the objective lens. Polarized light may be used to determine crystal orientation of metallic objects. Phase-contrast imaging can be used to increase image contrast by highlighting small details of differing refractive index.

5.3 Biological microscope: Observes objects that transmit light, such as thin sections of living tissues, cells, and bacteria. There is a light underneath the subject and it is observed through the light. Easily switches between transmission and epi-illumination.

III OTHER IRRADIATION SOURCES

1. Neutron microscope

Neutrons have no electric charge, enabling them to penetrate substances to gain information about structure that is not accessible through other forms of microscopy. Neutrons interact with atomic nuclei via the strong force. This interaction can scatter neutrons from their original path and can also absorb them. Thus, a neutron beam becomes progressively less intense as it moves deeper within a substance. In this way, neutrons are analogous to x-rays for studying object interiors. The behavior of samples composed of light elements can be observed.

After irradiating a neutron beam, the neutrons pass through the sample and enter a neutron-reactive material such as boron, lithium, or gadolinium on the opposite side of the irradiated surface, producing alpha rays, beta rays, gamma rays, and internal conversion electrons. There is also a method of emitting light from a fluorophore, amplifying it with a microchannel plate, and imaging it with a charge-coupled device (CCD). In order to improve resolution, it is necessary to thin the layer of neutron-reactive materials, and for this purpose boron-10, lithium-6, and gadolinium-157, which are highly reactive with neutrons, are used, but at the same time chemically stable materials are used. Since they also need to be compounds, they are used as stable compounds such as boron carbide, lithium fluoride, and gadolinium oxide. Boron, lithium, and gadolinium have different reaction cross sections and also vary depending on the energy of the neutron, so each can be used appropriately. Boron and lithium generate alpha rays that cause the phosphor to emit light, but gadolinium generates gamma rays that do not contribute to the phosphor's emission and internal conversion electrons that do. There is a suitable thickness for each phosphor layer, but in order to improve resolution, the phosphor layer needs to be made as thin as the reaction layer, but sensitivity is reduced, so there is a trade-off.

2. X-ray microscope

An X-ray microscope uses electromagnetic radiation in the X-ray band to produce magnified images of objects. Since X-rays penetrate most objects, there is no need to specially prepare them for X-ray microscopy observations. Unlike visible light, X-rays do not reflect or refract easily and are invisible to the human eye. Therefore, an X-ray microscope exposes film or uses a charge-coupled device (CCD) detector to detect X-rays that pass through the specimen. It is a contrast imaging technology using the difference in absorption of soft X-rays in the water window region (wavelengths: 2.34–4.4 nm, energies: 280–530 eV) between the carbon atom (main element composing the living cell) and the oxygen atom (an element of water).

3. Ultrasound microscope

3.1 Overview

It takes advantage of the fact that the propagation speed of ultrasound waves differs depending on the tissue and material. Conventional optical microscopes require staining treatments when examining biological tissue, but in many cases cells die during the staining process, whereas ultrasound microscopes allow samples to be treated in such a way. It can be visualized in vitro without any further treatment. Ultrasonic pulses with a frequency of 100MHz to 3GHz are irradiated onto a sample, and the reflected and transmitted waves are received by a piezoelectric element to observe the sample surface and optically invisible internal structure. Since spatial resolution is inversely proportional to frequency, high-frequency ultrasound enables high-resolution imaging of biological tissues and internal structures. The resolution depends on the ultrasonic frequency, and when a water coupler is used, the resolution is about 1 μm at 1 GHz.

3.2 Scanning Ultrasound Microscope

In an ultrasound microscope, a plane wave emitted from a plane transducer is focused by an acoustic lens such as sapphire, or ultrasound wave directly generated by concave surface-front shaped transducer is narrowed down to a focal point that is approximately the same as the wavelength. Focused ultrasound is irradiated onto the sample via a water coupler, and leaky surface acoustic waves (LSAW) excited on the sample surface are used for measurement. Currently, mechanical scanning ultrasound microscopes are the mainstream, and images are acquired by using a transducer to two-dimensionally scan a tissue section or cultured cell placed on a glass slide.

3.3 Ultrasonic Impedance Microscope

Although conventional ultrasound microscopes have made it possible to acoustically color unstained tissues, they require the samples to be cut into thin slices, which requires special equipment and the same effort as preparing samples for optical microscopy. However, unlike conventional ultrasound microscopy, which captures ultrasound waves that pass through tissue and reflect off a glass surface, an ultrasound impedance microscope captures ultrasound waves that pass through a plastic plate and reflect off a tissue surface.

3.4 Ultrasonic Force Microscope

Also known as an ultrasonic atomic force microscope, this microscope uses an ultrasonic vibrator to vibrate the sample perpendicular to the direction of the probe at an ultrasonic frequency higher than the resonance frequency of the cantilever, and then sequentially moves the cantilever for measurement. By applying an atomic force microscope, it is possible to image the shape of the sample surface as well as the distribution of elastic properties, and the spatial resolution of these images is significantly improved compared to conventional ultrasound microscopes that use acoustic lenses.

4 CHEMICAL MICROSCOPY

Abstract: Far-field chemical microscopy providing molecular electronic or vibrational fingerprint information opens a new window for the study of three-dimensional biological, material, and chemical systems. Chemical microscopy provides a nondestructive way of chemical identification without exterior labels. However, the diffraction limit of optics hindered it from discovering more details under the resolution limit. Recent development of superresolution techniques gives enlightenment to open this door behind far-field chemical microscopy. Here, we review recent advances that have pushed the boundary of far-field chemical microscopy in terms of spatial resolution. There are applications in biomedical research, material characterization, environmental study, cultural heritage conservation, and integrated chip inspection.

2.1.1 Optimal Detection Angle in Sub-diffraction Resolution Photothermal Microscopy

Application for High-sensitivity Imaging of Biological Tissues

1 INTRODUCTION

Photothermal microscopy is an efficient method for observing single molecules or nanomaterials using only their optical absorption [1–3]. One of the advantages of photothermal microscopy is that it is less affected by photobleaching, which is the major drawback in fluorescent measurement. By virtue of this fact, it has been used in several biological applications in recent years, such as imaging of live cells using 5-nm gold nanoparticles for arbitrarily long periods [4], evaluation of the hydrodynamic properties of cytosol by means of photothermal absorption correlation spectroscopy [5–8], label-free imaging of heme proteins with two-photon excitation [9], and the visualization or tracking of membrane proteins moving at the surface of live cells [10, 11].

Typically in photothermal microscopy, two laser beams with different wavelengths (pump and probe beams) are incident on the sample through a focusing lens. The pump beam increases the temperature around the focal point of the optical absorbing sample. This results in local refractive index variations and induces deflection of the probe beam. The variation of the refractive index with temperature $\partial n / \partial T$ is typically in the order of -10^{-4} K^{-1}, which is readily detected by a lock-in detection scheme as a change in probe beam transmissivity. The pump beam is modulated at 0.1 to 1 MHz and an integration time (the time constant of the lock-in amplifier) of 1–10 ms is used to attain a sufficient signal-to-noise ratio (SNR) [12].

 DOI: 10.1201/9781003476030-2

Thus, the image acquisition time is still slower than that achieved by conventional laser scanning fluorescence microscopy, with an integration time of down to ~1 μs per pixel.

Several technical improvements in photothermal microscopy have increased the SNR. As the relative change in the probe beam transmissivity is detected in a typical experimental setup, it is important to reduce the intensity noise of the beam. To reduce laser noise, a high frequency lock-in detection scheme of over 1 MHz is used, as laser noise is dominant at lower frequencies [13]. Alternatively, an auto-balanced detection scheme can be used for common mode noise reduction at a lower lock-in frequency (up to ~100 kHz) [14]. Using these techniques it is possible to reach shot noise limited sensitivity. It can also be useful to employ an embedding medium with large $\partial n / \partial T$, or media exhibiting a phase transition with large variations of refractive index around the critical temperature [15, 16]. Furthermore, it is important to select the optimal detection angle for the probe beam (the numerical aperture of the condenser lens NA_c) to maximize SNR. According to the literature on photothermal microscopy, NA_c is empirically set to be smaller than that of the focusing lens [5, 10, 12, 17, 18]. However, no systematic studies have been conducted to investigate this practice.

In this study, theoretical and experimental analyses were carried out to provide clear understanding of the relationship between the detection angle of the probe beam and the SNR.

High-sensitivity photothermal imaging of a mouse brain was also demonstrated with our sub-diffraction resolution photothermal microscopy, using intensity modulated laser diodes with a balanced detection scheme.

2 THEORY

Several models have been proposed to describe the nanoscopic photothermal signal. Selmke, Braun and Cichos calculated the difference in transmissivity of the focusing probe beam between two steady states, in the presence and absence of the pump beam, using generalized Lorenz-Mie theory [18–20]. They also suggested that the photothermal signal was analogous to Rutherford scattering [21, 22]. However, their model is based on a steady-state calculation and not applicable to the case when thermal diffusion length during one period of the modulating pump beam is comparable to or smaller than the focusing spot size of the probe beam. This condition is considered to be important to attain sub-diffraction resolution as detailed in Section 4. Berciaud et al. provided an analytical description of the photothermal signal, based on the theory of light scattering from a temporally modulating medium [17]. They introduced the time dependent refractive index profile induced by modulation of pump-beam intensity, based on the thermal diffusion equation, and derived the photothermal signal intensity from the interference term between the scattering field and an isotropic local oscillator. On the basis of this model, the dependence of the modulation frequency on the photothermal signal was discussed for both forward and backward configurations. However, their plane-wave approximation of the incident probe field did not include the effect of focusing (or transmitted diverging field) on the photothermal signal. As the transmitted field also interferes with the scattered field, it is important to take into account the effect of the focusing field on the photothermal signal.

In this study, we extend these models by considering both the effects of the focusing field and modulating heat sources on the photothermal signal. We consider the case when pump and probe beams are incident on the sample and only the pump beam is absorbed by the point absorber, which then causes an increase in temperature in a surrounding medium. The temperature rise induces an increase or decrease in the refractive index depending on the sign of the thermooptic index and thus the probe beam is deflected. As modulation frequency of the pump beam intensity is much lower than the frequency of the electromagnetic field, we consider the quasi-static Helmholtz equation [21]

$$U(\mathbf{r},t)+\mathrm{k}^2\left[n(\mathbf{r},t)/n_0\right]^2 U(\mathbf{r},t)=0, \tag{1}$$

where $U(\mathbf{r},t)$ is the scalar envelope electric field of the probe beam. The refractive index change is related to the temperature rise by $n(\mathbf{r},t) \cong n_0 + (\partial n / \partial T)\Delta T(\mathbf{r},t)$, where n_0 is the refractive index without temperature change. The excitation laser intensity is modulated with frequency of Ω and heat absorbed by the material is given by $\sigma I_0(1 + \cos\Omega t)$, where σ is the absorption cross-section at the pump laser wavelength and I_0 is the beam intensity at the focal point. Assuming that the size of the heat absorber is much smaller than the focal spot size, the temperature profiles can be described by using the heat point source model [11, 17]

$$\Delta T(\mathbf{r},t)=\frac{\sigma I_0}{4\pi\kappa}\left[\frac{1}{r}+\frac{1}{r}\exp\left(-\frac{r}{r_c}\right)\cos\left(\Omega t-\frac{r}{r_c}\right)\right]. \tag{2}$$

Here κ is the thermal conductivity, $r_c=\sqrt{2D/\Omega}$ is the characteristic length for thermal diffusion during one period with D, the thermal diffusion coefficient, and r is the distance from the point heat source. Substitution of Eq. (2) into Eq. (1) reads

$$\nabla^2 U(\mathbf{r},t)+\mathrm{k}^2[1+W(\mathbf{r},t)]^2 U(\mathbf{r},t)=0, \tag{3}$$

$$W(\mathbf{r},t)=\frac{\sigma I_0}{4\pi\kappa}\frac{1}{r}+\frac{1}{r}\exp\left(-\frac{r}{r_c}\right)\cos\left(\Omega t-\frac{r}{r_c}\right). \tag{4}$$

Equation (3) is equivalent to the quasi-static Schrödinger equations, where $W(\mathbf{r},t)$ corresponds to the potential field. The first term on the r.h.s. of Eq. (4) decays with the inverse of radial distance and corresponds to the Coulomb potential in quantum mechanical treatment. The Coulomb potential is effective over a greater distance. The second term is sinusoidally modulated and the amplitude $(1/r)\exp(-r/r_c)$ approaches zero at r_c. This factor is equivalent to the Yukawa potential (the screened

Coulomb potential), which was introduced to describe strong nuclear force between nucleons in particle physics [23]. The Yukawa potential coincides with the Coulomb potential in the limit $r_c \to \infty$. In photothermal microscopy, the first term can be regarded as the interaction with the temporally averaged (static) point heat source, whereas the second term is the interaction with the modulated point heat sources and thus the interaction length should be smaller than r_c.

The asymptotic solution of Eq. (3) in the far-field region for an incident plane wave reads

$$\lim_{r\to\infty} U_{\mathbf{k}}(\mathbf{r},t) \sim e^{i\mathbf{k}z} + \frac{e^{i|\mathbf{k}|r}}{r} f_{\mathbf{k}}(\theta,t) \tag{5}$$

where $\mathbf{k}$ denotes the wave vector of the incident plane wave and $f_{\mathbf{k}}(\theta, t)$ is the scattering amplitude, with θ the angle between $\mathbf{k}$ and the scattering field wave vector $\mathbf{k}_{sc}$ as pictured in Figure 2.1.1.1. The scattering amplitude is calculated under the first-order Born approximation as

$$\begin{aligned} f_k^{(1)}(\theta,t) &= -\frac{1}{4\pi}\int W(r',t)e^{i\Delta k\cdot r'}dr' \\ &= -\frac{\sigma I_0}{8\pi k n_0}\frac{\partial n}{\partial T}\left[\frac{2}{|\Delta k|^2} + \frac{2|\Delta k|^2}{|\Delta k|^4 + 4/r_c^4}\cos(\Omega t) + \frac{4/r_c^4}{|\Delta k|^4 + 4/r_c^4}\sin(\Omega t)\right] \end{aligned} \tag{6}$$

with $|\Delta\mathbf{k}| = |\mathbf{k} - \mathbf{k}_s| = (4\pi n_0/\lambda)\sin(\theta/2)$.

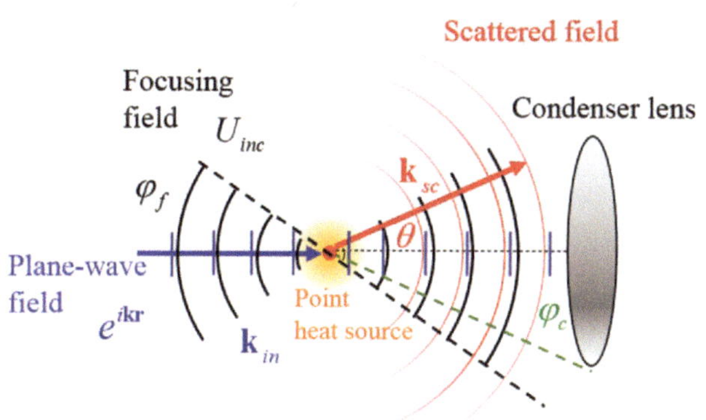

FIGURE 2.1.1.1 Schematic illustration of the principle of photothermal microscopy. A plane-wave field (blue lines) or a focusing field with NAf = n0sin φ f (black curves) is incident on the sample and is deflected (scattered) due to the local refractive index change around a point heat source. Transmitted and deflected photons are collected through a condenser lens with NAc = n0sin φ c. The copyright of the paper [26] we published is shared between ourselves and Optical Society of America which is now Optic (©2014 Optical Society of America).

The first term in the square bracket of Eq. (6) corresponds to the scattering amplitudes due to the Coulomb potential. The second and the third term correspond to the scattering amplitudes due to the temporary modulated Yukawa potential and their sum can be written as

$$2r_c^2 / \sqrt{u^2+1}\cos\left(\Omega t - \Phi\right)$$

where $\Phi = \tan^{-}(1/u)$ is the phase shift with $u = |\mathbf{\Delta k}|^2\, r_c^2/2$. The scattering cross-section, defined as the ratio of angularly integrated scattered field intensity to that of the incident field intensity, is given by

$$\begin{aligned}\sigma_{sc}(t) &= 2\pi \int_0^{\pi}\left[f_k^{(1)}(\theta,t)\right]^2 \sin\theta d\theta \\ &= \frac{4\pi}{k}\,\mathrm{Im} f_k^{(2)}(0,t).\end{aligned} \tag{7}$$

The conservation of the probability requires that σ_{sc}(t) is related to the imaginary part of the scattering amplitude under second-order Born approximation (optical theorem) [24]. When the incident and deflected (scattered) photons are collected through a condenser lens with $NA_c = n_0 \sin\varphi_c$ in the far field, as depicted in Fig. 2.1.1.1, the fraction of photons deflected outside the condenser lens (the residual scattering cross section) is expressed as

$$£\left(\varphi_{\omega a} t\right) = \sigma_{sc}(t) - 2\pi \int_0^{\varphi_c}\left[f_k^{(1b}(\theta,t)\right]^2 \sin\theta d\theta, \tag{8}$$

where the second term on the r.h.s of Eq. (8) is the fraction of photons deflected in the angle lower than φ_c. We note that the residual scattering cross-section is zero when photons are collected in all angles, i.e., $\Sigma(\pi,t) = 0$.

In photothermal microscopy, the focusing beam is incident on the sample through an objective lens and thus we consider a super-position of Eq. (5) at multiple incident angles to examine the effect of the focusing field on the photothermal signal

$$U(\mathbf{r},t) = \int d\mathbf{k} A(\mathbf{k}) U_{\mathbf{k}}(\mathbf{r},t) \sim U_{\mathrm{inc}}(\mathbf{r}) + U_{sc}(\mathbf{r},t) \tag{9}$$

Here,

$$U_{\mathrm{ina}}(\mathbf{r}) = \int d\mathbf{k} A(\mathbf{k}) e^{i\mathbf{k}\cdot\mathbf{r}} \tag{10}$$

denotes the incident focusing field and

$$U_{sc}(\mathbf{r},t) = \frac{e^{j\mathbf{k}|r}}{r}\int d\mathbf{k} A(\mathbf{k}) f_{\mathbf{k}}(\theta,t) \tag{11}$$

is the scattered field. For example, for an azimuthally symmetric and uniform intensity spherical wave focusing at $\mathbf{r}_0$, with $NA_f = n_0 \sin\varphi_f$, the amplitude factor $A(\mathbf{k})$ is expressed by

$$A(\mathbf{k}) = \begin{cases} e^{-i\mathbf{k}_0} / |\mathbf{r} - \mathbf{r}_0| & \text{for } \varphi' \leq \varphi_f \\ 0 & \text{for } \varphi' > \varphi_f \end{cases}, \tag{12}$$

$A(\mathbf{k})$ is expressed by where φ' is the polar angle of k. By assuming that f_k (θ, t)<<1, the square of the scattering field is negligibly small, and in this case, interference between the scattered field and the transmitted field is dominant. Thus, we approximate the scattering cross-section as an intensity ratio of the integrated interference signal to the incident beam intensity assuming that the field is negligibly small, and in this case, interference between the scattered field and the transmitted field is dominant. Thus, we approximate the scattering cross-section as an intensity ratio of the integrated interference signal to the incident beam intensity

$$\sigma'_{sc}(t) = J_{sc}(\pi, \mathrm{t}) / J_{\mathrm{inc}}(\pi) \tag{13}$$

Where

$$\begin{aligned} J_{sc}(\varphi, \mathrm{t}) &= (1/2)\left[2\pi \int_0^{\varphi} U_{sc}(\mathbf{r}, t) U_{\mathrm{inc}}(\mathbf{r})^* \sin\theta d\theta + c.c.\right. \\ J_{\mathrm{inc}}(\varphi) &= 2\pi \int_0^{\varphi} U_{\mathrm{inc}}(\mathbf{r}) U_{\mathrm{inc}}(\mathbf{r})^* \sin\theta d\theta. \end{aligned} \tag{14}$$

When the incident and deflected photons are collected through a condenser lens in far-field region, the photothermal signal is proportional to

$$\Sigma'(\varphi_c, t) = \sigma'_{sc}(t) J_{inc}(\varphi_c) - J_{sc}(\varphi_c, \mathrm{t}) / J_{\mathrm{inc}}(\pi), \tag{15}$$

where the first term on the r.h.s of Eq. (15) refers to the fraction of photons that have incident angle lower than φ_c and undergo scattering, while the second term is the fraction of the scattered photons in the angle lower than φ_c. Here, we assumed that the incident field is focused at the position of the point light absorber. The scattering amplitude is given by the sum of the three terms in Eq. (6), so Eq. (15) can be rewritten separately as

$$\Sigma'(\varphi_c, t) = \mathrm{C}(\varphi_c) + Y_{\mathrm{in}}(\varphi_c)\cos(\Omega \mathrm{t}) + Y_{\mathrm{out}}(\varphi_c)\sin(\Omega \mathrm{t}) \tag{16}$$

where $\mathrm{C}(\varphi_c)$ is attributable to the first term (the Coulomb potential), and $Y_{\mathrm{in}}(\varphi_c)$ and $Y_{\mathrm{out}}(\varphi_c)$ to the second and third terms (the temporally modulated Yukawa potential), respectively. In a typical photothermal experiment, the pump beam intensity is modulated and the resulting change in the probe beam transmissivity is detected

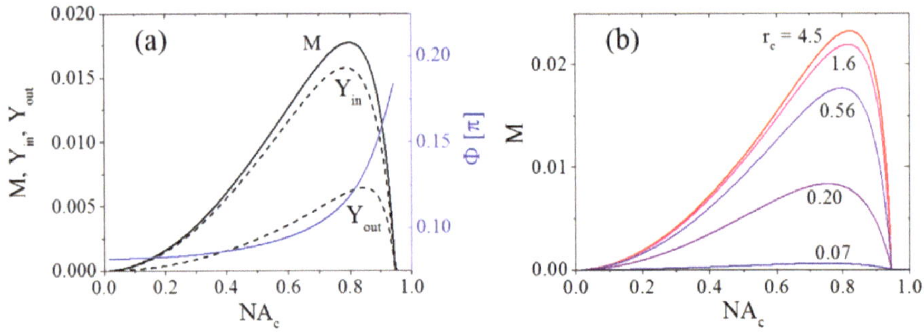

FIGURE 2.1.1.2 Detection angle dependence of the photothermal signal. (a) *Y*in and *Y*out as a function of numerical aperture of the condenser lens NAc. We set *r*c = 0.56. The vertical axis is normalized by $(\sigma I_0)/(8\pi\kappa n_0)\partial n/\partial T$. Their magnitude *M* and the phase shift Φ are shown as black and blue solid lines, respectively. (b) *M* as a function of NAc with various values of *r*c. The copyright of the paper [26] we published is shared between ourselves and Optical Society of America (©2014 Optical Society of America) which is now Optica.

by a dual-phase lock-in amplifier. In this case, $Y_{\text{in}}(\varphi_{\text{c}})$ and $Y_{\text{out}}(\varphi_{\text{c}})$ are measured as in-phase and π/2 as out-of-phase components, respectively.

Figure 2.1.1.2(a) shows Y_{in} and Y_{out} as a function of NA_c. Their magnitude $M = \text{sqrt}(Y_{in}^2 + X_{in}^2)$ and the phase $\Phi = \tan^{-1}[Y_{out}/Y_{in}]$ of the output are also shown for the set of $|\Delta\mathbf{k}| = 2$, $n_0 = 1$, and $\text{NA}_\text{f} = 0.95$. It is important to note that the magnitude has a peak at $\text{NA}_\text{c} = 0.73$ and is zero for $NA_c > NA_f$. Thus, NA_c should be smaller than that of the focusing lens to achieve optimal SNR. We also note that the sign of Y_{in} is negative for a negative ∂ / ∂, which suggests that transmissivity increases in the presence of the pump beam. This is because the wide-angled incident field component, with $\varphi' > NA_c$, is deflected to the angle lower than NA_c and thus contributes to the positive signal gain. We found that the peak position of Y_{in} is smaller than that of Y_{out}, and so Φ has a tendency to increase with φ_c. Figure 2.1.1.2(b) shows angular dependence of *M* for several values of r_c. We found that the peak value decreases and the peak position slightly shifts to the lower angular side as r_c decreases.

3 EXPERIMENTS

Detection angle dependence of the photothermal signal was examined experimentally in our pump-probe microscopy system, which uses intensity modulated laser diodes (LDs) and a balance detector. The details have been described previously [14]. We used a 488-nm LD for the pump, and a 640-nm LD for the probe beam. Each beam was collimated through a spatial filter and two beams were combined using dichroic mirrors. The probe and pump-beam intensities were typically modulated at frequencies of ω_1 and ω_2, respectively. A beat signal $|\omega_1 - \omega_2|$ is generated by the bilinear interaction of the pump field and the probe field in the sample. We used a lock-in technique to detect the signal, and the lock-in amplifier was referenced to the beat frequency. By changing the modulation frequencies (ω_1 and ω_2) while holding the difference frequency $|\omega_1 - \omega_2|$ constant, this dual modulation scheme has the potential to acquire time-resolved information without a high-speed detector. To counteract the intensity fluctuation of the probe laser, we split the probe beam into two and directed the two resulting beams to an auto-balanced photodetector (Nirvana, New Focus). This auto- balanced detector serves to cancel both the laser noise and the intensity imbalance due to refractive index variations during the sample scanning. We confirmed that when the probe beam power incident on the photodiode was above ~10 μW, the signal-to-noise ratio was mainly limited by shot noise. The pump and probe beams were linearly polarized. A polarizing beam splitter (PBS) was used to direct the combined beam to the objective lens (UPLSAPO 40X2, Olympus) with numerical aperture NA_f of 0.95. The beam size was adjusted to fill the back aperture of the objective lens. The sample position was raster scanned using a three-axis positioning stage driven by piezo actuators (MAX311D, Thorlabs or P-622.2, PI). A condenser lens with variable NA_c (U-AAC, Olympus) was used to collect the transmitted light. This lens is equipped with a diaphragm for varying NA_c. The maximum value of NA_c was 1.4 with immersion oil, while it was ~0.95 without immersion oil. In this study, immersion oil was not employed since a dry objective was used for focusing the probe beam. The distance between the glass substrate and the front side of the condenser lens was ~0.5 mm. A narrowband filter matching the laser line was placed in front of the detector to filter out the pump beam, so that only the probe beam was detected. The pump and probe beams were collimated so that they focused at the same position.

We measured the photothermal signal from 20-nm diameter gold nanoparticles dispersed in a polyvinyl alcohol (PVA) film on a glass slide with varying NA_c (Fig. 2.1.1.3). The wavelength of the pump beam overlapped the absorption spectrum of the gold nanoparticles as the plasmon resonance peaked at 524 nm. The pump and probe powers incident on the sample were 0.7 mW and 0.3 mW, respectively and were modulated at 1.015 MHz and 1.000 MHz, respectively. The time constant of the lock-in amplifier was 0.5 ms and the pixel dwell time was 1 ms. We found that the full width at half maximum (FWHM) value of a single particle is ~240 nm, which is 19% smaller than the diffraction-limited spot size of the pump beam and 32% smaller than that of the probe beam [25]. Furthermore, the axial spot size was 590 nm, which is 33% smaller than the diffraction-limited axial size of the probe beam. Upper and lower panels of Fig. 2.1.1.3 show magnitude, M, and phase shift, Φ, images, respectively, with several values of NA_c. Since the thickness of the PVA film is about 20 μm, gold nanoparticles are distributed in both lateral and axial directions. This leads to the signal intensity distribution of gold nanoparticles in the xy-image. To characterize the images, we calculated the integrated intensity of M and the average Φ for each image, summarized in Figure 2.1.1.4(a). It is important to note that with increasing NA_c, integrated intensity increases and then decreases when NA_c is larger than ~0.4, while Φ monotonically increases with NA_c. These tendencies are in accordance with the theoretical calculation as shown in Fig. 2.1.1.2(a). Furthermore, we evaluate the effect of NA_c on the signal-to-noise ratio (SNR) to ascertain the

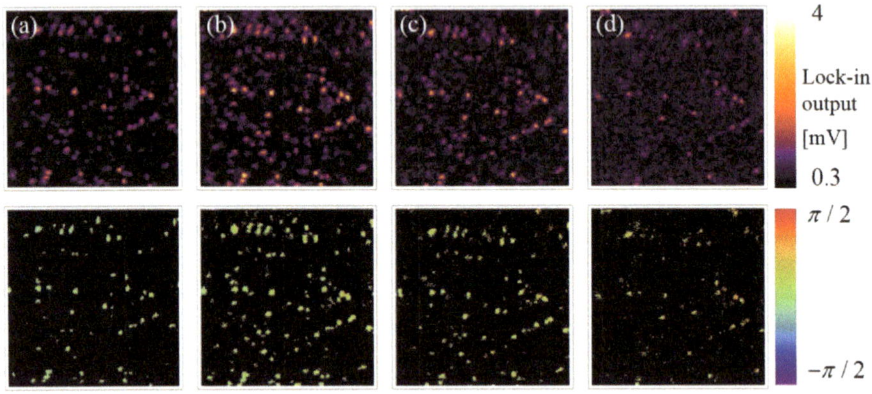

FIGURE 2.1.1.3 Magnitude (upper panel) and the corresponding phase shift images (lower panel) of gold nanoparticles dispersed in PVA film. The numerical aperture of the condenser lens NAc was (a) 0.3, (b) 0.42, (c) 0.54, and (d) 0.66. The image size is 19.6 x 19.6 μm with 200 x 200 pixels. The copyright of the paper [26] we published is shared between ourselves and Optical Society of America (©2014 Optical Society of America) which is now Optica.

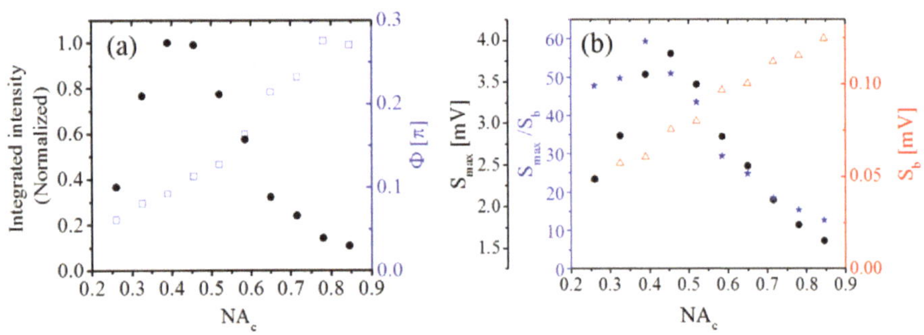

FIGURE 2.1.1.4 (a) The integrated intensity (filled black circles) and average phase shift, Φ, (open blue squares) as a function of the numerical aperture of the condenser lens (NAc). (b) Maximum intensities in the images Smax (filled black circles) and background noise levels Sb (open red triangles), and their ratios (blue stars) as a function of NAc. The copyright of the paper [26] we published is shared between ourselves and Optical Society of America (©2014 Optical Society of America) which is now Optica.

optimal NA_c. To this end, the maximum value S_{max} and the root mean square (r.m.s.) of background level S_b are calculated for each image. Figure 2.1.1.4(b) shows S_{max}, S_b, and their ratio S_{max}/S_b as a function of NA_c. We found that S_{max} has a peak at $NA_c = 0.45$, while S_b increases with increase in NA_c, as an increase in the beam power incident on the detector is accompanied by an increase in noise power. Consequently, S_{max}/S_b has a maximum value of 60 at NA_c ~0.4 and significantly decreases to 12 at $NA_c = 0.85$. We calculated maximum intensity S_{max} instead of the mean intensity S_{mean} because it is difficult to determine the latter due to the background noise that is prominent for a large NA_c. However, we found that S_{mean} gives a similar trend as S_{max}. We also checked the modulation frequency dependence on the signal intensity and found that the signal intensity decreases by approximately half when the modulation frequency reaches 2 MHz. This agrees with the experimental results in the literature [1, 2]. We also measured the incident beam power dependence and found that the signal is proportional to the product of the pump and probe power. These facts verify that the measured signal is to be attributed to the photothermal effect.

Biological photothermal imaging was demonstrated using mouse brain labeled with weakly fluorescent quantum dots (Figure 2.1.1.5). The brain of a two-month-old C57BL/6 mouse[1] was embedded in O.C.T. compound[2] (SAKURA Seiki Co. Ltd., Tokyo, Japan), frozen in liquid nitrogen, sliced at 14 μm, and air-dried. Sections were stained with anti-α-tubulin antibody labeled with eFluor 650NC (eBioscience, CA,

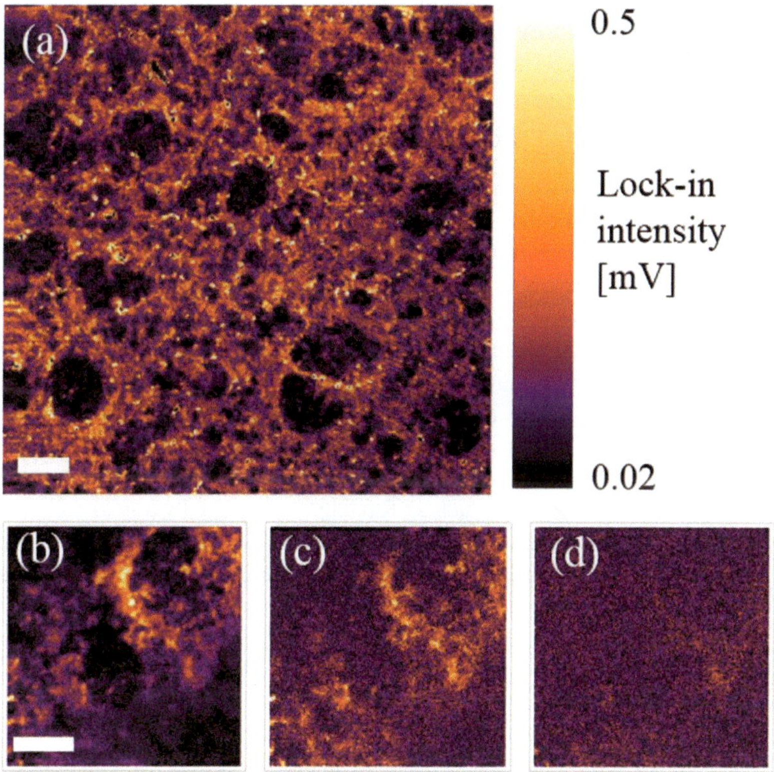

FIGURE 2.1.1.5 Photothermal imaging of a slice of mouse brain labeled with weakly fluorescent quantum dots. (a) Overview image and (b)–(d) images of the same area with different values of the numerical aperture of the condenser lens NAc. The image sizes are (a) 200 x 200 μm with 500 x 500 pixels and (b)–(d), 19.6 x 19.6 μm with 200 x 200 pixels. NAc was set at (a) 0.42, (b) 0.36, (c) 0.72, and (d) 0.9, respectively. The scale bars indicate (a) 20 μm and (b) 5 μm. The copyright of the paper [26] we published is shared between ourselves and Optical Society of America (©2014 Optical Society of America) which is now Optica.

San Diego), mounted on coverslips, and imaged. The samples were placed under atmospheric conditions for several months to induce surface oxidization of quantum dots, where rapid non-radiative exciton decay due to charge recombination dominates the relaxation pathway and photo excitation energy is mainly converted to heat. The pump and probe beam power were 4 mW and 0.6 mW, respectively. The time constant of the lock-in amplifier was 0.5 ms and the pixel dwell time was 1 ms. Pump and probe beams were modulated at 1.1 MHz and 1.0 MHz, respectively. Fig. 2.1.1.5(a) shows an overview image with the image size of 200 x 200 μm. Figs. 2.1.1.5(b)–(d) show a set of images with different values of NA_c. It can be clearly seen in Figs. 2.1.1.5(b)–(d) that the largest SNR is attained for $NA_c = 0.36$ [Fig. 2.1.1.5(b)]. The calculated values of S_{max}/S_b were (b) 40, (c) 17, and (d) 6. We confirmed that there was minimal damage to the sample after measurements. It is possible that stimulate emission contributed to the lock-in signal, as the laser wavelength overlapped the fluorescence of the quantum dot peak at 650 nm [14]. However, we consider the photothermal signal is a majority component because the detected signal significantly decreases for a large NA_c.

4 DISCUSSION AND CONCLUSION

In this study, theoretical and experimental analyses were carried out to ascertain the optimal detection angle for the probe beam to maximize SNR. We found that signal intensity significantly depends on NA_c. The magnitude of the photothermal signal has a maximum value when $NA_c = 0.4$, which is less than half that of the focusing lens of $NA_f = 0.95$. We also confirmed that the phase shift, Φ, increases as NA_c increases. These trends qualitatively agree with the theoretical calculation in which the photothermal signal has a maximum value in the region $NA_c < NA_f$, while it is zero for $NA_c > NA_f$. However there are several discrepancies between the experimental data and theoretical calculations. In the experimental data, the photothermal signal remains even when $NA_c \sim NA_f$, although the SNR is small. This is probably attributable to the static scattering field, such as Rayleigh scattering by a gold nanoparticle or light scattering at the surface of the glass plate, which serves as a local oscillator and causes interference with the temporary modulated scattering field [17]. In the present study, signal was detected in the forward direction, and thus we approximated that the photothermal signal mainly comes from the interference between the temporally modulated scattering field and the diverging transmitted probe field. This makes the signal vanish for $NA_c > NA_f$ in the theoretical calculation. On the other hand, for the backward detection, the photothermal signal may be attributable to the interference between the static scattering field and the temporally modulated scattering field. In that case, signal intensity should be proportional to the square of NA [6]. The peak position of the photothermal signal also differs between the experimental data and the theoretical calculations. In this regard, we examined how the theoretical curve of M depends on the parameters, and found that the optimal value of NA_c (position of the maximum M) decreases as λ or the focal spot size increases, while it increases with NA_f. Furthermore, we assume a uniform probe beam at the pupil of the focusing lens in theoretical calculations. However, experimentally, the incident beam is a Gaussian beam truncated by the limiting aperture of the objective pupil. For a Gaussian beam focusing at the sample position $A(\mathbf{k}) = \exp(-\varphi'^2/2\sigma^2)$ with $\sigma = 0.1\pi$, the optimal value appears at $NA_c \sim 0.3$. However, Φ decreases as NA_c increases. Further study is needed in regard to this point including the effect of the focal position on the signal.

Theoretical analysis and experiments were carried out for a particle whose diameter is much smaller than wavelength of visible light (20-nm Au nanoparticle and 5-nm quantum dot), since a smaller particle is usually preferred for a probe molecule in biological imaging. For a large particle, it is expected that the forward static scattering field also interferes with the modulated scattering field, which gives a similar result as in the case of the transmitted probe field. In this regard, in images of the slice of mouse brain, we observed several spots with the sizes from several hundred nm to ~1 μm due to the aggregates of quantum dots. We found that signal intensities from these spots also decrease for a large NA_c. This suggests that the optimal NA_c is smaller than NA_f regardless of the size of nanoparticles.

In a conventional thermal lens model, change in probe beam transmissivity is explained by considering the deflection by a concave lens induced by the temperature rise. This picture may be valid when r_c is larger than the focal spot size. In such a case, the temporally modulated Yukawa potential in Eq. (4) is approximated by the Coulomb potential, and consequently the photothermal signal is described in a similar manner to Rutherford scattering [21, 22] or by a lens-like model [18, 20]. In the present experiment, by assuming the thermal diffusivity of 0.1 mm^2/s and the modulation frequency of the pump beam of 1 MHz, r_c is estimated to be ~0.1 μm, which is about four times smaller than the calculated focal spot size of the probe beam ($0.61\lambda/NA_f = 410$ nm). In this regard, it is interesting to examine the spatial resolution with changing modulation frequencies, because when r_c is much smaller than the focal spot size, the spatial resolution (the point spread function) should be determined by the product of the pump and probe beam intensity at the focal point. In this case, spatial resolution improves [14, 25]. The improvement of the spatial resolution is attributed to the fact that the photothermal signal is based on the nonlinear interaction between the two laser beams and the sample. In contrast, it is expected that the spatial resolution is determined by a pump beam intensity distribution when r_c is much larger than the focal spot size. It is also important to examine the z-dependence of the photothermal signal with varying r_c. In a thermal lens model, signal intensity exhibits two peaks that correspond to focusing or defocusing when the sample position is moved in the axial direction [26]. However, in the present study, the amplitude of the lock-in signal exhibited only a single peak and the phase was nearly flat and did not exhibit the π shift at the origin. Further investigation is needed to clarify the modulation frequency dependence on the photothermal signal.

The abstract of this subsection is summarized as follows.

We evaluated the optimal detection angle for maximizing the signal-to-noise ratio (SNR) in sub-diffraction resolution photothermal microscopy. The angular dependent photothermal signal was calculated based on scattering theory using the temporally modulated Yukawa potential, and its detection angle and modulation frequency dependencies were analyzed [27]. We verified the theoretical findings by imaging gold nanoparticles using laser diode-based photothermal microscopy with balanced detection scheme. High-sensitivity (SNR ~40) photothermal biological imaging of a mouse brain was also demonstrated. The contents of this subsection are the outcome products of collaborative research study by the following people: J. Miyazaki, H. Tsurui, K. Kawasumi, and T. Kobayashi [27]. The copyright of the paper [26] we published is shared between ourselves and Optical Society of America (©2014 Optical Society of America) which is now Optica.

NOTES

1. C57BL/6 mouse: a common inbred strain of laboratory mouse. It is the most widely used "genetic background" for genetically modified mice for use as models of human disease. They are the most widely used and best-selling mouse strain due to the availability of congenic strains, easy breeding, and robustness.
2. O.C.T. compound: (O.C.T stands for optimal cutting temperature.) Compound is used to quickly embed fresh tissue specimens for frozen sections using a cryostat. It is a formulation of clear, water-soluble glycols and resins, providing a solid matrix to encapsulate tissue specimens and bond to the Specimen Holder for consistent sectioning in a cryostat working temperature of -10°C and below.

REFERENCES

1. D. Boyer, P. Tamarat, A. Maali, B. Lounis, and M. Orrit, "Photothermal imaging of nanometer-sized metal particles among scatterers," *Science* **297**(5584), 1160–1163 (2002).
2. S. Berciaud, L. Cognet, G. A. Blab, and B. Lounis, "Photothermal heterodyne imaging of individual nonfluorescent nanoclusters and nanocrystals," *Phys. Rev. Lett.* **93**(25), 257402 (2004).
3. A. Gaiduk, M. Yorulmaz, P. V. Ruijgrok, and M. Orrit, "Room-temperature detection of a single molecule's absorption by photothermal contrast," *Science* **330**(6002), 353–356 (2010).
4. C. Leduc, S. Si, J. Gautier, M. Soto-Ribeiro, B. Wehrle-Haller, A. Gautreau, G. Giannone, L. Cognet, and B. Lounis, "A highly specific gold nanoprobe for live-cell single-molecule imaging," *Nano Lett.* **13**(4), 1489–1494 (2013).
5. V. Octeau, L. Cognet, L. Duchesne, D. Lasne, N. Schaeffer, D. G. Fernig, and B. Lounis, "Photothermal absorption correlation spectroscopy," *ACS Nano* **3**(2), 345–350 (2009).
6. P. M. R. Paulo, A. Gaiduk, F. Kulzer, S. F. G. Krens, H. P. Spaink, T. Schmidt, and M. Orrit, "Photothermal correlation spectroscopy of gold nanoparticles in solution," *J. Phys. Chem. C* **113**(27), 11451–11457 (2009).
7. R. Radünz, D. Rings, K. Kroy, and F. Cichos, "Hot Brownian particles and photothermal correlation spectroscopy," *J. Phys. Chem. A* **113**(9), 1674–1677 (2009).
8. C. Leduc, J. M. Jung, R. P. Carney, F. Stellacci, and B. Lounis, "Direct investigation of intracellular presence of gold nanoparticles via photothermal heterodyne imaging," *ACS Nano* **5**(4), 2587–2592 (2011).
9. S. Lu, W. Min, S. Chong, G. R. Holtom, and X. S. Xie, "Label-free imaging of heme proteins with two-photon excited photothermal lens microscopy," *Appl. Phys. Lett.* **96**(11), 113701 (2010).

10. D. Lasne, G. A. Blab, S. Berciaud, M. Heine, L. Groc, D. Choquet, L. Cognet, and B. Lounis, "Single nanoparticle photothermal tracking (SNaPT) of 5-nm gold beads in live cells," *Biophys. J.* **91**(12), 4598–4604 (2006).
11. L. Cognet, C. Tardin, D. Boyer, D. Choquet, P. Tamarat, and B. Lounis, "Single metallic nanoparticle imaging for protein detection in cells," *Proc. Natl. Acad. Sci. U.S.A.* **100**(20), 11350–11355 (2003).
12. P. Vermeulen, L. Cognet, and B. Lounis, "Photothermal microscopy: optical detection of small absorbers in scattering environments," *J. Microsc.* **254**(3), 115–121 (2014).
13. L. Wei and W. Min, "Pump-probe optical microscopy for imaging nonfluorescent chromophores," *Anal. Bioanal. Chem.* **403**(8), 2197–2202 (2012).
14. J. Miyazaki, H. Tsurui, A. Hayashi-Takagi, H. Kasai, and T. Kobayashi, "Sub-diffraction resolution pump-probe microscopy with shot-noise limited sensitivity using laser diodes," *Opt. Express* **22**(8), 9024–9032 (2014).
15. W.-S. Chang and S. Link, "Enhancing the sensitivity of single-particle photothermal imaging with thermotropic liquid crystals," *J. Phys. Chem. Lett.* **3**(10), 1393–1399 (2012).
16. A. N. G. Parra-Vasquez, L. Oudjedi, L. Cognet, and B. Lounis, "Nanoscale thermotropic phase transitions enhancing photothermal microscopy signals," *J. Phys. Chem. Lett.* **3**(10), 1400–1403 (2012).
17. S. Berciaud, D. Lasne, G. A. Blab, L. Cognet, and B. Lounis, "Photothermal heterodyne imaging of individual metallic nanoparticles: theory versus experiment," *Phys. Rev. B* **73**(4), 045424 (2006).
18. M. Selmke, M. Braun, and F. Cichos, "Photothermal single-particle microscopy: detection of a nanolens," *ACS Nano* **6**(3), 2741–2749 (2012).
19. M. Selmke, M. Braun, and F. Cichos, "Gaussian beam photothermal single particle microscopy," *J. Opt. Soc. Am. A* **29**(10), 2237–2241 (2012).
20. M. Selmke, M. Braun, and F. Cichos, "Nano-lens diffraction around a single heated nano particle," *Opt. Express* **20**(7), 8055–8070 (2012)
21. M. Selmke and F. Cichos, "Photothermal single particle Rutherford scattering microscopy," *Phys. Rev. Lett.* **110**(10), 103901 (2013)
22. M. Selmke and F. Cichos, "Photonic Rutherford scattering: a classical and quantum mechanical analogy in ray and wave optics," *Am. J. Phys.* **81**(6), 405 (2013).
23. H. Yukawa, "On the interaction of elementary particles. I," *Proc. Phys.-Math. Soc. Jpn.* **17**, 48–57 (1935).
24. K. Gottfried and T.-M. Yan, *Quantum Mechanics: Fundamentals, Second Edition* (Springer, 2003).
25. J. Miyazaki, K. Kawasumi, and T. Kobayashi, "Resolution improvement in laser diodes-based pump-probe microscopy with an annular pupil filter," *Opt. Lett.* **39**(14), 4219–4222 (2014).
26. K. Uchiyama, A. Hibara, H. Kimura, T. Sawada, and T. Kitamori, "Thermal lens microscope," *Jpn. J. Appl. Phys.* **39**(9A), 5316–5322 (2000).
27. J. Miyazaki, H. Tsurui, K. Kawasumi, and T. Kobayashi, "Optimal detection angle in sub-diffraction resolution photothermal microscopy: application for high sensitivity imaging of biological tissues," *Opt. Express* **22**(16), 18833–18842 (2014).

2.1.2 Resolution Enhanced Pump-probe Microscope with a Spatial Filter

1 INTRODUCTION

1.1 MICROSCOPY

Optical microscopy is essentially an important methodology in biological and biomedical studies. The study of correlations between structure of the bio system and physiological function can provide key points for understanding the mechanism of the system. From the very beginning of physiological study of the biosystem, the optical microscope has been a vitally important tool. Technical progress of microscopy has succeeded in the development of the electron microscope. Even though the spatial resolution of optical microscopy is in general limited by the diffraction of light and is three orders of magnitude lower than that of electron microscopes which can have resolution of about 0.1 Å, wide applicability of optical microscopes is still extremely powerful in many areas of bio-relevant and medical studies.

Various types of optical microscope have been developed, as well as a revolutionary light source. Because of the invention and subsequent very rapid development of lasers from the early 1960s and also a variety of LEDs, the incandescent lamp or discharge lamp used in the early stages of optical microscopes was replaced by lasers and/or LEDs.

1.2 MICROSCOPY

Optical microscopy is essentially an important methodology in biological and biomedical studies. The study of correlations between structure of the biosystem and physiological function can provide key points for understanding the mechanism of the system. From the very beginning of physiological study of the biosystem, the optical microscope has been a vitally important tool. The technical progress of microscopes has succeeded in the development of the electron microscope. Even though the spatial resolution of optical microscopy is in general limited by the diffraction of light and is three orders of magnitude lower than that of electron microscopes which can have resolution of ~0.1 Å, the wide applicability of optical microscopes is still extremely powerful in many areas of bio-relevant and medical studies.

DOI: 10.1201/9781003476030-3

Various types of optical microscope have been developed, as well as a revolutionary light source. Because of the invention and subsequent very rapid development of lasers from the early 1960s and also a variety of LEDs, the incandescent lamp or discharge lamp used in the early stages of optical microscope was replaced by lasers and/or LEDs.

1.2 Fluorescence Microscopy

Most of the commercially available and laboratory-used laser microscopes can be termed as a (spontaneous) fluorescence-detection microscope. A common fluorescence microscope in general has the following several drawbacks.

(1) Spontaneous fluorescence from commonly used organic dyes and pigments for the fluorescence microscope has a very broad spectrum (as broad as more than 100 nm in the visible and near infrared spectral regions). Therefore, in cases where several fluorescent probe dyes tagged to some components are needed to be used for obtaining the distributions of several different biomaterials, microorganisms, and/or organelles, fluorescence from these are apt to spectrally overlap among them, resulting in crosstalk. Then it is difficult to separate the information about species with the tags of fluorescent dyes. If narrow band-pass color filters or notch filters are used, the weak spontaneous fluorescence intensities will become even weaker.

(2) The angular intensity distribution of spontaneous fluorescence from dye molecules via a transition-dipole mechanism is given by $(\cos\theta)^2$ in terms of the angle θ between the normal line and the dipole and radiation direction. However, the angle dependence is smeared out in the case of integrated intensity of fluorescence from the randomly oriented dye molecules, resulting in spatially quasi-isotopic emission to all 4πdirections. Since the commonly used microscope lens system cannot effectively detect the 4π-direction fluorescence, this results in even lower collection efficiency of the fluorescence signal.

(3) To obtain higher spatial resolution it is needed to utilize an iris to separate off-center beam as in a confocal microscope. Then it may decrease the signal intensity resulting in the reduction of S/N ratio.

(4) Sometimes fluorescent dyes in biological systems are fragile and easily photodamaged by phototoxic effects or some kind of photochemical and/or thermal reaction with surrounding components of biomolecules and biopolymers. This is a contrast with the fluorescent dyes that are relatively robust in organic solvents or in aqueous solutions.

(5) For studying microscopic bioprocesses taking place in biosystems, sometimes it is needed to perform microscopic observation for long period. In such a case, due to the elongated exposure of excitation light to the sample, it may be damaged and may result in change in the fluorescence spectrum due to degradation or induced reabsorption by extra absorption by the photodamaged species.

In the present study, we could overcome all the above five problems by using pump-probe microscopies, for example stimulated-emission and photothermal methodologies.

Other than the above described common fluorescence microscope may be called "optical pump fluorescence detection microscope", which requires only one actinic beam. In this case it may also be termed as "one actinic microscope".

1.3 Pump-probe Microscopy

In contrast to one actinic beam microscopy, there are pump/probe microscopy methods, in which two beams i.e. "actinic and probing" beams are introduced to a target sample. The simplest cases of two beams are an actinic beam for excitation and photothermal effect or stimulated emission sensed by the probing beam.

Here a short comment is required on the non-fluorescent characteristic of materials. Actually so-called non-fluorescent materials just have extremely low fluorescent efficiency. All materials have electrons in their systems except in special cases of atoms and molecules under high power optical and radiation fields or strongly ionized atoms. The common materials interact with the (weak) radiation field and have absorption because of the interaction. Then due to the relation between Einstein A and B coefficients, any material with a finite B coefficient corresponding to optical absorption capability has a non-zero spontaneous emission coefficient usually represented by A. There is a way to avoid the poor image quality due to low fluorescence quantum efficiency by utilizing the stimulated emission as an imaging signal.

Actually W. Min et al. developed pump-probe microscopy using stimulated emission intensity and imaged distributions of chromoproteins in live *Escherichia coli* cells [5]. The pump and probe beams were supplied by high intensity, high repetition rate femtosecond lasers, and a lock-in detection scheme was used to reduce probe noise [6].

Optical pump-probe microscopy can provide images by detecting changes in probe light intensity induced by stimulated emission, photoinduced absorbance change (PA) or photothermal (PT) effect in either transmission or reflection geometrical configuration mode, or both if needed. Microscopes based on PA and PT are hereafter called PAM and PTM, respectively. Also for convenience they are used for mentioning PT and PA "microscopy" so that no confusion is experienced. Because of simplicity and higher sensitivity, PTM is more extensively used and is mainly discussed in this chapter. Thanks to the extra freedom of probe, pump-probe technique has various applications. They include the identification of pigments in historical artworks, measurements of carrier dynamics in a single quantum well, lifetime measurements of fluorescent samples, and non-invasive characterization of melanocytes [1–4]. Especially in bio-imaging, pump-probe microscopy can be applied in a wide range of biological tissues regardless of their condition, whether the sample is fluorescent or non-fluorescent, with sub-diffraction resolution.

2 PHOTOTHERMAL MICROSCOPY

2.1 ADVANTAGES OF PTM

PTM can visualize microscopic images of non-fluorescent chromophores with high sensitivity, high spatial resolution, applicability to non-fluorescent materials, and robustness against photobleaching and photoblinking of samples [7–10]. Among them, determination of extremely low concentration [as low as sub yoctomole (1 yoctomole = 10^{-24} mole = $6.02214076 x 10^{-1}$ molecules)] non-fluorescent molecules using a thermal lens microscope were reported [7]. Photothermal heterodyne imaging was performed for both individual non-fluorescent nanoclusters and nanocrystals [8]. The method was extended to applications to biological systems, such as single nanoparticle photothermal tracking of 5-nm gold beads in live cells [9]. Photothermal heterodyne imaging of individual metallic nanoparticles was discussed theoretically for the experimental results obtained from the same group [10]. Thus far, it has been applied to various materials such as copper decorated grain boundary [11] and binary mixtures of liquid crystals to study phase transition [12]. Further applications are extended to biological systems. Photothermal microscopes can be used to visualize the distribution of endogenous chromoproteins in biological specimens as described later in this chapter.

One example is single-cell analysis by a scanning thermal lens microscope with a microchip performed for direct monitoring of cytochrome c[1] distribution during the apoptosis process [13]. Another example is photothermal confocal spectromicroscopy of multiple cellular chromophores and fluorophores [14]. Development of the photothermal microscope has succeeded in realization of ultrasensitive label-free imaging, spectral identification, and quantification of cytochrome c in mitochondria, live cells, and solutions [15]. Label-free optical imaging of mitochondria in live cells was performed by photothermal microscopy [16]. Photothermal microscopy was extended to the nonlinear optical microscope utilizing two-photon absorption excitation of biomaterials such as hemoproteins [17]. Examples of other useful applications of the photothermal microscope are "extension to simultaneous non-fluorescent tissues with 3D subdiffraction imaging microscope" [18], "imaging of ultrasharp nonlinear photothermal and photoacoustic resonances and holes beyond the spectral limit" [19], and "photoacoustic and photothermal detection of circulating tumor cells, bacteria and nanoparticles in cerebrospinal fluid in vivo and ex vivo" [20].

2.2 SIMULTANEOUS MULTI-WAVELENGTH IMAGING BY PTM

Simultaneous multi-wavelength imaging of non-fluorescent species was achieved with the use of a photothermal microscope. Dual wavelength sub-diffraction imaging of biological tissues stained with hematoxylin and eosin was demonstrated. Three-dimensional label-free imaging of mouse melanoma tissue section is also presented to demonstrate the effectiveness of the enhanced spatial resolution. This technique can be implemented using cost-effective and compact laser diodes and is applicable for various types of both fluorescent and non-fluorescent tissues [18].

Dual wavelength sub-diffraction imaging of biological tissues stained with hematoxylin[2] and eosin[3] is demonstrated. Three-dimensional label-free imaging of mouse melanoma tissue section is also presented to demonstrate the effectiveness of the enhanced spatial resolution. Our technique can be implemented using cost-effective and compact laser diodes and is applicable for various types of both fluorescent and non-fluorescent tissues. A laser spectroscopy based on nonlinear nanobubble-related photothermal and photoacoustic phenomena was reported [19]. It shows ultrasharp resonances and dips up to a few nanometers wide in broad plasmonic spectra of nanoparticles. It also demonstrates narrowing of absorption spectra of dyes and cellular chromophores, as well as an increase in the sensitivity and resolution of the spectral hole burning technique. This approach can permit the study of nonlinear plasmonics at a level of resolution beyond the spectral limits, the identification of weakly absorbing spectral holes, spectral optimization of photothermal nanotherapy, measurements of tiny red and blue resonance shifts in nanoplasmonic sensors, the use of negative contrast in photoacoustic technique, multispectral imaging, and multicolor cytometry [19].

2.3 High-sensitive Imaging by PTM

Circulating cells, bacteria, proteins, microparticles, and DNA in cerebrospinal fluid (CSF)[4] can be excellent biomarkers of many diseases, including cancer and infections. However, the sensitivity of existing methods is limited in their ability to detect rare CSF biomarkers at the treatable, early stage of diseases. Here, we introduce novel CSF tests based on *in vivo* photoacoustic flow cytometry[5] (PAFC) and ex vivo photothermal scanning cytometry. In the CSF of tumor-bearing mice, we molecularly detected *in vivo* circulating tumor cells (CTCs) before the development of breast cancer brain metastasis with 20-times higher sensitivity than with current assays. For the first time, we demonstrated assessing three pathways (i.e., blood, lymphatic, and CSF) of CTC dissemination, tracking nanoparticles in CSF *in vivo* and their imaging *ex vivo*. In label-free CSF samples, we counted leukocytes, erythrocytes, melanoma cells, and bacteria and imaged intracellular cytochromes, hemoglobin, melanin, and carotenoids, respectively. Taking into account the safety of PAFC, its translation for use in humans is expected to improve disease diagnosis beyond conventional detection limits [20].

These examples show that photothermal microscopic imaging applicable in such biosystems mentioned above, which do not have intense absorption, is an advantage is thanks to the high sensitivity.

Furthermore, photothermal imaging using pump-probe technique was further extended to biosystems with weak absorption in invisible near infrared spectral range, due to the spectral feature and/or small amount of pigments, if any, in the target system. In such a case it can also be applicable to biological tissues doped with metallic nanoparticles which induce the deflection of the probe beam by refractive index change due to photothermal effect [21–24]. As gold nanoparticles are not only physiologically inert and chemically stable but also do not suffer from photobleaching or photoblinking, they are highly useful in live cell imaging. Resolution enhancement is one of the great factors of study of PT microscope imaging. In our earlier works, we demonstrated pump-probe microscopy without using Ti:sapphire laser based femtosecond pulse lasers, optical parametric amplifiers, or high speed detectors [18, 25–29].

Especially we achieved sub-diffraction limited resolution with shot-noise-limited sensitivity using inexpensive laser diodes [25], details of which are described in [25]. An auto-balanced photoreceiver (New Focus, Nirvana) was used to cancel out the intensity-fluctuation noise of the probe laser. A beam splitter placed in front of the focusing lens reflected the probe beam, which was then directed to the reference port of the balanced detector through a multimode fiber. The output of the balanced detector can be expressed as $V_{out} = G\gamma\ [p_{sig} - gp_{ref}]$, where G and g are the trans-impedance gain[6] of the detector and the reference-channel gain, respectively, γ is the photodiode sensitivity and $p_{sig(ref)}$ is the beam power incident on the signal (reference) photodiode. A neutral density filter was used to set the reference-beam intensity to be about twice the probe intensity for optimal common-mode noise rejection. The fluorescence signal from the sample was detected by a photo-multiplier tube through the objective lens, a PBS and a multimode fiber with a core diameter of 100 μm. A signal generator (Rigol DG4162) and an amplifier (Mini-circuit ZHL-6A +) were used to sinusoidally modulate the light intensity of each LD by adjusting the injection current from the kHz to the MHz range. The light intensity of the diode-pump solid-state (DPSS) laser was externally modulated using an electro-optic modulator (EOM) (LINOS LM020). A lock-in amplifier (Signal Recovery 7270) was used to detect the pump-probe signal. The output of the balanced detector was connected to the lock-in amplifier through a low-pass filter (LPF; Mini-circuit LPF-B0R3). The reference signal for the lock-in amplifier was delivered from the two signal generators through a frequency mixer (Mini-circuit ZAD-1) and the low-pass filter. As it was crucial to avoid signal interference between the pump and the probe channels, a 30-dB attenuator and an amplifier (Mini-circuit ZFL-500LN) were placed between the signal generator and the frequency mixer. Furthermore, to reduce interference between the two channels due to electromagnetic (EM) waves, the LDs and the DPSS laser were placed in separate EM-shielding boxes.

2.4 Resolution Enhanced PTM

2.4.1 Resolution Enhancement by a Spatial Filter

We implemented a filter with an annular pupil into a microscopic system before the objective lens to improve the spatial resolution [27]. A donut-shaped beam created by this spatial filter is focused through an objective lens and forms a sharp main peak with unwanted side lobes in the focal plane [30, 31].

A very simple brief explanation on resolution is given as follows.

The spatial resolution of a pump-probe microscope can be approximated with the product of intensity profiles of pump and probe beams because the pump-probe signal is based on the bilinear interaction between two beams mediated by the sample [25]. Because of the effective nonlinearity induced by the bilinearity, the pump-probe microscope is in the nonlinear regime. It may be considered to be nonlinear providing higher resolution, but it is not the case. The PT signa is normalized by probe (= reference) intensity signal which is only linear to the pump.

The reasoning of higher spatial resolution of the PT microscope due to nonlinearity is explained as follows. The complex optical field, including phase change induced by the so-called thermal lensing, is given by the field with exponent of the exponential function in the form of exp(-iln) when no absorption effect takes place by the pump. Here l is the geometrical pathlength in the target material. Then the probe field is affected by refractive index change. It is given by n = no+δ Δ T with δbeing the temperature sensitivity of the refractive index and Δ T being temperature increment by the pump beam after it reaches equilibrium due to balancing between the energy deposition by the pump and the thermalization of the absorbing medium in the probe area being sensed by the probe laser.

The nonlinearity itself is not the true origin of the enhanced resolution better than the diffraction limit. This is induced by the lateral distribution of the temperature-dependent refractive index, which is determined by the pump beam cross-sectional distribution of intensity. By expanding the exponential function in the small exponent range case, given by the linear term and second order term, the latter is already in nonlinear term. This is to be discussed further in more detail later.

An annular pupil filter performed further improvement of the lateral resolution over the diffraction limit of conventional microscope lenses [27]. Although strong side ring lobes appeared surrounding the main peak with a common center of the intensity distribution, they can be suppressed because of difference in the peak intensity of radius of the side lobe. This is due to the different wavelength of pump and probe beams [27]. The annular pupil filter is an effective method to increase the lateral resolution and very easy to introduce into the microscope, however, the axial resolution becomes more than twice lower than that without the pupil filter [27, 32]. Therefore, it is not expected for the construction of three-dimensionally well-resolved images of thick samples. In some cases biological samples can transmit laser light deep enough to study the 3D structure information in principle but low resolution due to much longer Rayleigh length than the spot size at(near) the focal point.

2.4.2 Resolution Enhancement by Nonlinear Methods

Nonlinear detection imaging is an effective method to improve the resolution of the PT microscope. Further resolution improvement from the simple nominally linear (but effectively nonlinear component included as mentioned above) PT microscope by the pump-probe microscope can be obtained. It was realized by a nonlinear detection scheme using a supercontinuum white light laser taking advantages of its high intensity with the piezo-driven stage system [33, 34].

Some nonlinear methodologies based on plasmonic response are described in [35–37]. For example, Kildishev *et al.* developed an enhanced method for analysis of third-order nonlinearities in optical nanostructures with a scalar magnetic field frequency-domain formulation. It was shown to produce fast and accurate results for 2D problems without a superfluous vector electric field formalism [35]. While a standard TM representation using cubic nonlinear susceptibility results in an intractable implicit equation, their technique alleviates this problem. In addition to a universal approach, simpler, more efficient solutions are proposed for media having solely either a real (lossless Kerr-type medium) or an imaginary (nonlinear absorbing medium) nonlinearity. Combining these solutions with a finite-element method, they showed simulation examples validated with alternative approaches.

Surface plasmon, which is the collective oscillation of electrons in metal under light excitation, is capable of significantly enhancing the local field around the metal nanostructures and thus boosting up the efficiency of nonlinear optical interactions of the surrounding materials or of the metal itself [35–37]. Chu *et al.* showed that scattering from a single gold nanoparticle is saturable for the first time. Wavelength-dependent study reveals that the saturation behavior is governed by depletion of surface plasmon resonance, not the thermal effect [36]. We observed interesting flattening of the point spread function of scattering from a single nanoparticle due to saturation. By extracting the saturated part of the scattering signal obtained by temporal modulation experiment, it was claimed that they achieved $\lambda/8$ point spread function in far-field imaging with unambiguous separation of adjacent particles.

In 2016, Deka reported that plasmonic[7] scattering in an isolated nanostructure is indeed saturable [37]. The required excitation intensity for scattering saturation is lower when the excitation wavelength is closer to the surface plasmon (SP) scattering peak indicating that the saturation is governed by SP resonance. This is one of the resonant third-order nonlinear processes. By extracting the saturated part, sub-70-nm full width at half maximum (FWHM) was achieved, demonstrating the first superresolution far-field imaging based on scattering. There was no bleaching of the scattering after prolonged observation and the reversibility of saturable scattering is demonstrated.

Tzang *et al.* proposed advanced nonlinear measurement techniques [38, 39]. They reported resolution better than 100 nanometers in the nonlinear photo-modulated reflectivity and spatial modulation microscopy using an ultrashort pulse laser and 3D piezo stage. However, ultrashort pulse laser systems are much more expensive than diode lasers and require high-level technique to manipulate properly to take advantage of the high capability of the laser system. Sometimes the ultrashort pulse laser is very unstable both in terms of second-to-10seconds short-term stability but also minutes-to-hours long-term stability.

2.5 Fast PTM Imaging

Another important advantageous factor of PT imaging is its speed. Usually, a sample or a focused laser beam is scanned sequentially point-by-point and the pixel information is assembled into an image. Previously, PT imaging typically used a piezo stage to scan a sample sequentially point-by-point, but stage scanning requires a longer time with integration times per pixel ranging from 1 to 10 ms. In this case, the acquisition time of an image was 4 to 40 minutes for 500 x 500 pixels. One of the advantages of PT microscopy is its 3D imaging capability like confocal microscopy [14, 29]; however, it takes too long to acquire a stack of images at different focal plane depths (a volume scan) using a stage scanning PT microscope. Photoacoustic microscopy (PAM) [40, 41] and PT optical coherent tomography (PTOCT) [42, 43] are powerful techniques for the 3D imaging of optical absorbers *in vivo*. For example, deep *in vivo* photoacoustic imaging of mammalian tissues using a tyrosinase-based genetic reporter [40], and high-speed label-free functional photoacoustic microscopy of mouse brain in action [41]. But spatial resolution in PT microscopy is usually better than that of PAM and PTOCT by an order of magnitude.

Besides 3D imaging, PT imaging with fast image capturing has potential applications for live cell imaging and the tracking of physiological process such as biological transport in cells for example. PT microscopy equipped with a scanning mirror is a straightforward approach for fast imaging and there have been several reports of laser scanning PT microscopy [17, 44], which allows image acquisition with an 80 μs pixel dwell time so that 500 x 500 pixel images are acquired in 20 s.

To achieve fast imaging, it is also important to improve the signal-to-noise ratio (SNR) which inevitably reduces as it decreases measurement time. The PT signal increases with the pump power, but lower pump power is preferable to avoid photo and/or thermal damage to the sample. Several techniques have been proposed to improve SNR in PT microscopy. As the PT signal is proportional to the change in refractive index, chromophores embedded in liquid crystal, which exhibits a large refractive index change with temperature, produce larger PT signals than those in water [45, 46]. This technique is especially useful for single-molecule absorption spectroscopy.

3 FURTHER IMPROVEMENT

As mentioned above, a new scheme for photothermal microscopes systems with high detection sensitivity and with low noise by a balance detection method has been reported [25, 28, 29]. A spatially segmented balanced detection (SBD) method was proposed to improve SNR in forward detection [28]. In the SBD system, the center and peripheral parts of the transmitted probe beam are separated and detected by a balanced detector. This scheme was incorporated in the stage scanning PT microscope and improved the PT signal by nearly two times, while also canceling the intensity noise of the probe beam to reach nearly "the shot" noise limit.

New techniques for resolution enhancement and fast imaging in photothermal microscope (PTM) to leverage the capability of PTM.

3.1 3D Resolution Enhancement

First, to increase not only the lateral but also the axial resolution, we introduced an inverse-annular pupil filter rather than an annular pupil filter. Further improvement of the image quality can be attained by combination of the nonlinearity in the PT imaging and cancelation of the fringe patterns appearing in the lateral and axial intensity distributions of both pump and probe beams. It has been claimed theoretically that an inverse annular filter achieved a significant enhancement of the resolution of the fluorescence microscopy both in the lateral and axial directions

[47], but this claim has not actually been proven experimentally. Furthermore, the theoretically presented enhancement in the paper has been only discussed for the conventional fluorescence microscopy, not for pump-probe microscopy such as PT imaging. We also theoretically calculate the expected improvements in resolution. The resolutions are evaluated experimentally using gold nanoparticles (GNP) with a 20-nm diameter. The resolution evaluated in the X (perpendicular to the common pump and probe polarization direction), Y (parallel to the polarization direction), and Z (axial) directions are found to be improved by 15±6, 8±8, and 21±2 % relative to those without a pupil filter. The standard deviation is obtained by the properly estimated background signal inevitable in the imaging. The experimental results, which are even better than our theoretical predictions, are discussed. We also demonstrate PT imaging of thick biological samples (cells from rabbit intestine[8] and kidney[9]) stained with hematoxylin and eosin dye with enhanced resolution.

3.2 Higher Sensitivity

Second, we report the development of a fast, high-sensitivity PT microscope by implementing the SBD (segmented balance detection) scheme in a laser scanning microscope. For this purpose, a custom-made bifurcated fiber bundle was incorporated in a laser diode (LD)-based multimodal laser scanning microscope. Simultaneous PT and fluorescence (FL) imaging of gold nanoparticles and FL beads was performed to characterize the performance of the system. 3D visualization of endogenous mouse brain signals was also demonstrated.

3.3 High-speed Imaging

Third, we combine the nonlinear signal detection scheme enabling superresolution and high-speed imaging using a Galvano mirror with inexpensive laser diodes (LDs) as pump-probe light sources. The combined system can attain fast biological imaging with superresolution reaching 160 nm corresponding to the 61% improvement from the diffraction limit of probe wavelength enabling the clarification fine structures of biological cells.

4 INVERSE-ANNULAR FILTER

As mentioned before, we introduced a spatially blocking filter before a focusing objective lens of the microscope to enhance spatial resolution. Here detailed discussion of the methodology using one of spatial filters, i.e. "inverse annular filter", is described. To discuss spatial filters we first theoretically evaluated the spatial resolution of the pump-probe signal using a vector diffraction model with three illumination modes produced using three different sets of a pupil and filters, including inverse-annular illumination. To investigate the differences in spatial resolution among the three modes, we used the following three diameters for the filter, as illustrated in Figure 2.1.2.1(a): pupil diameter Φ_1, internal diameter of the annular beam Φ_2, and pupil diameter of the objective lens Φ_3. The values selected for these parameters are listed

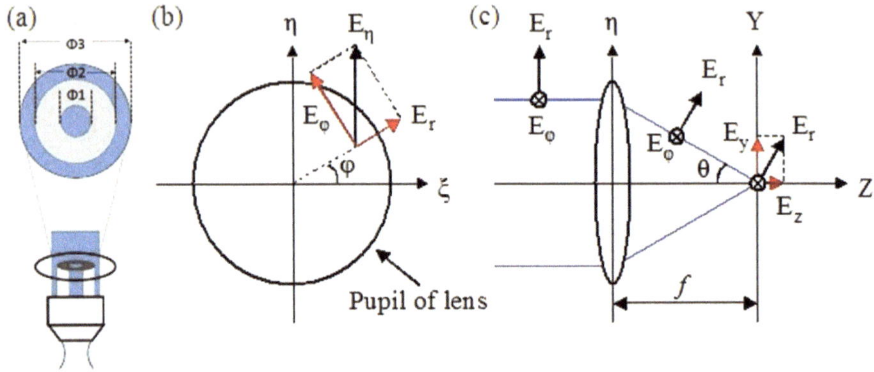

FIGURE 2.1.2.1 (a) Structure of inverse-annual filter in front of the objective pupil. The structures of circular, annular, and inverse annular filters are described with variable parameter sets of (Φ_1, Φ_2, Φ_3) listed in Table 2.1.2.1. This spatial filter can transmit light beam through the center and annular parts of the beams (blue-area). Diameters of the central and annular parts are expressed by Φ_1 and Φ_2, respectively. Φ_3 is the pupil diameter, which is determined by the objective lens. (b), (c) Electric field components in a lateral plane (b) and around a focal point (c). In this figure, the linear polarized laser is illuminated with polarization direction of η-axis. The figure was reported in the journal *Optical Review* and copyright is shared between the corresponding author (Takayoshi Kobayashi, TK) and the publisher Springer Verlag.

in Table 2.1.2.1. The radial coordinate is represented in terms of numerical aperture (NA).

Inverse-annular illumination produces both circular and annular beams [blue area in Fig. 2.1.2.1 (a)]. These beam areas determine the 3D spatial resolution in microscopy based on point-scanning systems [47, 48]. While Φ_3 depends on the focal length and NA of the objective lens, Φ_1 and Φ_2 can be set optimized by varying the beam-blocking ring structured area (shown in gray color). In this study, we set $\Phi_1 = 3.1$ mm, $\Phi_2 = 6.3$ mm and used the objective lens parameter $\Phi_3 = 8.55$ mm. By using vector diffraction theory we determined the 3D spatial resolution of the inverse-annular filter system before the pupil of the objective lens with much higher accuracy in comparison with the scalar diffraction theory. It is because the former theoretical treatment includes polarization dependence in each direction. Both pump

TABLE 2.1.2.1
Aperture Parameter Sets of the Filters for the Three Illumination Modes

Illumination mode	Φ_1 [mm]	Φ_2 [mm]	Φ_3 [mm]
Circular	0	0	8.55
Annular	0	5.8	8.55
Inverse annular	3.1	6.3	8.55

and probe fields in the present microscope system are linearly polarized parallel to Y-direction, so the resulting pump-probe signal intensity distribution on the focal plane is different between the parallel and perpendicular directions to the polarization [49, 50]. This situation differs from the case of scalar-approximation calculation, which has isotropic signal intensity distribution on the lateral plane perpendicular to the pump-probe beam.

Fig. 2.1.2.1 (b) shows the electric field components in a lateral plane (defined as XY). The illuminating laser is linearly polarized along the η-axis (or Y-axis), and the electric fields in the radial and declination directions (E_r and E_φ, respectively) in the microscopic region after the lens described schematically by an ellipsoidal shape are described as:

$$E_r = E_\eta sin\varphi \tag{1}$$

$$E_\varphi = E_\eta cos\varphi, \tag{2}$$

where E_η is the electric field component of the laser along the η-axis. E_r and E_φ can be resolved into X, Y, and Z components, as illustrated in Fig. 2.1.2.2(b) and given as:

$$E_{rx} = E_r cos\theta cos\varphi \tag{3}$$

$$E_{ry} = E_r cos\theta sin\varphi \tag{4}$$

$$E_{rz} = E_r sin\theta \tag{5}$$

and

$$E_{\varphi x} = -E_\varphi sin\varphi \tag{6}$$

$$E_{\varphi y} = E_\varphi cos\varphi. \tag{7}$$

From Eqs. (1) to (7), the components of the electric field amplitude in the X, Y, and Z directions in the microscopic region can be given by:

$$A_x = E_\eta \left(cos\theta sin\varphi cos\varphi - sin\varphi cos\varphi\right) \tag{8}$$

$$A_y = E_\eta \left(cos\theta sin^2\varphi - cos^2\varphi\right) \tag{9}$$

$$A_z = E_\eta sin\theta sin\varphi. \tag{10}$$

Point spread functions of these electric field components are determined in the following way. By the Fourier transform of the above equations [*Eqs.* (8), (9), and (10)] with proper integration range, the Point Spread Functions (PSF) are written by:

$$E_x\left(x,\ y,\ z\right)=C\iint A_x\left(cos\theta\right)^{\frac{1}{2}}exp\left\{-ik\left(xsin\theta cos\varphi+ysin\theta sin\varphi+zcos\theta\right)\right\}sin\theta d\theta d\varphi \tag{11}$$

$$E_y\left(x,\ y,\ z\right)=C\iint A_y\left(cos\theta\right)^{\frac{1}{2}}exp\left\{-ik\left(xsin\theta cos\varphi+ysin\theta sin\varphi+zcos\theta\right)\right\}sin\theta d\theta d\varphi \tag{12}$$

$$E_z\left(x,\ y,\ z\right)=C\iint A_z\left(cos\theta\right)^{\frac{1}{2}}exp\left\{-ik\left(xsin\theta cos\varphi+ysin\theta sin\varphi+zcos\theta\right)\right\}sin\theta d\theta d\varphi. \tag{13}$$

Here, k is the wave number and is equal to $2\pi\ n/\lambda$, θ is determined by the polar coordinates, and $r=n\sin\theta$ is the numerical aperture (NA), and Eqs. (11), (12) and (13) can be rewritten as follows:

$$E_x\left(x,\ y,\ z\right)=C\iint A_x exp\left\{-ik\left(rxcos\varphi+rysin\varphi+z\left(1-\left(r/n\right)^2\right)^{\frac{1}{2}}\right)\right\}drd\varphi \tag{14}$$

$$E_y\left(x,\ y,\ z\right)=C\iint A_y exp\left\{-ik\left(rxcos\varphi+rysin\varphi+z\left(1-\left(r/n\right)^2\right)^{\frac{1}{2}}\right)\right\}drd\varphi \tag{15}$$

$$E_z\left(x,\ y,\ z\right)=C\iint A_z exp\left\{-ik\left(rxcos\varphi+rysin\varphi+z\left(1-\left(r/n\right)^2\right)^{\frac{1}{2}}\right)\right\}drd\varphi \tag{16}$$

with

$$A_x=\frac{r}{n}\left[\left\{1-\left(\frac{r}{n}\right)^2\right\}^{\frac{1}{4}}-\left\{1/\left(1-\left(\frac{r}{n}\right)^2\right)\right\}^{\frac{1}{4}}\right]sin\varphi cos\varphi, \tag{17}$$

$$A_y=\frac{r}{n}\left[\left\{1/\left(1-\left(\frac{r}{n}\right)^2\right)\right\}^{\frac{1}{4}}sin^2\varphi+\left\{1-\left(\frac{r}{n}\right)^2\right\}^{\frac{1}{4}}cos^2\varphi\right], \tag{18}$$

and

$$A_z = \frac{r^3}{n}\left\{1/\left(1-\left(\frac{r}{n}\right)^2\right)^{\frac{1}{4}}\right\} sin\varphi. \tag{19}$$

The intensity distribution within the image volume range is given by the equation,

$$I(x, y, z) = |E_x(x, y, z)|^2 + |E_y(x, y, z)|^2 + |E_z(x, y, z)|^2 \tag{20}$$

This is calculated by integration over the open aperture area represented by pitch coordinate, r, of the NA and the declination angle φ . The integration ranges in the radius direction are $0 < r < 0.95$ for no filter (circular illumination), $0.68 < r < 0.95$ for the annular filter and $0 < r < 0.34$ and $0.70 < r < 0.95$ for the inverse annular filter in the unit of numerical aperture NA. The integration range in the declination angle is $0 < \varphi < 2\pi$ for all calculation.

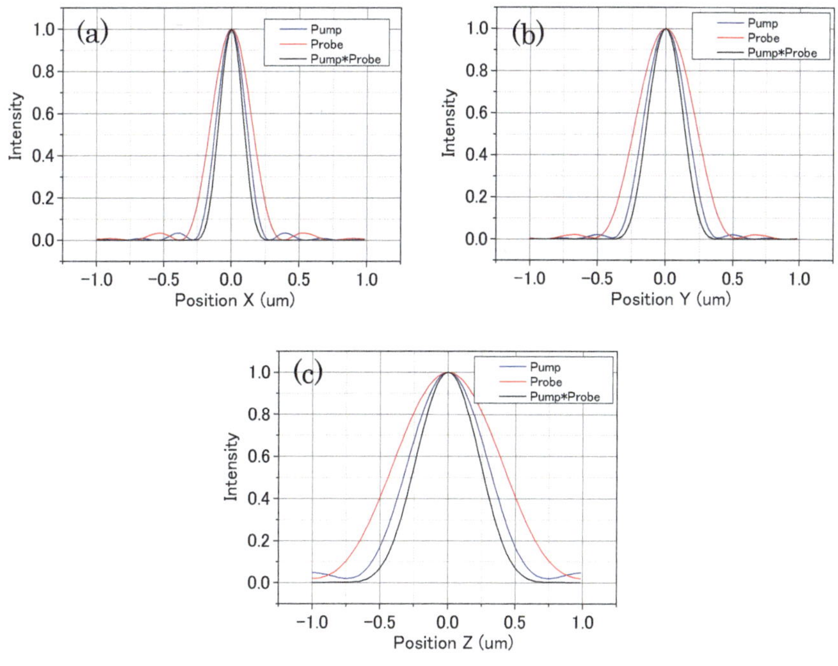

FIGURE 2.1.2.2 Calculated results of the intensity distribution to show the resolution improvement in pump-probe microscopy in X-axis (a), Y-axis (b) and Z-axis (c). The resolution of the intensity for the pump probe microscopy (black lines) is better than the resolution for the pump (blue lines) and the probe (red line) laser. In the calculation, the wavelengths of the pump and probe laser are 488 nm and 660 nm, respectively. The figure was reported in the journal *Optical Review* and copyright is shared between the corresponding author (Takayoshi Kobayashi, TK) and the publisher Springer Verlag.

TABLE 2.1.2.2
FWHMs of Lateral Intensity Distributions of the Pump Beam, the Probe Beam, and the Pump-probe Signal in Figure 2.1.2.2; Improvement of PP Results with Two Colors from the Cases of Single Color Experiment

Image direction	Pump (Blue) [nm]	Probe (Red) [nm]	Pump Probe (PP) (Black) [nm]	Improvement PP / Pump	Improvement PP / Probe
X	234	315	197	16 %	37 %
Y	337	456	285	15 %	37 %
Z	635	863	527	17 %	39 %

The laser wavelengths of pump and probe are 488 nm and 660 nm, respectively. The refractive index used in the calculation is set to be 1 (air).

Calculated results of the resolution enhancement in pump-probe microscopy with circular illumination (Figure 2.1.2.1) are shown in Figure 2.1.2.2. Since signal intensity distribution of the pump-probe signal, $I_{sig}(x, y, z)$ is given approximately by $I_{sig}(x, y, z) = I_{pu}(x, y, z) \times I_{pr}(x, y, z)$ [10], the peripheral ring structures introduced by diffraction in both the pump and probe beams can be canceled out because the ring radii of the pump and probe are different, because the two beams are of different wavelength. This can be confirmed in Figure 2.1.2.2, which shows intensity distribution improvement in the pump probe measurement by the similar calculation by the PSF. The improvement of the resolution in XY and XZ planes are 37% and 39%, respectively, as shown in Table 2.1.2.2 based on Fig. 2.1.2.2. Thus, in the case of PT imaging, inverse-annular illumination is even more advantageous.

The calculated results of intensity distribution are depicted in Figure 2.1.2.3, which shows the distributions in the XY (a1, b1, c1) and XZ (a2, b2, c2) planes of the pump-probe signal by circular illumination (a, without the filter, $\Phi_1 = \Phi_2 = 0$ mm), annular illumination (b, $\Phi_1 = 0$ mm, $\Phi_2 = 5.8$ mm), and inverse-annular illumination (c). The polarization direction is taken to be along Y-axis. Figure 2.1.2.4 shows cross-section profiles and Table 2.1.2.3 lists the calculated resolution of lateral (X- and Y-) directions and axial (Z-) direction in the case of Z-polarization probe field in terms of the full width a half maximum (FWHM) values of the PSF. As shown in these figures and the table, the improvements of resolutions with annular filter from the circular case in the lateral direction are 15% and 19%, respectively, in parallel (X-axis) and perpendicular (Y-axis) directions to the laser field (X-axis). However, the enhancement is at the expense of lowering the resolution in the axial direction by nearly 50% reduction indicating that the distribution is increased to 1.5 times in the spatial spread width of the image of delta-function like object. In contrast to the annular filter case, the improvements gained using the inverse-annular filter compared with the circular aperture case in the directions parallel and perpendicular to the laser field are 8% and 11%, respectively, and 6% improvement can be obtained even in the axial direction.

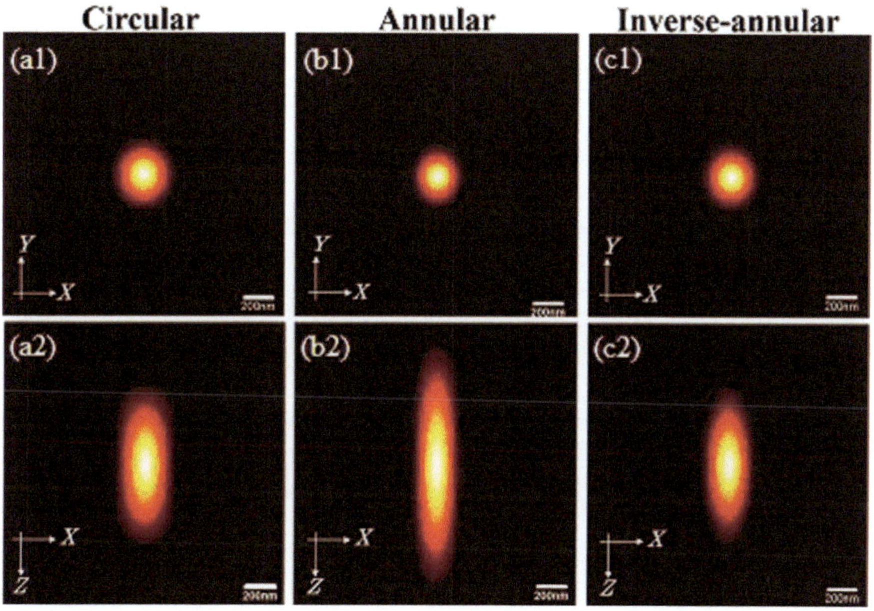

FIGURE 2.1.2.3 Intensity profiles in the planes of XY (a1, b1, and c1) and XZ (a2, b2, and c2), near the focal point of the objective lens in pump-probe calculation. The wavelengths of pump and probe beams are 488 and 660 nm, respectively. Numerical aperture (NA) of the objective lens is 0.95. The figure was reported in the journal *Optical Review* and copyright is shared between the corresponding author (Takayoshi Kobayashi, TK) and the publisher Springer Verlag.

3 EXPERIMENTAL SETUP

3.1 Pump-probe Microscope System

Figure 2.1.2.5 shows a schematic illustration of our pump-probe microscopy system, which employs two laser diodes (LDs) as the pump and probe light sources. A 488-nm LD (Nichia, NDS4216) and a 660-nm LD (Mitsubishi, ML101U29) are used for the pump and probe beams, respectively. The two beams are modulated at frequencies of $_{\omega 1}$ (pump) and $_{\omega 2}$ (probe) using signal generators (SGs) and are introduced to beam collimators that contain spatial filters. After the beams are combined using a beam-combiner made of a dichroic mirror, they irradiate the sample after passing through an objective lens (Olympus, UPLSAPO 40X2) with a NA of 0.95. A condenser lens collects the transmitted probe beam from the sample and introduces the light beam to a detector. To reduce the effect of probe noise, we use an

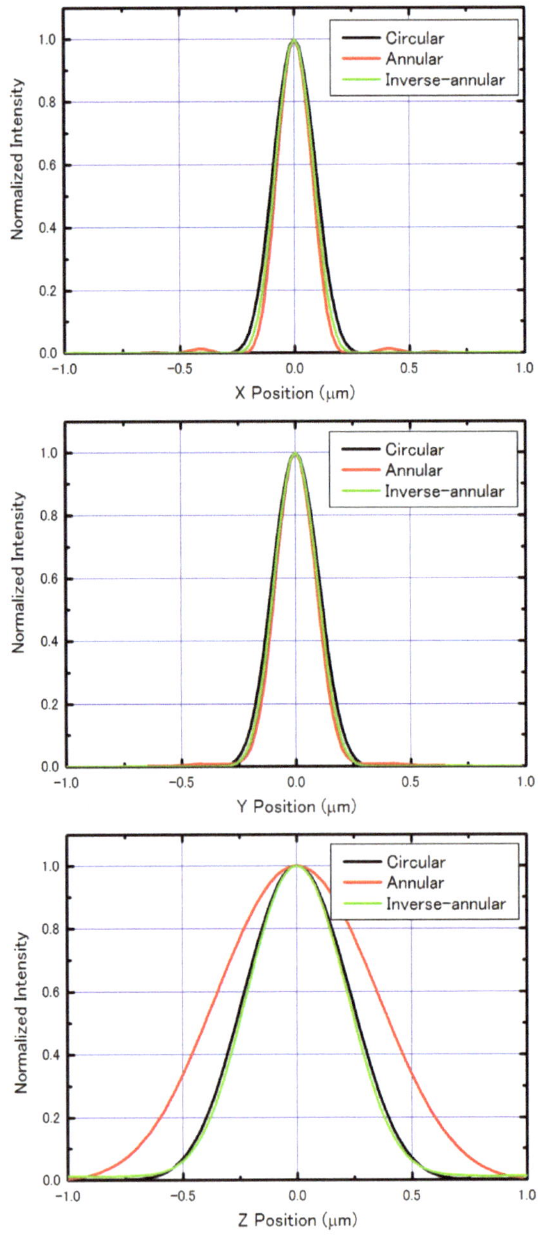

FIGURE 2.1.2.4 Cross-section profiles of the intensities by the pump-probe calculation in Figure 2.1.2.3 in X, Y, and Z directions. The width of the PSF in the X and Y-axis direction is decreased, but the width in the Z-axis direction is increased greatly in annular illumination (red lines) from the circular illumination (black lines). The resolution is also improved in the inverse-annular illumination (green lines) in any direction. The figure was reported in the journal *Optical Review* and copyright is shared between the corresponding author (Takayoshi Kobayashi, TK) and the publisher Springer Verlag.

TABLE 2.1.2.3
FWHMs in PSF by the Pump-probe Calculation in Figure 2.1.2.3

Imaging direction	Circular [nm]	Annular [nm]	Inverse-annular [nm]	Improvement (Annular filter)	Improvement (IA filter)
X-axis	207	167	185	19 %	11 %
Y-axis	232	198	214	15 %	8 %
Z-axis	527	793	494	-50 %	6 %

The Laser wavelengths of pump and probe are 488 nm and 660 nm, respectively. The refractive index used in the calculation is set to be 1.5 (PVA film employed as a matrix material in the experiment of resolution evaluation using the gold nanoparticles).

auto-balanced detector (Newfocus, Nirvana) containing two photodiodes; one is for the probe beam and the other is for a reference beam split off from the probe beam by a beam splitter (BS in Fig. 2.1.2.5) placed before a dichroic mirror (DM in Fig. 2.1.2.5). The intensity noise of these beams is canceled using this balance detection scheme. Thus, shot-noise-limited sensitivity can be achieved when the probe beam power is higher than ~about 10 μW measured on the photodiode surface [25]. A lock-in amplifier (Signal Recovery, 7270) detects the signal at the beat frequency $|_{\omega 1} - _{\omega 2}|$. In this study, we set modulation frequencies $_{\omega 1}$ and $_{\omega 2}$ to 1.000 MHz and 1.015 MHz,

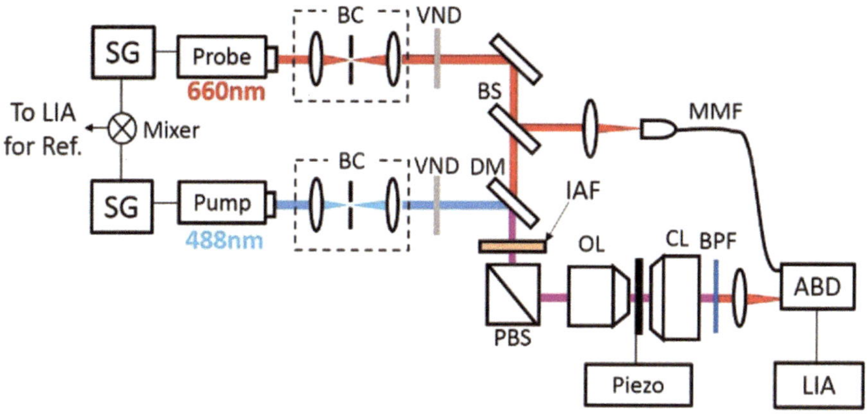

FIGURE 2.1.2.5 A scheme of pump-probe microscope system. SG: signal generator, Pump: 488 nm laser diode for pump light, Probe: 660 nm laser diode for probe light, BC: beam collimator, VND: variable neutral density filter, BS: probe beam splitter, DM: dichroic mirror, IAF: inverse-annular filter, PBS: polarized beam splitter, OL: objective lens, CL: condenser lens, BPF: band-pass filter, ABD: auto-balanced detector, LIA: lock-in amplifier, MMF: multi-mode fiber. The figure was reported in the journal *Optical Review* and copyright is shared between the corresponding author (Takayoshi Kobayashi, TK) and the publisher Springer Verlag.

respectively. A frequency mixer (Mini-circuit, ZAD-1) generates $|_{\omega 1} - _{\omega 2}|$ component for a reference signal of the lock-in amplifier. Closed loop piezo-actuators (Thorlabs, MAX311D) scan the sample in the three axial directions. An auto-balance detector (ABD) is used to improve the signal to noise ratio [18, 25–29].

To improve the 3D spatial resolution, an inverse-annular filter (IAF) is installed before the polarization beam splitter (PBS). The beam-blocking area of this spatial filter is made by affixing a metal washer (clip plate, facing ring) onto an optical window. A holder mounting the IAF is adjusted to produce a symmetric inverse-annular beam in the objective pupil plane.

3.2 Galvano Mirror Setup

A laser scanning PT microscope equipped with Galvano mirrors was developed based on a LD-based multi-wavelength PT microscope previously developed in our group (Figure 2.1.2.6) [10]. A 405-nm (Nichia NDV4B16) and a 520-nm (Osram PL-520-B1) LDs were used for pumping, with their intensity modulated at different frequencies ($f_1$and f_2). The two pump beams were combined using dichroic mirrors and collimated through polarization-maintaining single-mode fiber (PMSMF)

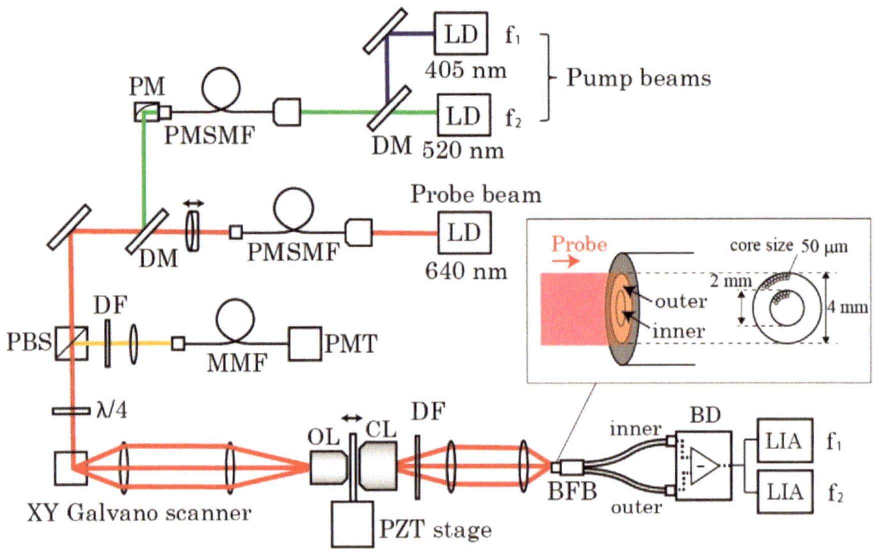

FIGURE 2.1.2.6 Schematic illustration of a photothermal microscope equipped with a Galvano scanner and spatially segmented balanced detection. LD: laser diode, PMSMF: polarization-maintaining single-mode fiber, PM: parabolic mirror, DM: dichroic mirror used for beam coupling, PBS: polarizing beam splitter, OL: objective lens, CL: condenser lens, DF: dielectric filter, BFB: bifurcated fiber bundle, BD: balanced detector, LIA: lock-in amplifier, MMF: multimode fiber, PMT: photomultiplier tube. The figure was reported in the journal *Optical Review* and copyright is shared between the corresponding author (Takayoshi Kobayashi, TK) and the publisher Springer Verlag.

and an off-axis parabolic mirror (Thorlabs RC08FC-P01) for spatial-mode filtering. A 640-nm LD (Opnext HL63133DG) was used for probing, which was collimated through another PMSMF and a doublet lens. The offset between the pump and probe beams at the focus was controlled by adjusting the position of the doublet lens to reduce the distortion of the PT signal caused by the twin peaks corresponding to the focusing or defocusing of the probe beam [29]. The combined pump and probe beams were directed to an XY Galvano scanner (GSI VM500PLUS) and focused on a sample through an oil immersion objective lens [(Olympus PLAPON60X) numerical aperture (NA) = 1.4]. To acquire images in different focal planes to construct 3D z-stacked image, the sample position was controlled in the axial direction using a positioning stage driven by a piezo actuator (MESS-TEK MA-140XLCa).

The SBD method was demonstrated to be useful for high-sensitivity PT imaging in forward detection as it improves signal intensity by nearly two times while also canceling the intensity noise of the probe beam [28]. In PT microscopy, the SNR is usually higher in forward detection than in backward (epi-)detection as in the report of previous works [10, 45]. In our previous stage-scanning PT microscope (not reported openly), a small mirror was used to separate the center and peripheral parts of the transmitted probe beam. However, it is impractical to incorporate this setup in a laser scanning microscope as it causes complexity in the optical setup making the alignment difficult and time-consuming. Therefore we decided not to use this configuration. To implement the SBD method, which is in principle an excellent method to improve S/N ratio of the detected image by a laser scanning microscope, we took the following scheme. A custom-made bifurcated fiber bundle with a core fiber size of 50 μm was used to separate the center and peripheral part of the transmitted probe beam. The diameters of the inner and outer circles were 2 and 4 mm, respectively. The incident port of the fiber bundle was placed on a plane conjugate to the pupil of the condenser lens. The center and peripheral parts of the beams separated by the fiber bundle were guided to a balanced photodetector (Newfocus 2307) to cancel the intensity fluctuation of the probe laser. The ratio of the laser powers between the two separated beams depends on the beam spot size at the incident port of the fiber. Thus, the focal lengths of the two lenses after the condenser lens were properly selected so that the laser powers of the two separated beams were nearly equal for optimal common mode noise rejection. PT signals (normal mode) at frequencies of f_1 and f_2 were demodulated by two lock-in amplifiers of the same model (Signal Recovery model 7270). The output from the lock-in amplifiers was recorded by a multichannel date acquisition board (NI PCIe-6321), which was also used to control the speed and range of the Galvano scanner. The time constant t_c of the lock-in amplifier was set at 10 μs and the pixel dwell time was 20 μs. In this case, image acquisition time was about 6 s for 500 x 500 pixels. The modulation frequencies (f_1 and f_2) were typically set at 150 kHz and 250 kHz, respectively. We confirmed that the channel interference effect was negligibly small since the bandwidth of the lock-in amplifier $1/(2t_c)$ was sufficiently narrower than the difference in modulation frequencies $|f_1 - f_2|$.

A confocal FL detection scheme was also set up for simultaneous PT and FL imaging, in which the FL signal from the sample was detected by a photomultiplier through a multimode fiber.

To prepare a slice of mouse brain, we fixed adult transgenic mice (two months old, thy1-GFP H-line, Jackson Lab.) by cardiac perfusion with PBS and paraformaldehyde, cut 100-μm slices of the frontal cortex using a vibratome (Leica VT1200), and embedded the slices in UNICRYL resin (Funakoshi). To remove the glycogen, the slices of mouse brain were incubated with 10 mg/ml α-amylase (Tokyo Chemical Industry) in PBS for one hour at 37°C. Then the slices were immunostained with anti-glycogen antibody (a gift from Dr. Otto Baba) to confirm the removal of glycogen. The slices were stained first with the anti-glycogen antibody (12 hours, dilution 1:10 at 4°C), followed by Alexa-Fluor-488 conjugated secondary antibody (Life Technologies) staining (2 hours, dilution 1:1000 at 20°C).

In the Galvano scanner system for photothermal microscopy experiments, an objective lens of 40x and NA = 0.95 is used. The wavelengths of the pump and probe laser diodes are 405 nm and 638 nm, respectively. Noise from the probe laser is reduced by the scheme of balance detection between probe beam transmitted through the sample and the reference beam split by a polarized beam splitter (PBS). Modulation frequency of the pump LD is 100 kHz. Images of 20-nm gold nanoparticles (GNPs) in polyvinyl alcohol (PVA) film spin coated on a slide glass plate, the HE stained rabbit organs (Animal Tissue D prepared slides, Japan Artec), and HE stained mouse brain are obtained from the signal in linear (f) and second-order nonlinear (2f) lock-in detection modes, when modulation frequency of the pump laser is f. The sample thickness of the GNP film: biological samples are about 15 μm and 10 μm, respectively. The pixel number is 90,000 = 300 (x) × 300 (y) points and the frame time of each linear and nonlinear images is 4.5 sec (= 50 μs × 90,000 points). The frame rate in the Galvano mirror system is 20 times shorter than the frame time when the piezo stage is used. The resolution of the bright-field image depends on the resolution of the detector (~5 μm, Thorlabs DCC1645, ⅓ inch CMOS, 1280 × 1024) and the image is taken with incoherent light irradiated from the back side. Details of the difference of the resolution between the f and 2f scheme were described in the references [33, 34].

4 THEORY

For the support of experimental study of photothermal microscope development, we used a model by considering both the effects of the focusing field and modulating heat sources on the photothermal signal. We consider the case when pump and probe beams are incident on the sample and only the pump beam is absorbed by the point absorber, which then causes an increase in temperature in a surrounding medium. The temperature rise induces an increase or decrease in the refractive index depending on the sign of the thermo-optic index and thus the probe beam is deflected and/or scattered. The schematic image of photothermal microscope is shown in Fig. 2.1.2.7. As modulation frequency of the pump beam intensity is 1 kHz to 10s kHz and much lower than the frequency of the electromagnetic field beyond petaHz, we consider the quasi-static Helmholtz equation [21].

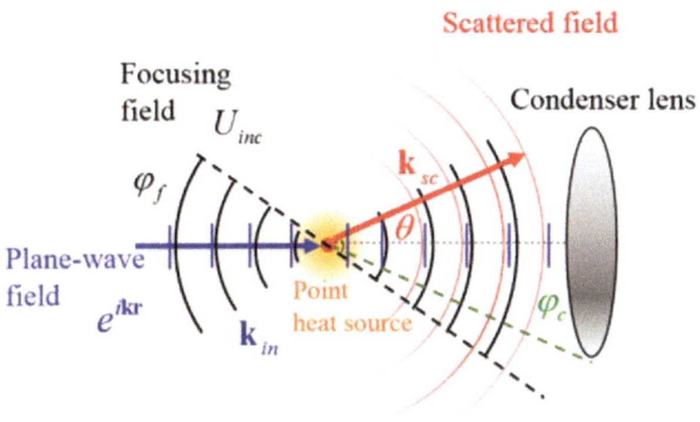

FIGURE 2.1.2.7 Schematic illustration of the principle of photothermal microscopy. A plane-wave field (blue lines) or a focusing field with NAf = $n_0 \sin \varphi_f$ (black curves) is incident on the sample and is deflected (scattered) due to the local refractive index change around a point heat source. Transmitted and deflected photons are collected through a condenser lens with NAc = $n_0 \sin \varphi_c$. The figure was reported in the journal *Optical Review* and copyright is share between the corresponding author (Takayoshi Kobayashi, TK) and the publisher Springer Verlag.

(Requesting comment: removed (1)-(15) in the following by white boxes in case the white boxes are moved by accident from the positions.)

$$\nabla^2 U(\mathbf{r},t) + \mathrm{k}^2 \left[n(\mathbf{r},t) / n_0 \right]^2 U(\mathbf{r},t) = 0, \tag{21}$$

where $U(\boldsymbol{r}, t)$ is the scalar envelope electric field of the probe beam. The refractive index change is related to the temperature rise by $n(\boldsymbol{r}, t)$. The excitation laser intensity is modulated with frequency of Ω and heat absorbed by the material is given by $\sigma I_0(1 + \cos\Omega t)$, where σ is the absorption cross-section at the pump laser wavelength and I_0 is the beam intensity at the focal point. Assuming that the size of the heat absorber is much smaller than the focal spot size, the temperature profiles can be described by using the heat point source model by the following equation [11, 17].

$$\Delta T(\mathbf{r},t) = \frac{\sigma I_0}{4\pi\kappa}\left[\frac{1}{r} + \frac{1}{r}\exp\left(-\frac{r}{r_c}\right)\cos\left(\Omega t - \frac{r}{r_c}\right)\right]. \tag{22}$$

Here κ is the thermal conductivity, $r_c = (2D/\Omega)^2$ is the characteristic length for thermal diffusion length during one period with D, the thermal diffusion coefficient, and r is the distance from the point heat source. Substitution of Eq. (21) reads

$$\nabla^2 U(\mathbf{r},t) + \mathrm{k}^2 \left[1 + W(\mathbf{r},t) \right]^2 U(\mathbf{r},t) = 0, \tag{23}$$

where

$$W(\mathbf{r},t)=\frac{\sigma I_0}{4\pi\kappa}\left[\frac{1}{r}+\frac{1}{r}\exp\left(-\frac{r}{r_c}\right)\cos\left(\Omega t-\frac{r}{r_c}\right)\right]. \tag{24}$$

Equation (19) is equivalent to the quasi-static Schrödinger equations, where $W(\boldsymbol{r}, t)$ corresponds to the potential field. The first term on the r.h.s. of Eq. (20) decays with the inverse of radial distance and corresponds to the Coulomb potential in quantum mechanical treatment. The Coulomb potential is effective over a greater distance r than r_0. The second term is sinusoidally modulated and the amplitude $(1/r)\exp(\zeta\, r/r_c)$ approaches zero at large r. This factor is equivalent to the Yukawa potential (the screened Coulomb potential), which was introduced to describe strong nuclear force between nucleons in particle physics [23]. The Yukawa potential coincides with the Coulomb potential in the limit $r_c\,\zeta\,\zeta$. In photothermal microscopy, the first term can be regarded as the interaction with the temporally averaged (static) point heat source, whereas the second term is the interaction with the modulated point heat sources and thus the interaction length should be smaller than r_c. The asymptotic solution of Eq. (23) in the far-field region for an incident plane wave reads

$$\lim_{r\to\infty} U_{\mathbf{k}}(\mathbf{r},t)\sim e^{i\mathbf{k}z}+\frac{i|\mathbf{k}|z}{r}f_k(\theta,t), \tag{25}$$

where $\mathbf{k}$ denotes the wave vector of the incident plane wave and $f_k(\theta, t)$ is the scattering amplitude, with θ the angle between $\mathbf{k}$ and the scattering field wave vector $\mathbf{k}_{sc}$ as shown in Fig. 2.1.2.1. The scattering amplitude is calculated under the first-order Born approximation as

$$\begin{aligned}f_{\mathbf{k}}^{(1)}(\theta,t)&=-\frac{1}{4\pi}\int W(\mathbf{r}',t)e^{i\Delta\mathbf{k}\cdot\mathbf{r}'}d\mathbf{r}'\\&=-\frac{\sigma I_0}{8\pi\kappa n_0}\frac{\partial n}{\partial T}\left[\frac{2}{|\Delta\mathbf{k}|^2}+\frac{2|\Delta\mathbf{k}|^2}{|\Delta\mathbf{k}|^4+4/r_c^4}\cos(\Omega t)+\frac{4/r_c^4}{|\Delta\mathbf{k}|^4+4/r_c^4}\sin(\Omega t)\right]\end{aligned} \tag{26}$$

with $|\Delta\mathbf{k}|=|\mathbf{k}-\mathbf{k}_s|=(4\pi n_0/\lambda)\sin(\theta/2)$. The first term in the square is given by the following

$$\begin{aligned}\sigma_{sc}(t)&=2\pi\int_0^{\pi}\left[f_k^{(1)}(\theta,t)\right]^2\sin\theta d\theta\\&=\frac{4\pi}{k}\operatorname{Im} f_k^{(2)}(0,t).\end{aligned} \tag{27}$$

$$\Sigma(\varphi_c,t)=\sigma_{sc}(t)-2\pi\int_0^{\varphi_c}\left[f_k^{(1)}(\theta,t)\right]^2\sin\theta d\theta, \tag{28}$$

In photothermal microscopy, the focusing beam is incident on the sample through an objective lens and thus we consider a super-position of Eq. (5) at multiple incident angles to examine the effect of the focusing field on the photothermal signal

$$U(\mathbf{r},t)=\int d\mathbf{k}A(\mathbf{k})U_{\mathbf{k}}(\mathbf{r},t)\sim U_{inc}(\mathbf{r})+U_{sc}(\mathbf{r},t). \tag{29}$$

Here,

$$U_{inc}(\mathbf{r})=\int d\mathbf{k}A(\mathbf{k})e^{i\mathbf{k}\cdot\mathbf{r}} \tag{30}$$

denotes the incident focusing field and

$$U_{sc}(\mathbf{r},t)=\frac{e^{i|\mathbf{k}|r}}{r}\int d\mathbf{k}A(\mathbf{k})f_{\mathbf{k}}(\theta,t) \tag{31}$$

is the scattered field. For example, for an azimuthally symmetric and uniform intensity spherical wave focusing at $\mathbf{r}_0$, with sin $NA_f = n_0$ sin $_{\zeta f}$, the amplitude factor $A(\mathbf{k})$ is expressed by

$$A(\mathbf{k})=\begin{cases} e^{-i\mathbf{k}r_0}/|\mathbf{r}-\mathbf{r}_0| & \text{for } \varphi'\le\varphi_f \\ 0 & \text{for } \varphi'>\varphi_f \end{cases}, \tag{32}$$

where φ' is the polar angle of **k**. By assuming that $f_{\mathrm{k}}(\zeta, t,) << 1$, the square of the scattering field is negligibly small, and in this case, interference between the scattered field and the transmitted field is dominant. Thus, we approximate the scattering cross-section as an intensity ratio of the integrated interference signal to the incident beam intensity

$$\sigma'_{sc}(t)=J_{sc}(\pi,t)/J_{inc}(\pi), \tag{33}$$

$$\begin{aligned} J_{sc}(\varphi,t)&=(1/2)\left[2\pi\int_0^{\varphi}U_{sc}(\mathbf{r},t)U_{inc}(\mathbf{r})^{*}\sin\theta d\theta+c.c.\right] \\ J_{inc}(\varphi)&=2\pi\int_0^{\varphi}U_{inc}(\mathbf{r})U_{inc}(\mathbf{r})^{*}\sin\theta d\theta. \end{aligned} \tag{34}$$

When the incident and deflected photons are collected through a condenser lens in far-field region, the photothermal signal is proportional to

$$\Sigma'(\varphi,t)=\left[\sigma'_{sc}(t)J_{inc}(\varphi_c)-J_{sc}(\varphi_c,t)\right]/J_{inc}(\pi), \tag{35}$$

where the first term on the r.h.s of Eq. (15) refers to the fraction of photons that have incident angle lower than φ_{c} and undergo scattering, while the second term is the fraction of the scattered photons in the angle lower than φ_c. Here, we assumed that the incident field is focused at the position of the point light absorber. The scattering amplitude is given by the sum of the three terms in Eq. (6), so Eq. (15) can be rewritten separately as

$$\zeta' (\varphi) \zeta C(\varphi_c) \zeta Y_{in} (\varphi_c) \cos(\zeta t) \zeta Y_{out} \sin (\zeta t) \quad (36).$$

Here C(φc) is attributable to the first term (the Coulomb potential), and $Y_{in}(\varphi_c)$ and $Y_{out}(\varphi_c)$ to the second and third terms (the temporally modulated Yukawa potential), respectively. In a typical photothermal experiment, the pump beam intensity is modulated and the resulting change in the probe beam transmissivity is detected by a dual-phase lock-in amplifier. In this case, $Y_{in}(\varphi_c)$ and $Y_{out}(\varphi_c)$ are measured as in-phase and π/2 as out-of-phase components, respectively.

4 RESULTS

4.1 Evaluation of 3D Spatial Resolution by Imaging Gold Nanoparticle (GNP) with Pupil Filters

To evaluate the improvement in the 3D spatial resolution, we capture PT images of 20-nm gold nanoparticles (GNPs) dispersed in a polyvinyl alcohol (PVA) film. Although the GNPs have a large absorption cross section due to their plasmon resonance, the luminescence yield is extremely small [10]. Thus, the pump beam causes an increase in temperature around the GNPs due to the PT effect, which in turn results in decrease in the local refractive index of the PVA at the modulation frequency $_{\omega 1}$. This refractive index change forming a thermal concave lens leads to a deflection of the probe beam and is detected by the ABD. The beat signal at $|_{\omega 1} - _{\omega 2}|$ is generated by the bilinear interaction of the pump and probe fields in the sample [27], which is detected using the lock-in scheme.

The sample is scanned in the focal plane [XY-plane, Figures 2.1.2.8(a) and (b)] and optical axis plane [XZ-plane, Figs. 2.1.2.8(c) and 8(d)] with and without the IAF (i.e., inverse-annular and circular illumination), respectively. The pump and probe powers are respectively set to 3.8 mW and 0.3 mW for inverse-annular illumination case and to 1.6 mW and 0.2 mW for circular illumination case. The time constant of the lock-in amplifier is 0.5 ms, and the dwell time is 1.0 ms. The pixel number in the image of the gold nanoparticle is 300 x 300 and the frame time of the images is 90 seconds. The motor driven stage step is 50 nm. The signal to noise ratios of the data of inverse-annular (black points) and circular (red points) illumination cases in Fig. 2.1.2.8 are 7.2 ± 2.8 and 11.5 ± 1.5 (e), 8.2 ± 3.5 and 12.8 ± 3.4 (f), 11.0 ± 6.3 and 6.5 ± 2.4 (g), respectively. The signal to background ratios of the data of inverse-annular (black points) and circular (red points) illumination in Fig. 2.1.2.7 are 12.1 ± 1.4 and 12.4 ± 1.1 (e), 10.8 ± 0.8 and 11.2 ± 0.9 (f), 19.5 ± 1.8 and 11.9 ± 0.7 (g), respectively. Table 2.1.2.4 shows the full width at half-maximum (FWHM) of the fitted Gaussian curves shown in Fig. 2.1.2.7. In the X and Y-axis directions, FWHMs for the inverse-annular illumination are smaller than in the circular illumination. In addition, each FWHM value agrees well with its corresponding value calculated

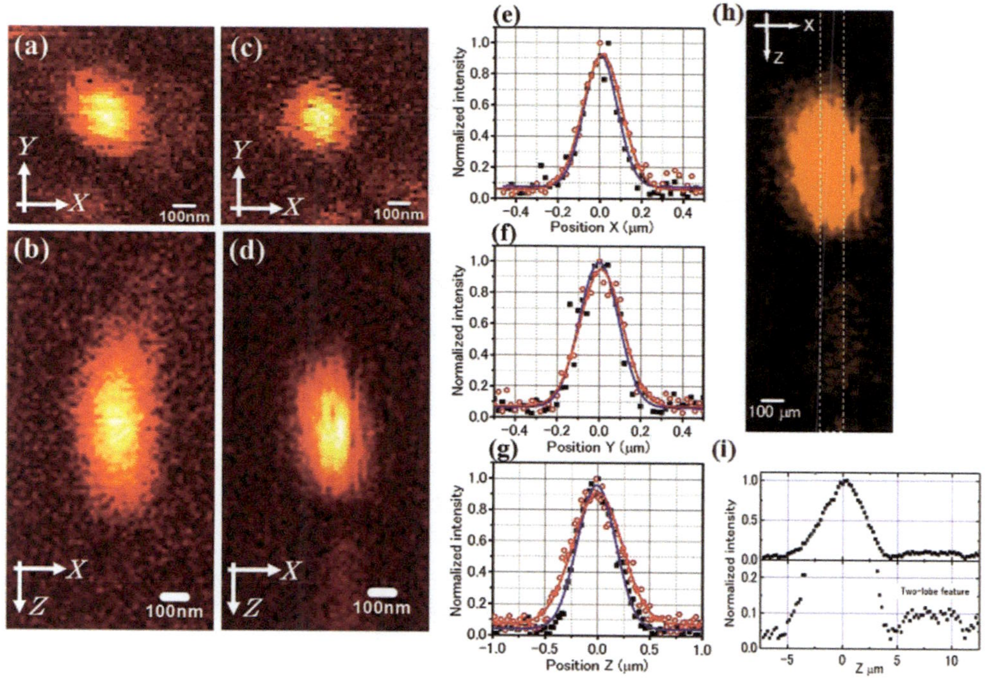

FIGURE 2.1.2.8 Photothermal images of a single GNP in PVA film and their intensity profiles. Normal illumination configuration in XY (a) and XZ-plane (b). Inverse-annular illumination configuration in XY (c) and XZ-plane (d). XZ-plane images are obtained directly by XZ scanning with piezo stage. Intensity cross section profiles of normal (red open dots) and inverse-annular illumination (black filled squares) in X, Y, and Z-axis directions are shown in (e), (f), and (g), respectively. Red and blue solid lines are Gaussian fitting curves in normal and inverse-annular illumination, respectively. Image with the two-lobe feature is shown in (h) and intensity along the axial direction integrated between two lines in (h) is shown in (i). The figure was reported in the journal *Optical Review* and copyright is shared between the corresponding author (Takayoshi Kobayashi, TK) and the publisher Springer Verlag.

TABLE 2.1.2.4
FWHMs of the Photothermal Intensity by the Pump-probe Calculation and Peaks of GNP Images in Figure 2.1.2.8

Imaging direction	Calculated Circular [nm]	Observed Circular [nm]	Calculated Inverse-annular [nm]	Observed Inverse-annular [nm]	Improvement by calculation	Improvement by experiment
X-axis	208.0	219±6	186.1	187±8	11 %	15±6 %
Y-axis	232.9	239±10	214.9	219±11	8 %	8±8 %
Z-axis	527.4	550±10	494.4	430±8	6 %	21±2 %

using vector diffraction theory, in Table 2.1.2.3. The resolution enhancement in the X and Y-axis directions are evaluated to be 15±6 %, 8±8 %, and 21±2 %, respectively. Difference in resolution enhancement between the X and Y-axis directions is due to the linear polarizations of the pump and probe beams being both along Y-axis. In the Z-axis direction, the resolution obtained using inverse-annular illumination is also better than the conventional circular illumination. Surprisingly, the enhancements in resolution obtained experimentally are even better than those predicted theoretically, which is discussed in the following.

Selmke and coworkers [51] discussed the dependence of the image resolution on the relative position of the pump and probe focal points. They found that the apparent lateral resolution depends sensitively on the relative focal positions of the two beams. In some cases, the intensity (= squared amplitude) image along the axial direction had two lobes in the amplitude distribution along the axial direction with a π phase jump, resulting in an apparent reduction in the lateral signal distribution caused by a cancelation between the two lobes. Even though this effect may distort the image, the location of the signal-generating particles does not change and therefore can be used in actual imaging when a precise lateral distribution of the target system is not required. By removing one of the lobes by shifting the relative position of the pump or probe focal plane, the two-lobe feature is detected as shown in Figs. 2.1.2.8(h) and (i). Because of this effect, we observe small changes in the size of the image structure when we move the focal position in the axial direction as expected and observed previously in reference [51] and in our experiment [26, 27].

4.2 Biological Imaging by a Microscope with Spatial Filters

We also performed PT imaging of biological systems using a rabbit small intestine and kidney as samples. The resolution was evaluated for tissues stained with hematoxylin and eosin (HE). The thickness of the small intestine sample is about 10 μm. HE (hematoxylin and eosin) staining has been routinely used in path morphology and histopathology, e.g., in the diagnosis of tumor progression [52]. Since HE can be used to stain anionic and cationic structures (e.g., nuclei containing DNA and basic protein) [53], these structures can be distinguished using a bright-field microscope. While eosin has a fluorescence peak around 540 nm in water [54], hematoxylin is an essentially non-fluorescent pigment [52, 53], and the energy from the excitation laser is instead converted into heat, resulting in the photothermal effect. However, we utilize both H and E as the source materials of the PT signal. Even in the case of eosin it is much more advantageous to utilize the PT effect for imaging because of much higher robustness than fluorescence. Our pump-probe microscope can be used to capture PT images in both regions of a sample stained with eosin and hematoxylin at the same time. The time constant of the lock-in amplifier and the dwell time are 500 μs and 1 ms, respectively, in the biological sample imaging. The pixel number is 800 x 800 and hence the frame time of the images is 640 seconds and the stage step is 18.75 nm.

Figures 2.1.2.9(a) and (b) show the photothermal images of a small intestine from a rabbit captured using the circular (a) and the inverse-annular (b) illumination in the focal (XY) plane. The pump and probe power are 140 μW and 120 μW for (a), and

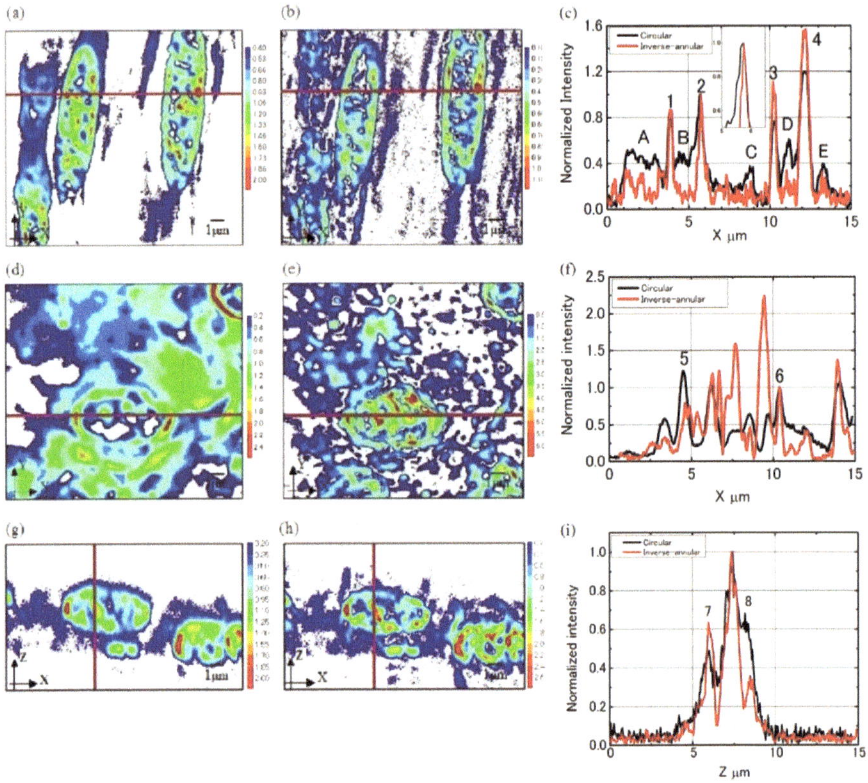

FIGURE 2.1.2.9 Images of several biological samples obtained without inverse annular filter (namely with circular illumination) and with inverse annular filter, and comparison in the lateral and axial resolutions between the two illuminations. (a) and (b) Photothermal images of cells from rabbit small intestine for the circular and inverse-annular illumination, respectively, in the focal (XY) plane. The pump and probe power are 140 μW and 120 μW for (a), 150 μW and 420 μW for (b), respectively. (c) The intensity profiles along the red lines in (a) and (b). The red and black solid lines are intensity profiles with the inverse-annular and circular illumination, respectively. Inset shows enlarged view of the peak No. 2. (d) and (e) photothermal images of cells form rabbit kidney for the circular and inverse-annular illumination in the focal (XY) plane. The pump and probe power are 150 μW and 200 μW for (d), 700 μW and 650 μW for (e), respectively. (f) The intensity profiles along the red lines in (d) and (e). (g) and (h) Photothermal images of cells form rabbit kidney for the circular and inverse-annular illumination in optical axial (XZ) plane. The pump and probe power are 100 μW and 200 μW for (h), 780 μW and 260 μW for (h), respectively. (i) The intensity profiles along the red lines in (g) and (h). All scale bars are 1 μm. The most intense peak in (i) is not used for the resolution discussion because it suffers from heavy overlaps with the neighboring peaks. The figure was reported in the journal *Optical Review* and copyright is shared between the corresponding author (Takayoshi Kobayashi, TK) and the publisher Springer Verlag.

TABLE 2.1.2.5
FWHMs of the PT Intensity Peaks in XY-plane in Figure 2.1.2.8

Peak number	Circular [nm]	Inverse-annular [nm]	Difference of width	$(\varepsilon_c^2+\varepsilon_I^2)^{0.5}$	Improvement
1	344±54	271±32	73	62.8	21 %
2	478±63	320±35	158	72.1	33 %
3	399±19	263±17	136	25.5	34 %
4	735±61	585±27	150	66.7	20 %
5	458±4	341±67	117	67.1	26 %
6	314±15	271±8	43	17.0	14 %
Average					25(±8) %

ε_c and ε_I: Standard deviations of the peak intensity profiles using circular and inverse annular illumination, respectively.

150 μW and 420 μW for (b), respectively. The signals originating from the elliptical structures in (a) and (b) are particularly well detected. It can be concluded that these regions are cell nuclei because they are primarily stained with hematoxylin and also because of their shapes and sizes. To evaluate the lateral resolution of the two images obtained using the two illumination modes, we compare the intensity profiles [Fig. 2.1.2.9(c)] along the red lines drawn in Figs. 2.1.2.9(a) and (b). The intensities of peaks 1–4 in Fig. 2.1.2.9(c) are clearly separated for both illumination types. Table 2.1.2.5 shows the FWHM values for each peak obtained by Gaussian fit. All the FWHMs of the Gaussian curves are smaller for inverse-annular illumination than for circular illumination by 14–34%. Peaks A–E for circular illumination in Fig. 2.1.2.9(c) are missing for inverse-annular illumination, which can be explained as follows. They are considered to originate from signal produced far from the focal plane and thus do not appear when inverse-annular illumination is employed. This is proved by moving the location of the focal planes of pump and probe together. This result indicates that the resolution is improved not only in the lateral directions but also in the axial direction. In the case of inverse-annular illumination, the signals produced in the vicinity of the focal plane are more efficiently detected, resulting in the observed improvement in the axial resolution over the case of circular illumination.

Figs. 2.1.2.9(d) and (e) show the photothermal images of a rabbit kidney captured using the circular and the inverse-annular illumination in the focal, respectively, in the lateral (XY) plane. The pump and probe power are 150 μW and 200 μW for (d), 700 μW and 650 μW for (e), respectively. As with the small intestine sample, the intensity profiles [Fig. 2.1.2.9(f)] along the red lines also indicate improvement in the lateral resolution. Figs. 2.1.2.9(g) and 9(h) show the PT images of a sample of rabbit kidney captured using circular and inverse-annular illumination, respectively, in the optical axis (*XZ*) plane. We have defined the Z-axis as the optical axis in Figs. 2.1.2.9(g) and 9(h). The pump and probe power are 100 μW and 200 μW for

TABLE 2.1.2.6
FWHMs of the Photothermal Intensity Peaks in XZ-plane in Figure 2.1.2.9

Peak number	Circular [nm]	Inverse-annular [nm]	Difference of width	(εc2+εl2)0.5	Improvement
7	828±57	552±24	276	61.8	33 %
8	1004±39	774±40	230	55.9	23 %
Average					28 (±5)%

ε_c and ε_I: Standard deviations of the peak intensity profiles using circular and inverse annular illumination, respectively.

Fig. 2.1.2.9(g), 780 μW and 260 μW for (h), respectively. The inner structures of the cell nuclei can be observed clearly with inverse-annular illumination compared to without it. The intensity profiles [Fig. 2.1.2.8(i)] along the red lines in Figs. 2.1.2.9(g) and (h) also indicate an improvement in the axial resolution. Table 2.1.2.5 and Table 2.1.2.6 show the width of the peaks in Fig. 2.1.2.9(c, f, XY plane) and (i, XZ plane), respectively. The biological sample has thickness of about 10 μm, hence there is a possibility of the existence of neighboring structures on the planes above or below the focal plane causing the modification of the observing image of the target. This may be the reason for the variation of the enhancement depending on the position of the target structure. Thus, it is hard to remove the image deformation completely. Under such situation some peaks may have large error values, but yet resolution is improved. To properly evaluate the effects with and without the inverse-annular filter, we take the root of the sum of squared standard deviations (corresponding to the widths of the two cases = $(\varepsilon_c^2+\varepsilon_I^2)^{0.5}$). For peak number 1 in the Table 2.1.2.5, the square root of $(54^2 + 32^2)$ is approximately 63, which is smaller than the "differences between the cases of with and without" listed in the table to be "73" as in Table 2.1.2.5. In all other cases, the same inequality condition is satisfied as shown in Tables 2.1.2.5 and 2.1.2.6. The most intense peak in Fig. 2.1.2.9(i) is not used for the resolution discussion because it suffers from overlap with the neighboring peaks and separation between the two peaks is too small. The averages of the fractional increment of resolutions by photothermal imaging by using inverse annular illumination are 25 ± 8 % in the XY planes and 28 ± 5 % in the XZ plane, respectively.

4.3 Evaluation of the Specially Segmented Balanced Detection Method

The PT image measured with SBD using the bifurcated fiber bundle was compared with that measured by the conventional detection scheme. Fig. 2.1.2.9 shows the PT images of hematoxylin- and eosin-stained tissue with and without SBD. Here only a 520-nm beam was used for pumping. To obtain a conventional PT image without SBD, the iris diaphragm in the collimator lens was adjusted so that only the center part of the probe beam passed through to maximize the SNR [17, 22], and the transmitted beam was detected using a single photodiode. In contrast, in the SBD, the iris

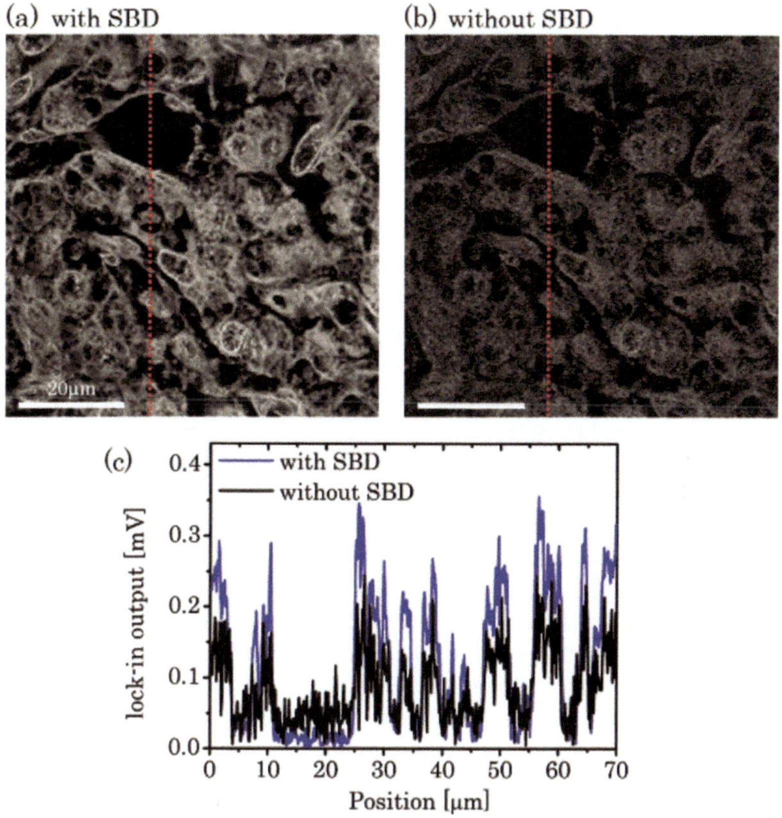

FIGURE 2.1.2.10 (a) Photothermal images of a hematoxylin and eosin stained rabbit ovary (a) with and (b) without spatially segmented balanced detection (SBD). Pump and probe beam power incident on the sample were 8 μW and 4 mW, respectively. Image acquisition time was 6 s at 500 x 500 pixels. (c) Intensity profiles along the vertical lines in (a) (blue) and (b) (black). The figure was reported in the journal *Optical Review* and copyright is shared between the corresponding author (Takayoshi Kobayashi, TK) and the publisher Springer Verlag.

diaphragm of the collimator lens was fully opened and the entire transmitted beam was directed to the incident port of the bifurcated fiber bundle separating the center and peripheral area of the beam. It was confirmed that signal intensity in the SBD was about 1.8 times higher than that with conventional detection. Furthermore, the intensity noise of the LD was reduced by 37%. Thus, the SNR was improved by a factor of 4.9. This indicates that image acquisition speed with SBD is 24 times faster than with the conventional scheme to obtain the same SNR using the same pump power, because the SNR is proportional to the square root of the measurement time. In the present setup, imaging speed was limited by the time constant of the lock-in amplifier (10 μs at minimum), which must be shorter than the pixel dwell time. However, because the time response of the PT signal is in principle determined by the heat conductivity (~1 MHz at -6 dB point for tightly focused laser beam) [10, 55], it is possible to further decrease the pixel dwell time to a few microseconds by using a fast lock-in amplifier that works with a shorter time constant. To evaluate the effect of light scattering on the lock-in signal, we measured images without pump beam and confirmed that light scattering is negligible (less than 5% of the PT signal).

4.4 Comparison between Photothermal and Fluorescence Imaging

FL imaging is a well-established methodology for identifying specific biomolecules in a biological specimen, and the combination of PT and FL imaging can provide further information on the tissues because both of the non-fluorescent and fluorescent molecules can be visualized at the same time. To characterize the performance of present imaging modality, the spatial displacement between the PT and FL images was examined by measuring 1 μm FL beads stained with Nile red [Figures 2.1.2.11(a)–(d)]. Here, only the 520-nm beam was used for pumping. PT and FL signals were measured simultaneously during the beam scan. The central position of an individual bead was determined by calculation using the Gaussian fitting function. From the displacement between the determined peaks of PT and FL images was estimated to be 0.18 μm in the lateral plane. In the same way PT and FL images of a single bead were simultaneously acquired by changing the sample position in the axial direction, and the PT and FL displacement in the axial direction was estimated to be 0.22 μm. These displacements are considered to be due to misalignment of the confocal pinhole. Since the displacement vector in the lateral direction is nearly uniform in the scanning area, it can be corrected by the pixel shift scheme. The PT signal from the beads suffer from the axial position dependence of PT effect intensity

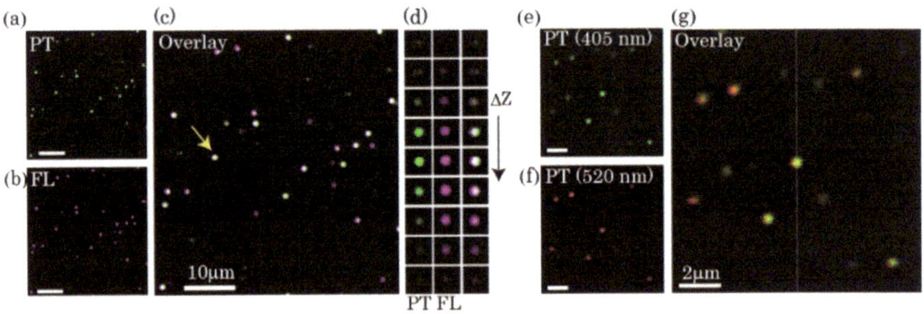

FIGURE 2.1.2.11 (a) Photothermal, (b) fluorescence, and (c) overlay images of 1 μm Nile red beads dispersed in PVA on a glass slide. Pump and probe beam power incident on the sample were 0.5 and 4 mW, respectively. Imaging area was 48 x 48 μm. Scale bars in (a)–(c): 10 μm. (d) Images of a single bead with a different focal plane for the bead designated by the arrow in (c). The step size was 0.27 μm. Simultaneous dual-wavelength photothermal images with excitation at (e) 405 nm and (f) 520 nm, and (c) overlay images. Pump beam power for both was 2 mW and probe beam power was 12 mW. Scale bars in (e)–(g): 2 μm. The figure was reported in the journal *Optical Review* and copyright is shared between the corresponding author (Takayoshi Kobayashi, TK) and the publisher Springer Verlag.

with twin peaks when the focal offset of the pump and probe beam was close to zero [29]. The ratio of the signal intensities between FL and PT images is therefore different among individual beads depending on the image position in the axial plane.

The spatial resolution was evaluated by PT imaging of 20-nm gold nanoparticles dispersed in PVA on a glass slide [Figs. 2.1.2.11(e)–(g)]. The average of the full width at half maximum (FWHM) values for the 61 measured spots was 149±11 nm and 204 ±10 nm with excitation at 405 nm and 520 nm with NA = 1.4, respectively. This is slightly larger than the diffraction-limited spot sizes of the pump beams ($0.51\lambda/\mathrm{NA}$ = 147 nm and 189 nm for λ = 405 nm and 520 nm). The point spread function (PSF) of PT microscopy is considered to be dependent on the intensity profiles of the pump and probe beams and the refractive index profile at the focal point, and the width of the PSF should be smaller than both of the diffraction-limited spot sizes of the pump and probe beams [29]. We consider that wavefront distortion induced by optical elements may have caused degradation in the resolution and that further optimization of the optics should provide better resolution. From a stack of images at different focal planes, the FWHM for the axial plane was estimated to be 0.6 μm and 0.7 μm for the 405-nm and 520-nm beams, respectively.

Multi-wavelength PT imaging is invaluable for identifying the components observed in the PT image especially in the case of biological imaging. Since the magnitude of the PT signal is proportional to the absorption spectrum [14, 18] it can show the spatial distribution of the components of the system studied. To perform multi-wavelength PT imaging, it is important to investigate the spatial displacement between the images obtained by different colors, as chromatic aberration of the optical elements may cause distortion of the multi-wavelength images. PT images of gold nanoparticles were obtained simultaneously at the pumping wavelengths of 405 and 520 nm [Figs. 2.1.2.11(e)–(g)]. The spatial image displacement between the two colors was estimated to be smaller than 50 nm in the lateral direction and 0.2 μm in the axial direction, which are about one-third and one-fourth of the PSF width, respectively.

4.5 Imaging of Endogenous Brain Signals

Various linear and nonlinear interactions of light with matter have been applied for label-free brain imaging, such as autofluorescence [56], photoacoustic [44], spontaneous Raman scattering [57, 58], infrared absorption [57, 59], second harmonic generation [56, 60], third harmonic generation (THG) [61], coherent anti-Stokes Raman scattering [62], and stimulated Raman scattering [63]. PT microscopy is a powerful tool for visualizing endogenous signals in biological tissue, yet little brain imaging has been reported. We performed simultaneous PT and FL imaging of a fixed slice of transgenic mouse brain expressing yellow fluorescent protein (YFP). Figure 2.1.2.12 shows the results of simultaneous PT and confocal FL imaging of different layers of the mouse brain cortex. FL images visualize neurons expressing YFP at the dendrite terminals [Fig. 2.1.2.12(a)], a dendrite [Fig. 2.1.2.12(b)], and cell bodies [Fig. 2.1.2.12(c)], respectively. The PT images show distinct structures and organization. Red blood cells and vessels can be identified morphologically. They produce strong

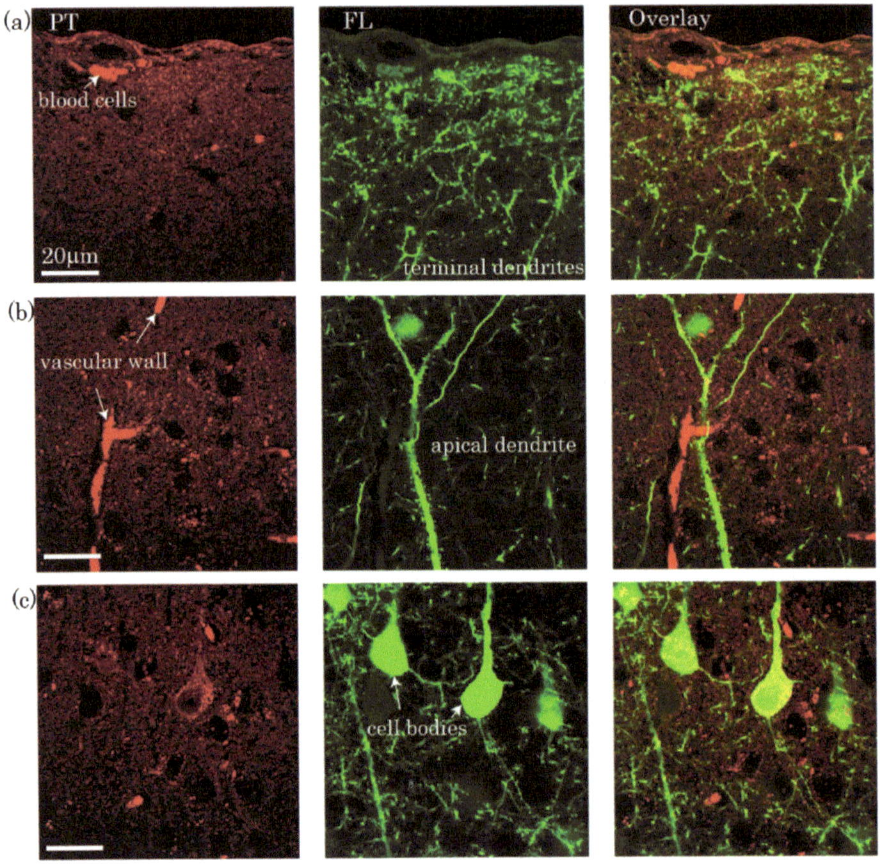

FIGURE 2.1.2.12 Simultaneous photothermal (PT) and fluorescence (FL) imaging of mouse brain cortex at (a) the surface, (b) layer II/III, and (c) layer IV. Left and center panels show the PT and FL images, respectively, and their overlay is shown in the right panel. The pump and probe beam power incident on the sample were 0.5 and 4 mW, respectively. Image acquisition time was 6 s at 500 x 500 pixels. The figure was reported in the journal *Optical Review* and copyright is shared between the corresponding author (Takayoshi Kobayashi, TK) and the publisher Springer Verlag.

PT signals as they may contain heme proteins [Figs. 2.1.2.12(a) and (b)]. A Y-shaped junction of a vessel is also observed in Fig. 2.1.2.12(b). Although cardiac perfusion with PBS is conducted in the sample preparation, there is a possibility that red blood cells are fragmented or stacked in a vessel and result in the PT signal of a vessel.

The cell bodies of the neurons appear as dark shadows. It should be noted that fine structures are observed in all areas of the brain cortex. Figures 2.1.2.13(a) and (b) shows close-up images around an axon and the dendrites of neurons. Bright specks with diameters of around 1 μm appear surrounding the cell bodies. These are attributable to the lipofuscins in lysosomes as weak autofluorescence is detected [56].

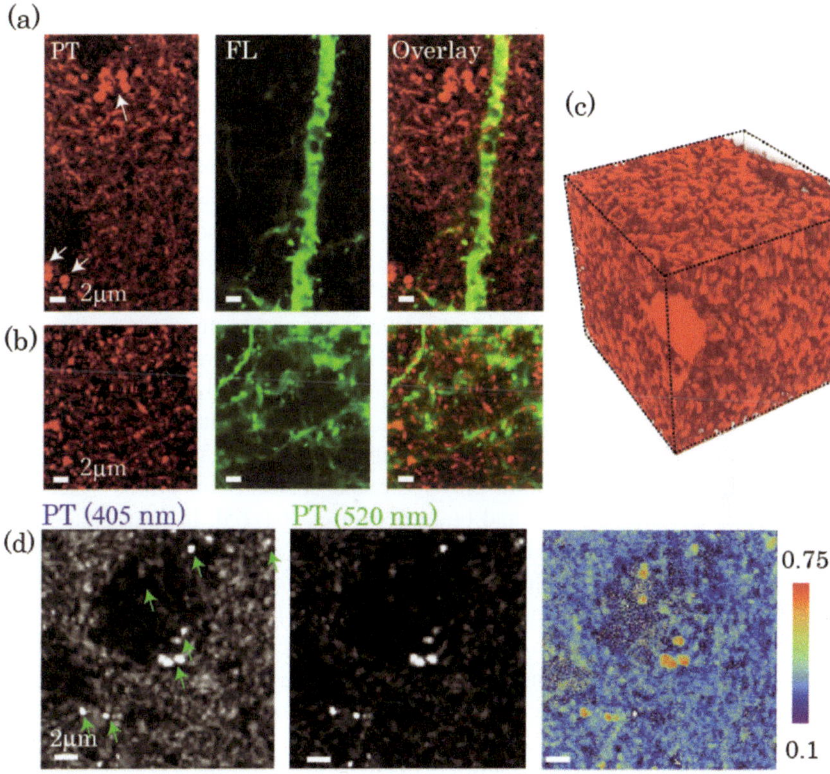

FIGURE 2.1.2.13 Close-up photothermal (PT) and fluorescence (FL) images of mouse brain cortex. PT and FL images around (a) a dendrite and (b) axons. Arrows shows bright specks localized around cell bodies, which are attributable to lipofuscins in lysosomes. (c) Reconstructed 3D model of PT signal in cortex. Image size is 23 x 23 x 21 μm^3 with 400 x 400 x 80 voxels and acquisition time was about 9 minutes. A stack of 30 images of brain cortex was acquired by changing the sample position in the axial direction with a step size of 0.68 μm. Each image size is 70 x 70 μm^2 with 600 x 600 pixels. Acquisition times were 5 minutes (d) dual-wavelength PT images around a cell body with excitation at 405 nm (left) and 520 nm (center), and their ratio image (right). The figure was reported in the journal *Optical Review* and copyright is shared between the corresponding author (Takayoshi Kobayashi, TK) and the publisher Springer Verlag.

Sponge-like structures of sub-micrometer size cover large areas of the tissue. The combined PT and FL images show that the axon and dendrites visualized in the FL image do not overlap with the fine structures in the PT images. This indicates that the contrasting agent of the sponge-like structure is outside the neuron. A stack of images in discrete z-steps clearly shows the depth profile of the shadow contrast of the cell bodies and the fine structures. The 3D-imaging capability is also demonstrated [Fig. 2.1.2.13(c)] by the reconstructed 3D model of a volume scan.

To obtain further information on the fine structures observed in the PT image, simultaneous dual-wavelength imaging was performed using two pump beams at wavelengths of 405 and 520 nm [Fig. 2.1.2.13(d)]. From the dual-wavelength PT imaging of the sponge structures, the absorbance ratio between 520 and 405 nm is calculated to be about 0.3. The brain slice shows a faint yellowish color. Thus, the light absorption of the brain tissue is predominantly attributed to the species contained in the sponge structures. In contrast, the intensity ratio of the bright specks around the dark shadows is ~0.7, which suggests that the bright specks and the sponge structures are composed of different species. To exclude the possibility that transgenic expression of YFP affects the PT signal, we acquired PT images of a wild-type mouse brain and found identical fine sponge structures. This indicates that the PT signal originates mainly from the endogenous chromophores in the brain.

We presume that the fine sponge-like structures can be attributed to the lipids comprising 30–60% of the brain. Hydrophobic dyes such as lutein and zeaxanthin that can cross the blood-brain barrier and/or lipid peroxides may serve as PT contrast agents. Lutein and zeaxanthin are carotenoid contained in grain, which are concentrated in the neural retina and brain neocortex [64, 65]. It is suggested that lipid has a steeper refractive index/temperature slope ($\sim 1 \times 10^{-3}$ K^{-1}) [66], and thus the chromophores in the lipids result in a significant PT signal. It is interesting to note that the PT image of the brain cortex is similar to the THG image [61]. Lipid bodies are a major source of contrast in THG microscopy, which is used to visualize lipid structures in the brain. Interestingly, however, unlike the THG signal, the PT signal was not seen in neurons and myelin sheaths, but rather found in intracellular organelles of non-neuronal cells (Fig. 2.1.2.13). In this regard, in THG imaging, a partial phase match of third-order susceptibility at the focal produces the TH signal and the signal is most intense when a structure is half the size of the focal volume. This can result in the signal intensity distribution different from the concentration distribution and may work as a kind of spatial filtering function to coarse grain the distribution, reducing the intensity of finer structure. This may be the reason why THG imaging provides more uniform distribution than the PT method [61]. On the other hand, since the contrast mechanisms are different between the THG and PT images, they may provide complemental information on the structure or composition of the lipid.

We also evaluated the possibility that the fine sponge-like structures represent glycogen granules, which are abundant in astrocytes. However, the PT signal did not change after treatment with amylase, which digests glycogen in the sample. Moreover, the PT signals rarely overlapped with glycogen immune-reactive granules.

Previous study showed that the PT signal of live cultured neurons originated from cytochrome c [16]. Cytochrome c is water soluble and the refractive index/temperature slope of water is 1×10^{-4} K^{-1}, while lipid has a steeper slope [66]. Therefore, the PT signal of lipid may be larger than that of cytochrome c in imaging of brain slice if they have comparable optical absorption and concentration with cytochrome c. Careful study is needed to precisely separate the signals due to lipid and cytochrome c.

Lipid metabolism in the brain is closely related to neurodegeneration illnesses such as Alzheimer's disease [59, 67]. Hence, the ability to image both the neuron and the lipid structure at high spatial resolution will be a powerful technique for studying the physiology of such diseases.

4.6 Non-linear PT Imaging

We are developing PT imaging with even higher special resolution. To do so, we used gold nanoparticle (GNP) at first and then biological sample. Figures 2.1.2.14 (a) and (b) show the photothermal image of a 20-nm GNP dispersed in a PVA film in an XY-plane obtained by the Galvano mirror photothermal microscope system with the linear (a) and the non-linear (b) detection schemes. The powers of pump and probe laser diodes are 4.8 mW and 6.3 mW, respectively. In the GNP measurement case, temperature rise of 16.5 K is calculated for sample thickness = 50 μm, pump power = 2.3 MW/cm^2, heat capacity C = 1.62 × 10^6 J/Km^3 [68], and absorbance at pump energy = 0.6 %. The distribution temperature increase is considered to be Gaussian whose size is determined by the diffusion process. The FWHMs of the intensity profile curves of the GNP along the lines in Figs. 2.1.2.14 (a) and (b) obtained by the linear and nonlinear modes are 269.5 nm and 160 nm with NA = 0.95, respectively. The FWHMs are better than the diffraction limit (410 nm) calculated from the NA = 0.95 of the 40x objective lens and the probe wavelength 638 nm. The result shows that the no-linear detection scheme provides superresolution by 61% over the diffraction limit, and by 42% enhancement from the linear scheme. The instrument function determined and observed as the convolution of point spread function of the optical system and resolution function of the detector system of the photothermal microscope using the linear detection scheme obtained by using the GNP can be calculated to be 268.8 nm [= (FWHM of detected image of GNP)^2 - (True size of GNP)^2)$^{0.5}$] from the observed size (FWHM = 269.5 nm) and true GNP size (20 nm). The instrument function of the photothermal microscope by nonlinear detection scheme is calculated in the same way to be 159 nm (see Table 2.1.2.7). As expected, the size of the GNP can be neglected, because the difference between the FWHM of the detected image and that of the deconvolution calculation is negligibly small. Figs. 2.1.2.14 (d) and (e) are the photothermal images in the XZ plane by the linear and the nonlinear detection modes, respectively. In the figures, the substantially large elongation of images along Z axis (= axial direction) is found. The FWHMs of the intensity curves of the GNP obtained by the linear and nonlinear modes are 292 nm [center line in Fig. 2.1.2.14 (d)] and 226 nm [left line in Fig. 2.1.2.14 (e)], respectively. It is confirmed by the intensity profiles along the axial direction [see Fig. 2.1.2.14 (f)]. The result shows that the nonlinear detection scheme provides resolution enhancement by 23% from the linear scheme in the axial direction.

Thermal diffusion length is given by $r_{th} = (2D/2\pi f)^{0.5}$ with thermal diffusivity D and modulation frequency f. In the present experiment, r_{th} ~ 837 nm is obtained for the system of GNP / polyvinyl alcohol film with D_{PVA} ~ 2.2 × 10^{-7} m^2/s [69], and f = 100 kHz. The calculated r_{th} is about three times larger than the results of linear measurement. With the value of r_{th} and the probe light spatial distribution, the expected FWHM of GNP is estimated to be 311 nm. Even with such a large temperature-increased area, high resolutions of about 269 nm are obtained, which is better than the expectation by ~15 %. The difference may be explained in terms of even higher nonlinearity due to the saturation of laser absorption, temperature-dependent heat capacity and thermal conductivity.

To evaluate the size of the biological systems from the observed images, we estimate the "real size" of the structures of the rabbit kidney, and discuss the improvement

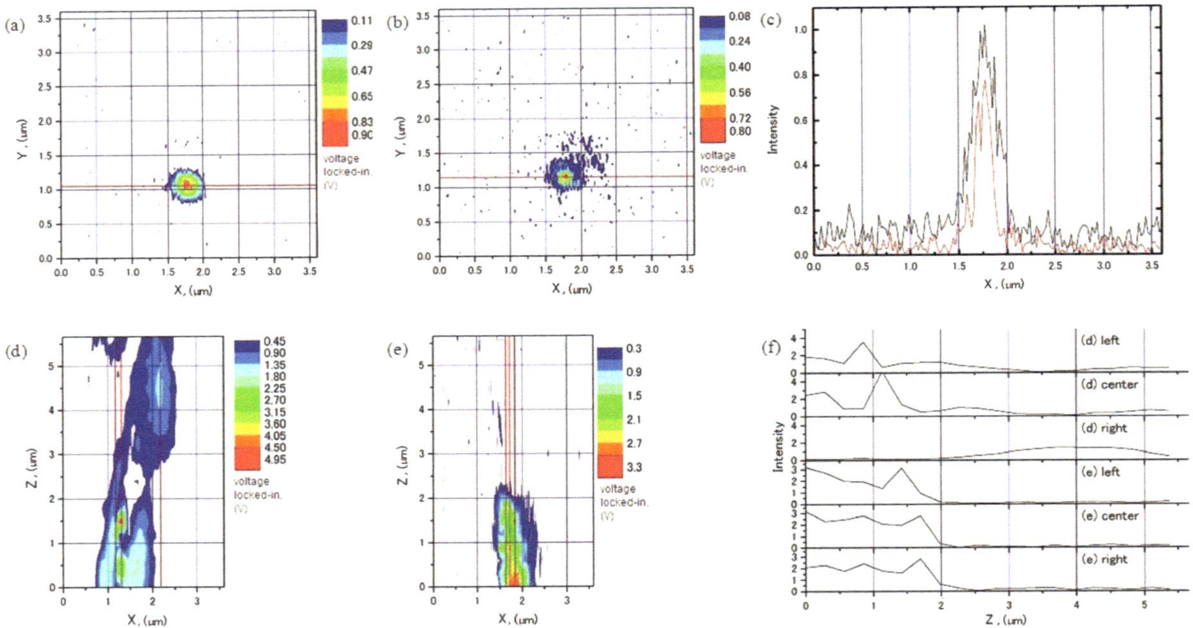

FIGURE 2.1.2.14 (a) and (b), The images of a 20-nm gold nanoparticle (GNP) in an XY-plane obtained by the Galvano mirror photothermal microscope system with the linear (a) and the nonlinear (b) detection schemes. The pixel number is 90,000 = 300 (x) × 300 (y) and the frame time for both linear and non-linear images is 4.5 seconds (= 50 μs × 90,000 points). (c) Black and red curves are the intensity distribution profiles along the red lines shown in the linear (a) and the non-linear (b) images, respectively. (d) and (e), The images of a 20-nm GNPs in XZ-plane obtained by the Galvano mirror system with the linear (d) and the nonlinear (e) detection schemes. (f) Intensity profiles along the red lines in the frame images (d) and (e). The values in the color scales in 2D photothermal figures (a, b, d, and e) indicate the signal output intensity in the voltage of the locked-in detections. The figure was reported in the journal *Optical Review* and copyright is partially held between the corresponding author (Takayoshi Kobayashi, TK) and the publisher Springer Verlag.

TABLE 2.1.2.7
Characteristic Resolution Parameters Evaluated by Gold Nanoparticle Images in the Present Microscope. FWHM by the f-detection and FWHM by the 2f-detection are determined by Gaussian fitting to Figure 2.1.2.13 (c). Instrument function (f) and (2f) are calculated by the equation $((\text{FWHM})^2 - (\text{true size})^2)^{0.5}$ and FWHMs by f- and 2f-detection, respectively

	FWHM by f-detection [a]	FWHM by 2f-detection[a]	True size	FWHM of instrument function (f)	FWHM of instrument function (2f)
X	270±8 nm	160±5 nm	20 nm	269 nm	159 nm
Z	292±28 nm	226±28 nm	20 nm	291 nm	225 nm

(a) FWHM in X direction is evaluated by Gaussian fitting and that in Z direction is evaluated by the positions of half values of the intensity distribution data.

of resolution in the 2f-detection mode from the linear detection scheme. The "real size" is based on the assumption that the observed structure is given by the spatial convolution of the structure of the target object and the instrument function of the measurement including the optical system and the detection system.

Figures 2.1.2.15 (a) and (b) are the photothermal images in the XY plane of rabbit kidney obtained by the linear and the nonlinear lock-in detection schemes showing the structure of the tubules existing in the cell is clearly resolved. The laser diode power of pump in linear scheme, pump in nonlinear scheme and probe is 0.23 mW, 2.3 mW, and 2.1 mW, respectively. The intensity distribution of image profiles of linear (black curve) and non-linear (red curve) detection schemes along the red lines is shown in Fig. 2.1.2.15 (c). The real size of the biological structures is difficult to estimate by the ordinary bright-image microscopy, Fig. 2.1.2.15 (d), because of low contrast and resolution. To evaluate the resolution of the present photothermal microscope utilized for biological systems, we discuss the improvement of resolution in the 2f-detection mode, and estimate the real size of the structures of the rabbit kidney in this experiment.

In Fig. 2.1.2.15(f), the intensity distribution profiles of linear (red) and non-linear (black) detection schemes along the two red lines in (b). In (g) the intensity distribution profiles of linear detection scheme (black) and bright-field image (red) along the two red lines in (c). Upper box shows that along the Y = 23 μm line, lower box shows that along the Y = 21 μm line in (f) and (g). The values in the color scales in 2D photothermal figures (b, c) indicate the signal output intensity in the voltage of the locked-in detections. Therefore, the resolution is improved by 23% by the nonlinear detection scheme from the linear scheme. The FWHMs of the structures together with the other three structures are listed in Table 2.1.2.8.

When the reference frequency is f, the real size of structure in each rabbit kidney is estimated by deconvolution to be 200–320 nm as shown in Table 2.1.2.8 using the estimated instrument function of the linear detection scheme microscope. The ratio of the real size to each FWHM is 0.61–0.77 in the linear detection scheme. When the reference frequency is 2f, the real size of each rabbit kidney structure is calculated by deconvolution to be 200–320 nm using the estimated instrument function of the

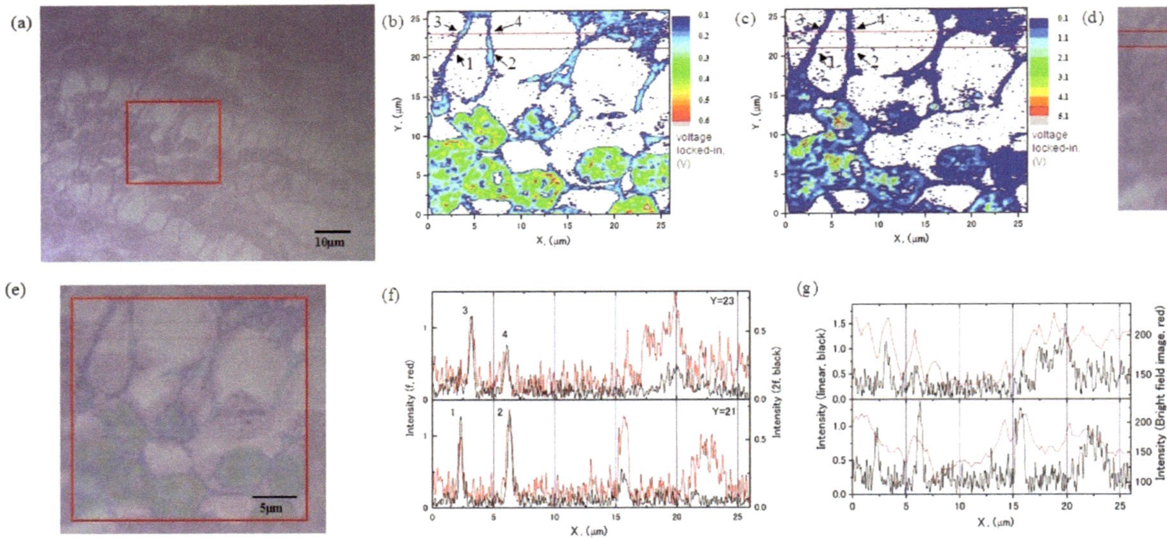

FIGURE 2.1.2.15 (a) Bright-field image of the rabbit kidney (110 μm × 88 μm). (b), (c) Images of the rabbit kidney obtained by the photothermal microscope equipped with Galvano mirror in the linear (a) and the nonlinear (b) detection schemes (26 μm × 26 μm). The image frame time of these images is 4.5 seconds. (d) Enlarged bright-field image of the rabbit kidney in a red square (26 μm × 26 μm) in (a). (e) Overlaid image of the bright field image (e) and the photothermal image (b) of linear detection scheme. The red squares in (a) and (e) show the area in the photothermal images of (b) and (c). (f) The intensity distribution profiles of linear (red) and non-linear (black) detection schemes along the two red lines in (b). (g) The intensity distribution profiles of linear detection scheme (black) and bright field image (red) along the two red lines in (c). Upper box shows that along the Y = 23 μm line, lower box shows that along the Y = 21 μm line in (f) and (g). The values in the color scales in 2D photothermal figures (b and c) indicate the signal output intensity in the voltage of the locked-in detections. The figure was reported in the journal *Optical Review* and copyright is partially held between the corresponding author (Takayoshi Kobayashi, TK) and the publisher Springer Verlag.

TABLE 2.1.2.8
Estimated Values of the Size of Imaged Structure of the Biological Sample of Rabbit Kidney. FWHMs of the Gaussian fit curves to Figure 2.1.2.14 (c) by f-detection mode and 2f-detection mode. "Ratio of FWHM" means FWHM by 2f-detection / FWHM by f-detection. "Real size (linear)" and "Real size (non-linear)" (see the context) are calculated using instrument function (f) and (2f) in Table 7, respectively

	FWHM by f-detection	FWHM by 2f-detection	Ratio of FWHM	Real size (linear)	Real size (non-linear)
1	339±31 nm	262±8 nm	0.77±0.10	206 ±52 nm	208 ±10 nm
2	416±21 nm	337±11 nm	0.81±0.07	314 ±28 nm	297 ±12 nm
3	378±34 nm	284±16 nm	0.75±0.11	261 ±49 nm	235 ±19 nm
4	422±73 nm	351±18 nm	0.83±0.19	322 ±97 nm	313 ±20 nm

nonlinear detection scheme of the microscope. The ratio of the real size to each of FWHM is 0.79–0.89 in the nonlinear detection scheme.

These results show the image size obtained by the nonlinear detection mode is much closer to real size than the linear detection mode. Nearly the same real size of the rabbit kidney structures is determined from the observed FWHMs by both the linear and the nonlinear detection scheme using corresponding instrument functions. This proves that the deconvoluted size in terms of FWHM can be concluded to be very close to the real size. It is also confirmed from the results that the difference of the resolution between the linear and the non-linear detection scheme is due to the difference in the instrument function. Thus, real size of the biological structures can be estimated from the photothermal images using the instrument function of the microscope determined by the linear and the nonlinear detection scheme of sufficiently small objects as GNPs.

The ratio of the width of the instrument function between 2f- and f-detection modes is 159 / 269 = 0.59. The instrument function of a microscope system during 2f-measurement is 58% for the linear detection mode, and this indicates that the diffraction images obtained by the nonlinear detection scheme are very close to the actual size. Estimated relative accuracy is given by (0.59–0.58)/0.59+0.58)/2 = 0.017 corresponding to 1.7%. Figures 2.1.2.16, 17, and 18 show the photothermal images of ovary, brain, and small intestine of rabbit, respectively. The laser diode power of the pump in the linear scheme, and probe is 0.23 mW and 2.1 mW, respectively. The power of the pump in the non-linear scheme is 2.3 mW in Fig. 2.1.2.16 and 17 and it is 1.5mW in Fig. 2.1.2.18. The photothermal images obtained by using the Galvano mirror show equivocal structures within the cell in the bright-field image.

Figure 2.1.2.19 shows the images of structures in the mouse brain sample. The image in Fig. 2.1.2.19(b) obtained by the nonlinear lock-in measurement has better resolution than the linear detection image, Fig. 2.1.2.19(a). Figure 2.1.2.19(c) shows

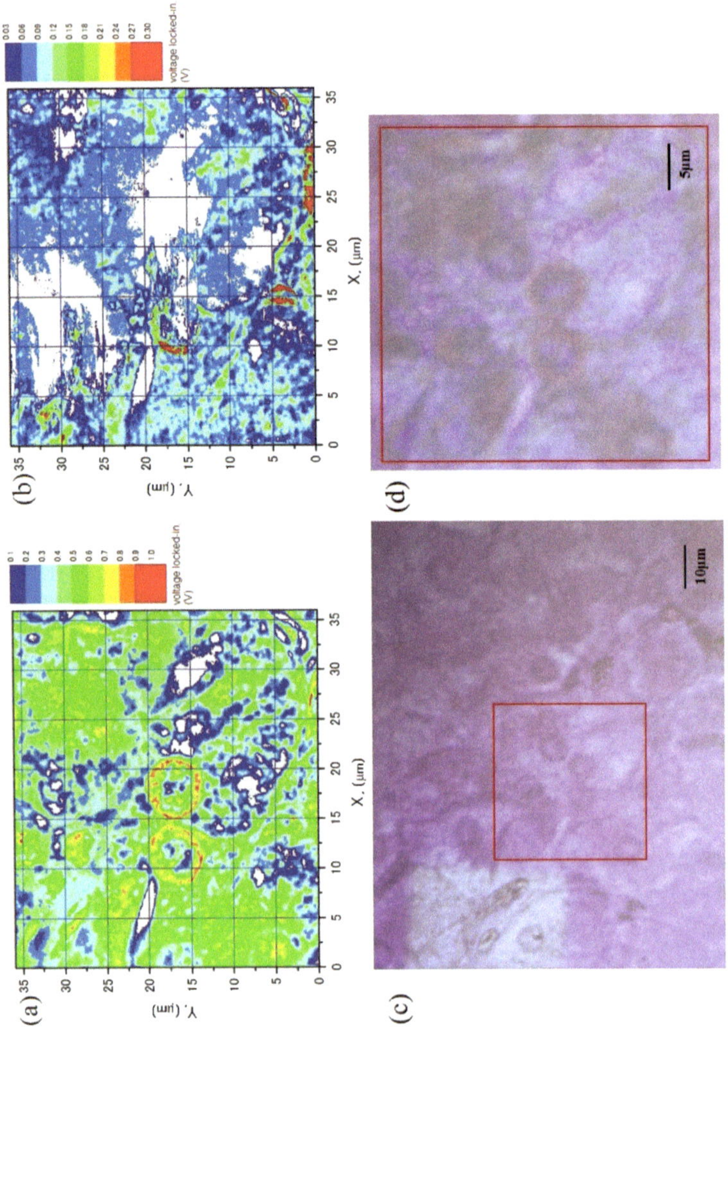

FIGURE 2.1.2.16 (a) and (b) The photothermal images of the rabbit ovary (36 μm × 36 μm) obtained in the f- and 2f- detection schemes, respectively. (c) Bright-field image of the rabbit ovary sample (110 μm × 88 μm). (d) Overlapped image of the bright-field image and the photothermal image of linear detection scheme. Red squares in (c) and (d) show an area in the photothermal images of (a) and (b). The values in the color scales in 2D Photothermal figures (a and b) indicate the signal output intensity in the voltage of the locked-in detections. The figure was reported in the journal *Optical Review* and copyright is shared between the corresponding author (Takayoshi Kobayashi, TK) and the publisher Springer Verlag.

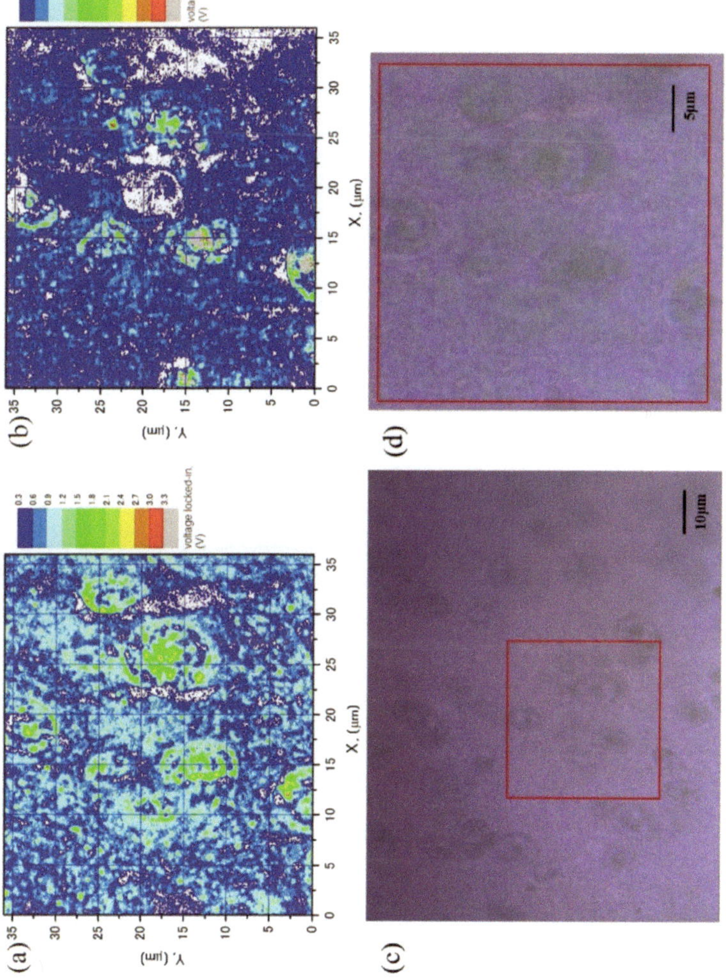

FIGURE 2.1.2.17 (a) and (b) The photothermal images of the rabbit brain (36 μm × 36 μm) obtained in the linear and the nonlinear detection schemes, respectively. (c) Bright-field image of the rabbit brain sample (110 μm × 88 μm). (d) Overlapped image of the bright-field image and the photothermal image of linear detection scheme. Red squares in (c) and (d) show an area in the photothermal images of (a) and (b). The values in the color scales in 2D Photothermal figures (a and b) indicate the signal output intensity in the voltage of the locked-in detections. The figure was reported in the journal *Optical Review* and copyright is shared between the corresponding author (Takayoshi Kobayashi, TK) and the publisher Springer Verlag.

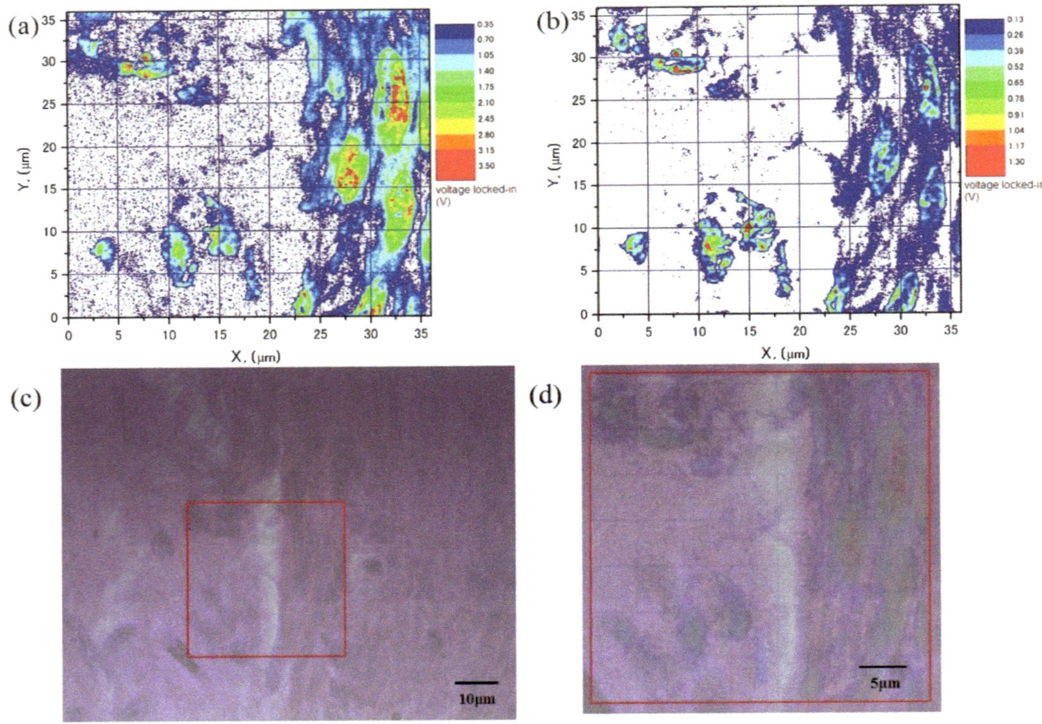

FIGURE 2.1.2.18 (a) and (b) The photothermal images of the rabbit small intestine (36 μm × 36 μm) obtained in the linear and the nonlinear detection schemes, respectively. (c) Bright-field image of the rabbit small intestine sample (110 μm × 88 μm). (d) Overlapped image of the bright-field image and the photothermal image of linear detection scheme. Red squares in (c) and (d) show an area in the photothermal images of (a) and (b). The values in the color scales in 2D Photothermal figures (a and b) indicate the signal output intensity in the voltage of the locked-in detections. The figure was reported in the journal *Optical Review* and copyright is shared between the corresponding author (Takayoshi Kobayashi, TK) and the publisher Springer Verlag.

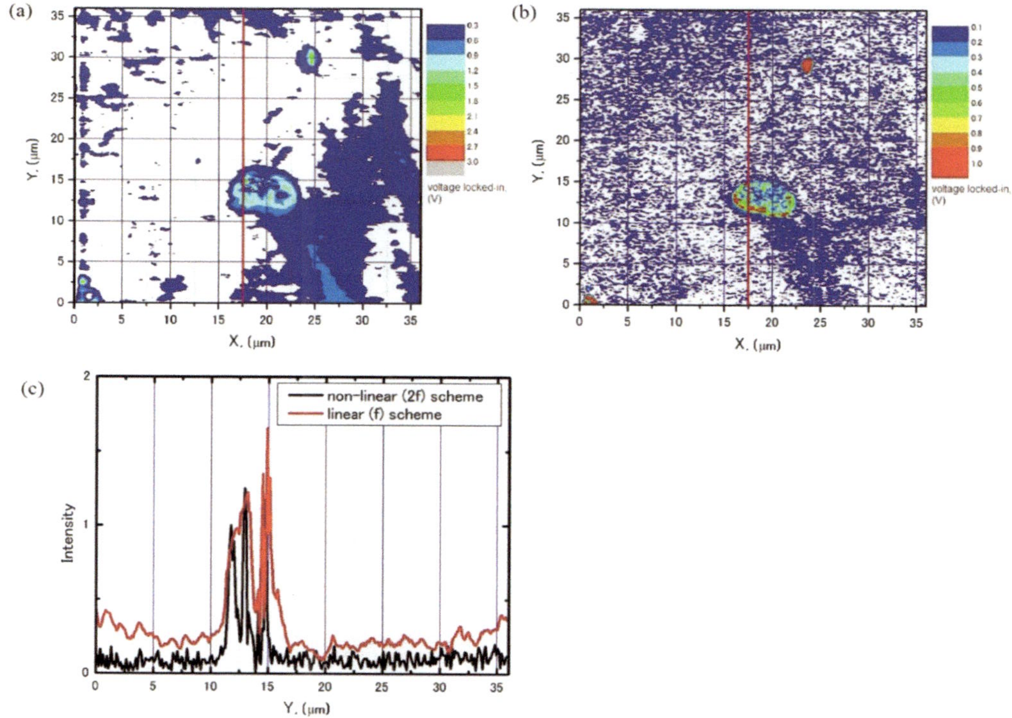

FIGURE 2.1.2.19 Images of the mouse brain sample in linear (a) and non-linear (b) detection by the Galvano mirror microscope. Intensity distribution of the image is shown in (c). The values in the color scales in 2D Photothermal figures (a and b) indicate the signal output intensity in the voltage of the locked-in detections. The figure was reported in the journal *Optical Review* and copyright is shared between the corresponding author (Takayoshi Kobayashi, TK) and the publisher Springer Verlag.

well-resolved fine structures in mouse brain cells that cannot be resolved by the linear detection scheme.

These results show that the microscope can improve the resolution of the photothermal images by the second-order nonlinear lock-in detection mode with fast frame time of 4.5 seconds using a Galvano mirror and simple laser diodes avoiding usage of an intense and expensive ultrashort pulse laser.

5 CONCLUSION

In conclusion, a new scheme of fast photothermal imaging microcopy for resolution enhancement is developed by adding an inverse-annular filter, second-order nonlinear lock-in detection scheme, and Galvano mirror.

First, we introduce a new pupil filter, an inverse-annular pupil filter in a pump-probe photothermal microscope, which provides resolution enhancement in three dimensions. The resolutions are proved to be enhanced in both lateral and axial directions by imaging experiments using commercially available gold nanoparticles of 20-nm diameter The resolution of a pump-probe photothermal microscope has been significantly improved compared to that of a conventional microscope simply by adding an inverse-annular pupil filter to a conventional microscope configuration. The improvement in X (perpendicular to the common pump and probe polarization direction), Y (parallel to the polarization direction), and Z (axial direction) are by 15±6, 8±8, and 21±2 %, respectively, evaluated from the resolutions with and without a pupil filter. The method has been applied to the photothermal imaging of thick biological samples (rabbit small intestine and kidney cells) stained with hematoxylin and eosin dye.

Second, fast, high-sensitivity PT microscopic imaging methodology was developed and an imaging experiment was conducted by incorporating the SBD scheme into a laser scanning multi-wavelength and multimodal microscope. It was confirmed that SNR improves by a factor of 4.9; thus, image acquisition time becomes 24 times faster than that of conventional detection with the same SNR and pump power. Simultaneous PT and confocal FL imaging of transgenic mouse brain tissue expressing YFP was demonstrated. The PT method detected endogenous non-fluorescent chromophores in brain tissue and various organizations such as blood cells, lipofuscin in lysosomes, and fine sponge structures, which are presumably attributable to lipids, all visualized with high 3D resolution.

Third, we introduced a new pupil filter, an inverse-annular pupil filter in a pump-probe photothermal microscope, which provides resolution enhancement in three dimensions. An imaging experiment was performed and the resolutions were proved to be enhanced in both lateral and axial directions by using commercially available gold nanoparticles of 20-nm diameter. The improvement in X (perpendicular to the common pump and probe polarization direction), Y (parallel to the polarization direction), and Z (axial direction) are by 15±6, 8±8, and 21±2 %, respectively, evaluated from the resolutions with and without a pupil filter. We obtained the specification of the superresolution reaching 160 nm and fast frame time of 4.5 seconds in the photothermal microscope using the Galvano mirror and laser diodes by the second-order non-linear lock-in measurement. From the measurement of the gold nanoparticle using the photothermal microscope, resolution of the images by the nonlinear detection scheme improved by 42% from the linear detection scheme corresponding to 61% of the diffraction limit. Fine structures of the biological samples unresolvable by the linear detection mode are well resolved by the non-linear detection scheme.

As this imaging modality is based on compact, low-cost laser diodes, it will be widely useful in the life and medical sciences, e.g., for lipid metabolism in the brain and related neurodegenerative diseases.

The research output presented in this subsection is based on the collaborative activity of T. Kobayashi, K. Kawasumi, J. Miyazaki, and K. Nakata [70]. The above study was reported in the journal *Optical Review* and copyright is shared between the corresponding author (Takayoshi Kobayashi, TK) and the publisher Springer Verlag. TK would like to express thanks to the company for allowing use of the revised paper in the book *Photothermal and Raman Microscopy* (Taylor & Francis).

NOTES

1. cytochrome c: Redox-active proteins containing a heme, with a central Fe atom (cofactor) at its core. They are involved in the electron transport chain and redox catalysis. They are classified according to the type of heme and its mode of binding cytochromes a, cytochromes b, cytochromes c and cytochrome d. Cytochrome function is linked to the reversible redox change from ferrous [Fe(II)] to the ferric [Fe(III)] oxidation state of the iron found in the heme core.
2. hematoxylin: naturally derived dye has been used as a histologic (microscopic anatomy or microanatomy) stain and is commonly followed by counterstaining with eosin. When the process is paired, this staining procedure is known as H&E staining and is one of the most commonly used combinations in histology.
3. eosin: the name of several groups of fluorescent acidic molecules which bind to and form salts with basic, or eosinophilic, compounds like proteins containing amino-acid residues e.g. arginine and lysine, and stains them dark red or pink as a result of the interaction of bromine with eosin. It can be used to stain proteins in the cytoplasm, collagen and muscle fibers for microscopic examination. Structures that stain readily with eosin are termed eosinophilic. In the field of histology, Eosin Y is the form of eosin used most often as a histologic stain.
4. cerebrospinal fluid: clear liquid produced in the ventricles (one of two large chambers toward the bottom of the heart that collect and expel blood towards the peripheral beds within the body and lungs) of the brain; fills and protects cavities in the brain and spinal cord.
5. flow cytometry: a technique used to detect and measure physical and chemical characteristics of a population of cells or particles. In this process, a sample containing cells or particles is suspended in a fluid and injected into the flow cytometer instrument. The sample is focused to ideally flow one cell at a time through a laser beam, where the light scattered is characteristic to the cells and their components. Cells are often labeled with fluorescent markers so light is absorbed and then emitted in a band of wavelengths. Tens of thousands of cells can be quickly examined and the data gathered are processed by a computer. Flow cytometry is routinely used in basic research, clinical practice, and clinical trials.

6. trans-impedance gain: Gain is the amplification factor of an amplifier. A trans-impedance amplifier (TIA) is a current to voltage converter, almost exclusively implemented with one or more operational amplifiers. The TIA can be used to amplify the current output of Geiger–Müller tubes, photo multiplier tubes, accelerometers, photo detectors and other types of sensors to a usable voltage. Current to voltage converters are used with sensors that have a current response that is more linear than the voltage response. This is the case with photodiodes where it is not uncommon for the current response to have better than 1% nonlinearity over a wide range of light input.
7. Plasmon: an oscillation of electron density with respect to the fixed positive ions in a metal. To visualize a plasma oscillation, imagine a cube of metal placed in an external E field pointing to the right. Electrons will move to the left side (uncovering positive ions on the right side) until they cancel the field inside the metal. If the electric field is removed, the electrons move to the right, repelled by each other and attracted to the positive ions left bare on the right side. They oscillate back and forth at the plasma frequency until the energy is lost in some kind of resistance or damping. Plasmons are a quantization of this kind of oscillation with at the plasma frequency .
8. intestine: intestine is an organ in the gastrointestinal tract where most of the absorption of nutrients from food takes place. It receives bile and pancreatic juice through the pancreatic duct to aid in digestion.
9. kidney: kidneys are two reddish-brown bean-shaped blood-filtering organs that are located on the left and right in the retroperitoneal space, and in adult humans are about 12 cm in length. They receive blood from the paired renal arteries; blood exits into the paired renal veins. Each kidney is attached to a ureter, a tube that carries excreted urine to the bladder.

REFERENCES

1. Samineni, P., deCruz, A., Villafana, T.E., Warren, W.S., Fischer, M.C.: Pump-probe imaging of historical pigments used in paintings. *Opt. Lett.* 37, 1310–1303 (2012)
2. Nechay, B.A., Siegner, U., Achermann, M., Bielefeldt, H., Keller, U.: Femtosecond pump-probe near-field optical microscopy. *Rev. Sci. Instrum.* 70, 2758–2764 (1999)
3. Dong, C.Y., Buehler, C., So, P.T., French, T., Gratton, E.: Implementation of intensity-modulated laser diodes in time- resolved, pump-probe fluorescence microscopy. *Appl. Opt.* 40, 1109–1115 (2001)
4. Matthews, T.E., Wilson, J.W., Degan, S., Simpson, M.J., Jin, J.Y., Zhang, J.Y., Warren, W.S.: In vivo and ex vivo epi-mode pump-probe imaging of melanin and microvasculature. *J. Biomed. Opt.* 2, 1576–1583 (2011)
5. Min, W., Lu, S., Chong, S., Roy, R., Holtom, G.R., Xie, X.S.: Imaging chromophores with undetectable fluorescence by stimulated emission microscopy. *Nature.* 461, 1105–1109 (2009)
6. Wei, L., Min, W.: Pump-probe optical microscopy for imag- ing nonfluorescent chromophores. *Anal. Bioanal. Chem.* 403, 2197–2202 (2012)
7. Tokeshi, M., Uchida, M., Hibara, A., Sawada, T., Kitamori, T.: Determination of sub-yoctomole amounts of nonfluorescent molecules using a thermal lens microscope: subsingle-molecule determination. *Anal. Chem.* 73, 2112–2116 (2001)
8. Berciaud, S., Cognet, L., Blab, G.A., Lounis, B.: Photothermal heterodyne imaging of individual nonfluorescent nanoclusters and nanocrystals. *Phys. Rev. Lett.* 93, 257402 (2004)
9. Lasne, D., Blab, G.A., Berciaud, S., Heine, M., Groc, L., Choquet, D., Cognet, L., Lounis, B.: Single nanoparticle photothermal tracking (SNaPT) of 5-nm gold beads in live cells. *Biophys. J.* 91, 4598–4604 (2006)

10. Berciaud, S., Lasne, D., Blab, G.A., Cognet, L., Lounis, B.: Photo- thermal heterodyne imaging of individual metallic nanoparticles: theory versus experiment. *Phys. Rev. B.* 73, 045424 (2006)
11. Inglehart, L.J., Broniaotowski, A., Fournier, D., Boccara, A.C., Lepoutre, F.: Photothermal imaging of copper-decorated grain boundary in silicon. *Appl. Phys. Lett.* 56, 1749 (1990)
12. Sahraoui, A.H., Kolinsky, C., Delenclos, S., Daoudi, A., Buisine, J.M.: Photothermal imaging as a tool to study phase transitions in binary mixtures of liquid crystals. *J. Appl. Phys.* 82, 6209 (1997)
13. Tamaki, E., Sato, K., Tokeshi, M., Sato, K., Aihara, M., Kitamori, T.: Single-cell analysis by a scanning thermal lens microscope with a microchip: direct monitoring of cytochrome c distribution during apoptosis process. *Anal. Chem.* 74, 1560–1564 (2002)
14. Nedosekin, D.A., Galanzha, E.I., Ayyadevara, S., Shmookler Reis, R.J., Zharov, V.P.: Photothermal confocal spectromicroscopy of multiple cellular chromophores and fluorophores. *Biophys. J.* 102, 672–681 (2012)
15. Brusnichkin, A.V., Nedosekin, D.A., Galanzha, E.I., Vladimirov, Y.A., Shevtsova, E.F., Proskurnin, M.A., Zharov, V.P.: Ultrasensi- tive label-free photothermal imaging, spectral identification, and quantification of cytochrome c in mitochondria, live cells, and solutions. *J. Biophoton.* 3, 791–806 (2010)
16. Lasne, D., Blab, G.A., Giorgi, F.D., Ichas, F., Lounis, B., Cognet, L.: Label-free optical imaging of mitochondria in live cells. *Opt. Express.* 15(21), 14184–14193 (2007)
17. Lu, S., Min, W., Chong, S., Holtom, G.R., Xie, X.S.: Label-free imaging of heme proteins with two-photon excited photothermal lens microscopy. *Appl. Phys. Lett.* 96, 113701 (2010)
18. Miyazaki, J., Tsurui, H., Kawasumi, K., Kobayashi, T.: Simultaneous dual-wavelength imaging of nonfluorescent tissues with 3D subdiffraction photothermal microscopy. *Opt. Express.* 23(3), 3647–3656 (2015)
19. Zharov, V.P.: Ultrasharp nonlinear photothermal and photoacoustic resonances and holes beyond the spectral limit. *Nat. Photonics.* 5, 110–116 (2011)
20. Nedosekin, D.A., Juratli, M.A., Sarimollaoglu, M., Moore, C.L., Rusch, N.J., Smeltzer, M.S., Zharov, V.P., Galanzha, E.I.: Photoacoustic and photothermal detection of circulating tumor cells, bacteria and nanoparticles in cerebrospinal fluid in vivo and ex vivo. *J. Biophotonics.* 6, 523–533 (2013)
21. Cognet, L., Tardin, C., Boyer, D., Choquet, D., Tamarat, P., Lou- nis, B.: Single metallic nanoparticle imaging for protein detection in cells. *Proc. Natl. Acad. Sci. USA.* 100, 11350–11355 (2003)
22. Vermeulen, P., Cognet, L., Lounis, B.: Photothermal microscopy: optical detection of small absorbers in scattering environments. *J. Microsc.* 254, 115–121 (2014)
23. Leduc, C., Si, S., Gautier, J., Soto-Ribeiro, M., Wehrle-Haller, B., Gautreau, A., Giannone, G., Cognet, L., Lounis, B.: A highly specific gold nanoprobe for live-cell single-molecule imaging. *Nano Lett.* 13, 1489–1494 (2013)
24. Leduc, C., Jung, J.M., Carney, R.P., Stellacci, F., Lounis, B.: Direct investigation of intracellular presence of gold nanoparticles via photothermal heterodyne imaging. *ACS Nano.* 5, 2587–2592 (2011)
25. Miyazaki, J., Tsurui, H., Hayashi-Takagi, A., Kasai, H., Kobayashi, T.: Sub-diffraction resolution pump probe microscopy with shot-noise limited sensitivity using laser diodes. *Opt. Express.* 22, 9024–9032 (2014)
26. Miyazaki, J., Tsurui, H., Kawasumi, K., Kobayashi, T.: Opti- mal detection angle in sub-diffraction resolution photothermal microscopy: application for high sensitivity imaging of biological tissues. *Opt. Express.* 22, 18833–18842 (2014)

27. Miyazaki, J., Kawasumi, K., Kobayashi, T.: Resolution improvement in laser diodes-based pump probe microscopy with an annular pupil filter. *Opt. Lett.* 39, 4219–4222 (2014)
28. Miyazaki, J., Tsurui, H., Kawasumi, K., Kobayashi, T.: Sensitivity enhancement of photothermal microscopy with radially segmented balanced detection. *Opt Lett.* 40(4), 479–482 (2015)
29. Miyazaki, J., Tsurui, H., Kobayashi, T.: Reduction of distortion in photothermal microscopy and its application to the high resolution three-dimensional imaging of nonfluorescent tissues. *Biomed. Opt. Exp.* 6(9), 3217–3224 (2015)
30. Sick, B., Hech, B.: Orientational imaging of single molecules by annular illumination. *Phys. Rev. Lett.* 85, 4482–4485 (2000)
31. Kim, J., Kim, D.C., Back, S.H.: Demonstration of high lateral resolution in laser confocal microscopy using annular and radially polarized light. *Microsc. Res. Techniq.* 72, 441–446 (2009)
32. Kitamura, K., Sakai, K., Noda, S.: Sub-wavelength focal spot with long depth of focus generated by radially polarized, narrow-width annular beam. *Opt. Express.* 18, 4518–4525 (2010)
33. He, J., Miyazaki, J., Wang, N., Tsurui, H., Kobayashi, T.: Label- free imaging of melanoma with nonlinear photothermal micros- copy. *Opt. Lett.* 40(7), 1141–1144 (2015)
34. He, J., Miyazaki, J., Wang, N., Tsurui, H., Kobayashi, T.: Bio- logical imaging with nonlinear photothermal microscopy using a compact supercontinuum fiber laser source. *Opt. Express.* 23(8), 9762–9771 (2015)
35. Kildishev, A.V., Sivan, Y., Litchinitser, N.M., Shalaev, V.M.: Frequency-domain modeling of TM wave propagation in optical nanostructures with a third-order nonlinear response. *Opt. Lett.* 34(21), 3364–3366 (2009)
36. Chu, S.W., Su, T.Y., Oketani, R., Huang, Y.T., Wu, H.Y., Yone- maru, Y., Yamanaka, M., Lee, H., Zhuo, G.Y., Lee, M.Y., Kawata, S., Fujita, K.: Measurement of a saturated emission of optical radiation from gold nanoparticles: application to an ultrahigh resolution microscope. *Phys. Rev. Lett.* 112, 017402 (2014)
37. Deka, G., Sun, C.K., Fujita, K., Chu, S.W.: Nonlinear plasmonic imaging techniques and their biological applications. *Nanophotonics* (2016) https://doi.org/10.1515/nanoph-2015-0149
38. Tzang, O., Cheshnovsky, O.: New modes in label-free super resolution based on photo-modulated reflectivity. *Opt. Express.* 23(16), 20926–20932 (2015)
39. Tzang, O., Pevzner, A., Marvel, R.E., Haglund, R.F., Cheshnovsky, O.: Super-resolution in label-free photomodulated reflectivity. *Nano Lett.* 15, 1362–1367 (2015)
40. Jathoul, A.P., Laufer, J., Ogunlade, O., Treeby, B., Cox, B., Zhang, E., Johnson, P., Pizzey, A.R., Philip, B., Marafioti, T., Lythgoe, M.F., Pedley, R.B., Pule, M.A., Beard, P.: Deep in vivo photoacoustic imaging of mammalian tissues using a tyrosinase- based genetic reporter. *Nat. Photon.* 9, 239–246 (2015)
41. Yao, J., Wang, L., Yang, J.M., Maslov, K.I., Wong, T.T., Li, L., Huang, C.H., Zou, J., Wang, L.V.: High-speed label-free functional photoacoustic microscopy of mouse brain in action. *Nat. Methods.* 12, 407–410 (2015)
42. Pache, C., Bocchio, N.L., Bouwens, A., Villiger, M., Berclaz, C., Goulley, J., Gibson, M.I., Santschi, C., Lasser, T.: Fast three- dimensional imaging of gold nanoparticles in living cells with photothermal optical lock-in optical coherence microscopy. *Opt. Express.* 20, 21385–21399 (2012)
43. Tucker-Schwartz, J.M., Meyer, T.A., Patil, C.A., Duvall, C.L., Skala, M.C.: In vivo photothermal optical coherence tomography of gold nanorod contrast agents. *Biomed. Opt. Express.* 3, 2881–2895 (2012)

44. Arunkarthick, S., Bijeesh, M.M., Varier, G.K., Kowshik, M., Nandakumar, P.: Laser scanning photothermal microscopy: fast detection and imaging of gold nanoparticles. *J. Microsc.* 256, 111–116 (2014)
45. Chang, W.S., Link, S.: Enhancing the sensitivity of single-particle photothermal imaging with thermotropic liquid crystals. *J. Phys. Chem. Lett.* 3, 1393–1399 (2012)
46. Parra-Vasquez, A.N.G., Oudjedi, L., Cognet, L., Lounis, B.: Nanoscale thermotropic phase transitions enhancing photothermal microscopy signals. *J. Phys. Chem. Lett.* 3, 1400–1403 (2012)
47. Corral, M.M., Andres, P., Rodriguez, C.J.Z., Kowalczyk, M.: Three-dimensional superresolution by annular binary filters. *Opt. Commun.* 165, 267–278 (1999)
48. Corral, M.M., Andres, P., Castaiieda, J.O., Saavedra, G.: Tun- able axial superresolution by annular binary filters: application to confocal microscopy. *Opt. Commun.* 119, 491–498 (1995)
49. Richards, B., Wolf, E.: Electromagnetic diffraction in optical systems. II. Structure of the image field in an aplanatic system. *Proc. R. Soc. Lond. A.* 253, 358–379 (1959)
50. Youngworth, K.S., Brown, T.G.: Focusing of high numerical aperture cylindrical-vector beams. *Opt. Express.* 7, 77–87 (2000)
51. Selmke, M., Braun, M., Cichos, F.: Photothermal single-particle microscopy: detection of a nanolens. *Nano Lett.* 6. 2741–2749 (2012)
52. Conklin, M.W., Provenzano, P.P., Eliceiri, K.W., Sullivan, R., Keely, P.J.: Fluorescence lifetime imaging of endogenous fluorophores in histopathology sections reveals differences between normal and tumor epithelium in carcinoma in situ of the breast. *Cell Biochem. Biophys.* 53, 145–157 (2009)
53. Apgar, J.M., Juarranz, A., Espada, J., Villanueva, A., Canete, M., Stockert, J.C.: Fluorescence microscopy of rat embryo sections stained with haematoxylin–eosin and Masson's trichrome method. *J. Microsc.* 191, 20–27 (1997)
54. de Carvalho, H.F., Taboga, S.R.: Fluorescence and confocal laser scanning microscopy imaging of elastic fibers in hematoxylin- eosin stained sections. *Histochem. Cell Biol.* 106, 587–592 (1996)
55. Boyer, D., Tamarat, P., Maali, A., Lounis, B., Orrit, M.: Photothermal imaging of nanometer-sized metal particles among scatterers. *Science.* 297, 1160–1163 (2002)
56. Kwan, A.C., Duff, K., Gouras, G.K., Webb, W.W.: Optical visualization of Alzheimer's pathology via multiphoton-excited intrinsic fluorescence and second harmonic generation. *Opt. Express.* 17, 3679–3689 (2009)
57. Krafft, C., Kirsch, M., Beleites, C., Schackert, G., Salzer, R.: Methodology for fiber-optic Raman mapping and FTIR imag- ing of metastases in mouse brains. *Anal. Bioanal. Chem.* 389, 1133–1142 (2007)
58. Krafft, C.: Bioanalytical applications of Raman spectroscopy. *Anal. Bioanal. Chem.* 378, 60–62 (2004)
59. Benseny-Cases, N., Klementieva, O., Cotte, M., Ferrer, I., Clad- era, J.: Microspectroscopy (mu FTIR) reveals co-localization of lipid oxidation and amyloid plaques in human Alzheimer disease brains. *Anal. Chem.* 86, 12047–12054 (2014)
60. Dombeck, D.A., Kasischke, K.A., Vishwasrao, H.D., Ingelsson, M., Hyman, B.T., Webb, W.W.: Uniform polarity microtubule assemblies imaged in native brain tissue by second-harmonic generation microscopy. *Proc. Natl. Acad. Sci. USA.* 100, 7081–7086 (2003)
61. Witte, S., Negrean, A., Lodder, J.C., de Kock, C.P., Testa Silva, G., Mansvelder, H.D., Louise Groot, M.: Label-free live brain imaging and targeted patching with third-harmonic generation microscopy. *Proc. Natl. Acad. Sci. USA.* 108, 5970–5975 (2011)

62. Evans, C.L., Xu, X., Kesari, S., Xie, X.S., Wong, S.T., Young, G.S.: Chemically-selective imaging of brain structures with CARS microscopy. *Opt. Express.* 15, 12076–12087 (2007)
63. Freudiger, C.W., Min, W., Saar, B.G., Lu, S., Holtom, G.R., He, C., Tsai, J.C., Kang, J.X., Xie, X.S.: Label-free biomedical imag- ing with high sensitivity by stimulated Raman scattering micros- copy. *Science.* 322, 1857–1861 (2008)
64. Hammond, B.R. Jr., Renzi, L.M.: Carotenoids. *Adv. Nutr.* 4, 474–476 (2013)
65. Johnson, E.J.: A possible role for lutein and zeaxanthin in cognitive function in the elderly. *Am. J. Clin. Nutr.* 96, 1161S–1165S (2012)
66. van der Meer, F.J., Faber, D.J., Cilesiz, I., van Gemert, M.J., van Leeuwen, T.G.: Temperature-dependent optical properties of individual vascular wall components measured by optical coherence tomography. *J. Biomed. Opt.* 11, 041120 (2006)
67. Liu, J.P., Tang, Y., Zhou, S., Toh, B.H., McLean, C., Li, H.: Cholesterol involvement in the pathogenesis of neurodegenerative diseases. *Mol. Cell. Neurosci.* 43, 33–42 (2010)
68. Xie, X., Li, D., Tsai, T.H., Liu, J., Braun, P.V., Cahill, D.G.: Ther- mal conductivity, heat capacity, and elastic constants of water- soluble polymers and polymer blends. *Macromolecules.* 49, 972–978 (2016)
69. Bama, G.K., Devi, P.I., Ramachandran, K.: Structural and thermal properties of PVDF/PVA blends. *J. Mater. Sci.* 44, 1302–1307 (2009)
70. Kobayashi, T., Kawasumi, K., Miyazaki, J., Nakata, K., Resolution enhancement of pump–probe microscope with an inverse-annular filter. *Opt. Rev.* 25, 271–294 (2018)

2.1.3 Reduction of Distortion in Photothermal Microscopy and its Application to the High-resolution Three-dimensional Imaging of Nonfluorescent Tissues

1 INTRODUCTION

Photothermal (PT) microscopy's capability to measure light-absorbing molecules at high sensitivity and spatial resolution has made it useful in detecting endogenous non-fluorescent chromoproteins in biological tissues. Thus far, it has been applied to visualizing the distribution of cytochrome[1] in mitochondria[2] [1–3], the three-dimensional (3D) imaging of hemoglobin[3] in blood vessels [4], and the high-resolution 3D imaging of melanin[4] pigments in skin cancer [5]. Furthermore, 5-nm-diameter gold nanoparticles have been used to identify biomolecules with antibody labeling techniques that have been applied to cellular imaging by means of PT microscopy [6–8]. As gold nanoparticles[5] are physiologically inert and do not suffer from photobleaching or photoblinking, they are useful in biological imaging and potentially applicable to immunoanalysis and gene analysis at the cellular level.

In PT microscopy, two laser beams (pump and probe) are incident on the sample through a focusing lens. The pump beam increases the temperature in the region

DOI: 10.1201/9781003476030-4

around the light absorbing sample within the focal spot volume. This results in local refractive index changes and induces the deflection of the probe beam. In the forward detection scheme, the nano-lens (thermal lens) effect produces positive and negative peaks corresponding to the focusing or defocusing of the probe beam when a sample is scanned in the axial direction [9–11]. This signal shape would be useful in the tracking or positioning of nanomaterials, but it causes a serious problem in biological imaging, as the twin peaks result creates distortion when reconstructing an image of a structured sample.

To eliminate this drawback, a confocal forward detection scheme has been proposed in previous studies [12, 13] and the experimental results have shown that a single peak with a 450-nm axial resolution is achieved when the sample is a single layer thin film and a water immersion objective lens is used. Confocal PT microscopy can be conducted with the same detection schema as the classical thermal lens setup, but the pinhole position must be carefully adjusted to avoid an unwanted photothermal signal.

In contrast, it has been demonstrated that either of the twin peaks is suppressed by an offset between the focal planes of the pump and probe beams resulting in the very small noise level [9]. Because the focal plane's position can be easily controlled by the divergence of the incident beam, this scheme would be a simple approach for reducing distortion without using a confocal pinhole.

The purpose of this study is to demonstrate the validity and usefulness of the offset scheme for reducing distortion in PT imaging. To this end, the relationship between the focal plane offset and the spatial resolution is examined in detail in both the lateral and axial planes, as there is concern about the degradation in the spatial resolution prompted by the focal plane offset. Experimentally, it is confirmed that both the lateral and axial resolutions are better than the conventional optical microscopy, even with the offset. A procedure for reducing image distortion in PT microscopy, which has not been discussed previously, is presented and the high-fidelity 3D imaging of thick biological tissues is demonstrated.

2 EXPERIMENT

PT imaging is conducted using an LD-based pump-probe microscopic system. The details of the experimental setup and sample preparation have been described in previous papers [5, 14–16]. A 488-nm LD was used for pumping and a 640-nm LD was used for probing (Figure 2.1.3.1). Each beam was collimated through a polarization-maintaining single-mode (SM) fiber and a doublet lens for spatial mode filtering. The combined beams were directed to an objective lens [40X, numerical aperture (NA) = 0.95]. The sample position was raster scanned using a three-axis positioning stage driven by piezo actuators. An iris diaphragm was used to adjust the effective NA of the condenser lens in optimizing the signal-to-noise ratio. An auto-balanced photodetector (Newfocus, Nirvana) was used to cancel the intensity fluctuation of the probe laser, achieving nearly shot noise limited sensitivity. The probe and pump beam intensities were set to frequencies of ω_1 and ω_2, respectively. A beat signal at $|\omega_1 - \omega_2|$ was generated by the PT effect in the sample, and a dual-phase lock-in

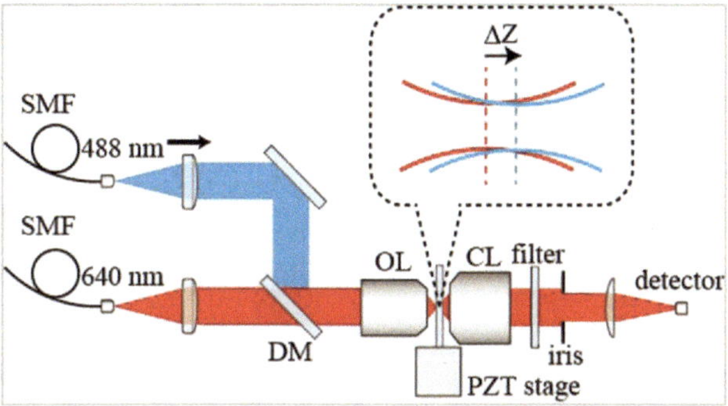

FIGURE 2.1.3.1 Schematic illustration of laser diode-based photothermal microscopy. SMF: single mode fiber, DM: dichroic mirror, OL: objective lens, CL: condenser lens. The distance Δz between the focal planes of the pump and probe beams is controlled by positioning the doublet lens after the SMF in the pump beam path. The figure was reported in the journal *Optical Review* and copyright is partially held between the representative author (Takayoshi Kobayashi, TK) and the AIP Advances.

amplifier was used to demodulate the beat signal. In this study, ω_1 and ω_2 were set at 500 kHz and 600 kHz, respectively.

Results and Discussion

To evaluate the usefulness of the present scheme, it is important to understand how the PT signal depends on the beam overlap. For this purpose, the intensity profiles of the pump and probe beams at focus were measured using a knife edge method. A glass slide with chrome-coated lines (Edmond Optics) was mounted on the sample stage and used as a knife edge. A cover slip was attached to the glass slide to avoid spherical aberration, which can be induced by using a dry objective lens. In the knife edge method, the intensity profile is given by the differentiation of the transmitted beam power with respect to the edge position. In our experiment, the differential of the transmitted beam power was directly measured in such a way that the edge position was modulated in the lateral direction at a frequency of 50Hz with an amplitude of 20 nm. The resulting variation in the transmitted beam power was detected with a lock-in amplifier.

Figure 2.1.3.2(a) shows the intensity profiles of the pump and probe beams in the focal plane. Fitting with a Gaussian provides the full width at half maximum (FWHM) values of the pump and probe beams (282 and 368 nm, respectively). They are only 1.07–1.08 times larger than the diffraction limited spot size (262 and 344 nm, respectively). The displacements between the peak positions of the pump and probe beams in the lateral plane are maintained within ~20 nm. Figure 2.1.3.2(b) shows that the intensity distribution in the axial direction exhibits long tails and fitting with a Lorentzian produces FWHM values of 1.19 and 1.55 μm for the pump and probe beams, respectively.

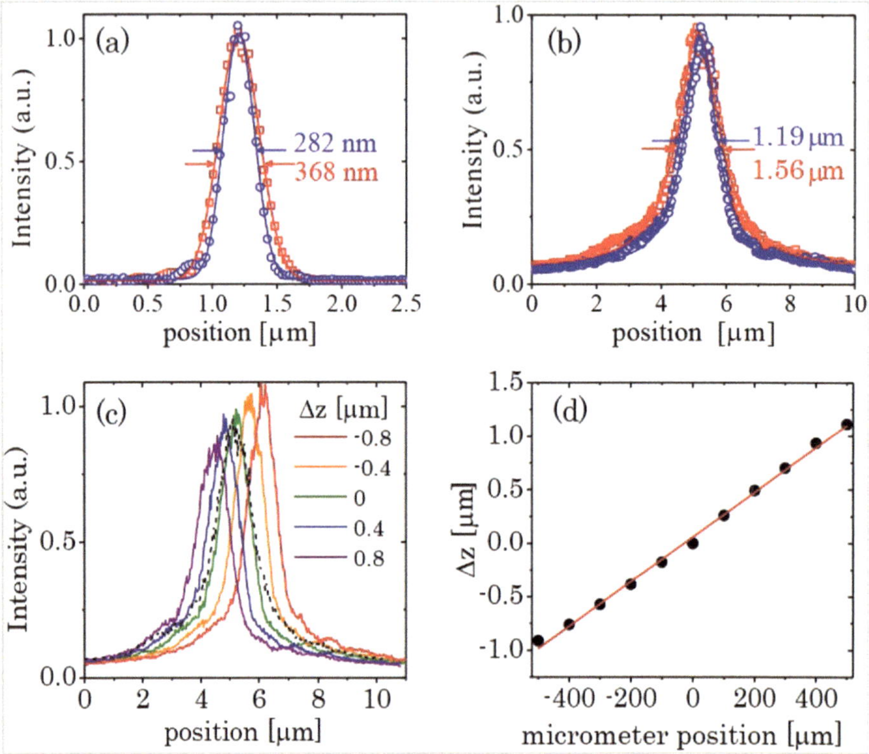

FIGURE 2.1.3.2 Intensity profiles of pump (blue) and probe (red) beams in the (a) lateral and (b) axial planes, as measured using a knife edge method. (c) Axial intensity distribution of the pump beam with several offset values Δz from the focal plane location. The broken line is the intensity profile of the probe beam. (d) Relation between Δz and the position of the doublet lens, which is mounted on a translation stage with a micrometer. The figure was reported in the journal *Optical Review* and copyright is partially held between the representative author (Takayoshi Kobayashi, TK) and the AIP Advances.

The spatial offset Δz along the optical axis (z-axis) between the focal planes of the pump and probe beams was controlled by the divergence of the incident pump beam, which was adjusted by positioning the doublet lens after the SM fiber, as depicted in Fig. 2.1.3.1. Figure 2.1.3.2(c) shows the intensity profiles of the pump beam in the axial plane as the position of the doublet lens changed. The spatial offset Δz is nearly proportional to the position of the doublet lens [Figure 2.1.3.2(d)]. The peak intensity depends somewhat on the beam divergence because the incident beam is expanded to overfill the back aperture of the objective lens to maximize the spatial resolution.

The spatial resolution in PT microscopy is evaluated by changing Δz in the imaging procedure of a single 20-nm-diameter gold nanoparticle embedded in polyvinyl alcohol film. The pump and probe beams' power incident on the sample are both 0.2 mW. The time constant of the lock-in amplifier is 0.5 ms and the pixel dwell time is 1 ms.

The PT imaging of individual nanoparticles showed that some particles exhibited twin peaks in the axial direction when Δz was close to zero while others exhibited a single peak, regardless of Δz in the same sample.

Figure 2.1.3.3 shows a typical PT image of a gold nanoparticle that exhibited twin peaks when Δz was close to zero. The amplitude and phase images show that the sign of the probe beam transmissivity (phase of the lock-in amplifier output) changes depending on the axial position of the sample, and either of the peaks is suppressed with increasing or decreasing Δz [Figure 2.1.3.3(a)]. This can be explained by the lens-like model, where the probe beam is focused or defocused depending on the sample position and the twin peaks become asymmetric in the presence of the focal plane offset [9–11]. The positive peak (in-phase) is larger than the negative (180° out of phase) peak at Δz = 0 μm.

In the case when the axial intensity profile exhibited a single peak with an offset, the resolution was defined as its FWHM value [Figure 2.1.3.3(b)]. The axial resolution deteriorated with increasing or decreasing |Δz|, thus a small |Δz| is preferable for high resolution imaging. However, when Δz was close to zero, the twin peaks emerged, distorting the image of the structured nanomaterials. Hence, it is necessary to suppress the peaks with an appropriate offset. When Δz = −0.4 μm, the intensity

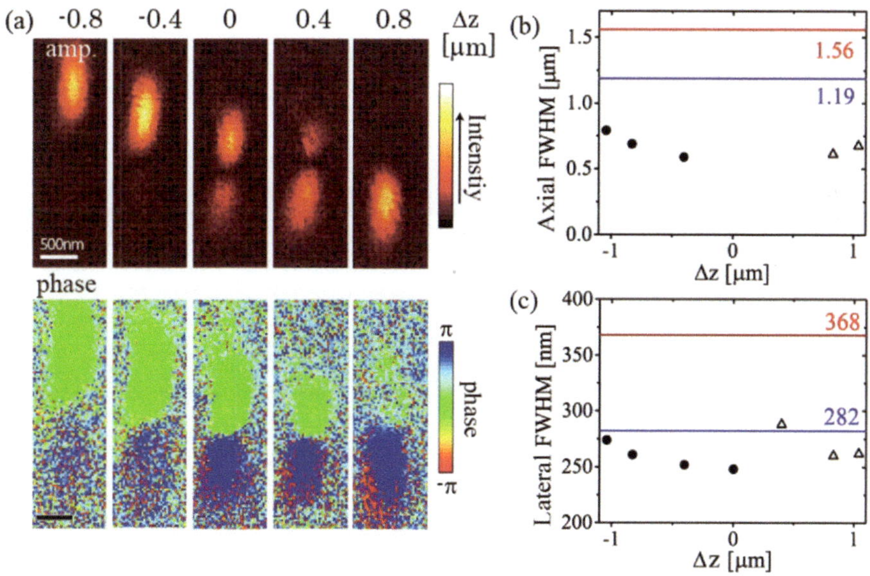

FIGURE 2.1.3.3 Photothermal images of a single gold nanoparticle exhibiting twin peaks when the focal offset Δz was close to zero. (a) Intensity distribution in the axial plane with several values of Δz. (b) Axial and (c) lateral FWHM values as a function of Δz. The filled circles and open triangles are the FWHM values for positive (in-phase) and negative (180° out of phase) peaks, respectively. The figure was reported in the journal *Optical Review* and copyright is partially held between the representative author (Takayoshi Kobayashi, TK) and the AIP Advances.

profile exhibited a single peak in the axial plane with an FWHM width of 590 nm—smaller than the focal spot size of the pump (probe) beam by 50% (62%). In contrast, the lateral FWHM value at Δz = −0.4 μm was 252 nm [Figure 2.1.3.3(c)] which is smaller than the focal spot size of the pump (probe) beam by 11% (32%). Thus, by providing a small focal plane offset, it was possible to reduce distortion and obtain a high fidelity image with a high spatial resolution. The peak intensity of the PT signal at Δz = −0.4 μm is about 2.1 times larger than that without the offset.

In contrast, some other nanoparticles exhibited single peaks regardless of Δz value, as shown in Figure 2.1.3.4(a). The peak position shifted in the axial direction as Δz changed. The phase images show that the transmissivity of the probe beam was in phase with the pump beam intensity, and the phase value was less affected by Δz. The spatial resolutions in the axial and lateral directions are summarized in Figures 2.1.3.4(b) and 2.1.3.4(c), respectively. As long as Δz was within 1 μm, the axial and lateral FWHM values were less sensitive to Δz, with averages of 609 and 236 nm, respectively. The lateral FWHM value was close to that of the product of the pump and probe beam intensities (231 nm).

It should be noted that the axial FWHM value was much smaller than that of the product of the pump and probe intensities (890 nm), in Figs 2.1.3.3 and 2.1.3.4.

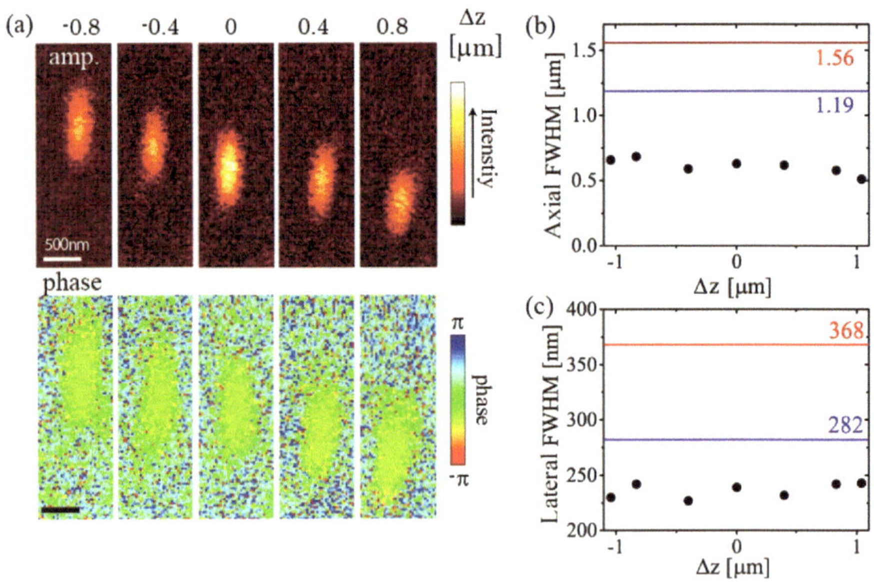

FIGURE 2.1.3.4 Photothermal images of a single gold nanoparticle that exhibits a single peak in the axial direction, regardless of the focal plane offset Δz. (a) Intensity distribution in the axial plane with several values of Δz. The upper and lower panels are amplitude and phase images, respectively. The scale bar is 0.5μm. (b) Axial and (c) lateral FWHM values as a function of Δz The figure was reported in the journal *Optical Review* and copyright is partially held between the representative author (Takayoshi Kobayashi, TK) and the AIP Advances.

Furthermore, the intensity of tails in the axial PSF was dramatically reduced, suggesting strong sectioning capability in thick specimens, as the optical sectioning properties resulted from the characteristics of the integrated intensity point spread function.

It is unclear why there were two different types in the same sample. One possible reason is an incomplete thermal lens due to the non-uniformity of the surrounding medium affected by local conditions. It is expected that incomplete thermal lens also occurs in imaging biological tissues since they have microscopically non-uninform temperature coefficient of refractive index d*n*/d*T* and thermal diffusivity. The results did not depend on the modulation frequencies and pump beam intensity.

Based on these experimental observations, the procedure for obtaining a PT image with minimized distortion can be summarized as follows. First, it is recommended to measure a sectional image without offset to check for the presence/absence of the distortion. Twin peaks created by the lens-like effect produce a double image and the resulting distortion can be identified from the sign of the signal (or the phase of the lock-in amplifier output). If there are noticeable twin peaks, provide a small offset to suppress either the positive or the negative signal. With decreasing (increasing) Δz, it is possible to enhance the positive (negative) signal while suppressing the negative (positive) signal. Consequently, image distortion is suppressed and a high-resolution image can be obtained. A small distortion can be eliminated by simply setting the pixel intensity values to zero. The phase image is useful to check the presence of the distortion due to the dual lobe but the distortion cannot be avoided just by considering the phase.

The 3D imaging of a slice of biological tissue is demonstrated based on this procedure. Figure 2.1.3.5 shows PT images of a slice of a rabbit ovary stained with hematoxylin[6] and eosin[7] (HE). The sample thickness is ~10 μm. Figure 2.1.3.5(a) shows a bright field image of the HE stained tissues. Both the hematoxylin and eosin molecules were detected in the PT imaging as they absorbed light at a wavelength of 488 nm [5]. The pump and probe beams' power incident on the sample were 0.1 and 0.2 mW, respectively. The focal plane offset was set at Δz = 0 μm. The amplitude and phase images in the lateral [Figure 2.1.3.5(b)] and axial [Fig. 2.1.3.5(c)] planes show that the positive (in-phase) signal was dominant while the negative (180° out of phase) signal was buried in noise. In this case, it was not necessary to provide an offset. Figure 2.1.3.5(d) and (Visualization 1 reduction of distortion in photothermal microscopy and its application to the high-resolution three-dimensional imaging of nonfluorescent tissues (figshare.com)) show 3D images of HE-stained biological tissue. To reconstruct the 3D model, a set of 30 image slices was acquired by changing the sample position in the axial direction with a step size of 0.28 μm. Each image size is 19.6 x 19.6 μm at 200 x 200 pixels. The time constant of the lock-in amplifier was 0.5 ms and the dwell time per point was 1 ms. The complex structures of the cell nucleus and cytoplasm were clearly visualized. One of the advantages of PT imaging is that it dramatically improves the axial resolution over conventional bright field imaging, allowing the production of high-resolution 3D models or high-contrast transverse images for a thick specimen [5].

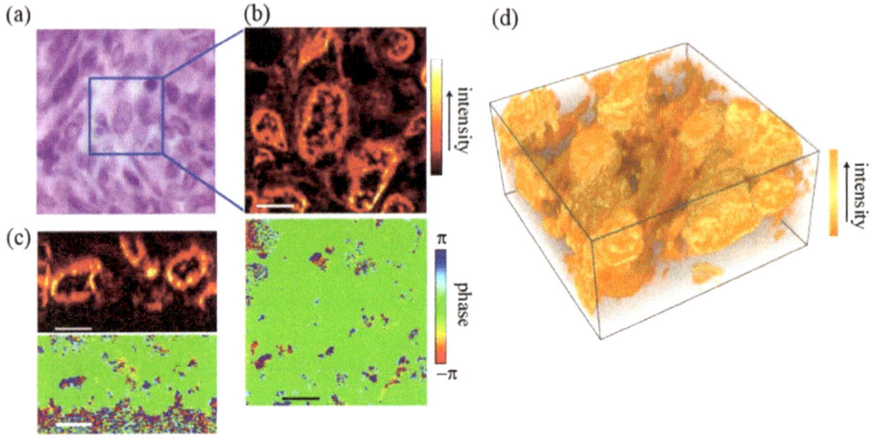

FIGURE 2.1.3.5 Photothermal (PT) images of a hematoxylin and eosin (HE)-stained tissue. (a) Bright field image of HE-stained tissue. A doublet lens and the same objective lens are used to produce the image on a CMOS camera with a white LED backlight. PT images of the tissue in the (b) lateral and (c) axial planes without focal plane offset. The upper and lower panels indicate amplitude and phase images, respectively. Scale bars are 4 μm. (d) 3D image of He-stained tissue (see Visualization 1). The figure was reported in the journal *Optical Review* and copyright is partially held between the representative author (Takayoshi Kobayashi, TK) and the AIP Advances.

In contrast, some of the samples exhibited noticeable distortion in the case of no offset. Figure 2.1.3.6 shows PT images of a slice of mouse melanoma in the lateral plane [Figs. 2.1.3.6(a) and 2.1.3.6(b)] and axial plane [Figs. 2.1.3.6(c) and 2.1.3.6(d)]. Although the positive signal is dominant with $\Delta z = 0$ μm, the PT images clearly include large negative signals whose intensity is 20–30% of that of the positive signal [Figs. 2.1.3.6(a) and (c)]. This negative signal causes image distortion and can be eliminated by setting the offset at $\Delta z = -0.5$ μm [Figs. 2.1.3.6(b) and 6(d)]. Figure 2.1.3.6(e) and Visualization 2: Reduction of distortion in photothermal microscopy and its application to the high-resolution three-dimensional imaging of nonfluorescent tissues (figshare.com) show the reconstructed 3D image of a slice of mouse melanoma with $\Delta z = -0.5$ μm, in which the 3D distribution of melanin granules is clearly visualized.

In the transmission-mode confocal PT microscopy, the pinhole position should be carefully adjusted so that it lies on a plane one Rayleigh range away from the focal plane [12, 13]. This scheme is useful for measuring a thin specimen. However, because the confocal position is sensitive to the sample thickness, care must be taken with samples of non-uniform thickness or refractive index for the measurement of biological tissues. In contrast, the present scheme was less affected by the non-uniformity of the sample because the confocal pinhole was not used. Furthermore, it can easily be incorporated into the spatially segmented balanced detection method, which increases the signal-to-noise ratio in PT microscopy [16].

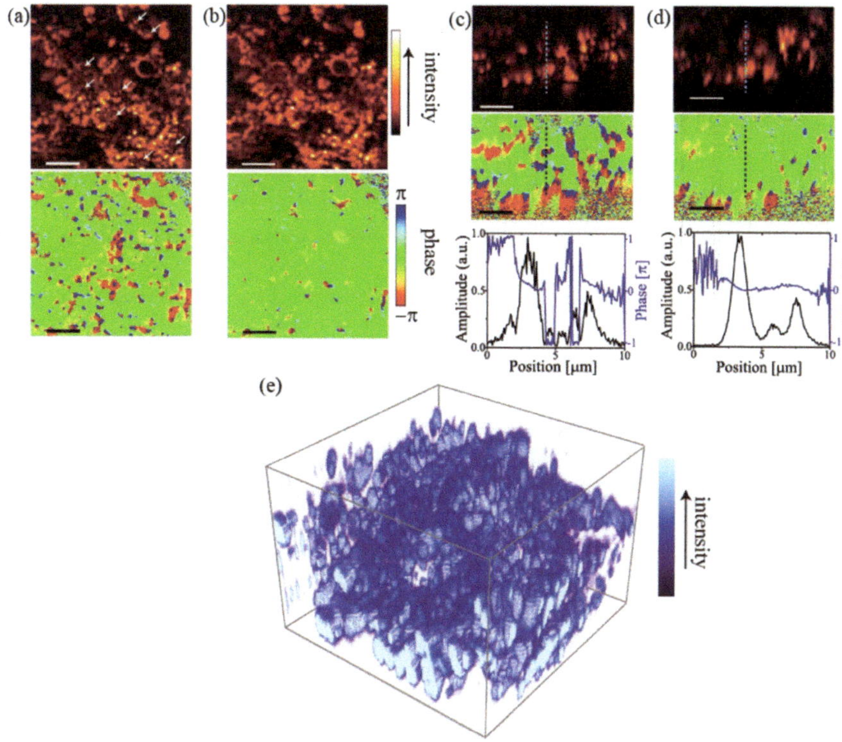

FIGURE 2.1.3.6 Photothermal (PT) images of a slice of mouse melanoma in the lateral plane with (a) Δz = 0 μm and (b) Δz = −0.5 μm. The upper and lower panels indicate amplitude and phase images, respectively. The arrows in (a) indicate a negative (180° out of phase) signal that causes image distortion. The PT images in the axial plane with (c) Δz = 0 μm and (d) Δz = −0.5 μm. Lower panels show intensity and phase profiles along the broken lines. Scale bars are 4 μm. (e) 3D rendering of a set of 50 image slices (see Visualization 2). Image size is 19.6 x 19.6 x 13.9 μm at 200 x 200 x 50 voxel. The time constant of the lock-in amplifier is 0.5 ms and the dwell time per point is 1 ms. The figure was reported in the journal *Optical Review* and copyright is partially held between the representative author (Takayoshi Kobayashi, TK) and the AIP Advances.

3 CONCLUSION

In this subsection a scheme is demonstrated for reducing distortion in photothermal microscopic imaging. In photothermal microscopy, the signal shape exhibits twin peaks corresponding to the focusing or defocusing of the probe beam when a sample is scanned in the axial direction. This causes a distortion when imaging a structured sample in the axial plane. Image distortion due to the lens-like effect can be reduced by providing a small focal offset between the pump and probe beams. This scheme does not cause serious degradation in the spatial resolution. When a dry objective lens with an NA of 0.95 was used, the axial and lateral resolution were 0.6 and 0.25

μm in full width at half maximum, respectively, and they were 50% (62%) and 11% (32%) smaller than the focal spot sizes of the pump (probe) beam. Accordingly, the three-dimensional structures of nonfluorescent biological tissues were clearly visualized. Given that the focal offset can be controlled simply by the divergence of the incident beams, this method is easy and simpler than the confocal detection setup, which makes it highly useful in obtaining high-fidelity images with a high spatial resolution [17].

The research described in this subsection is based on collaborative work of the following people: J. Miyazaki, H. Tsurui, and T. Kobayashi [17]. The above study was reported in Biomedical Optical Express and copyright is partially held between the corresponding author (Takayoshi Kobayashi, TK) and Optica (formally Optical Society of America). The figure was reported in the journal *Optical Review* and copyright is partially held between the representative author (Takayoshi Kobayashi, TK) and the AIP Advances. TK would like to express thanks to the company for understanding the revised paper to be published as one section in the book *Photothermal and Raman Microscopy* (Taylor & Francis).

NOTES

1. cytochrome: Redox-active proteins containing a heme, with a central Fe atom (cofactor) at its core. They are involved in the electron transport chain and redox catalysis. They are classified according to the type of heme and its mode of binding. cytochromes a, cytochromes b, cytochromes c and cytochrome d. Cytochrome function is linked to the reversible redox change from ferrous (Fe(II)) to the ferric (Fe(III)) oxidation state of the iron found in the heme core.
2. mitochondria: mitochondria (plural form of mitochondrion. is an organelle): Mitochondria exist in the cells of most eukaryotes, such as animals, plants and fungi. They have a double membrane structure and use aerobic respiration to generate adenosine triphosphate (ATP), which is used throughout the cell as a source of biochemical energy.
3. haemoglobin or hemoglobin: (Hb or Hgb) is a protein containing iron that facilitates the transport of oxygen in red blood cells. Almost all vertebrates contain hemoglobin. Hemoglobin in the blood carries oxygen from the respiratory organs (lungs or gills) to the other tissues of the body, where it releases the oxygen to enable aerobic respiration which powers the animal's metabolism.
4. melanin: consist of oligomers or polymers arranged in a disordered manner which among other functions provide the pigments of many organisms. Melanin pigments are produced in a specialized group of cells known as melanocytes.
5. gold nanoparticle: has been the topic of extensive research and a great deal of interest among researchers in recent years (nanotechnology). The manufacturing and utilization of nanoparticles (NPs) has expanded dramatically as a result of the rapid development of nanotechnology. Gold nanoparticles (AuNPs) are one of the most important nanoparticles, and they have been widely used for medical, biological, and non-medical and non-biological applications as ideal material because of their unique distinct features: inert, biocompatible, and especially due to low toxicity comparing with semiconductor nanoparticle such as CdSNps and CdSeNPs.

6. hematoxylin, haematoxylin: is a compound extracted from heartwood of the logwood tree with a chemical formula of $C_{16}H_{14}O_6$. This naturally derived dye has been used as a histologic stain, as an ink and as a dye in the textile and leather industry.
7. eosin: is the name of several fluorescent acidic compounds which bind to and from salts with basic, or eosinophilic, compounds like proteins containing amino acid residues such as arginine and lysine, and stains them dark red or pink as a result of the actions of bromine on eosin

REFERENCES

1. E. Tamaki, K. Sato, M. Tokeshi, K. Sato, M. Aihara, and T. Kitamori, "Single-cell analysis by a scanning thermal lens microscope with a microchip: direct monitoring of cytochrome c distribution during apoptosis process," *Anal. Chem.* 74(7), 1560–1564 (2002).
2. D. A. Nedosekin, E. I. Galanzha, S. Ayyadevara, R. J. Shmookler Reis, and V. P. Zharov, "Photothermal confocal spectromicroscopy of multiple cellular chromophores and fluorophores," *Biophys. J.* 102(3), 672–681 (2012).
3. A. V. Brusnichkin, D. A. Nedosekin, E. I. Galanzha, Y. A. Vladimirov, E. F. Shevtsova, M. A. Proskurnin, and V. P. Zharov, "Ultrasensitive label-free photothermal imaging, spectral identification, and quantification of cytochrome c in mitochondria, live cells, and solutions," *J. Biophotonics* 3(12), 791–806 (2010).
4. S. Lu, W. Min, S. Chong, G. R. Holtom, and X. S. Xie, "Label-free imaging of heme proteins with two-photon excited photothermal lens microscopy," *Appl. Phys. Lett.* 96(11), 113701 (2010).
5. J. Miyazaki, H. Tsurui, K. Kawasumi, and T. Kobayashi, "Simultaneous dual-wavelength imaging of nonfluorescent tissues with 3D subdiffraction photothermal microscopy," *Opt. Express* 23(3), 3647–3656 (2015).
6. C. Leduc, S. Si, J. Gautier, M. Soto-Ribeiro, B. Wehrle-Haller, A. Gautreau, G. Giannone, L. Cognet, and B. Lounis, "A highly specific gold nanoprobe for live-cell single-molecule imaging," *Nano Lett.* 13(4), 1489–1494 (2013).
7. C. Leduc, J. M. Jung, R. P. Carney, F. Stellacci, and B. Lounis, "Direct investigation of intracellular presence of gold nanoparticles via photothermal heterodyne imaging," *ACS Nano* 5(4), 2587–2592 (2011).
8. L. Cognet, C. Tardin, D. Boyer, D. Choquet, P. Tamarat, and B. Lounis, "Single metallic nanoparticle imaging for protein detection in cells," *Proc. Natl. Acad. Sci. U.S.A.* 100(20), 11350–11355 (2003).
9. M. Selmke, M. Braun, and F. Cichos, "Photothermal single-particle microscopy: detection of a nanolens," *ACS Nano* 6(3), 2741–2749 (2012).
10. M. Selmke, M. Braun, and F. Cichos, "Nano-lens diffraction around a single heated nano particle," *Opt. Express* 20(7), 8055–8070 (2012).
11. K. Uchiyama, A. Hibara, H. Kimura, T. Sawada, and T. Kitamori, "Thermal lens microscope," *Jpn. J. Appl. Phys.* 39(Part 1, No. 9A), 5316–5322 (2000).
12. J. Moreau and V. Loriette, "Confocal dual-beam thermal-lens microscope: model and experimental results," *Jpn. J. Appl. Phys.* 45(9A), 7141–7151 (2006).
13. J. Moreau and V. Loriette, "Confocal thermal-lens microscope," *Opt. Lett.* 29(13), 1488–1490 (2004).
14. J. Miyazaki, K. Kawasumi, and T. Kobayashi, "Frequency domain approach for time-resolved pump-probe microscopy using intensity modulated laser diodes," *Rev. Sci. Instrum.* 85(9), 093703 (2014).

15. J. Miyazaki, H. Tsurui, A. Hayashi-Takagi, H. Kasai, and T. Kobayashi, "Sub-diffraction resolution pump-probe microscopy with shot-noise limited sensitivity using laser diodes," *Opt. Express* 22(8), 9024–9032 (2014).
16. J. Miyazaki, H. Tsurui, K. Kawasumi, and T. Kobayashi, "Sensitivity enhancement of photothermal microscopy with radially segmented balanced detection," *Opt. Lett.* 40(4), 479–482 (2015).
17. J. Miyazaki, H. Tsurui, and T. Kobayashi Biomed. "Reduction of distortion in photothermal microscopy and its application to the high-resolution three-dimensional imaging of nonfluorescent tissues," *Opt. Exp.* 6(9), 3217–3224 (2015).

2.2.1 Numerical Study of the Subtraction Threshold for Fluorescence Difference Microscopy

1 INTRODUCTION

Various types of far-field optical microscopes with superresolution have been proposed and realized for sub-diffraction-limit fluorescence imaging during the last decades, such as photoactive localization microscopy (PALM) [1], stochastic optical reconstruction microscopy (STORM) [2], structured illumination microscopy (SIM) [3], stimulated emission depletion (STED) microscopy [4], and others [5–9]. In addition, mathematically calibrated microscopes such as fluorescence emission difference (FED) microscopy [10], scanning laser mode microscopy (SLAM) [11], and others [12] have attracted much interest recently for the ease with which they achieve superresolution. The essential key point of the latter methods is to subtract two fluorescence images captured by two excitation lasers with the same wavelength but separate solid and donut spots. However, the factors used in these microscopes to subtract the one from the other of two images are experiential. If the factor is too large, an exaggerated size reduction leading to even smaller than the real sample size may result. Meanwhile, some of the sample's important information may be missed due to over-processing. Therefore, a quantitative study of the parameters related to the subtraction factor must be conducted to obtain a definite criterion for determining the appropriate subtraction factor. Previous reports [10–12] have mainly focused on the subtractions of the point spread function (PSF) without taking the sample structure into consideration. In fact, due to the removal of the negative values in the subtracted images, the data processing involved is nonlinear. As such, investigating only the PSF subtractions cannot sufficiently describe the real PSF of the subtraction microscope. In another case, due to the complexity of the convolution between the

DOI: 10.1201/9781003476030-5

donut spot and different sample sizes, it is difficult to obtain an analytical description of the resolution in subtraction microscopes.

In this chapter, a numerical investigation is carried out based on vector beam diffraction theory (VBDT) and two-dimensional convolution calculations. Using the calculated focal PSFs, the resolutions of the samples in the convoluted and subtracted images are analyzed and discussed along with the sample size, sample shape, numerical aperture (NA) of the microscope objective, and excitation laser field polarizations. Combined with the subtraction threshold line obtained from the simulation, the peak intensity after subtraction and resolution derivative versus the subtraction factor are found to be two useful parameters for evaluating the appropriate subtraction.

2 PRINCIPLE AND NUMERICAL MODELS

Figure 2.2.1.1 shows the schematic principle of the fluorescence subtraction microscope. The sample is excited with the same laser wavelength but with two different focal spot shapes, which are usually a Gaussian distribution and a donut intensity

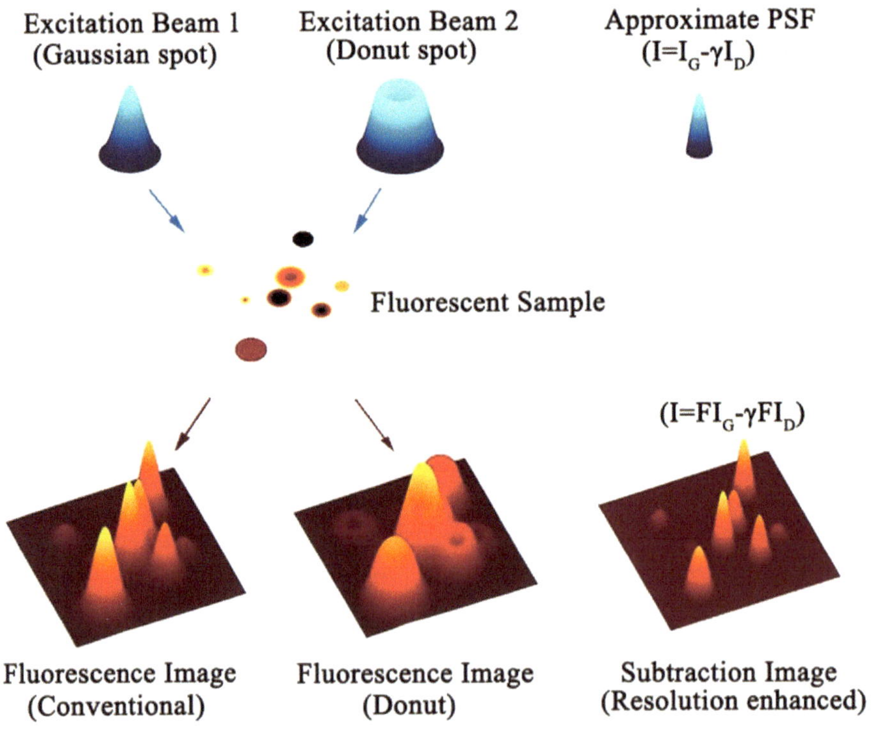

FIGURE 2.2.1.1 Schematic of subtraction microscope. The figure was reported in the journal *Optica* and copyright is shared between the representative author (Takayoshi Kobayashi, TK) and Optica.

distribution. The fluorescence image obtained by the latter is subtracted from that by the former as follows: $I = FI_G - \gamma\, FI_D$, where FI_G and FI_D are the intensities of the two fluorescence images and γ is the subtraction factor. The produced image has a higher resolution compared with the conventional fluorescence image FI_G. The most important point of this method is to obtain two suitable spot shapes and define a reasonable factor γ. Several combinations of spot shapes can be used for the subtraction microscope, such as the circular polarized Gaussian spot and vortex-phase-plate-generated donut spot [10, 11], or the vector beam radially polarized solid spot and azimuthally polarized donut spot [12]. The subtraction factor γ, which should be related to both the parameters of the two excitation lasers and the properties of the samples, plays a critical role in improving the resolution power. However, it can also introduce uncertainties.

In the chapter, an approximate numerical model is applied to investigate the image subtraction procedure and the factor γ. Because the sample is assumed to be very thin and excited without saturation, its structural shape along the axial direction can be disregarded. Thus, the intensity of the fluorescence is linearly proportional to the excitation laser intensity and to the density distribution of the fluorophores in the sample. The fluorescence excited by the two spots is assumed to have the same emission luminosity and collection efficiencies, and the detection PSFs and emission Stokes shifts of the fluorophores are not considered. Therefore, the fluorescence images can be described as the two-dimensional convolution of the excitation PSFs with the density distribution of the fluorophores in the sample. After the subtraction and the removal of negative values, the final intensity $I(x,y)$ of the subtracted image can be expressed as follows:

$$
\begin{aligned}
I_{RAW}(x,y) &= I_G(x,y) * Ob(x,y) - \gamma \cdot I_D(x,y) * Ob(x,y) \\
I(x,y) &= \begin{cases} I_{RAW}(x,y) & I_{RAW}(x,y) \geq 0 \\ 0 & I_{RAW}(x,y) < 0 \end{cases}
\end{aligned}
\tag{1}
$$

where I_G and I_D are the intensity-normalized PSFs of the excitation lasers, Ob is the shape (fluorophore density distribution) of the sample, γ is the subtraction factor, and symbol $*$ is the convolution. The PSFs are derived from the focal spots of the objective lens calculated by the VBDT [13–20] and expressed as follows:

$$
I_{G,D}(x,y,0) = |E_x|^2 + |E_y|^2 + |E_z|^2 \tag{2}
$$

$$
E_{x,y,z}(x,y,0) = \frac{iC}{\lambda} \int_0^{2\pi} \int_0^{\arcsin(NA/n)} A_{AMP} \cdot A_{\mathrm{Phase}} \cdot A_L \begin{pmatrix} P_x \\ P_y \\ P_z \end{pmatrix} \sin\theta \cdot e^{ikn\sqrt{x^2+y^2}\sin\theta\cos(\varphi - \arctan(y/x))} d\theta\, d\varphi \tag{3}
$$

TABLE 2.2.1.1
Setting values of I_G and I_Dd

I_{PSF}	Polarization	A_{AMP}	A_{PHASE}	Polarization	A_{AMP}	A_{PHASE}
I_G	Right Circular	1	1	Radial	1	1
I_D	Left Circular	1	$e^{i\varphi}$	Azimuthal	1	1

where C is the normalization constant, NA = 1.4 for the oil-immersion type objective, n is the refractive index of the immersion medium of the objective, which usually has the value of 1.518 for oil and 1 for air, respectively, and A_{AMP} and A_{phase} are the real amplitude and phase terms of the input beams, respectively. If the plane wave is used, then $A_{AMP} = 1$. Furthermore, $A_{phase} = e^{i\varphi}$ or $A_{phase} = 1$ where a vortex phase plate is present or absent, respectively. Two groups of I_G and I_D are calculated according to the setting values listed in Table 2.2.1.1. The second group is used only in Section 3.5.

A_L is the vector weight matrix of the lens, calculated as follows:

$$A_L = \sqrt{\cos\theta}\begin{pmatrix} 1+(\cos\theta-1)\cos^2\varphi & (\cos\theta-1)\cos\varphi\sin\varphi & -\sin\theta\cos\varphi \\ (\cos\theta-1)\cos\varphi\sin\varphi & 1+(\cos\theta-1)\cos^2\varphi & -\sin\theta\sin\varphi \\ \sin\theta\cos\varphi & \sin\theta\sin\varphi & \cos\theta \end{pmatrix} \quad (4)$$

The $[P_x, P_y, P_z]$ in equation 3 is the polarization unit vector matrix, and has the values shown in Table 2.2.1.2.

The fluorophore density of the sample Ob is assumed to have a Gaussian distribution, and the densities with semicircular and rectangular types are also calculated for comparison. Table 2.2.1.3 describes their numerical models. C_0 is the intensity normalization constant, and r is the full width at half maximum (FWHM), radius, and half width of the fluorophore density distributions ("Gauss", "Circ", and "Rect"), respectively.

TABLE 2.2.1.2
Polarization unit vector matrix

	Right Circular	Left Circular	Radial	Azimuthal
$\begin{pmatrix} P_x \\ P_y \\ P_z \end{pmatrix}$	$\frac{1}{\sqrt{2}}\begin{pmatrix} 1 \\ -i \\ 0 \end{pmatrix}$	$\frac{1}{\sqrt{2}}\begin{pmatrix} 1 \\ i \\ 0 \end{pmatrix}$	$\begin{pmatrix} \cos\varphi \\ \sin\varphi \\ 0 \end{pmatrix}$	$\begin{pmatrix} -\sin\varphi \\ \cos\varphi \\ 0 \end{pmatrix}$

TABLE 2.2.1.3
Numerical models for the sample in Equation (1)

	Gauss	Circ	Rect
$Ob(x,y)$	$C_0 e^{-\left(4\ln 2\frac{x^2+y^2}{r^2}\right)}$	$\begin{cases} C_0\sqrt{x^2+y^2} & \sqrt{x^2+y^2}\leq r \\ 0 & \sqrt{x^2+y^2}\geq r \end{cases}$	$\begin{cases} C_0 & \sqrt{x^2+y^2}\leq r \\ 0 & \sqrt{x^2+y^2}\geq r \end{cases}$

The subtraction image I of the sample can be obtained by numerically calculating the PSFs and determining the convolution. The obtained FWHM(r,γ) of the sample in the subtraction images can be compared with the preset sample width r to evaluate the subtraction factor. In addition, the density types of the fluorophore distributions, the NA of the objective lens, and the polarization combinations of the excitation lasers can also be investigated.

3 SIMULATION RESULTS

3.1 Sample Size

Figure 2.2.1.2 shows the convolution and subtraction results with three different sample sizes. Due to the convolution, the widths of the fluorescent distributions are broadened and the dark center of the donut spot is flattened to be non-zero compared with the initial PSFs, as shown in the 2D convoluted images in Fig. 2.2.1.2(a) and their central cut lines in Figs. 2.2.1.2(c)–2(d). The sample is illustrated in Fig. 2.2.1.2(b). This illustrates that the donut center is sensitive to the sample sizes, and that the subtraction factor cannot be infinitely large because the FI_G can be totally erased with some specific subtraction factors. This conclusion conflicts with the situation where only the PSFs are subtracted [10–12], in which case the subtraction factor γ can be infinitely large. Subtracting the PSFs and setting the negative values to be zero introduces additional nonlinearity and derivation, making the situation inappropriate. When subtracting with factors γ = 0, 0.3, and 1.0, the obtained FWHM resolutions are 0.44λ, 0.33λ, and 0.2λ with r = 0.2λ; and 0.49λ, 0.35λ and 0.13λ with r = 0.3λ, respectively, as shown in Figs. 2.2.1.2(e) and (f). This not only simply shows that a large subtraction factor leads to a better resolution, but also indicates that the FWHM decreases more rapidly as the subtraction factor increases in case of a large-radius sample. This can also be seen when the evolution of the FWHM resolution is compared with that of the subtraction factors in Fig. 2.2.1.2(i). The resolution with a sample size of r = 0.3λ has a sharper slope than those with smaller sample sizes. Figs. 2.2.1.2(e)–2(i) also show that with a smaller γ the sample may be resolved properly or below-subtraction. When the γ value is larger, the subtracted FWHM resolution of a large-size sample has an even smaller value than that of a small-size sample, which represents obvious over-subtraction.

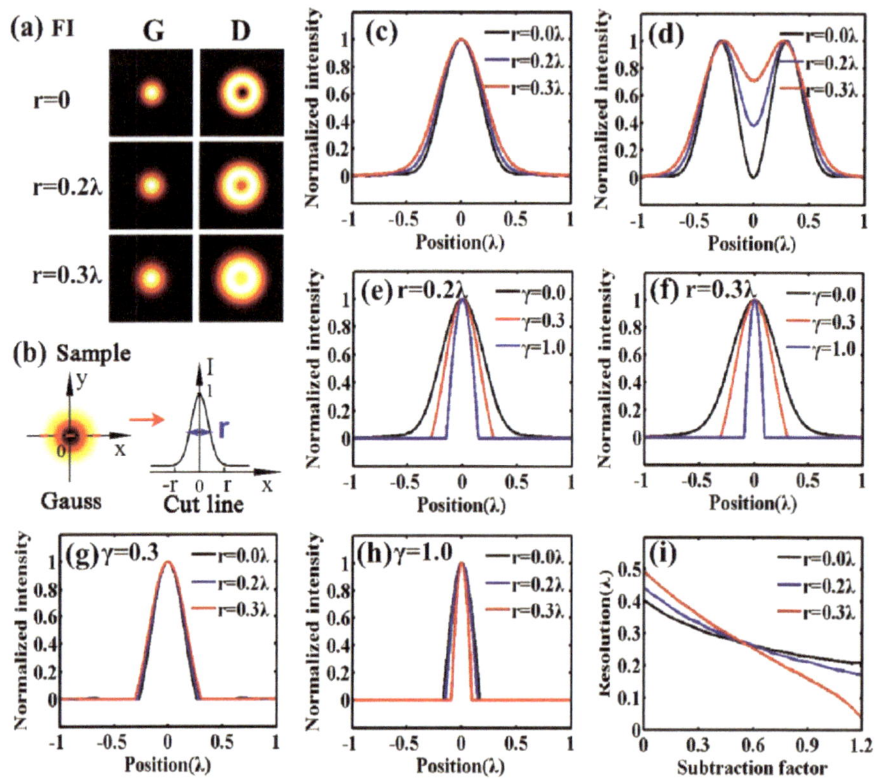

FIGURE 2.2.1.2 (a) Convoluted images with sample sizes r = 0, 0.2λ, 0.3λ. (b)The sample illustration of Gaussian shape. The cross-sections of the convoluted (c) Gaussian, (d) donut distributions, subtracted images with (e) r = 0.2λ and (f) r = 0.3λ while γ= 0, 0.3 and 1.0, and subtracted images with (g) γ = 0.3 and (h) γ = 1.0 while r = 0, 0.2λ, and 0.3λ. (i) FWHM resolutions of the sample versus the subtraction factor with r = 0, 0.2λ, 0.3λ. The figure was reported in the journal *Optica* and copyright is shared between the representative author (Takayoshi Kobayashi, TK) and Optica.

3.2 Sample Shape

The simulations with fluorophore densities of semicircular and rectangular types are carried out in comparison with the Gaussian type. The convoluted images and their cut lines using three sample shapes are shown in Figs. 2.2.1.3(a), (b), (d), (e), (h), and (i), and the sample sizes are r = 0.15λ and 0.25λ, respectively. The samples with semicircular and rectangular shapes are illustrated in Fig. 2.2.1.3(c). They indicate that the contribution of three shapes to the FWHM resolution is considered to be approximately the same as each other if the sample widths are small. When a sample size of r = 0.15λ and factors of 0.3 and 1.0 are applied, the FWHM resolutions after subtraction with three different shapes are all close to 0.32λ and 0.21λ, respectively, as shown in Figs. 2.2.1.3(f) and 3(g). This synchronized evolution of the FWHM

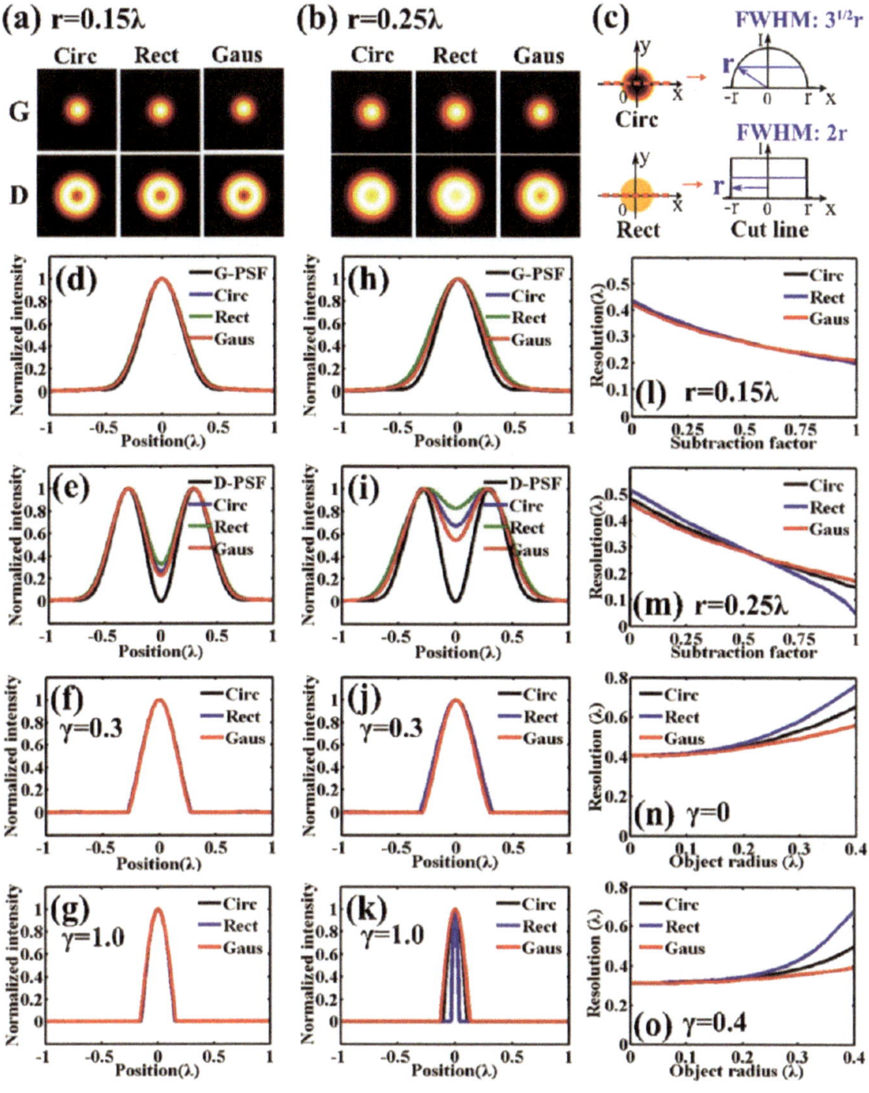

FIGURE 2.2.1.3 The (a), (d)–(g) and (l) are simulated with sample r = 0.15λ, while (b), (h)–(k) and (m) with sample r = 0.25λ. The (a) and (b) are convoluted images of PSFs of Gaussian and donut shape with sample shapes of "Circ", "Rect" and "Gauss". (c)The sample illustration of semicircular shape (Circ) and rectangular shape (Rect). The cross-sections of (d) and (h) Gaussian-PSF convolutions, (e) and (i) Donut-PSF convolutions, (f) and (j) subtracted images with γ= 0.3, (g) and (k) subtracted images with γ=1.0. The FWHM resolutions versus (l) and (m) sample *r*, versus (n) and (o) γwith three sample shapes. The figure was reported in the journal *Optica* and copyright is shared between the representative author (Takayoshi Kobayashi, TK) and Optica.

resolution versus subtraction factors for three different types can similarly be found in Fig. 2.2.1.3(l). When the sample width increases to r = 0.25λ, the FWHM curves of the three sample shapes are divergent, as shown in Figs. 2.2.1.3(j), (k), and (m). The one with "Gauss" shape shows a smoother slope and the one with the "Rect" shape shows a sharp decrease in resolution with the increase of the subtraction factors. Figures. 2.2.1.3(n) and (o) give the FWHM resolution evolutions with three sample shapes versus the sample sizes with $\gamma = 0$ and 0.4. They indicate that the FWHM resolution change with the subtraction factor is more sensitive to the sample sizes than to the sample shapes. For the sample sizes of $r < 0.2\lambda$ in particular, the influence of the sample shapes on the resolution and subtraction factor can be ignored.

3.3 Subtraction Factor

The principal criterion for evaluating the appropriate subtraction factor is that the obtained FWHM resolution in the subtracted images can be matched with the sample's real size. Figures. 2.2.1.4(a)–(c) give the 2D FWHM resolutions versus different sample sizes and subtraction factors with three sample shapes. In terms of the Gaussian-type samples, the obtained FWHM should be equal to the preset sample FWHM according to the subtraction criterion, which is displayed as the dotted black line in Fig. 2.2.1.4(a). The subtraction factors on this line are critical values for obtaining the right subtracted resolutions. The separate left and right regions are the below- and over-subtraction regions, respectively. This line can be picked out and plotted in Fig. 2.2.1.4(d) with the solid blue curve. We call it the subtraction threshold line or γ_0–r_0 line. With a specific subtraction factor γ_0, the sample structures with sizes below the corresponding value r_0 can be imaged properly, and the sizes of $r > r_0$ are over-subtracted. The dotted blue threshold line in Fig. 2.2.1.4(d) is estimated simply by enhancing the FWHM with a factor of 1.4 for the confocal microscope case. However, using the unknown sample size to define the subtraction factor suitable for resolving the sample is not rational. Therefore, additional parameters combined with the threshold line are needed to evaluate the proper subtraction.

On the one hand, when Fig. 2.2.1.4(a) is coordinate transformed to Fig. 2.2.1.4(e), which is similar to Fig. 2.2.1.2(i), the FWHM slopes for the different sample sizes are clearly different and could be useful for identifying the real sample sizes. The absolute values of the slopes, defined as $\Delta FWHM/\Delta\gamma$, are plotted with the threshold line (dotted black curve) in Fig. 2.2.1.4(f) with sample sizes of r = 0.1λ, 0.2λ, 0.3λ, 0.4λ, and 0.5λ. Although the threshold line is not systematic in the $\Delta FWHM/\Delta\gamma$ figure, relative sample sizes can still be compared among them. The larger the absolute value of $\Delta FWHM/\Delta\gamma$, the larger the sample size is. On the other hand, the peak intensities of the sample in subtracted images decrease as the subtraction factor increases. Fig. 2.2.1.4(g) shows the evolution map of the peak intensities, which are normalized to the initial peak intensities. The shape of the sample is also assumed to be Gaussian. The γ_0–r_0 threshold line is plotted in Fig. 2.2.1.4(g). It is interesting to see that the γ_0–r_0 line can horizontally and almost smoothly separate the below- and over-subtraction regions. The critical value associated with the peak intensity status for appropriate subtraction is around 0.6 even for different object sizes. This means that if the fluorescence peak intensity of the sample after subtraction drops

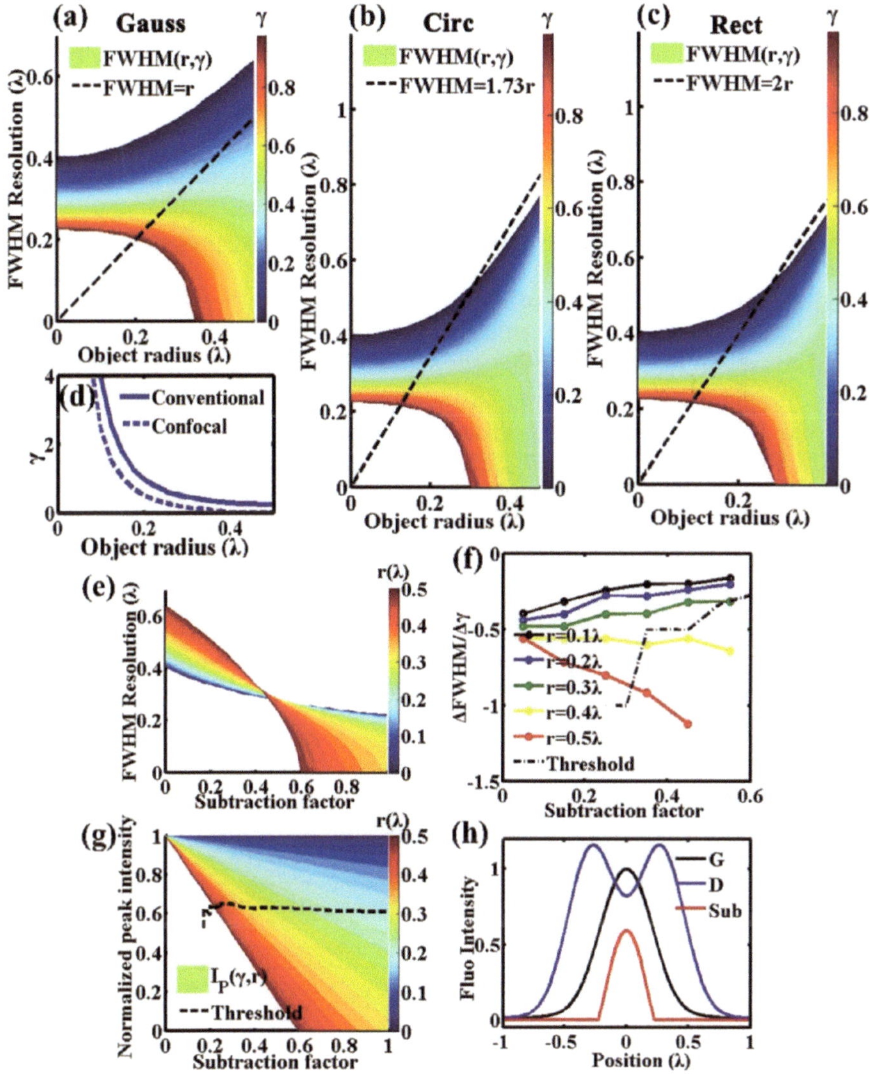

FIGURE 2.2.1.4 (a)–(c) FWHM (r, γ) with sample shape of "Gauss", "Circ" and "Rect". (d) Value of subtraction factors on the line of FWHM = r in (a). (e) FWHM (r, γ) with shape of "Gauss". (f) Derivative of FWHM resolution to the subtraction factor. (g) Peak intensity after subtraction versus subtraction factor and object radius. (h) PSF intensity when r = 0.3λ and γ = 0.5. The figure was reported in the journal *Optica* and copyright is shared between the representative author (Takayoshi Kobayashi, TK) and Optica.

below 60% of its initial value, the possibility that it will be over-subtracted increases. It can be simply called the I_P factor. Although the best subtraction factor for the specific sample size varies, the decrease in peak intensity can be used to determine whether the factor is proper. Fig. 2.2.1.4(h) gives an example to show that, when γ= 0.5 is used for r = 0.3λ, the fluorescence peak intensity slightly drops below 0.6

and a preliminary over-subtraction appears. The corresponding fluorescence peak intensity and FWHM resolution after subtraction is 0.59 and 0.28λ, respectively. Similarly, the FWHM values after subtraction with semicircular and rectangular sample shapes should approach their preset FWHM sizes of (SQRT3)*r* and 2*r* [illustrated in Fig. 2.2.1.3(c)]. Their threshold lines are shown in Figs. 2.2.1.4(b) and 4(c) as the dotted black lines. Because these two types are rarely used in the imaging analysis, their results are not going to be shown in detail. Their resolution slopes are also identical for different sample sizes while their I_P factors are found around 0.75 and 0.83, respectively.

To inspect the effectiveness of the I_P factor and the $\Delta\ FWHM/\Delta\ \gamma$ factor, further simulations are carried out with a laser wavelength of 488 nm, NA = 1.4, and bead FWHMs of 200, 150, 100, and 50 nm with the Gaussian shape, respectively, shown in Figure 2.2.1.5. After the 2D convolution with an intensity-normalized PSF, their FWHMs in the fluorescence image become 278, 244, 219, and 204 nm as shown in Fig. 2.2.1.5(a); the FI_D and the sample is shown in Figs. 2.2.1.5(b) and 5(c), respectively. With five different subtraction factors of γ=0.2, 0.4, 0.6, 0.8, and

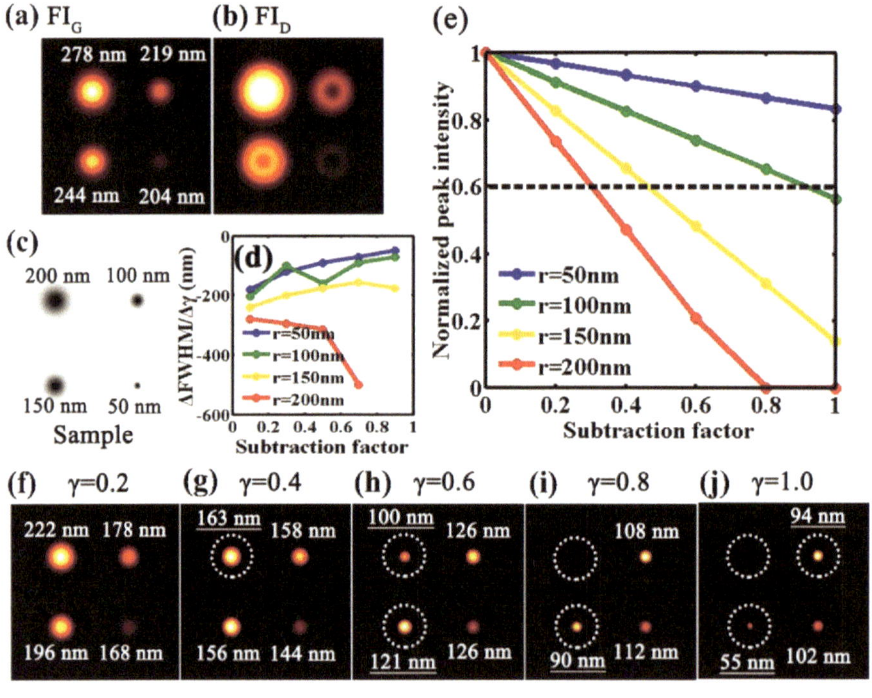

FIGURE 2.2.1.5 (a) Gaussian and (b) donut spot convoluted images with (c) sample. (Laser wavelength: 488 nm; NA=1.4). (d) $\Delta\ FWHM/\Delta\ \gamma$. (e) Normalized peak intensity of four beads versus the subtraction factor. (f)–(j) Subtracted images with γ = 0.2, 0.4, 0.6, 0.8, and 1.0. The figure was reported in the journal *Optica* and copyright is shared between the representative author (Takayoshi Kobayashi, TK) and Optica.

1.0, the FWHM values of the four beads in the subtracted images decrease nonlinearly. Figure 2.2.1.5(d) displays the evolutions of their Δ *FWHM/*Δ γ factors. That the sample with a larger absolute Δ *FWHM/*Δ γfactor gets a larger size is in agreement with the prediction. Fig. 2.2.1.5(e) shows the evolutions of their peak intensities after subtraction with the variation of the subtraction factor. The imaging with a sample width of 200 nm can be taken as an example. Its I_P after subtraction drops below 0.6 when the subtraction factor goes further than 0.4. The FWHM resolution in the image reaches 163 nm as shown in Fig. 2.2.1.5(g) and this value decreases even further in Figs. 2.2.1.5(h)– (j) where they are all below the real width of this model sample. The resolutions of the other three samples are over-subtracted when their peak intensities drop below 60%. Therefore, the over-subtraction of the sample is predicted and demonstrated based on the peak intensity after subtraction. The over-subtracted areas are marked with dotted white circles in Figs. 2.2.1.5(f)–(j).

3.4 Numerical Aperture

As shown in Figure 2.2.1.6, the simulations with an objective lens of NA = 0.9 are also carried out using the same procedure. Because of the coarse resolution, the maps of the FWHM(r, γ) are upward shifted, and the threshold lines are also right shifted compared with Fig. 2.2.1.4(d). The value of the threshold line is also around 0.6. Therefore, the value of 0.6 could also be used as a criterion to inspect the over-subtraction. Figure 2.2.1.6(h) describes a critical subtraction with a Gaussian shape, sample radius of 0.3λ, and subtraction factor of 0.5. After subtraction, the peak intensity drops to 60% of the initial intensity, and the achieved resolution is around 0.3λ.

3.5 Polarization

The combination of radial and azimuthal polarizations is demonstrated to have fewer negative sidebands for subtraction microscopes [12]. Simulations with this polarization combination and NA = 1.4 are also realized here, as shown in Figure 2.2.1.7. The FWHM resolution maps and ΔFWHM/Δ γ factor are similar to those in Fig. 2.2.1.4 and are not discussed further. The IP factor is a little smaller than 0.6 in Fig. 2.2.1.7(h) but can still be considered flat. Figure 2.2.1.7(i) shows the right-subtraction status and implements the subtraction factor of 0.5 and sample FWHMs of 0.3λ. Although the polarization combination has changed, the criterion of the IP factor and ΔFWHM/Δ γ still works.

4 DISCUSSION

The simulations in this paper are carried out under several assumptions. First, the assumption of non-saturation allows the fluorescence obtained to be linearly proportional to the intensity of the excitation laser and the density of the fluorophores in the sample. In this case, the direct convolution of laser PSFs and the sample can be used to describe the conventional laser scanning fluorescence imaging. When a high excitation power is used, the fluorescence signals are intensity saturated, at which

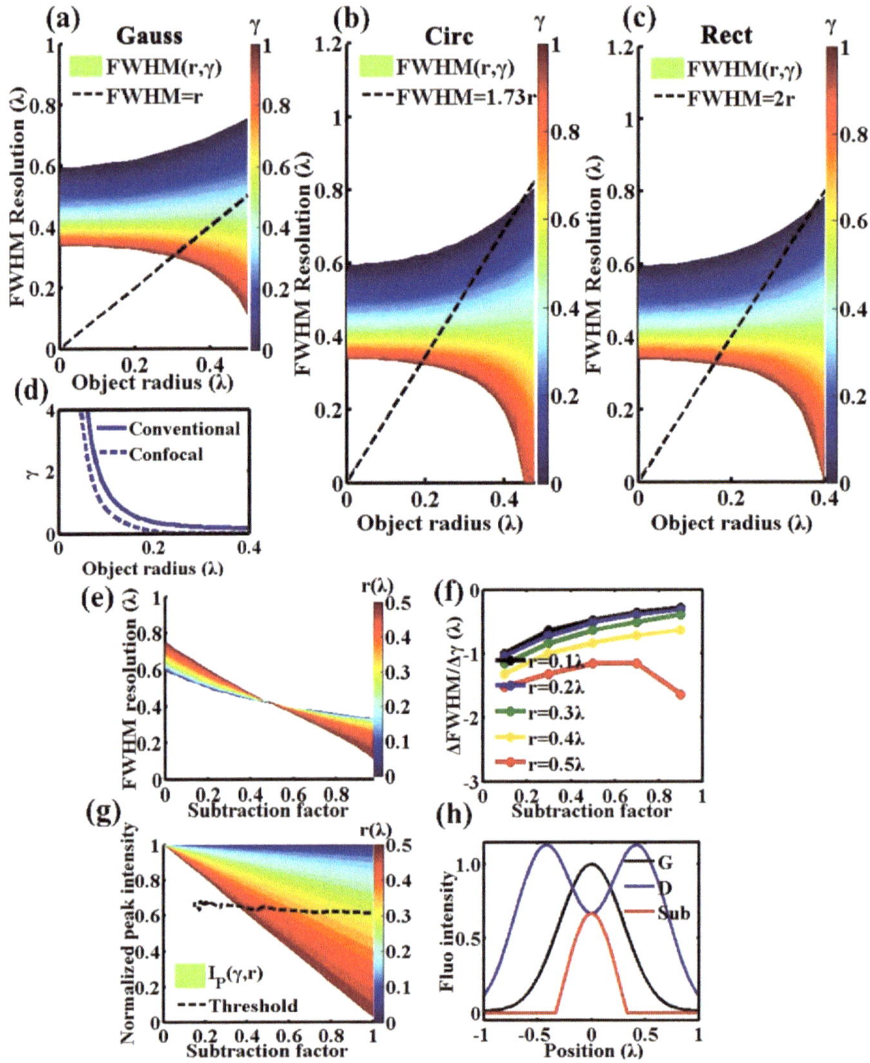

FIGURE 2.2.1.6 Simulated results with NA = 0.9. (a)–(c) FWHM(r,γ) with sample shape of "Gauss", "Circ" and "Rect". (d) Value of subtraction factors on the line of FWHM = r in (a). (e) FWHM(γ,r) with shape of "Gauss". (f) Derivative of FWHM resolution to the subtraction factor. (g) Peak intensity after subtraction versus subtraction factor and object radius. (h) Fluorescence intensity when $r = 0.3\lambda$, $\gamma = 0.5$. The figure was reported in the journal *Optica* and copyright is shared between the representative author (Takayoshi Kobayashi, TK) and Optica.

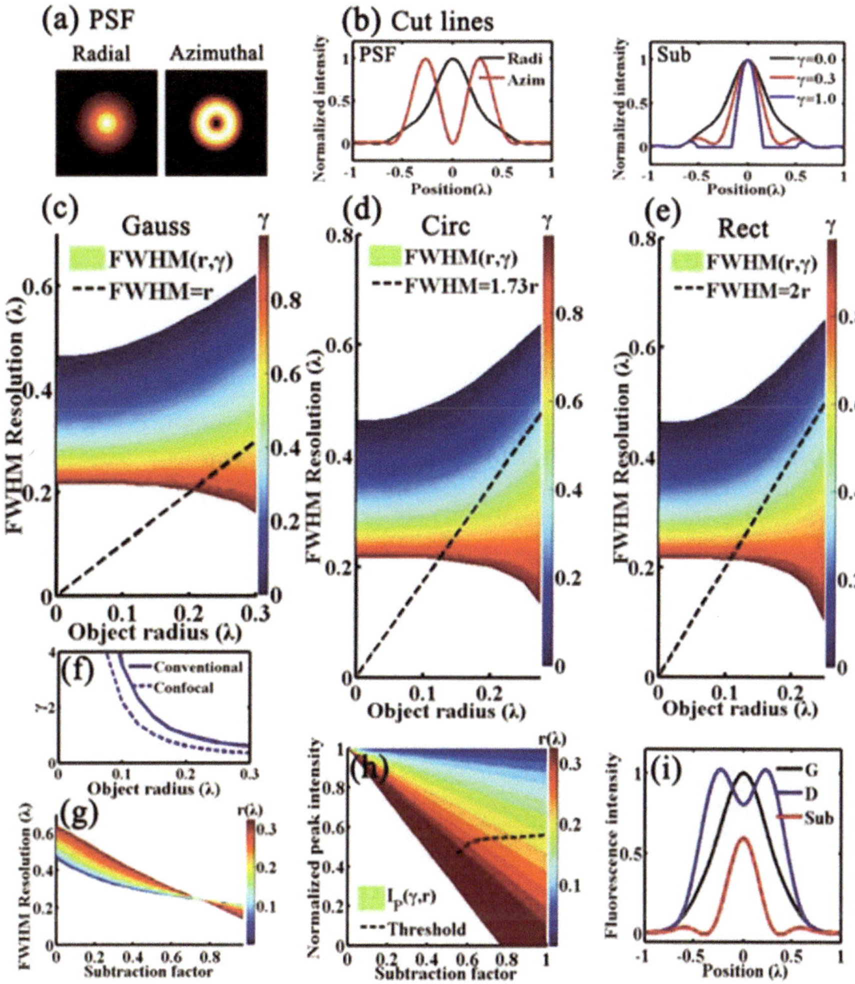

FIGURE 2.2.1.7 Simulated results with radial and azimuthal polarizations for excitation. (a) focal PSFs and (b) cut lines of PSFs and subtractions with factor γ = 0, 0.3, 1.0. (c) – (e) FWHM (r, γ) with sample shape of "Gauss", "Circ" and "Rect". (f) Value of subtraction factors on the line of FWHM = r in (c). (g) FWHM (r, γ) with shape of "Gauss". (h) Peak intensity after subtraction versus subtraction factor and object radius. (i) PSF intensity when r = 0.3λ, γ=0.5. The figure was reported in the journal *Optica* and copyright is shared between the representative author (Takayoshi Kobayashi, TK) and Optica.

point the additional saturation factor should be included. Second, the detection PSF is assumed to be constant having the value of unity, which means that all of the fluorescence in one scanning step is collected. For confocal microscopy with different detection focal Airy disks and different fluorophores with different emission Stokes shifts, additional detection PSFs could be included to obtain more specific and accurate subtraction factors. Third, the samples are simulated with simple structures. For the biological samples with complex structures, the subtraction factors for the overlap regions still need to be addressed. In addition, dynamic subtraction factors could be a good measure for obtaining an ideal resolving power, but obtaining an accurate peak positioning requires the development of a complicated algorithm. However, the numerical simulation in this paper can provide information of the FWHM resolutions as a function of the subtraction factor relevant to the sample size, shapes, NA, and polarizations in the subtraction microscope and to derive the appropriate subtraction threshold line. Meanwhile, this chapter demonstrates that combined with the threshold line, the peak intensity after subtraction and the derivative of FWHM resolution are two useful parameters for evaluating the subtraction factor and estimating the real size of a sample. The samples with larger sizes are found to be sensitive to the subtraction factor and easily over-subtracted while the ones with smaller sizes can endure a larger subtraction factor without over-processing. For the sample sizes of $r < 0.2\lambda$, the sample shapes do not play an important role in affecting the resolutions and determining the subtraction factors. When the fluorescence peak intensity in the subtracted image drops to 60% of the initial value, the corresponding subtraction factor is found to be properly chosen for the imaging even with different sample sizes. This criterion is also valid when a smaller NA is used, and can also roughly work when the radial and azimuthally polarized lights are used. The sample sizes can also be relatively compared according to the resolution derivative versus the subtraction factor. In summary, based on vector beam diffraction theory and two-dimensional convolution, we have numerically investigated the parameters that may affect the selection of subtraction factors in the subtraction microscopes and proposed that the fluorescence peak intensity after subtraction and resolution derivative can possibly be used as the key parameters for determining the appropriate subtraction factors and obtaining the sample sizes in the subtraction microscopes.

5 CONCLUSION

Subtraction microscopy has recently been promoted for its compactness and simplicity of enhancing spatial resolution. However, until now, the subtraction factors used in such microscopes to process raw images have been chosen experientially, and it has been impossible to determine whether the resolved structures after subtraction are appropriate or over-processed. Based on vector diffraction theory and two-dimensional convolution, this subsection presents numerical investigations of the parameters that may offer the selection criterion of subtraction factors used in subtraction microscopy. We proposed in the present subsection two essential parameters for appropriately evaluating the subtraction factor: the fluorescence peak intensity after subtraction and resolution derivative.

The research discussed in this subsection was performed collaboratively of N. Wang and T. Kobayashi. The above study was reported in *Biomedical Optical Express* and copyright is partially held between the corresponding author (Takayoshi Kobayashi, TK) and *Optica* (formally Optical Society of America). TK would like to express thanks to Optica for permit us to publish the revised paper as one section in the book *Photothermal and Raman Microscopy* (Taylor & Francis).

REFERENCES

1. E. Betzig, G. H. Patterson, R. Sougrat, O. W. Lindwasser, S. Olenych, J. S. Bonifacino, M. W. Davidson, J.L. Schwartz, and H. F. Hess, "Imaging intracellular fluorescent proteins at nanometer resolution," *Science* 313, 1642–1645 (2006).
2. M. J. Rust, M. Bates, and X. W. Zhuang, "Sub-diffraction-limit imaging by stochastic optical reconstruction microscopy (STORM)," *Nat. Meth.* 3, 793–796 (2006).
3. M. G. L. Gustafsson, "Surpassing the lateral resolution limit by a factor of two using structured illumination microscopy," *J. Microscopy* 198, 82–87 (2000).
4. S. W. Hell and J. Wichmann, "Breaking the diffraction resolution limit by stimulated emission: stimulated emission-depletion fluorescence microscopy," *Opt. Lett.* 19(11), 780–782 (1994).
5. M. P. Gordon, T. Ha, and P. R. Selvin, "Single-molecule high-resolution imaging with photobleaching," *P. Natl. Acad. Sci. USA* 101(17), 6462–6465 (2004).
6. A. Sharonov and R. M. Hochstrasser, "Wide-field sub-diffraction imaging by accumulated binding of diffusing probes," *Proc. Natl. Acad. Sci. USA* 103(50), 18911–18916 (2006).
7. T. Dertinger, R. Colyer, G. Iyer, S. Weiss, and J. Enderlein, "Fast, background-free, 3D super-resolution optical fluctuation imaging (SOFI)," *P. Natl. Acad. Sci. USA* 106(52), 22287–22292 (2009).
8. J. Miyazaki, H. Tsurui, A. Hayashi-Takagi, H. Kasai, and T. Kobayashi, "Sub-diffraction resolution pump-probe microscopy with shot-noise limited sensitivity using laser diodes," *Opt. Express* 22(8), 9024–9032 (2014).
9. J. Miyazaki, K. Kawasumi, and T. Kobayashi, "Resolution improvement in laser diode-based pump-probe microscopy with an annular pupil filter," *Opt. Lett.* 39(14), 4219–4222 (2014).
10. C. F. Kuang, S. Li, W. Liu, X. Hao, Z. T. Gu, Y. F. Wang, J. H. Ge, H. F. Li, and X. Liu, "Breaking the diffraction barrier using fluorescence emission difference microscopy," *Sci. Rep.* 3(1441) (2013).
11. H. Dehez, M. Piche, and Y. D. Koninck, "Resolution and contrast enhancement in laser scanning microscopy using dark beam imaging," *Opt. Express* 21(13), 15912–15925 (2013).
12. S. Segawa, Y. Kozawa, and S. Sato, "Resolution enhancement of confocal microscopy by subtraction method with vector beams," *Opt. Lett.* 39(11), 3118–3121 (2014).
13. B. Richards and E. Wolf, "Electromagnetic diffraction in optical systems. II. structure of the image field in an aplanatic system," *Proc. R. Soc. Lond. A* 253, 358–379 (1959).
14. D. Ganic, X. Gan, and M. Gu, "Focusing of doughnut laser beams by a high numerical-aperture objective in free space," *Opt. Express* 11(21), 2747–2752 (2003).
15. R. Dorn, S. Quabis, and G. Leuchs, "Sharper focus for a radially polarized light beam," *Phys. Rev. Lett.* 91, 233901 (2003).
16. G. M. Lerman and U. Levy, "Effect of radial polarization and apodization on spot size under tight focusing conditions," *Opt. Express* 16(7), 4567–4581 (2008).
17. S. Deng, L. Liu, Y. Cheng, R. Li, and Z. Xu, "Investigation of the influence of the aberration induced by a plane interface on STED microscopy," *Opt. Express* 17(3), 1714–1725 (2009).
18. X. Hao, C. Kuang, T. Wang, and X. Liu, "Effects of polarization on the de-excitation dark focal spot in STED microscopy," *J. Opt.* 12(115707) (2010).

19. Y. Xue, C. Kuang, S. Li, Z. Gu, and X. Liu, "Sharper fluorescent super-resolution spot generated by azimuthally polarized beam in STED microscopy," *Opt. Express* 20(16), 17653–17666 (2012). http://en.wikipedia.org/wiki/Circular polarization.
20. N. Wang and T. Kobayashi, "Numerical study of the subtraction threshold for fluorescence difference microscopy," *Opt. Express* 22(23), 28819–28830 (2014).

2.2.2 Sub-Diffraction-Limit Imaging Using Mode Multiplexing

1 INTRODUCTION

Over the last 20 years, there has been increasing interest in improving the resolving power of far-field optical microscopy [1–4]. Based on various innovative concepts, several types of superresolution microscopic methods have been proposed and realized, such as stimulated emission depletion (STED) microscopy [5], structured illumination microscopy (SIM) [6], photoactive localization microscopy (PALM) [7], stochastic optical reconstruction microscopy (STORM) [8], and others [9–14]. However, due to the complexity of these methods, researchers continue to try to break the diffraction barrier by mathematically improving the optical point spread function (PSF) [15–20]. One of the most convenient and effective solutions is to subtract two confocal fluorescence images that are captured with the same excitation wavelength but with different spots of solid and donut intensity distributions. Kuang et al. realized frame-by-frame fluorescence emission difference (FED) microscopy [18] by alternately acquiring then subtracting one set of images. Dehez et al. demonstrated line-by-line switching laser mode (SLAM) microscopy [19] by mechanically switching the beam paths when the sample was scanned line-by-line. Nevertheless, a repeating scan, whether by frame or by line, creates uncertainties due to mechanical drift or changes in the properties of the fluorescent probes, even fast scanning by modern mechanism is utilized.

Especially for moving targets or live cell imaging, scans by frame or by line may introduce errors because of the motion of the sample or the diffusion of the components in biological organisms. Therefore, the ability to simultaneously acquire and process the images excited by the two beams is essential for subtraction microscopes. Although the dual path programmable array microscope (PAM) [15] and multiplexed two-color imaging [21] have been demonstrated, imaging with resolution enhancement using multiplexed solid and donut spots for excitation has not been reported. In this study, pixel-by-pixel processed fluorescence difference microscopy is realized by multiplexing the excitation laser spots with Gaussian and donut shapes and then demultiplexing the fluorescent signals using lock-in amplifiers (LIAs). Using a homemade microscope, fluorescent polystyrene beads and a slice of brain tissue of a mouse are imaged with sub-diffraction-limit resolution. A "moving" sample is also imaged to compare the present method with the frame-by-frame and line-by-line scanning methods. In the experiment, the subtraction weight

DOI: 10.1201/9781003476030-6

of the two images, which is not a numerical factor but the ratio of excitation laser powers, simplifies the subtraction process compared with previous FED and SLAM methods. With the same excitation power of two beams, the peak intensity ratio of the PSFs of donut beam path (DBP) and Gaussian beam path (GBP) is investigated to be around 0.3. The corresponding spatial resolution improvement is about 26%. The time resolution of the scan can also be improved by a factor of 2 resulting from the simultaneous measurement. Furthermore, by using a commercially available off-the-shelf laser diode and lock-in detections, the cost of this imaging system is further reduced compared to the previous FED and SLAM setups, while a high signal–noise ratio can be maintained or improved.

2 EXPERIMENTAL SETUP

The experimental setup is shown in Figure 2.2.2.1. The excitation laser source is a continuous-wave laser diode (LD) emitting at 445 nm (NDB4116, NICHIA). After passing through a beam splitter, each of the beam paths is spatially filtered by a polarization-maintained single mode fiber (PM-SMF) (P1-488PM-FC-2, Thorlabs). By mechanically rotating the end adapter of the PM-SMF, the polarizations of the two beams from the two fibers are set to be perpendicular to each other. Then one of the excitation spots is kept constant, while the other passes through a vortex phase plate (VPP-1a, RPC Photonics) to generate a donut-shaped intensity distribution. After the beams are combined using a polarization beam splitter to avoid interference, the polarizations of the beams are converted to a circular form by a quarter-wave plate

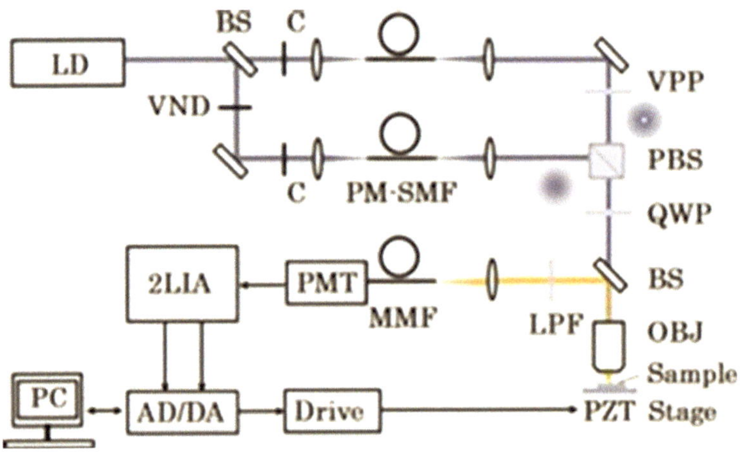

FIGURE 2.2.2.1 Experimental setup of the microscope. LD, laser diode; BS, beam splitter; VND, variable neutral density filter; PM-SMF, polarization maintained single mode fiber; C, Chopper; VPP, vortex phase plate; PBS, polarization beam splitter; QWP, quarter wave plate; OBJ, objective lens; MMF, multi-mode fiber; LPF, long pass filter; PMT, photomultiplier; LIA, lock-in amplifier; PZT, piezoelectric ceramic transducer; AD/DA, analog digital signal converter; PC, personal computer. The figure was reported in the *Journal of Optics* and copyright is partially held between the representative author (Takayoshi Kobayashi, TK) and Elsevier.

(AQWP05M-600, Thorlabs) to improve the focal performance of the donut spots [22, 23]. An objective lens of the oil-immersion type (UPlanFLN100X/1.30, Olympus) is used to focus the excitation beams and simultaneously collect the fluorescence. The sample is fixed on a three-axis piezoelectric controlled stage (MAX311D/M, Thorlabs) with a minimum scanning step of 5 nm using a closed-loop feedback drive (BPC303, Thorlabs). The collimated fluorescence from the objective is partly reflected by a 50% beam splitter (EBS1, Thorlabs), filtered after passing through a long-pass filter (LP03-458RU-25, Semrock), and focused into a multi-mode fiber (M68L, Thorlabs) with the focal length of 100 mm and a fiber core diameter of 25 μm—corresponding to ~0.8 Airy units. Then the signal is detected by the combined system of a current-output type photomultiplier (H11460-1, Hamamatsu Photonics), an electrical converter (C9999), and two LIAs (M7265, Signal Recovery). Two chopper blades are installed to modulate the excitation beam paths; they encode the signals and then the trigger outputs of the chopper controllers (MC2000, Thorlabs), which are sent to the reference inputs of the two LIAs. The modulation frequency is 1.2 kHz for the GBP and 1.8 kHz for the DBP. The time constant of the LIAs is set as 5 ms, and the dwell time is 10 ms per pixel for imaging the fluorescent beads; when imaging biological samples, they are set as 10 ms and 20 ms, respectively. Both of the LIAs operate at the same voltage sensitivity scale of 5 mV.

3 DATA PROCESSING

The images can be processed during the stage scanning or after they are acquired. The data processing is carried out in the way as $I_{Sub} = I_G(P, \lambda, h_G, \eta) - I_D(P, \lambda, h_D, \eta)$, where I_{Sub} is the target image I_G and I_D are the confocal images excited by GBP and DBP, respectively; P is the excitation power; λ is the laser wavelength; h is the total PSF, and η is the electronic detection efficiency. The same parameters P, λ, and η in the I_G and I_D are implemented in such a way that the two excitation laser beams have the same power, wavelength, and detection efficiency. The optical PSF, polarization response efficiency, and optical collection efficiency are all combined together in the total PSF h_G and h_D for the two beams. However, by studying the images with conventional confocal configuration without the phase plate, the resolutions and peak intensities of the PSFs of the two beams are quite similar (not shown here). This means that the differences between the polarization responses and fluorescence collection efficiencies of the two beams play only minor roles in the peak-to-peak intensity ratio of two PSFs. The images I_G and I_D are directly subtracted without a subtraction factor and generated negative values are removed as shown in [18–20].

4 EXPERIMENTAL RESULTS

To check the spatial resolution and to investigate the peak-to-peak intensity ratio of the PSFs of the two beams with the same excitation power, fluorescent beads with radius of 100 nm (T14792, Invitrogen) are imaged. One group of images scanned with a step size of 20 nm and a total scale of 3 μm is shown in Figure 2.2.2.2(a)–(c). The measured laser powers of the GBP and DBP for excitation before an objective lens are 0.8 μW and 0.7 μW, respectively. By making the cut-through lines on the

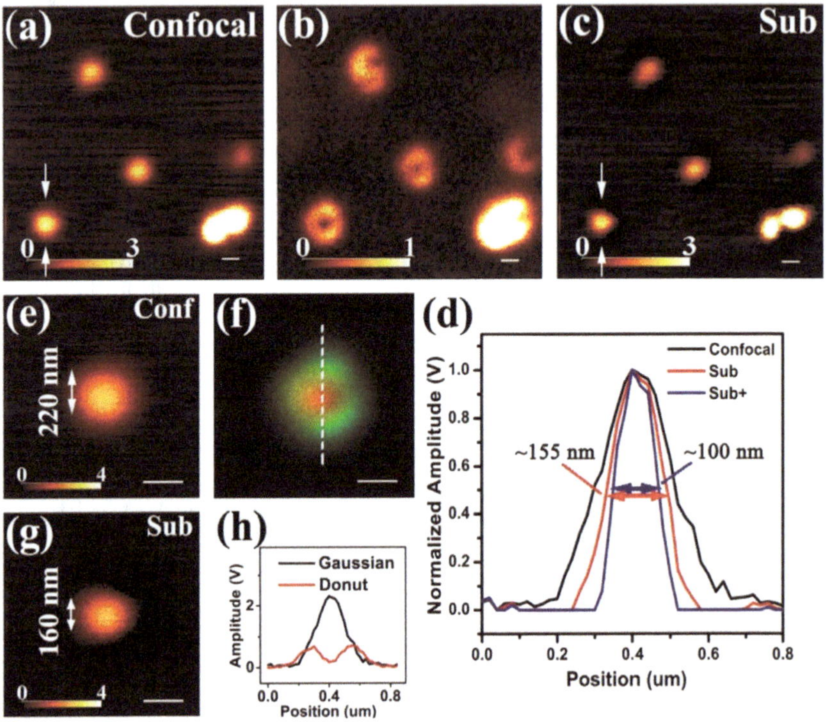

FIGURE 2.2.2.2 (a)-(c) The confocal images of fluorescent beads excited by GBP, DBP and their subtraction. (d) The cut-through lines of a bead in (a) and (c). See text for blue curve in (d). (e)-(g) The confocal image of a single fluorescent bead, overlap of images excited by GBP (red) and DBP(green),and the image obtained after the subtraction. (h)The intensity profiles of the dotted line in (f). White scale bar is 200nm. Images (a)-(c), and (e)-(g) have scales of 3 μm × 3 μm and 1 μm × 1 μm, respectively. (For interpretation of the references to color in this figure caption, the reader is referred to the web version of this paper.)

fluorescent bead in Fig. 2.2.2.2(a) and (b), the peak-to-peak intensity ratio of the PSFs of the DBP and GBP is found to be 0.31. The image after their direct subtraction is shown in Fig. 2.2.2.2(c); the corresponding cut-through line is shown in Fig. 2.2.2.2(d) as a red curve and has a full width at half maximum (FWHM) value of 155 nm. A spatial resolution enhancement of 26%, over the confocal resolution of 210 nm (see the black curve in Fig. 2.2.2.2(d)) is achieved. By further multiplying the results shown in Fig. 2.2.2.2(b) by a factor of 2.2, the peak-to-peak intensity ratio of DBP and GBP becomes 0.68. The FWHM value of the cut-through line in the subtracted image reaches 100 nm, as shown by the blue curve in Fig. 2.2.2.2(d), which illustrates a spatial resolution improvement of over 52%. Another group of measurements of a single bead with a stage-scanning step of 10 nm is shown in Fig. 2.2.2.2(e)–(h). The measured excitation powers of the GBP and DBP are 3.38 μW and 3.33 μW, respectively. The black and red curves in Fig. 2.2.2.2(h) corresponding to the PSFs of the GBP and DBP, respectively, depict the intensities of two spots along the white dotted line in Fig. 2.2.2.2(f). The peak-to-peak intensity ratio of the PSFs of DBP and

GBP is 0.25. The image of this single bead after direct subtraction is shown in Fig. 2.2.2.2(g), which illustrates enhanced resolution compared to the confocal image. The FWHM values of the image widths shown in Fig. 2.2.2.2(e) and (g) are 220 nm and 160 nm, respectively. Thus the enhancement of spatial resolution is about 27% when the direct subtraction method is implemented under the same excitation power.

Next, a biological sample of mouse brain tissue labeled with yellow fluorescent protein (YFP) is imaged. The excitation laser wavelength with peak efficiency for YFP is around 515 nm, whereas the excitation efficiency drops to 5% of the peak at 445 nm [24]. To prevent the sample from damage and ensure sufficient brightness, the GBP and DBP laser powers used are measured to be 0.3 μW and 0.27 μW, respectively. Several groups of measurements are carried out and no sample damage is observed in the experiment. Two groups of images of mouse brain tissue labeled with YFP are shown in Figure 2.2.2.3. Fig. 2.2.2.3(a) displays the confocal image of the tissue at the scale of 20 μm with a pixel size of 50 nm. The two rectangular box areas marked by white lines in Fig. 2.2.2.3(a) are imaged separately and shown in Fig. 2.2.2.3(b)–(e) and (f)–(h). In the first group, the confocal image of the GBP shown in Fig. 2.2.2.3(b) is overlaid with red color in Fig. 2.2.2.3(c), whereas the confocal

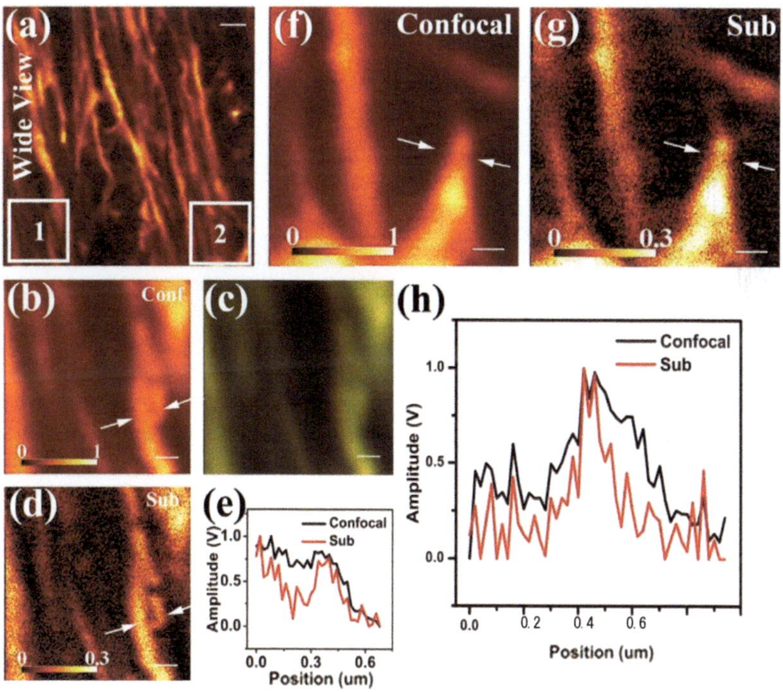

FIGURE 2.2.2.3 Imaging of a slice of mouse brain tissue. (a) Wide view (20 μm) confocal image. (b)–(d) The confocal image, overlapped image of GBP and DBP, and the image obtained after the subtraction of areal in (a). (e) The cut-through lines in between the arrows in (b) and (d). (f)–(h) The confocal image, image obtained after subtraction, and corresponding cut through lines of area 2 in (a). The white scale bar in (a) is2 μm; white scale bars in (b)–(d), (f) and (g) are 500nm.

image of the DBP is set as green. The direct subtraction image is shown in Fig. 2.2.2.3(d). By comparing the confocal and subtraction images in Fig. 2.2.2.3(b) and (d), we can see visually that both the resolution and contrast are improved, as clearly seen graphically in Fig. 2.2.2.3(e); the FWHM value of the curve after subtraction is 150 nm. In the second group, the confocal image and the image after subtraction are shown in Fig. 2.2.2.3(f) and (g), respectively. The black and red curves in Fig. 2.2.2.3(h) display the cut-through line profiles of Fig. 2.2.2.3(f) and (g) of which the FWHM values are 320 nm and 140 nm, respectively. The results demonstrate that sub-diffraction limited imaging of both the fluorescent beads and biological samples can be obtained with this mode multiplexed microscope. When similar excitation powers of the two beams are used, the spatial resolution improvement is better than 26% compared with confocal microscopes.

The figure was reported in the journal *Optica* and copyright is partially held between the representative author (Takayoshi Kobayashi, TK) and Elsevier. To inspect the robustness of the present imaging method for moving samples, fluorescent beads are imaged with specially programmed scanning parameters to mimic "live" cells or microorganisms. The images acquired by the frame-by-frame scan, line-by-line scan and pixel-by-pixel scan methods are then compared. The scanning algorithms and one group of acquired images are shown in Figure 2.2.2.4. (See also Media 1 for frame-by-frame scan, Media 2 for line-by-line scan and Media 3 for pixel-by-pixel scan.) A frame-by-frame scan is implemented by alternating the data acquisition channels after the scan of each frame is finished. After one set of image frames excited by the GBP and DBP is obtained, the scanning goes to the second group to take the image of the moving beads.

Similarly, a line-by-line scan is carried out. In each scan group with the line scan, the images of the GBP and DBP are acquired by alternately changing the channel after scanning one line. Then, scanning the sample pixel-by-pixel is performed with the present mode-multiplexed measurement. Each frame is in the range of 2 μm × 2 μm, and the beads are set to move downward by 1 μm within six frames. As shown in the videos and Fig. 2.2.2.4, the confocal and donut images scanned by the three methods can trace the moving beads. However, due to the asynchronous data acquisition of the frame-by-frame and line-by-line scans, the target images with subtraction are substantially distorted. Furthermore, the resolution cannot be improved with these two methods if the shifts of the beads' absolute position in the GBP- and DBP-excited images are too large during the scanning. By using the multiplexing of excitation spots method, resolution-enhanced imaging of moving objects can be obtained, as shown in Fig. 2.2.2.4(i). The time resolution can also be improved by a factor of 2 compared to the alternative frame or line scans Video S1, Video S2, Video S3.

Video S1. Confocal, overlapped, and subtracted images acquired using frame-by-frame scan method (version of this paper). The figure was reported in the *Journal of Optics* and copyright is partially held between the representative author (Takayoshi Kobayashi, TK) and Elsevier. Supplementary data associated with this paper can be found in the online version at http://dx.doi.org/10.1016/j.optcom.2014.12.072

Video S2. Confocal, overlapped and subtracted images acquired using the line-by-line scan method. The figure was reported in the *Journal of Optics* and copyright

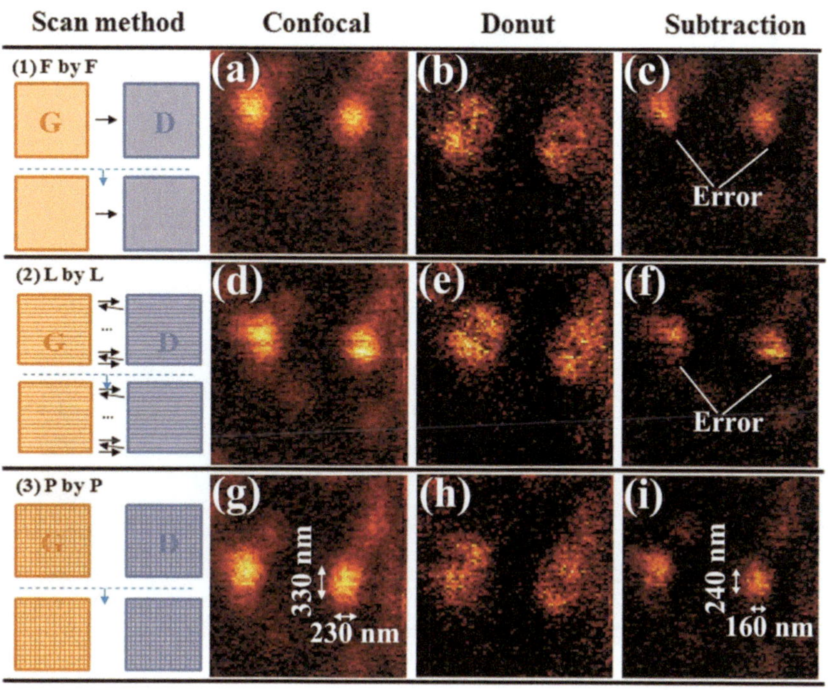

FIGURE 2.2.2.4 Images from sequential frames show the comparison of the subtraction imaging of moving beads with three methods. The frame-by-frame scan (seeMedial) is shown in (1); the line-by-line scan (seeMedia2) is shown in (2); and the pixel-by-pixel scan (seeMedia3) is shown in (3). The four columns, from left to right, show the scanning algorithm, confocal images of GBP, images of DBP, and images after subtraction, respectively.

is partially held between the representative author (Takayoshi Kobayashi, TK) and Elsevier.

Supplementary data associated with this paper can be found in the online version at http://dx.doi.org/10.1016/j.optcom.2014.12.072

Video S3. Confocal, overlapped and subtracted images acquired using the pixel-by-pixel scan method. The figure was reported in the *Journal of Optics* and copyright is partially held between the representative author (Takayoshi Kobayashi, TK) and Elsevier.

Supplementary data associated with this paper can be found in the online version at http://dx.doi.org/10.1016/j.optcom.2014.12.072

5 DISCUSSION

If the excitation power of DBP is improved or the image of DBP is multiplied by a factor with a value larger than 1, corresponding to a high performing subtraction, the resolution can definitely be improved over present results. However, the possibility of oversubtraction would increase. Meanwhile, the removal of negative values in the

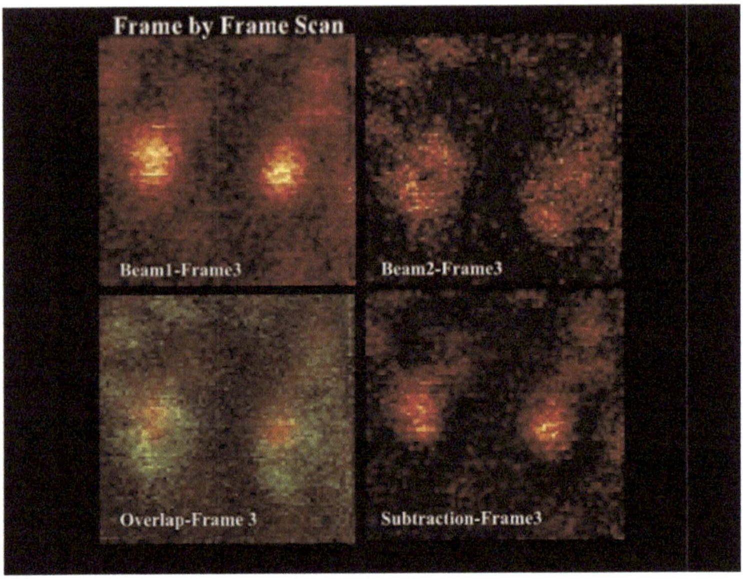

Video S1. Confocal, overlapped and subtracted images acquired using frame-by-frame scan method. Supplementary data associated with this paper can be found in the online version at http://dx.doi.org/10.1016/j.optcom.2014.12.072.

subtracted images may also bring nonlinearities. In this case, a subtraction threshold should be defined for an appropriate subtraction. We have proposed and shown a subtraction threshold numerically elsewhere [25] and will not discuss it in detail here. In this paper, the corresponding subtraction factor of around 0.3, resulting from the similar excitation powers of the GBP and DBP, is much smaller than the values used in previous reported FED and SLAM, leading to a smaller amount of resolution enhancement but a lower possibility for oversubtraction. In addition, this microscope may seem quite similar with FED and SLAM. However, due to the photon counting detection or direct fluorescence detection, it is almost impossible to obtain pixel-by-pixel image processing with previous FED and SLAM setups.

Although the present microscope is able to resolve a moving sample with enhanced resolution, a further upgrade is necessary for the following live cell imaging. The mechanical choppers are planned to be replaced with fast optical modulators, such as an acousto-optic modulator (AOM) or Electro-optic modulator (EOM); a Galvano scanning system is preferable to the present scanning stages to reduce the image acquisition time, and thus make live-cell imaging [26] possible. The image quality can be also improved by removal of the nonlinearity of the cross-sectioning of choppers.

This mode multiplexed subtraction microscopy can be very useful for sub-diffraction-limit multi-color imaging since common wavelength is used for the GBP and DBP in this method and the number of the wavelengths is readily increased compared to STED microscopy. By introducing a third excitation beam installed with a central disk 0–π phase plate, three-dimensional superresolution imaging of this microscopy can also be achieved with resolution enhancement both in lateral and axial directions. Furthermore, by using various laser sources and fluorescent probes,

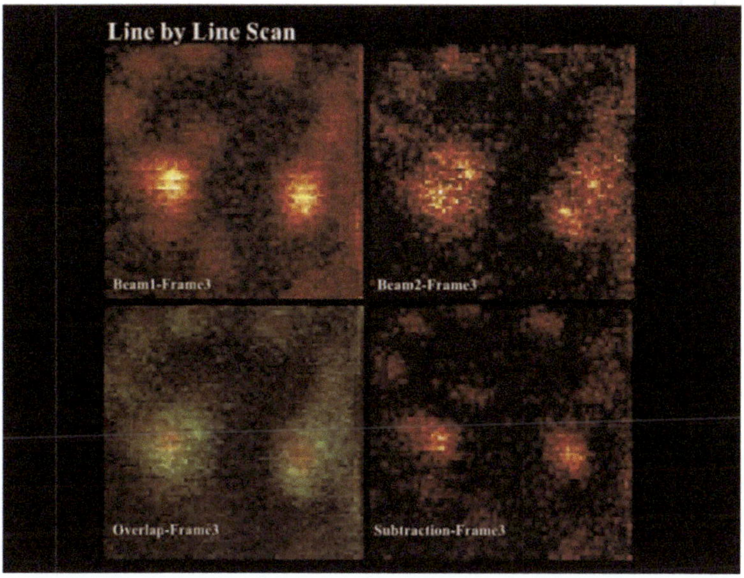

Video S2 Images from sequential frames show the comparison of the subtraction imaging of moving beads with three methods. The frame-by-frame scan (see Media 1) is shown in (1); the line-by-line scan (see Media 2) is shown in (2); and the pixel-by-pixel scan (see Media 3) is shown in (3). The four columns, from left to right, show the scanning algorithm, confocal images of GBP, images of DBP, and images after subtraction, respectively. The figure was reported in the *Journal of Optics* and copyright is partially held between the representative author (Takayoshi Kobayashi, TK) and Elsevier.

both two-photon excitation and harmonic generation could also be carried out by this microscopy, which is useful for deep-locating microorganisms and refractive index imaging with superresolution.

6 CONCLUSION

In conclusion, pixel-by-pixel processed fluorescence difference microscopy is demonstrated experimentally to extend the possible applications of mathematically calibrated microscopes. By simultaneously acquiring the images excited by laser spots with Gaussian and donut intensity distributions, fluorescent spheres and mouse brain tissue are imaged with sub-diffraction-limit resolution. The same excitation power of the two beams is applied for illumination, allowing the experimental investigation of the peak-to-peak intensity ratio of the PSF of the GBP and DBP, which is around 1:0.3. In this case, the achieved resolution enhancement is about 26%. An artificially moving sample is also imaged and three scanning methods are compared; the mode multiplexing is shown to have good features for imaging moving samples with improved spatial and temporal resolution. With further transformations, several different forms of this microscope could be built and used for various applications. The research activities to produce the results described in this subsection are obtained by the cooperation of the following people: N. Wang, J. Miyazaki, J. He, K. Seto, and T. Kobayashi [27].

The above study was reported in *Biomedical Optics Communications* published by Elsevier and copyright is partially held between the corresponding author (Takayoshi Kobayashi, TK) and Elsevier. TK would like to express thanks to Elsevier for permitting us to publish the revised paper as one section in the book *Photothermal and Raman Microscopy* (Taylor & Francis). http://dx.doi.org/10.1016/j.optcom.2014.12.072

REFERENCES

1. S.W. Hell, *Science* 316 (2007) 1153.
2. A.R. Kherlopian, T. Song, Q. Duan, M.A. Neimark, M.J. Po, J.K. Gohagan, A.F. Laine, *BMC Syst. Biol.* 2 (74) (2008).
3. B. Huang, H. Babcock, X. Zhuang, *Cell* 143 (2010) 1047.
4. B.O. Leung, K.C. Chou, *Appl. Spectrosc.* 65 (9) (2011) 967.
5. S.W. Hell, J. Wichmann, *Opt. Lett.* 19 (11) (1994) 780.
6. M.G.L. Gustafsson, *J. Microsc.* 198 (2000) 82.
7. E. Betzig, G.H. Patterson, R. Sougrat, O.W. Lindwasser, S. Olenych, J.S. Bonifacino, M.W. Davidson, J.L. Schwartz, H.F. Hess, *Science* 313 (2006) 1642.
8. M.J. Rust, M. Bates, X.W. Zhuang, *Nat. Methods* 3 (2006) 793.
9. M.P. Gordon, T. Ha, P.R. Selvin, *Proc. Natl. Acad. Sci. USA* 101 (17) (2004) 6462.
10. A. Sharonov, R.M. Hochstrasser, *Proc. Natl. Acad. Sci. USA* 103 (50) (2006) 18911.
11. T. Dertinger, R. Colyer, G. Iyer, S. Weiss, J. Enderlein, *Proc. Natl. Acad. Sci. USA* 106 (52) (2009) 22287.
12. J. Miyazaki, H. Tsurui, A. Hayashi-Takagi, H. Kasai, T. Kobayashi, *Opt. Express* 22 (8) (2014) 9024.
13. K. Fujita, M. Kobayashi, S. Kawano, M. Yamanaka, S. Kawata, *Phys. Rev. Lett.* 99 (2007) 228105.
14. K. Isobe, H. Kawano, T. Takeda, A. Suda, A. Kumagai, H. Mizuno, A. Miyawaki, K. Midorikawa, *Biomed. Opt. Express* 3 (7) (2012) 1594.
15. R. Heintzmann, Q.S. Hanley, D. Arndt-Jovin, T.M. Jovin, J. Microsc. 204 (2001) 119.
16. R. Heintzmann, V. Sarafis, P. Munroe, J. Nailon, Q.S. Hanley, T.M. Jovin, *Micron* 34 (2012) 293.
17. E. Sanchez-Ortiga, C.J.R. Sheppard, G. Saavedra, M. Martnez-Corral, A. Doblas, A. Calatayud, *Opt. Lett.* 37 (7) (2012) 1280.
18. C.F. Kuang, S. Li, W. Liu, X. Hao, Z.T. Gu, Y.F. Wang, J.H. Ge, H.F. Li, X. Liu, *Sci. Rep.* 3 (2013) 1441.
19. H. Dehez, M. Pich, Y.D. Koninck, *Opt. Express* 21 (13) (2013) 15912.
20. S. Segawa, Y. Kozawa, S. Sato, *Opt. Lett.* 39 (11) (2014) 3118.
21. N. Aslund, K. Carlsson, *Micron* 24 (6) (1993) 603.
22. S. Galiani, B. Harke, G. Vicidomini, G. Lignani, F. Benfenati, A. Diaspro, P. Bianchini, *Opt. Express* 20 (7) (2012) 7362.
23. X. Hao, C. Kuang, T. Wang, X. Liu, *J. Opt.* 12 (2010) 115707.
24. http://www.lifetechnologies.com/us/en/home/life-science/cell-analysis/labeling-chemistry/fluorescence-spectraviewer.html.
25. N. Wang, T. Kobayashi, *Opt. Express* 22 (2014) 28819.
26. S.S. Gorthi, D. Schaak, E. Schonbrun, *Opt. Express* 21 (4) (2013) 5164.
27. N. Wang, J. Miyazaki J. He, K. Seto, T. Kobayashi, *Opt. Commun.* 343 (2015) 28.

2.2.3 Polarization Modulation for Fluorescence Emission Difference Microscopy

1 INTRODUCTION

Various types of superresolution techniques have achieved great success in improving the resolving power of optical far-field microscopes during the last 20 years, such as stimulated emission depletion (STED) microscopy [1], structured illumination microscopy (SIM) [2], photoactive localization microscopy (PALM) [3], stochastic optical reconstruction microscopy (STORM) [4], and others [5–10]. Recently, mathematically calibrated microscopes, such as fluorescence emission difference (FED) microscopy [11], scanning laser mode microscopy (SLAM) [12], and others [13, 14], have attracted interest for being able to provide a simple approach to break the diffraction barrier. The concept is to subtract two fluorescence images captured by two excitation lasers with the same wavelength but with different spatial intensity distributions, e.g., solid and donut shapes. Because the quality of subtracted images strongly depends on the spot shapes of the two excitation lasers and the implemented subtraction factors, the possible problems of nonlinearities and over-process resulting from the subtraction could be generated. To reduce the distortion and negative sidebands brought by improper illumination, excitation beams with radial and azimuthal polarizations and with extended solid and hollow spot shapes have been proposed [15, 16]. Further, to achieve appropriate subtraction, the concept of the subtraction threshold has also been constructed [17]. However, the imaging quality can be easily deteriorated by imperfect polarizations in coherent optical microscopes [18–21]. Especially for fluorescence subtraction microscopy, the resolution enhancement is obtained from direct subtraction of fluorescence images, in which the quality of the polarization for excitation should be considered as significant. However, there has been no report on the detailed study of the effects of the imperfect polarization on the subtracted parameters in such microscopes until now. In this chapter, a numerical model is built based on the vector beam diffraction theory (VBDT) and it is applied to investigate the importance of the polarizations in fluorescence

DOI: 10.1201/9781003476030-7

difference microscopy. The resolution, peak intensity, and further parameters in subtracted images are quantitatively evaluated by altering the polarization described by the Jones matrices. It is found that when the wave plate for polarization conversion deviates from normal incidence by 10°, the resolution deterioration is about 5%; peak intensity decreases within 3%; the corresponding amplitude of negative sidebands is reduced by about 5%. The deviation from appropriate polarization conditions leads to easier over-subtraction and the threshold peak intensity decreases for sample sizes of smaller than 0.1λ. The subtraction factor of 0.6 is found to be able to eliminate the spot deformation after subtraction caused by the deviation of the polarizations. Simulation results with combined radially and azimuthally polarized spots for excitation are also given.

2 PRINCIPLE AND NUMERICAL MODELS

The schematic of polarization conversion in the subtraction microscope is depicted in Figure 2.2.3.1(a). To avoid the interference between each other, two orthogonally linear-polarized excitation beams are combined with a polarization beam splitter

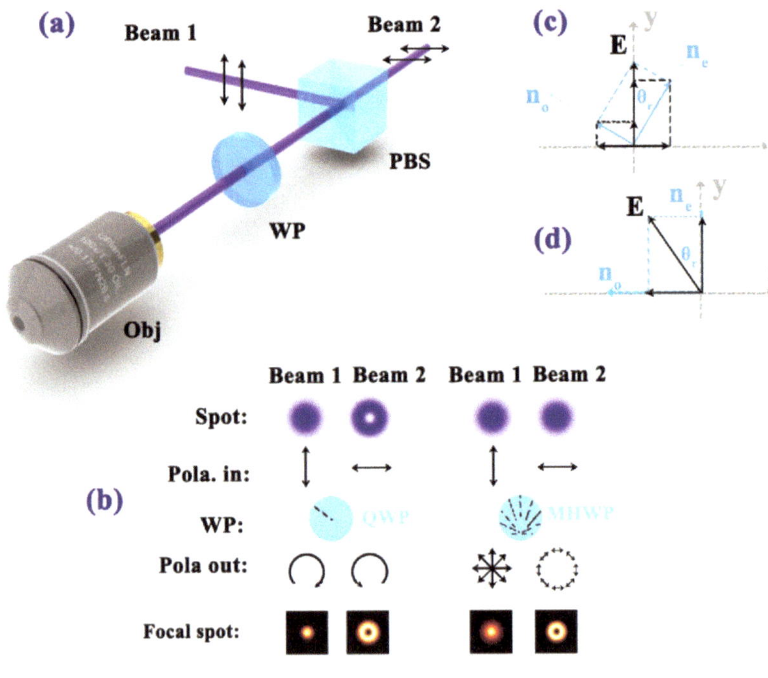

FIGURE 2.2.3.1 (a) Schematic of polarization conversion for excitation beams in subtraction microscopy. (b) Two groups of polarization combinations for excitation. (c) Electric field components with rotated wave plate. (d) Electric field components with fixed wave plate. PBS: polarization beam splitter; WP: wave plate; Obj: objective lens; QWP: quarter wave plate; MHWP: mosaic half wave plate. Copyright of this figure is partially held between the representative author (Takayoshi Kobayashi) and Optica (formally Optical Society of America).

and converted into desired polarization combinations with corresponding wave plates. Two combinations of spots and polarizations are usually used for excitation, as are shown in Fig. 2.2.3.1(b). The first group is the combination of a solid-shape spot and a donut-shape spot generated with a vortex phase plate. Then they are converted into right-handed and left-handed circular polarization after passing through a quarter-wave plate (QWP). The second group is the combination of radially and azimuthally polarized spots converted from two solid-shape spots using a mosaic half wave plate (MHWP). With each of the polarization combinations, the solid and donut focal spots for excitation can be obtained.

The mathematical formula of subtracted image can be expressed as [17, 22]:

$$I_{RAW}\left(x,y,z,\theta_r,r,\gamma\right)=I_S\left(x,y,z,\theta_r\right)*Ob\left(x,y,r\right)-\gamma\cdot I_D\left(x,y,z,\theta_r\right)*Ob\left(x,y,r\right)$$

$$I\left(x,y,z,\theta_r,r,\gamma\right)=\begin{cases} I_{RAW}\left(x,y,z,\theta_r,r,\gamma\right) & I_{RAW}\left(x,y,z,\theta_r,r,\gamma\right)\geq 0 \\ 0 & I_{RAW}\left(x,y,z,\theta_r,r,\gamma\right)<0 \end{cases} \tag{2.2.3.1}$$

where γ is the subtraction factor, $*$ is the convolution, I_S and I_D are the focused solid and donut spots, the point spread functions (PSFs) of which can be derived from VBDT [23–25]. Ob is the sample. They are expressed as:

$$E_{x,y,z}\left(x,y,z,\theta_r\right)=\frac{iC}{\lambda}\int_0^{2\pi}\int_0^{\arcsin(NA/n)}A_{AMP}\cdot A_{\text{Phase}}\cdot A_L\begin{pmatrix}P_x\left(\theta_r\right)\\ P_y\left(\theta_r\right)\\ P_z\left(\theta_r\right)\end{pmatrix} \sin\theta\cdot e^{ikn\sqrt{x^2+y^2}\sin\theta\cos\left(\varphi-\arctan\left(y/x\right)\right)}d\theta d\varphi \tag{2.2.3.2}$$

$$I_{S,D}\left(x,y,z,\theta_r\right)=\left|E_x\right|^2+\left|E_y\right|^2+\left|E_z\right|^2 \tag{2.2.3.3}$$

$$Ob\left(x,y,r\right)=C_0 e^{-\left(4\ln 2\right)\frac{x^2+y^2}{r^2}} \tag{2.2.3.4}$$

Here C and C_0 are the normalization constants; λ is the wavelength of incident light; NA is the numerical aperture of the objective lens; n is the refractive index of the contact medium between the objective lens and the cover glass of the sample, the value of which is 1.518 for the immersion oil; k is the wave number; A_{AMP} is the amplitude factor; A_{phase} is the phase factor; r is the preset full width at half maximum (FWHM) of the sample envelope. For simplification of A_{AMP}, $A_{AMP} = 1$ is assumed. The $A_{AMP} = 1$ means the plane wave is used to simulate the intensity distribution of the incident excitation lasers; the $A_{phase} = e^{i\phi}$ is the phase distribution of the vortex phase plate. The set values of A_{AMP} and A_{phase} for I_S and I_D are shown in Table 2.2.3.1.

TABLE 2.2.3.1
Parameter sets for I_S and I_D

	Group 1			Group 2		
IPSF	**Polarization**	A_{AMP}	A_{phase}	**Polarization**	A_{AMP}	A_{phase}
IS	Right Circular	1	1	Radial	1	1
ID	Left Circular	1	$_ei\phi$	Azimuthal	1	1

A_L is the vector weight matrix of the aplanatic objective lens, calculated as follows:

$$A_L = \sqrt{\cos\theta}\begin{pmatrix} 1+(\cos\theta-1)\cos^2\varphi & (\cos\theta-1)\cos\varphi\sin\varphi & -\sin\theta\cos\varphi \\ (\cos\theta-1)\cos\varphi\sin\varphi & 1+(\cos\theta-1)\cos^2\varphi & -\sin\theta\sin\varphi \\ \sin\theta\cos\varphi & \sin\theta\sin\varphi & \cos\theta \end{pmatrix} \tag{2.2.3.5}$$

[*Px, Py, Pz*] is the unit vector matrix of the polarization. According to the Jones Calculus [26, 27], the electric field of s- or p-polarized light has components along the tilted crystal axis and then they are projected to the horizontal and vertical axes when rotating the wave plate. The phase delay between the two birefringent optical axes is $\delta = kd(n_o - n_e)$, where d is the thickness of wave plate; n_o and n_e are the refractive indices of ordinary and extraordinary optical axes, respectively. For a QWP of negative crystal, δ is $\pi/2$. From Fig. 2.2.3.1(c), Jones matrices of the horizontal and vertical electric field components for s- and p-polarized beams after a negative-type QWP can be written as:

$$s-p: \begin{bmatrix} P_x \\ P_y \\ P_z \end{bmatrix} = \begin{bmatrix} (1-i)\sin\theta_r\cos\theta_r \\ \cos^2\theta_r + i\sin^2\theta_r \\ 0 \end{bmatrix}, p-p: \begin{bmatrix} P_x \\ P_y \\ P_z \end{bmatrix} = \begin{bmatrix} \sin^2\theta_r + i\cos^2\theta_r \\ (1-i)\sin\theta_r\cos\theta_r \\ 0 \end{bmatrix} \tag{2.2.3.6}$$

where θ_r is the angle between the electric field of s-polarized beam and the fast axis. Then the intensity distributions of focused spots with different angles of the QWP can be calculated. Because the long and short axes of the focused spots are always along the fast and slow axes of the wave plates, it is inconvenient to analyze the resolutions along two perpendicular axes when the focused spots are tilted. Fixing the wave plate and rotating the polarization can be a calculation strategy to solve this problem. Then the long and short axes of the focused spots can be fixed to the horizontal or vertical axes. As is shown in Fig. 2.2.3.1(d), the Jones matrices used to calculate the sectioning lines can be written as:

$$s-p:\begin{bmatrix}P_x\\P_y\\P_z\end{bmatrix}=\begin{bmatrix}-i\sin\theta_r\\\cos\theta_r\\0\end{bmatrix},p-p:\begin{bmatrix}P_x\\P_y\\P_z\end{bmatrix}=\begin{bmatrix}\cos\theta_r\\-i\sin\theta_r\\0\end{bmatrix} \tag{2.2.3.7}$$

For the radially or azimuthally polarized beams generated from MHWP, the Jones matrices are

[28]:

$$s-p:\begin{bmatrix}P_x\\P_y\\P_z\end{bmatrix}=\begin{bmatrix}\cos(\varphi-\theta_r)\\\sin(\varphi-\theta_r)\\0\end{bmatrix},p-p:\begin{bmatrix}P_x\\P_y\\P_z\end{bmatrix}=\begin{bmatrix}\sin(\varphi-\theta_r)\\-\cos(\varphi-\theta_r)\\0\end{bmatrix} \tag{2.2.3.8}$$

Where ϕ is the azimuthal angle. Then the focal PSFs and subtracted spots related with the angles θ $_r$ for different polarizations can be obtained. In order to quantitatively analyze the quality of the image, the peak intensity and amplitude of negative sidebands after subtraction are defined as:

$$I_{\text{peak}}=\frac{MAX\left[I(x,y,z,\theta_r,r,\gamma)\right]}{\text{MAX}\left[I(x,y,z,\theta_r,r,0)\right]} \tag{2.2.3.9}$$

$$\left|I_{Neg}\right|=\begin{cases}0 & \text{MIN}\left[I_{RAW}(x,y,z,\theta_r,r,\gamma)\right]\geq 0\\ \dfrac{\left|MIN\left[I_{RAW}(x,y,z,\theta_r,r,\gamma)\right]\right|}{MAX\left[I(x,y,z,\theta_r,r,\gamma)\right]} & MIN\left[I_{RAW}(x,y,z,\theta_r,r,\gamma)\right]<0\end{cases} \tag{2.2.3.10}$$

The subtraction threshold conditions, which describe the appropriate subtraction [17], and normalized peak intensity at threshold under the condition, are expressed by Eqs. (11) and (12), respectively.

$$FWHM_{Ob(x,y,r_0)}=FWHM_{I(x,yz,\theta_r,r_0,\gamma_0)} \tag{2.2.3.11}$$

$$I_P=\frac{MAX\left[I(x,y,z,\theta_r,r_0,\gamma_0)\right]}{MAX\left[I(x,y,z,\theta_r,r_0,0)\right]} \tag{2.2.3.12}$$

Then the effects of the polarization on the FWHM resolution, ellipticity, peak intensity, negative sidebands, subtraction threshold, and peak intensity after subtraction can be carried out.

3 SIMULATION RESULTS

3.1 Linear and Circular Polarizations

The calculated fluorescence and subtracted spots with polarizations of different circularity are shown in Figure 2.2.3.2. The preset sample size is $r = 0.1\lambda$. From Fig. 2.2.3.2(a), it can be seen that the shape of the focused spot varies with different polarization status. The focused solid spot has elliptical shape when the incident light is linearly or elliptically polarized. Because of the phase distribution, the center of the donut spot is round and has zero intensity only if θ_r is $-45°$, where the polarization is left-handed circular. This is the appropriate angle for illumination in a subtraction microscope. As is shown in Fig. 2.2.3.2(b), the subtracted images with $\theta_r = -45°$ keep the same circularity while images with this angle ±10o have varying circularities after subtraction. Figures 2(c)–2(e) display the FWHM values of the cross sectioning profiles along the long and short axes of subtracted focal spots and corresponding ellipticities with factors of $\gamma = 0, 0.3, 0.6, 1.0$. It can be seen that the FWHM values of subtracted spots are fluctuating by about 0.1λ when rotating the QWP by π; the FWHM values change within 5% even when the angles of the QWP deviate from the right angle for 10°; the ellipticities of the spots after subtraction with the two critical angles of $-45°$ and 45° are both unity; with subtraction factors of larger than 0.6, the long and short axes of the subtracted spots are reversed. The ellipticity here is defined as the ratio of FWHM values on the primary short and long axes of the focused spot. What is more, it is interesting to find that the factor of 0.6 can almost eliminate the errors of ellipticity induced by the misalignments of the polarization. It also indicates that when the factor of 0.6 is implemented in the subtraction scheme, the problem of polarization aberration cannot be intuitively found out from the subtracted images.

The evolution of peak intensity, amplitude of negative sidebands, and the subtraction threshold are shown in Fig. 2.2.3.3. From Fig. 2.2.3.3(a), it can be seen that the subtracted peak intensities are maximal when $\theta_r = -45°$ even with different subtraction factors. With $\theta_r = 45°$, the center of donut spot is not completely dark, so the peak intensity drops significantly after the convolution with samples and the following subtraction. Fig. 2.2.3.3(b) describes that the decreased amount of peak intensity after subtraction is about 2.5% if the angle of QWP is displaced with 10° and subtraction factor of 1. Thus, the problem of signal loss due to the imperfect polarization for excitation within 10° is not serious. Besides, the negative sidebands are generated inevitably after subtraction. However, from Fig. 2.3.3.3(c), the values of negative sidebands introduced are similar in a wide range from $-90°$ to 0°. The deviation of the polarization may reduce the negative sidebands by 5% as is shown in Fig. 2.3.3.3(d). It describes that the amplitude of negative sidebands after subtraction is insensitive to the ellipticity of the polarization, as well. Similarly, the calculations on the resolution, peak intensity and negative sidebands—with different sample sizes— have been carried out. The curves give the similar evolutions as present ones which will not be illustrated in further detail. For a sample size of 0.4λ, the variation of FWHM values of subtracted spots is by about 0.1λ versus the rotation angles of

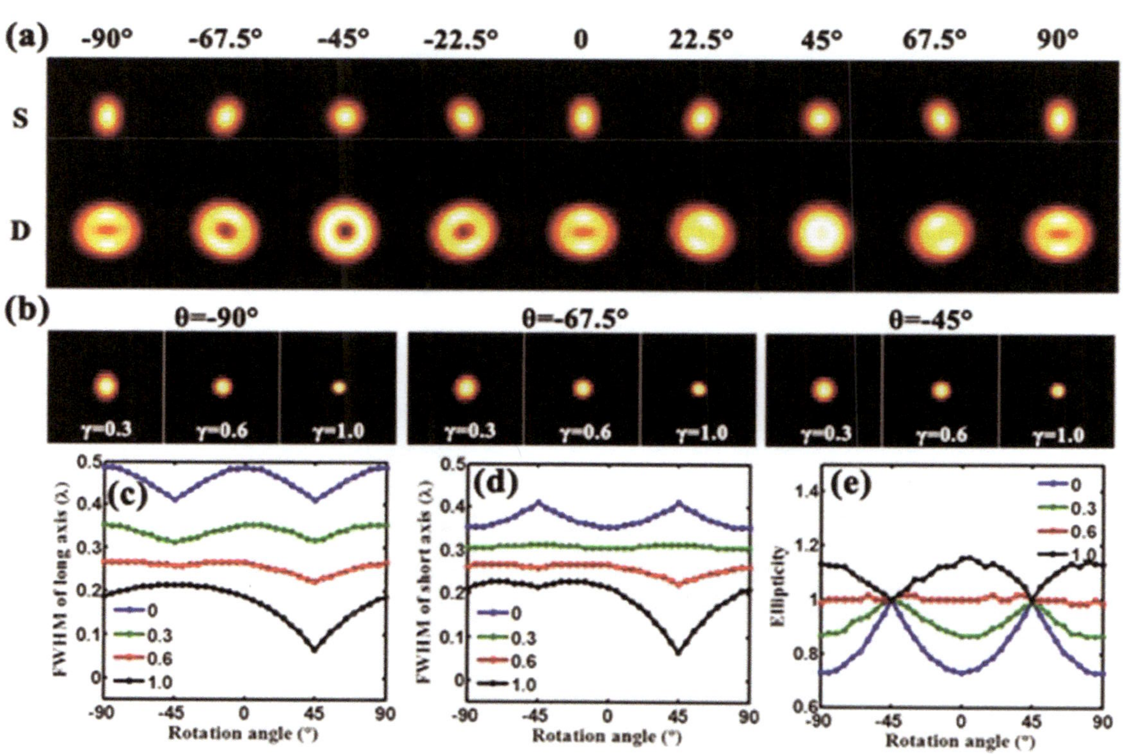

FIGURE 2.2.3.2 (a) Convoluted solid and donut fluorescence images with different rotation angles of the QWP. The sample has size of r = 0.1λ. (b) Subtracted images with factors of γ = 0.3, the long and short axes, and ellipticity with different subtraction factors of 0.6, 1.0 for θ r =−90°, −67.5°,−45°. (c)–(e) FWHM values of sectioning profiles along $\gamma_{=0, 0.3, 0.6,}$ 1.0 versus different rotation angles (θ_r). Copyright of this figure is partially held between the representative author (Takayoshi Kobayashi) and Optica (formally Optical Society of America).

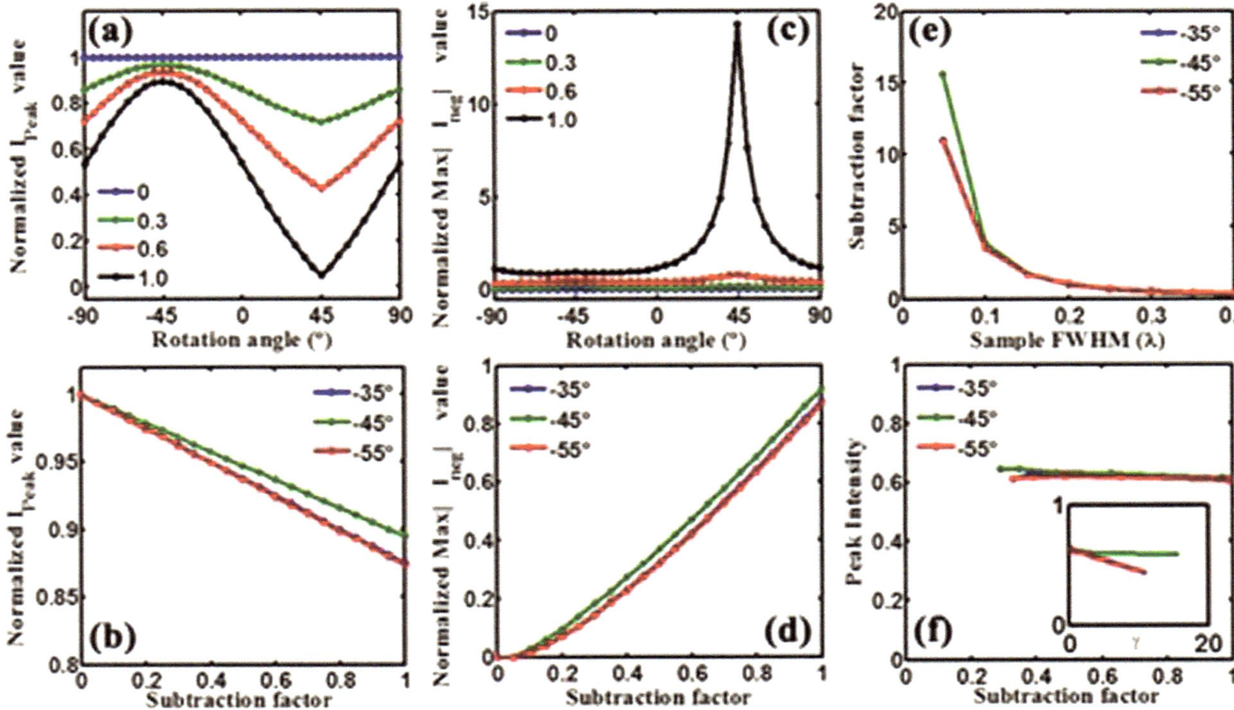

FIGURE 2.2.3.3 (a) Peak intensity and (c) Negative sidebands versus rotation angles of QWP with subtraction factors of $\gamma = 0, 0.3, 0.6, 1.0$. (b) Peak intensity and (d) Negative sidebands versus subtraction factors with rotation angles of $\theta_r = -35°, -45°, -55°$. (e) Subtraction threshold line. (f) Threshold peak intensity curves. Copyright of this figure is partially held between the representative author (Takayoshi Kobayashi) and Optica (formally Optical Society of America).

the QWP. The peak intensities or negative sidebands are affected by less than 1% with 10° deviation from the right angle. The subtraction threshold lines are shown in Fig. 2.3.3.3(e). It can be seen that the threshold subtraction factors for θr = −45° are a little larger than the cases of −35° and −55° with sample sizes of smaller than 0.1λ. It means the polarization-deviated excitation is vulnerable for over-subtraction and it mainly acts on the fine structures in the sample. Meanwhile, from Fig. 2.3.3.3(f), it can be seen that the threshold peak intensity factors slightly drop when increasing the subtraction factors. It means that the deviation of polarization will not introduce significant signal loss at the subtraction threshold for sample sizes larger than 0.1λ.

3.2 Radial and Azimuthal Polarizations

The calculation results with radially and azimuthally polarized excitation spots are shown in Figure 2.2.3.4. The focused PSFs with different rotation angles of MHWP are shown in Fig. 2.3.3.4(a). Because the evolutions of the subtracted spot shapes appear periodically, only the angles between −45° and 45° are analyzed here. From Fig. 2.3.3.4(b), it can be seen that the subtracted spot shapes are not deformed with displaced polarization angles while the resolutions are a little worse than the case of 0° for MHWP. Figures 2.3.3.4(c)–4(e) depict the FWHM values, peak intensities, and the values of negative sidebands when changing the angles of the MHWP. It can be found that when the angle of MHWP is 0° and subtraction factor keeps the same, the resolved FWHM values are smaller than those with other angles. Meanwhile, the peak intensity is also maximal at the angle of 0°. It is possible to use these two parameters to calibrate the angles of the used MHWP. When rotating the plate, the angle with maximal peak intensity and minimal FWHM values corresponds to the appropriate angle for MHWP in a subtraction microscope scheme. The amplitude of negative sidebands is insensitive to the rotation angles of the plate, as is shown in Fig. 2.3.3.4(e). From Fig. 2.3.3.4(f), it can be seen that the peak intensity decrease is about 5% when the angle is deviated with 10° and a subtraction factor of 1. The subtraction threshold lines and the threshold peak intensity lines are shown in Figs. 2.3.3.4(g) and 2.3.3.4(h), respectively. Similar to the results in the case of circular polarization, the threshold lines show that the right angle of 0° can stand the over-subtraction better than the incorrect polarizations. Meanwhile, the threshold peak intensity factors are a little smaller for the displaced angles. It means that when the appropriate subtraction is achieved, the peak intensity is slightly decreased, especially for the sample sizes of smaller than 0.1λ.

4 CONCLUSION

In summary, we have built a numerical model to evaluate the effects of polarization for a subtraction microscope based on the vector beam diffraction theory. By modulating the Jones matrices, the effects of different polarization status on the parameters of resolution, ellipticity, peak intensity, negative sidebands, and subtraction threshold in fluorescence emission difference microscopy are quantitatively investigated. The resolution deterioration, signal loss, and intensity reduction of negative

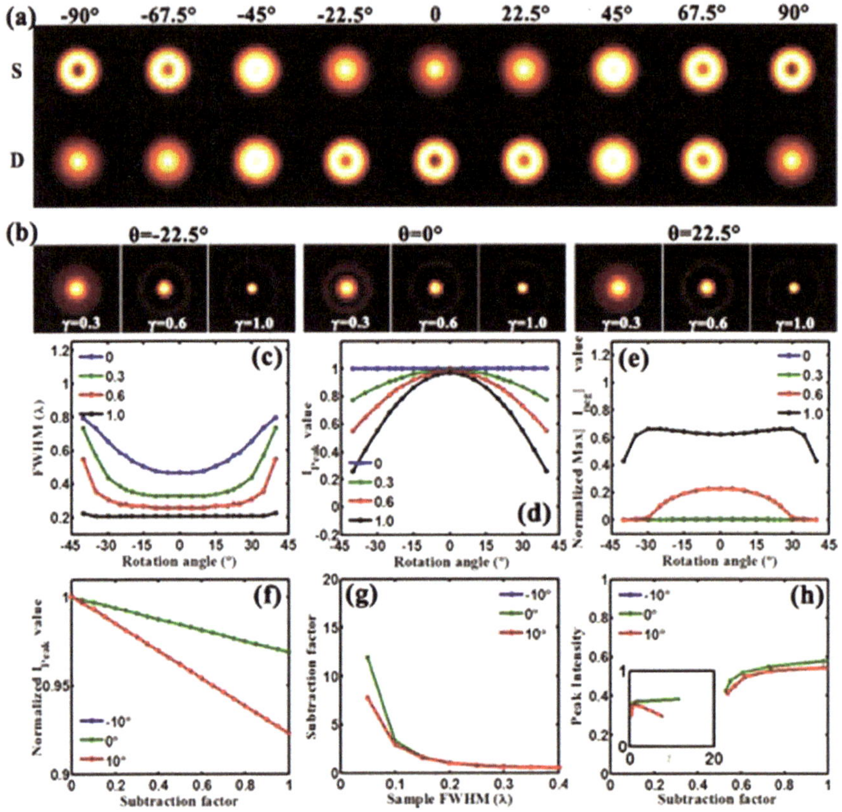

FIGURE 2.2.3.4 Simulation results with radial and azimuthal polarization. (a) Excitation spots versus different angles of wave plate. (b) Subtracted images with factors of 0.3, 0.6 and 1.0 for the rotation angles of -22.5 °, 0°, 22.5°. Copyright of this figure is partially held between the representative author (Takayoshi Kobayashi) and Optica (formally Optical Society of America).

sidebands are all within 5% for both of the polarization combinations for excitation, even when the angles of wave plates are deviating from the normal incidence by 10°.

The corresponding threshold peak intensity curve finds that the signal loss at appropriate subtraction is negligibly small for sample sizes of larger than 0.1λ. The rotating angle of polarizer being 45° must be critically avoided, since it can generate serious signal intensity loss and resolution deterioration. The subtraction factor of 0.6 is found to be a specially behaving parameter for the case when the polarization is modified with a quarter wave plate. With this factor, the ellipticity is always unity even when the excitation beams are linearly or elliptically polarized. To calibrate the polarization, ellipticity variation, or peak intensity after subtraction can be used. Furthermore, the simulation here is carried out based on the model in ideal conditions. However, calculations in this chapter can basically describe and evaluate the effects of polarization on the parameters for fluorescence subtraction microscopy even though the confocal detection parameter or noise effect is not included in the model. The improved model including them is to be presented in future work. Sensitive fluorescence detection with high signal-to-noise ratio is necessary for experimental demonstration of present calculations. Our related experiments are in progress.

The contents described in this subsection are based on the collaborative work of N. Wang and T. Kobayashi [29]. The above study was reported in *Optics Express* and copyright is partially held between the corresponding author (Takayoshi Kobayashi, TK) and Optica (formally Optical Society of America). TK would like to express thanks to Optica for permitting us to publish the paper properly revised to be suited to be one section in the book *Photothermal and Raman Microscopy* (Taylor & Francis).

REFERENCES

1. S. W. Hell and J. Wichmann, "Breaking the diffraction resolution limit by stimulated emission: stimulated emission-depletion fluorescence microscopy," *Opt. Lett.* **19**(11), 780–782 (1994).
2. M. G. L. Gustafsson, "Surpassing the lateral resolution limit by a factor of two using structured illumination microscopy," *J. Microscopy* **198**, 82–87 (2000).
3. E. Betzig, G. H. Patterson, R. Sougrat, O. W. Lindwasser, S. Olenych, J. S. Bonifacino, M. W. Davidson, J. L. Schwartz, and H. F. Hess, "Imaging intracellular fluorescent proteins at nanometer resolution," *Science* **313**, 1642–1645 (2006).
4. M. J. Rust, M. Bates, and X. W. Zhuang, "Sub-diffraction-limit imaging by stochastic optical reconstruction microscopy (STORM)," *Nat. Meth.* **3**, 793–796 (2006).
5. M. P. Gordon, T. Ha, and P. R. Selvin, "Single-molecule high-resolution imaging with photobleaching," *P. Natl. Acad. Sci. USA* **101**(17), 6462–6465 (2004).
6. A. Sharonov and R. M. Hochstrasser, "Wide-field sub-diffraction imaging by accumulated binding of diffusing probes," *P. Natl. Acad. Sci. USA* **103**(50), 18911–18916 (2006).
7. K. Fujita, M. Kobayashi, S. Kawano, M. Yamanaka, and S. Kawata, "High-resolution confocal microscopy by saturated excitation of fluorescence," *Phys. Rev. Lett.* **99**, 228105 (2007).
8. T. Dertinger, R. Colyer, G. Iyer, S. Weiss, and J. Enderlein, "Fast, background-free, 3D super-resolution optical fluctuation imaging (SOFI)," *P. Natl. Acad. Sci. USA* **106**(52), 22287–22292 (2009).
9. J. Yao, L. Wang, C. Li, C. Zhang, and L. V. Wang, "Photoimprint photoacoustic microscopy for three-dimensional label-free subdiffraction imaging," *Phys. Rev. Lett.* **112**, 014302 (2014).
10. J. Miyazaki, H. Tsurui, A. Hayashi-Takagi, H. Kasai, and T. Kobayashi, "Sub-diffraction resolution pump-probe microscopy with shot-noise limited sensitivity using laser diodes," *Opt. Express* **22**(8), 9024–9032 (2014).
11. C. F. Kuang, S. Li, W. Liu, X. Hao, Z. T. Gu, Y. F. Wang, J. H. Ge, H. F. Li, and X. Liu, "Breaking the diffraction barrier using fluorescence emission difference microscopy," *Sci. Rep.* **3**(1441) (2013).
12. H. Dehez, M. Piche, and Y. D. Koninck, "Resolution and contrast enhancement in laser scanning microscopy using dark beam imaging," *Opt. Express* **21**(13), 15912–15925 (2013).

13. S. Segawa, Y. Kozawa, and S. Sato, "Resolution enhancement of confocal microscopy by subtraction method with vector beams," *Opt. Lett.* **39**(11), 3118–3121 (2014).
14. N. Wang, J. Miyazaki, J. He, K. Seto, and T. Kobayashi, "Sub-diffraction-limit imaging using mode multiplexing," *Opt. Commun.* **343**, 28–33 (2015).
15. S. Segawa, Y. Kozawa, and S. Sato, "Demonstration of subtraction imaging in confocal microscopy with vector beams," *Opt. Lett.* **39**(15), 4529–4532 (2014).
16. S. You, C. Kuang, Z. Rong, and X. Liu, "Eliminating deformations in fluorescence emission difference microscopy," *Opt. Express* **22**(21), 26375–26385 (2014).
17. N. Wang and T. Kobayashi, "Numerical study of the subtraction threshold for fluorescence difference microscopy," *Opt. Express* **22**(23), 28819–28830 (2014).
18. X. Hao, C. Kuang, T. Wang, and X. Liu, "Effects of polarization on the de-excitation dark focal spot in STED microscopy," *J. Opt.* **12**, 115707 (2010).
19. C. Yuan, G. Situ, G. Pedrini, J. Ma, and W. Osten, "Resolution improvement in digital holography by angular and polarization multiplexing," *Appl. Opt.* **50**(7), B6–B11 (2011).
20. S. Galiani, B. Harke, G. Vicidomini, G. Lignani, F. Benfenati, A. Diaspro, and P. Bianchini, "Strategies to maximize the performance of a STED microscope," *Opt. Express* **20**(7), 7362–7374 (2012).
21. Y. Chen, H. Lin, M. J. Ludford-Menting, A. H. Clayton, M. Gu, and S. M. Russell, "Polarization of excitation light influences molecule counting in single-molecule localization microscopy," *Histochem. Cell Biol.* **143**(1), 11–19 (2015).
22. N. Wang and T. Kobayashi, "Numerical calibration of the spatial overlap for subtraction microscopy," *Opt. Express.* **23**(10), 13410 (2015).
23. B. Richards and E. Wolf, "Electromagnetic diffraction in optical systems. II. structure of the image field in an aplanatic system," *Proc. R. Soc. Lond.* A **253**, 358–379 (1959).
24. D. Ganic, X. Gan, and M. Gu, "Focusing of doughnut laser beams by a high numerical-aperture objective in free space," *Opt. Express* **11**(21), 2747–2752 (2003).
25. G. M. Lerman and U. Levy, "Effect of radial polarization and apodization on spot size under tight focusing conditions," *Opt. Express* **16**(7), 4567–4581 (2008).
26. R. C. Jones, "A new calculus for the treatment of optical systems," *J. Opt. Soc. Am.* **31**(7), 488–493 (1941).
27. W. H. Mcmaster, "Matrix representation of polarization," *Rev. Mod. Phys.* **33**(1), 8–28 (1961).
28. M. Stalder and M. Schadt, "Linearly polarized light with axial symmetry generated by liquid-crystal polarization converter," *Opt. Lett.* **21**(23), 1948–1950 (1996).
29. N. Wang and T. Kobayashi, Polarization modulation for fluorescence emission difference microscopy," *Opt. Express* **23**(10), 13704–13712 (2015).

2.2.4 Numerical Study of Superresolved Optical Microscopy with Partly Staggered Beams

1 INTRODUCTION

Optical microscopy has developed very fast in the last decades, and fluorescence microscopy is among the most powerful technologies with the capability of superresolution imaging of samples stained with fluorescent labels. Several well established superresolution fluorescence techniques are now being used in biological and material research fields, such as stimulated emission depletion (STED) [1, 2], stochastic optical reconstruction (STORM) [3] and photoactivated localization microscopy (PALM) [4]. However, fluorescence microscopy still has some limitations [5]: (1) Many tissues, chromophores, and molecules in biological systems absorb light but do not have efficient fluorescence emission, such as collagen and hemoglobin; (2) Many intracellular small molecules are intrinsically nonfluorescent and they are also too small to be labeled with fluorescent proteins or dyes, such as neurotransmitters and drugs. As an alternative, many fluorescence-free or label-free microscopies have been developed, such as pump-probe microscopy (PPM) [6–9], photothermal microscopy (PTM) [10–12], and nonlinear optical microscopy (NOM) [13–15]. However, resolution of previous imaging with these microscopies has nearly always been diffraction-limited, and superresolution technique is still very limited. Recently, photothermal microscopy with partly staggered quasi-super-Gaussian pump and probe beams has been demonstrated experimentally in the imaging of mouse melanoma, and clear resolution enhancement was observed [16]. This microscopic technology is suitable for superresolution imaging involving two or more beams, which is the case of PPM, PTM, and some of the NOM.

In this chapter, we give a detailed numerical study of the type of optical microscopy with partly staggered beams (OM-PSB) by using vector diffraction theory. Different polarizations of pump and probes are discussed, such as linear, circular, radial, and azimuthal polarizations. Compared with the traditional optical microscopy, the highest resolution improvement for OM-PSB can reach $> 45\%$ in lateral direction and $>35\%$ in longitudinal direction, with pump and probe beams of ideal plane waves.

DOI: 10.1201/9781003476030-8

2 PRINCIPLE AND NUMERICAL MODELS

The principle of OM-PSB is similar to PTM-PSB [16]. The pump and probe beams are partly staggered in a lateral direction at the focal point of the objective, and as a result, the interaction area of the two beams and the sample or the effective PSF of PTM can be reduced if the beam shapes of pump and probes are carefully chosen. The signal intensity in PPM and PTM is proportional to the change of the probe beam intensity ΔI_{probe}, while that in NOM involving two or more beams, such as sum frequency generation (SFG) and four-wave mixing (FWM) microscopies, is the intensity of a newly generated signal I_{new}. Both ΔI_{probe} and I_{new} are proportional to the product of the intensities of the two beams involved I_{pump} and I_{probe}, under small signal condition [5, 6]. In this paper, only the cases of PPM and PPM with partly staggered beams (PPM-PSB) are discussed. For simplification, $\Delta I_{probe}(\mathrm{r}) = I_{pump}(\mathrm{r}) * I_{probe}(\mathrm{r})$ is taken as the effective PSF of PPM, and $\Delta I_{probe}(\mathrm{r}, \Delta\mathrm{r}) = I_{pump}(\mathrm{r} - \Delta\mathrm{r}/2) * I_{probe}(\mathrm{r} + \Delta\mathrm{r}/2)$ is that for PPM-PSB with the spatial displacement of the pump and probe beams ofΔr.

Vectorial diffraction theory is used to simulate the propagation of input lights through a lens. The electric field vector near the focal spot can be obtained with generalized Debye integral [17] as follows,

$$\vec{E}(r_2, \varphi_2, z_2) = \mathrm{i}C \iint_\Omega \sin(\theta) \cdot A_1(\theta, \varphi) \cdot A_2(\theta, \varphi) \cdot \begin{bmatrix} p_x \\ p_y \\ p_z \end{bmatrix} \cdot e^{ikn(z_2 \cos\theta + r_2 \sin\theta \cos(\varphi - \varphi_2))} \mathrm{d}\theta \mathrm{d}\varphi \tag{2.2.4.1}$$

where $\vec{E}(r_2, \varphi_2, z_2)$is the electric field vector, $C$is the normalized constant, $A_1(\theta, \varphi)$is the amplitude of the input light, $A_2(\theta, \varphi)$is a 3×3 image-transfer matrix of the objective lens, $[p_x; p_y; p_z]$ is the unit matrix of input light polarization. $A_1(\theta, \varphi)$ is equal to 1 for ideal plane-wave. The polarization matrix for linear, left circular, right circular, radial, and azimuthal beams is shown in Table 2.2.4.1. During the simulation, 0-2π vortex phase plates are applied in both the pump and probe beam paths for the tight focusing of the two beams, with the pump and probe having azimuthal polarization.

With an aplanatic lens as the focal lens, $A_2(\theta, \varphi)$can be expressed as follows,

$$A_2(\theta, \varphi) = \sqrt{\cos\theta} \cdot \begin{bmatrix} 1 + (\cos\theta - 1)\cos^2\varphi & (\cos\theta - 1)\cos\varphi \sin\varphi & -\sin\theta\cos\varphi \\ (\cos\theta - 1)\cos\varphi\sin\varphi & 1 + (\cos\theta - 1)\sin^2\varphi & -\sin\theta\sin\varphi \\ \sin\theta\cos\varphi & \sin\theta\sin\varphi & \cos\theta \end{bmatrix} \tag{2.2.4.3}$$

The PSFs of pump and probe beams are calculated by $I(x,y,z) = |E_x|^2 + |E_y|^2 + |E_z|^2$. While the effective PSF of PPM-PSB is defined by

TABLE 2.2.4.1
Polarization Matrix Unit for Different Polarizations

Polarization	Polarization matrix unit
Linear	[1; 0; 0]; [0; 1; 0]
Left circular	$[1; i; 0]/\sqrt{2}$
Right circular	$[1; -i; 0]/\sqrt{2}$
Radial	[cosφ; sinφ; 0]
Azimuthal	[-sinφ; cosφ; 0]

$\Delta I(x,y,z)=I_{pump}(x,y,z)I_{probe}(x\text{-}\Delta x, y-\Delta y, z-\Delta z)$, where I_{pump} and I_{probe} are intensity distributions of pump and probe beams, Δ x, Δ y, and Δ z are the offset of pump and probe beams in x, y, and z directions, respectively. To fit the common experimental conditions, the numerical aperture (NA) of the focal lens is assumed to be 1.4 (100×, oil immersion), and the wavelengths of pump and probe beams are set as 488 and 632 nm respectively. By solving the above equations, PSFs of pump beam, probe beam, normally used pump-probe microscopy (no spatial offset between pump and probe) (NPPM), and PPM-PSB are obtained.

3 RESULTS AND DISCUSSIONS

To give a qualitative demonstration of the lateral resolution enhancement of PPM-PSB, the effective PSFs of PPM-PSB are given. With the offset of pump and probe only in the X axis, the PSFs of pump, probe, NPPM, and PPM-PSB in the XY plane for different polarizations are shown in Figure 2.2.4.1. The first column in Fig. 2.2.4.1 is the PSFs of pump (blue spots) and probe (red spots) beams, while the second and third columns are the effective PSFs of NPPM and PPM-PSB, respectively. In the case of PPM-PSB, the values of the offset Δ x are 350, 225, 290, 520, and 150 nm for linear polarization in the X direction (Lx), linear polarization in the Y direction (Ly), right circular polarization (Cr), radial polarization (Ra) and azimuthal polarization (Az), respectively. The resolving power of NPPM is higher than that of traditional optical microscopy, as illustrated by the spots in the first and second columns. It is also proved experimentally in the previous report [18]. The spots size is even smaller for PPM-PSB with Lx, Ly and Cr polarizations, compared with NPPM, as shown in the second and third columns in Fig. 2.2.4.1, although the resolving power decreases for Ra and Az polarizations for their long tails of the intensity distribution. From this, we can find that the polarization of pump and probe influences the resolving power of PPM-PSB greatly.

Figure 2.2.4.2 shows the axial resolution enhancement of PPM-PSB. The offset of pump and probe is set only in the Z axis with Δ z of 610, 615, 700, and 620 nm for Lx, Cr, Ra, and Az polarizations, respectively, with the intensity of side lobes of the

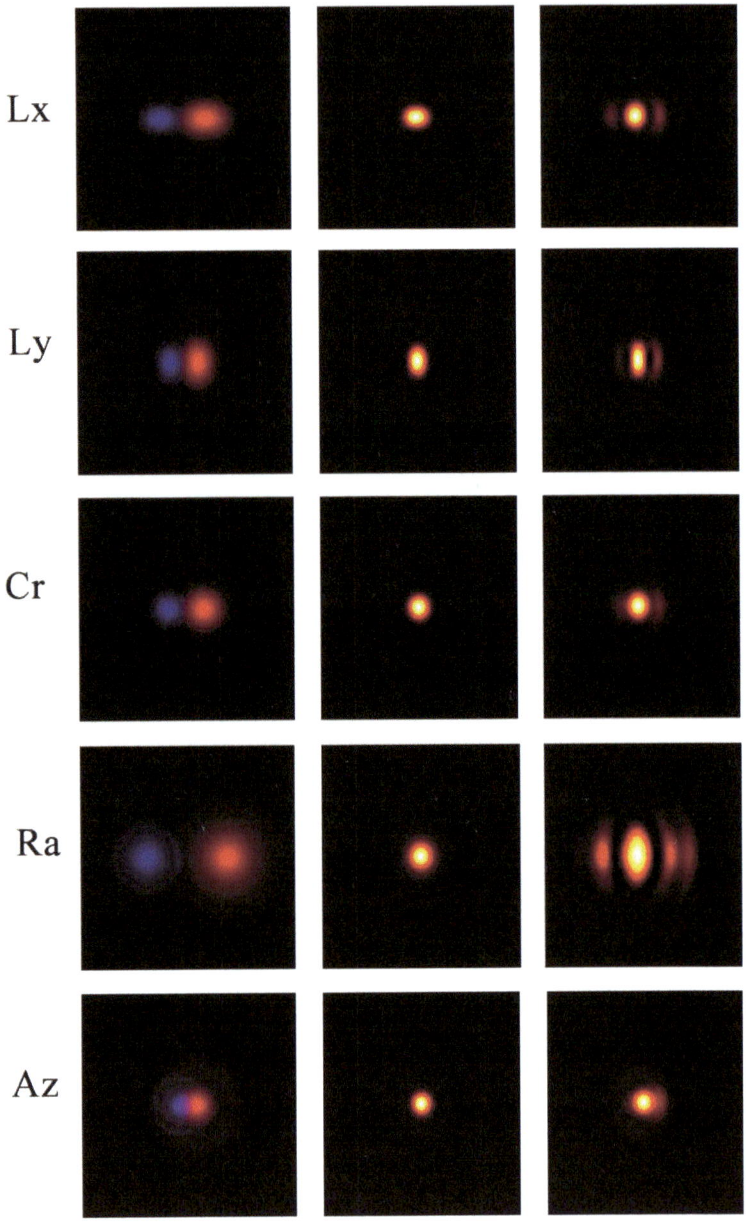

FIGURE 2.2.4.1 Effective PSFs of pump (blue spots in the first column), probe (red spots in the first column), NPPM (bright spots in the second column), and PPM-PSB (bright spots in the third column) in XY plane with different polarizations. Lx: linear polarization with polarization direction in X axis; Ly: linear polarization with polarization direction in X axis; Cr: right circular polarization; Ra: radial polarization; Az: azimuthal polarization. The values of Δx are 350, 237, 290, 520, and 150 nm for Lx, Ly, Cr, Ra, and Az polarizations, respectively. The figure was reported in the *Journal of Optics* and copyright is partially held between the representative author (Takayoshi Kobayashi, TK) and IOP Science.

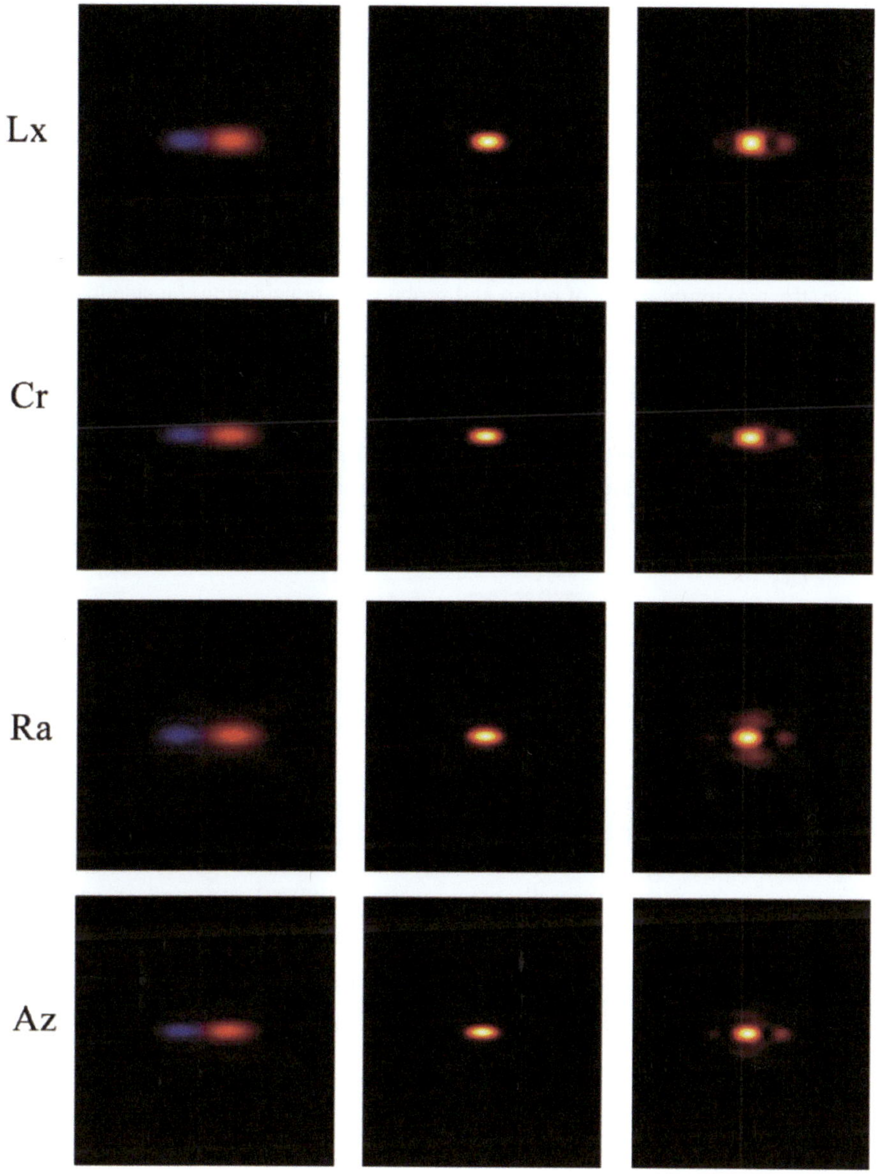

FIGURE 2.2.4.2 Effective PSFs of pump (blue spots in the first column), probe (red spots in the first column), NPPM (bright spots in the second column), and PPM-PSB (bright spots in the third column) in XZ plane with the offset of pump and probe only in Z axis. Lx: linear polarization with polarization direction in X axis; Cr: right circular polarization; Ra: radial polarization; Az: azimuthal polarization. The values of Δz are 610, 615, 700, and 620 nm for Lx, Cr, Ra, and Az polarizations, respectively. The figure was reported in the *Journal of Optics* and copyright is partially held between the representative author (Takayoshi Kobayashi, TK) and Optica.

effective PSFs smaller than 25% of the main peak intensity. The sizes of the effective PSFs of PPM-PSB are smaller than those of NPPM for all the four polarizations, showing higher rsolving power of PPM-PSB mode than NPPM mode.

To obtain the maximum enhancement of spatial resolving power with a PPM-PSB mode, we calculated the sizes of the effective PSFs for different offsets (Δx or Δz) and polarizations ((a) and (b) in Figure 2.2.4.3–2.2.4.9). Since the offset of pump and probe reduces the peak intensity of the effective PSFs and then brings large side lobes, the relationship between peak intensity reduction and the offsets is also given for different polarization of pump and probe ((c) in Fig. 2.2.4.3–2.2.4.9). In the calculation, the maximum offset of pump and probe is set considering the side lobes, the intensity of which is kept below 25% of the intensity of the main peak. The intensity profiles of the effective PSFs of PPM-PSB with the maximum offset are also shown ((d) in Fig. 2.2.4.3–2.2.4.9).

Fig. 2.2.4.3–2.2.4.5 demonstrate the characteristics of the lateral components of the effective PSFs for PPM-PSB with pump and probe of different polarizations. Fig. 2.2.4.3 shows the case of Lx polarized pump and probe beams. The FWHM of

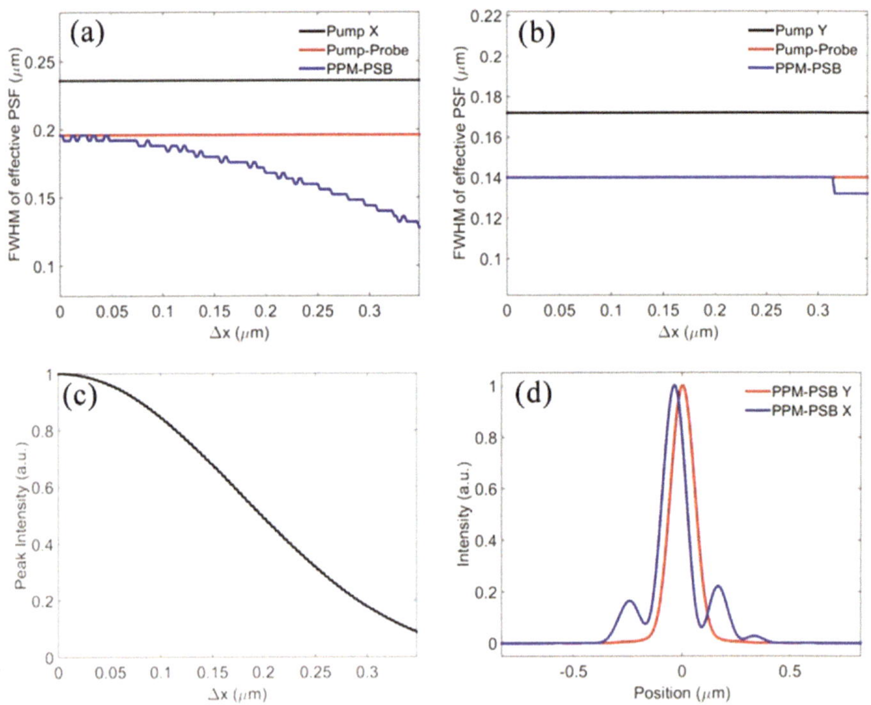

FIGURE 2.2.4.3 With the polarization of Lx, dependence of the FWHM of effective PSFs in X axis (a) and Y axis (b) on Δx. (c) The dependence of the peak intensity of the effective PSF for PPM-PSB on Δx. (d) The Intensity distributions of PPM-PSB along X and Y axis with Δx of 350 nm. The figure was reported in the *Journal of Optics* and copyright is partially held between the representative author (Takayoshi Kobayashi, TK) and IOP Science.

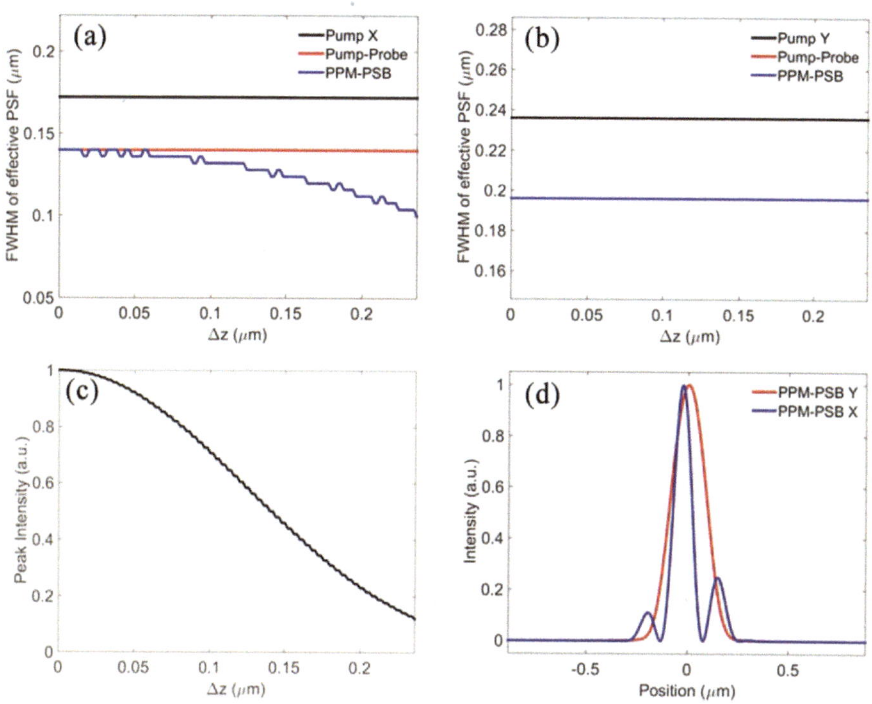

FIGURE 2.2.4.4 With the polarization of Ly, dependence of the FWHM of effective PSFs in X axis (a) and Y axis (b) on Δx. (c) The peak intensity of the effective PSF of PPM-PSB decreases with the increase of Δx. (d) The Intensity distributions of PPM-PSB along the X and Y axis with Δx of 237 nm. The figure was reported in the *Journal of Optics* and copyright is partially held between the representative author (Takayoshi Kobayashi, TK) and IOP Science.

the effective PSF of PPM-PSB reduces from 196 nm to 128 nm in the X axis with Δx increased from 0 to 350 nm (denoted by the blue curve in Fig. 2.2.4.3(a)), with the resolving power improvement of ~34.7% and 45.8% compared with NPPM (blue curve in Fig. 2.2.4.3(a)) and conventional optical microscopy with the excitation of the pump beam (black curve in Fig. 2.2.4.3(a)), respectively. The effective PSF size in the Y axis nearly does not change with Δx , shown by the blue and red curves in Fig. 2.2.4.3(b). With linear polarized pump and probe, the effective PSF of NPPM has different sizes in x and y axes, as shown in Fig. 2.2.4.1 and the red curves in Figs. 2.2.4.3 (a) and (b). Here, x and y axes are horizontal and vertical axes, respectively. The difference can be overcome by utilizing PPM-PSB with carefully chosen Δx, i.e., the FWHM values of the effective PSF in x and y axes are both eaqual to 132 nm with Δx of ~340 nm. As shown in Fig. 2.2.4.3(c), the peak intensity decreases with the increase of the offset Δx. With Δx of 350 nm, the peak intensity in the PPM-PSB mode is about 10% of that with NPPM mode. Fig. 2.2.4.3(d) shows the intensity profiles of the effective PSF of PPM-PSB mode in both x and y axes. The

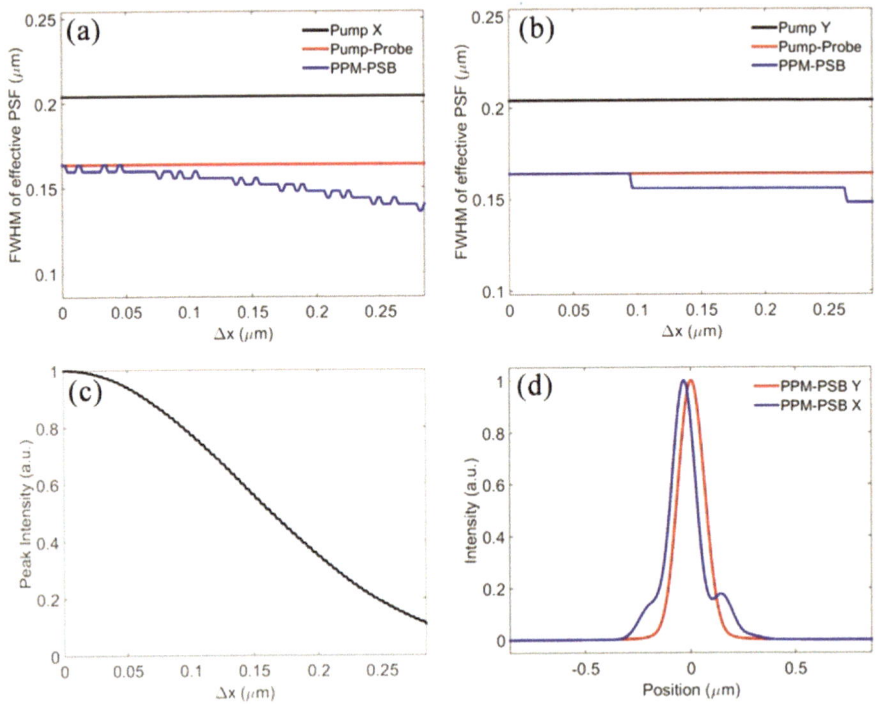

FIGURE 2.2.4.5 With the polarization of Cr, dependence of the FWHM of effective PSFs in X axis (a) and Y axis (b) on Δ *x*. (c) The peak intensity of the effective PSF of PPM-PSB decreases with the increase of Δ *x*. (d) The Intensity distributions of PPM-PSB along the X and Y axis with Δ *x* of 290 nm. The figure was reported in the *Journal of Optics* and copyright is partially held between the representative author (Takayoshi Kobayashi, TK) and IOP Science.

side lobe intensity is about 22% of the peak intensity, and the peak position of the effective PSF of PPM-PSB also shifts 36 nm in the x axis at *x* =of 350 nm. The cases of Ly and Cr polarized pump and probe are shown in Figures 2.2.4.4 and 2.2.4.5, and the detailed information is summarized in Table 2.2.4.2. From Fig. 2.2.4.4(a) and Table 2.2.4.2, we can find that the minimum size of the effective PSF in the X

TABLE 2.2.4.2
Lateral Properties of PPM-PSB with the Offset of Pump and Probe in X Axis

Polar.	Δx/nm	Pump-X/nm	NPPM-X/nm	PPM-PSB-X/nm	I_{peak}	$I_{side\ lobe}$
Lx	350	236	196	128	9.3%	22.2%
Ly	237	172	140	100	12.1%	24.9%
Cr	290	204	164	140	10.2%	19%

axis can reach 100 nm, with Ly polarized pump and probe, and the resolving power improvement is ~28.5% and 41.9% compared with NPPM and conventional optical microscopy with the excitation of the pump beam, respectively. Typically, the peak intensity is ~10 % of the maximum value with side lobes of ~20 % of the main peak, as shown in (c) and (d) in Fig. 2.2.4.3,–2.2.4.5, and Table 2.2.4.2. The cases of Ra and Az polarized pump and probe are not given here, for nearly no resolving power improvement can be observed for PPM-PSB with the two types of polarizations even with the optimized Δ *x*, as shown in Fig. 2.2.4.1.

Figure 2.2.4.6–2.2.4.9 demonstrate the characteristics of the longitudinal components of the effective PSFs for PPM-PSB with Lx, Cr, Ra, and Az polarized pump and probe, respectively. Here I would like to repeat the meanings of the terms as follows: Lx, Cr, Ra, and Az polarizations are linear, circular, radial, and azimuthal, respectively. A summarization of the calculated data is shown in Table 2.2.4.3. From Fig. 2.2.4.6–2.2.4.9 and Table 2.2.4.2, we can find that the typical size of the effective PSF in Z axis is ~300 nm for PPM-PSB, with the resolving power improvement of 20–30% and 30–40% compared with NPPM and conventional optical microscopy

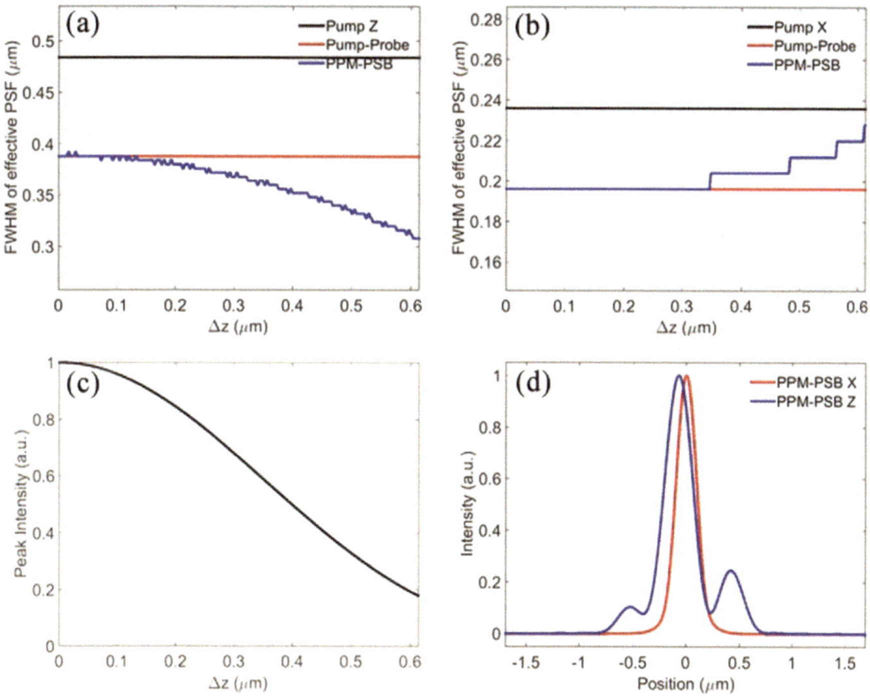

FIGURE 2.2.4.6 With the polarization of Lx, dependence of the FWHM of effective PSFs in X axis (a) and Z axis (b) on Δ *z*. (c) The peak intensity of the effective PSF of PPM-PSB decreases with the increase of Δ *z*. (d) The Intensity distributions of PPM-PSB along the X and Y axis with Δ *z* of 610 nm. The figure was reported in the *Journal of Optics* and copyright is partially held between the representative author (Takayoshi Kobayashi, TK) and IOP Science.

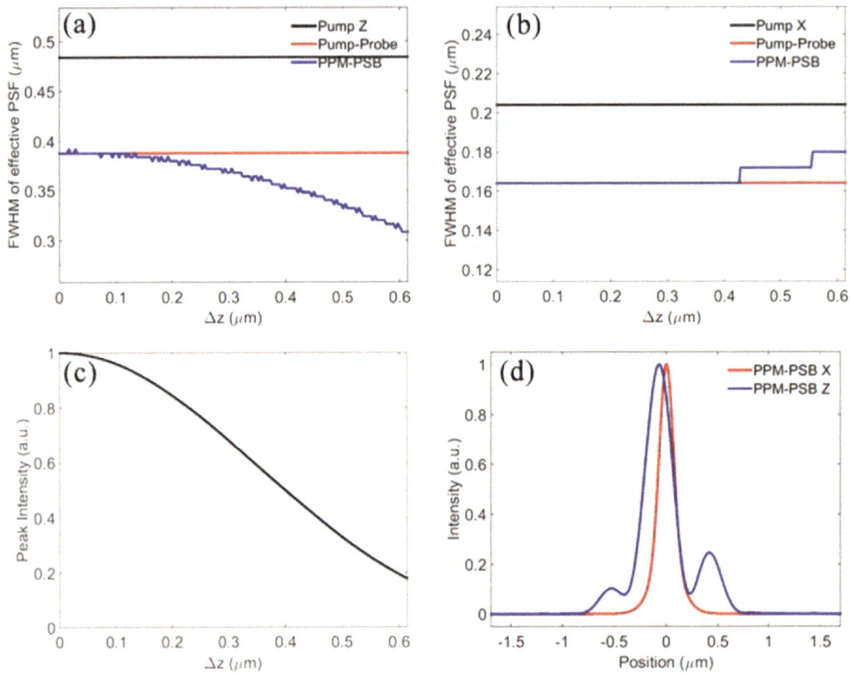

FIGURE 2.2.4.7 With the polarization of Cr, dependence of the FWHM of effective PSFs in X axis (a) and Z axis (b) on Δ *z*. (c) The peak intensity of the effective PSF of PPM-PSB decreases with the increase of Δ *z*. (d) The Intensity distributions of PPM-PSB along the X and Y axis with Δ *z* of 615 nm. The figure was reported in the *Journal of Optics* and copyright is partially held between the representative author (Takayoshi Kobayashi, TK) and IOP Science.

with excitation of the pump beam, respectively. The peak intensity reduction is typically ~85% with the side lobes smaller than 25% of the main peak. The PSF sizes in X direction will get worse with the increased Δ *z* for Lx and Cr polarized beams, as shown in Figs. 2.2.4.6(b) and 2.2.4.7(b). The minimum size of the effective PSF in the Z axis can reach 288 nm, with Az polarized pump and probe, and the resolving power improvement is ~24.2% compared with NPPM.

One of the main drawbacks of the superresolution imaging technology reported so far is the peak intensity reduction with the increased offset of pump and probe. One of the solutions for this problem is just increasing the pump and probe powers [16]. The resolving power improvement is also limited to ~30 % compared with NPPM, with pump and probe of ideal plane waves, and it is similar to the nonlinear PTM [11, 19]. However, even higher resolving power is achievable if both the PTM-PSB and nonlinear PTM are applied in the imaging process, and a multiplex imaging techonique, nonlinear PTM with partly staggered beams (NLPTM-PSB) is

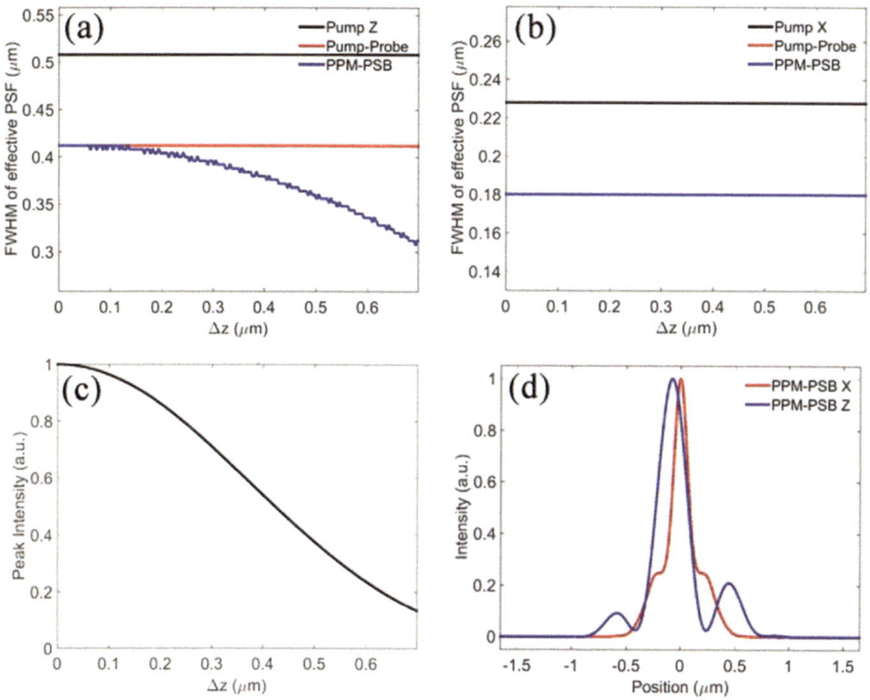

FIGURE 2.2.4.8 With the polarization of Ra, dependence of the FWHM of effective PSFs in X axis (a) and Z axis (b) on Δ *z*. (c) The peak intensity of the effective PSF of PPM-PSB decreases with the increase of Δ *z*. (d) The Intensity distributions of PPM-PSB along the X and Y axis with Δ *z* of 700 nm. The figure was reported in the *Journal of Optics* and copyright is partially held between the representative author (Takayoshi Kobayashi, TK) and IOP Science.

expected. The resolving power improvement can also be obtained simultaneously in lateral and longitude, with the offset of pump and probe in both X and Z axes. As a result, high-resolution, three-dimensional imaging is achievable, although the peak intensity will become even lower.

In this subsection, we have performed the simulation of OM-PSB with pump and probe of ideal plane wave. Other beam shapes may introduce even higher performance for OM-PSB. On the other hand, it is also not easy to obtain an ideal plane wave experimentally. SG beams or quasi-SG beams may be good candidates because of their easy generation and relatively good performance [16]. The detailed discussions about OM-PSB with pump and probe beams with SG distribution, such as the influence of the order of SG function, will be given elsewhere. Further work about the influence of beam shapes is still necessary.

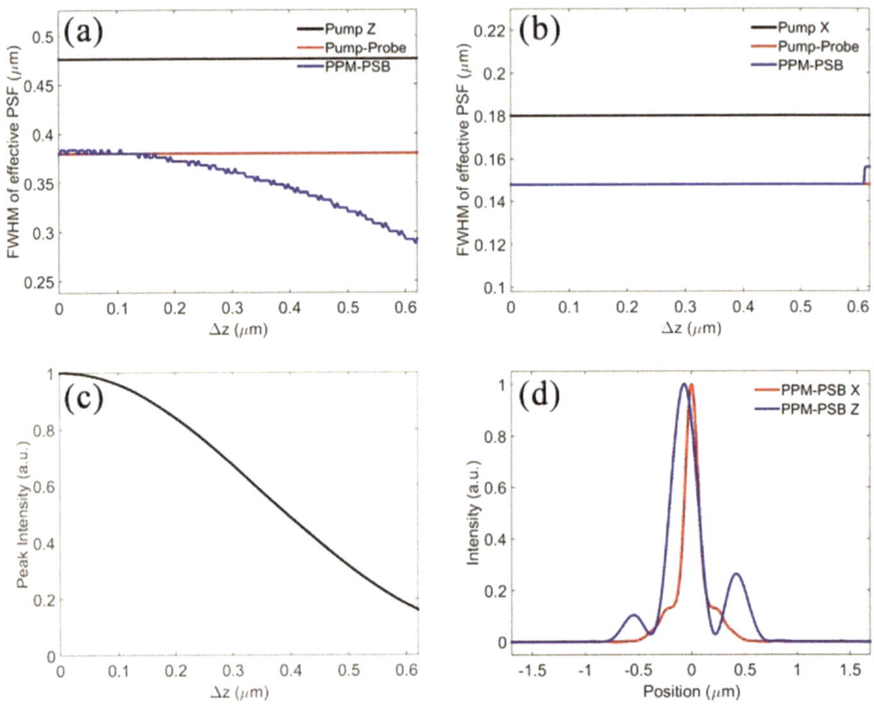

FIGURE 2.2.4.9 With the polarization of Az (Azimuthal polarization), dependence of the FWHM of effective PSFs in the X axis (a) and Z axis (b) on Δ *z*. (c) The peak intensity of the effective PSF of PPM-PSB decreases with the increase of Δ *z*. (d) The Intensity distributions of PPM-PSB along the X and Y axis with Δ *z* of 620 nm. The figure was reported in the *Journal of Optics* and copyright is partially held between the representative author (Takayoshi Kobayashi, TK) and IOP Science.

TABLE 2.2.4.3
Longitudinal Properties of PPM-PSB with the Offset of Pump and Probe in Z axis

Polar.	Δ*z*/nm	Pump-Z/nm	NPPM-Z/nm	PPM-PSB-Z/nm	I_{peak}	$I_{side\ lobe}$
Lx	610	484	388	308	18.3%	23.9%
Cr	615	484	388	308	17.9%	24.6%
Ra	700	508	412	308	13.5%	20.9%
Az	620	476	380	288	15.9%	26%

4 CONCLUSION

In the present subsection, a numerical study of a new superresolution imaging technique, OM-PSB, is performed with vector diffraction theory. Sub-100nm resolution in lateral direction and sub-300nm in longitudinal direction is achievable by this method, with the pump and probe wavelengths of 488 and 632 nm, respectively. The lateral resolving power improvement is typically ~30% and 40%, compared with NPPM and traditional optical microscopy, respectively. The longitudinal resolving power improvement is typically ~20% and 30%, compared with NPPM and traditional optical microscopy, respectively. The influence of polarization is discussed. The highest resolving power in lateral direction is achieved with linearly (Ly) polarized pump and probe, while that in longitudinal direction is obtained with azimuthal (Az) polarized beams. Sub-100nm resolution in lateral and sub-300 nm in longitudinal is achievable.

The research described in this subsection is performed collaborative work by J. He, N. Wang, and T. Kobayashi [20].

The above study was reported in the *Journal of Optics* and copyright is partially held between the corresponding author (Takayoshi Kobayashi, TK) and IOP Science. TK would like to express thanks to IOP Science for permitting us to publish the paper properly revised to be a section in the book *Photothermal and Raman Microscopy* (Taylor & Francis),

REFERENCES

1. S. W. Hell, and J. Wichmann, "Breaking the diffraction resolution limit by stimulated emission: Stimulated-emission-depletion fluorescence microscopy," *Opt. Lett.* **19**(11), 780–782 (1997).
2. K. I. Willig, B. Harke, R. Medda, and S. W. Hell, "STED microscopy with continuous wave beams," *Nat. Methods* **4**(11), 915–918 (2007).
3. M. J. Rust, M. Bates, and X. Zhuang, "Sub-diffraction-limit imaging by stochastic optical reconstruction microscopy (STORM)," *Nat. Methods* **3**(10), 793–795 (2006).
4. E. Betzig, G. H. Patterson, R. Sougrat, O. W. Lindwasser, S. Olenych, J. S. Bonifacino, M. W. Davidson, J. L. Schwartz, and H. F. Hess, "Imaging intracellular fluorescent proteins at nanometer resolution," *Science* **313**(5793), 1642–1645 (2006).
5. W. Lu, and W. Min, "Pump-probe optical microscopy for imaging nonfluorescent chromophores," *Anal. Bioanal. Chem.* **403**(8), 2197–2202 (2012).
6. W. Min, S. Lu, S. Chong, R. Roy, G. R. Holtom, and X. S. Xie, "Imaging chromophores with undetectable fluorescence by stimulated emission microscopy," *Nature* **461**(7267), 1105–1109 (2009).
7. S. Chong, W. Min, and X. S. Xie, "Ground-state depletion microscopy: Detection sensitivity of single-molecule optical absorption at room temperature," *J. Phys. Chem. Lett.* **1**(23), 3316–3322 (2010).
8. D. Fu, T. Ye, T. E. Matthews, and G. Yurtsever, "Two-color, two-photon, and excited-state absorption microscopy," *J. Biomed. Opt.* **12**(5), 054004 (2007).
9. M. Tokeshi, M. Uchida, A. Hibara, T. Sawada, and T. Kitamori, "Determination of subyoctomole amounts of nonfluorescent molecules using a thermal lens microscope: Subsingle-molecule determination," *Anal. Chem.* **73**(9), 2112–2116 (2001).

10. D. Boyer, P. Tamarat, A. Maaili, B. Lounis, and M. Orrit, "Photothermal imaging of nanometer-sized metal particles among scatterers," *Science* **297**(5584), 1160–1163 (2002).
11. J. He, J. Miyazaki, N. Wang, H. Tsurui, and T. Kobayashi, "Label-free imaging of melanoma with nonlinear photothermal microscopy," *Opt. Lett.* **40**(7), 1141–1143 (2015).
12. M. Han, G. Giese, and J. F. Bille, "Second harmonic generation imaging of collagen fibrils in cornea and sclera," *Opt. Express* **13**(15), 5791–5797 (2005).
13. P. Mahou, N. Olivier, G. Labroille, L. Duloquin, J. M. Sintes, N. Peyrieras, R. Legouis, D. Debarre, and E. Beaurepaire, "Combined third-harmonic generation and four-wave mixing microscopy of tissues and embryos," *Opt. Express* **2**(10), 2837–2849 (2011).
14. S. Kumar, T. Kamali, J. M. Levitte, O. Katz, B. Hermann, R. Werkmeister, B. Povazay, W. Drexler, A. Unterhuber, and Y. Silberberg, "Single-pulse CARS based multimodal nonlinear optical microscope for bioimaging," *Opt. Express* 23(10), 13082–13098 (2015).
15. Y. K. Tao, D. Shen, Y. Sheikine, O. O. Ahsen, H. H. Wang, D. B. Schmolze, N. B. Johnson, J. S. Brooker, A. E. Cable, J. L. Connolly, and J. G. Fujimoto, "Assessment of breast pathologies using nonlinear microscopy," *Proc. Natl. Acad. Sci. USA* **111**(42), 15304–15309 (2014).
16. J. He, N. Wang, and T. Kobayashi, ". Numerical study of super-resolved optical microscopy with partly staggered beams, J. Opt. **18**(18), 125303 (2016).
17. B. Richards, and E. Wolf, "Electromagnetic diffraction in optical systems. II. Structure of the image field in an aplanatic system," *Proc. Roy. Soc. Lond. Ser.* A **253**(1274), 358 (1959).
18. J. Miyazaki, H. Tsurui, A. H. Takagi, H. Kasai, and T. Kobayashi, "Sub-diffraction resolution pump-probe microscopy with shot-noise limited sensitivity using laser diodes," *Opt. Express* **22**(8), 9024–9032 (2014).
19. J. He, J. Miyazaki, N. Wang, H. Tsurui, and T. Kobayashi, "Biological imaging with nonlinear photothermal microscopy using a compact supercontinuum fiber laser source," *Opt. Express* **23**(8), 9762–9771 (2015).
20. J. He, N. Wang, and T. Kobayashi, "Numerical study of super-resolved optical microscopy with partly staggered beams," *J. Opt.* **18**, 125303 (2016).

2.2.5 Resolution Enhancement of Pump-probe Microscopy with an Inverse-annular Spatial Filter

I INTRODUCTION

Pump-probe microscopy can provide images by detecting changes in probe light intensity induced by stimulated emission, photoinduced absorbance change (PA) or photothermal (PT) effect in either transmission or reflection mode. This technique has various applications. They include the identification of pigments in historical artworks, measurements of carrier dynamics in single quantum well, lifetime measurements of fluorescent samples, and non-invasive characterization of melanocytes [1–4]. A single quantum well is composed of a trench structure where band gap is narrower than the 2D structured semiconductor. Especially in bio-imaging, pump-probe microscopy can be applied in a wide range of biological tissues regardless of whether the sample is fluorescent or not, with sub-diffraction resolution. Xie et al. developed pump-probe microscopy using stimulated emission and imaged distributions of chromoproteins in live Escherichia coli cells [5]. The pump and probe beams were supplied by high intensity, high repetition rate femtosecond lasers, and a lock-in detection scheme [5] was used to reduce probe noise [6]. Photothermal imaging using pump-probe technique was also applied to biological tissues doped with metallic nanoparticles, which induces the deflection of probe beam by refractive index change due to photothermal effect [7–9].

In our previous works, we demonstrated pump-probe microscopy without using Ti:sapphire laser-based femtosecond pulse lasers, optical parametric amplifiers, or high speed detectors [10–15]. Our pump-probe system is equipped with two intensity-modulated laser diodes (LDs), a lock-in amplifier, and an auto-balanced detector to remove the probe noise. It could achieve high sensitivity detection at a low cost. Especially the laser diodes are nearly 50 times less expensive than ultrashort pulse lasers. In addition, another advantage of using laser diodes is the capability of tuning

DOI: 10.1201/9781003476030-9

of the LD wavelength and modulation frequency. We can easily change the laser wavelength by changing the LD element, and can select the modulation frequency (~MHz) by modulating the laser diode (LD) current. Recently, we implemented an annular pupil filter into a microscope before the objective lens to improve the spatial resolution [12]. A donut-shaped beam created by this filter is focused through an objective lens and forms a sharp main peak with unwanted side lobes in the focal plane [16, 17]. Side lobes are the satellite ring peaks appearing in the diffraction pattern by a circular aperture. Its structure is given by the first-order Bessel function $J_1(r)$ which has a dumped oscillatory feature along the radial direction (r). The spatial resolution of a pump-probe microscope can be approximated with the product of intensity profiles of pump and probe beams because the pump-probe signal is based on the bilinear interaction between two beams mediated by the sample [10]. Because of the effective nonlinearity induced by the bi-linearity, the pump-probe microscope has spatial resolution better than conventional optical microscopes. This mechanism of bi-linearity of pump and probe intensity is especially powerful to the biological sample which does not have strong absorption in the visible range being used for probing. In the case of semiconductors, reflection mode can be utilized which is expected to detect the change in signal intensity of the Mie scattering in the backward direction. An annular pupil filter performed further improvement of the lateral resolution over the diffraction limit [12]. Although relatively strong side lobes appear near the main peak due to diffraction, they can be suppressed because of difference in the locations of the side lobe peaks due to the different wavelength of pump and probe beams [12]. The diffraction fringe intensities are decreased by the cancelation of them.

The annular pupil filter is an effective method to increase the lateral resolution and very easy to introduce into the microscope; however, the axial resolution becomes more than twice lower than that without pupil filter [12, 18]. Therefore, it is not suitable for the construction of three-dimensional (3D-) images of thick biological samples. In this chapter, to increase not only the lateral but also the axial resolution, we introduced a new pupil filter, an inverse-annular pupil filter. It has been claimed theoretically that an inverse annular filter achieved a significant enhancement of the resolution of the fluorescence microscopy region both in the lateral and axial directions [19], but this claim has not been proven experimentally. Furthermore, the theoretical enhancement in the chapter has been discussed only for the conventional fluorescence microscopy, not for pump-probe microscopy including PT imaging. It is expected that with the combination of the nonlinearity in the PT imaging and removal of the interference taking place in the structured illumination, further resolution enhancement is expected. Furthermore, an even clearer image can be expected by the aforementioned side lobe cancelation in both lateral and axial directions.

In this study, further improvement of the image quality can be attained by cancelation of the fringe patterns appearing in the lateral and axial intensity distributions of both pump and probe beams. We also theoretically calculate the expected improvements in resolution. To increase not only the lateral but also the axial resolution, we introduced an inverse-annular pupil filter rather than an annular pupil filter. The resolution is evaluated experimentally using gold nanoparticles (GNP) with a 20-nm diameter. The resolution powers evaluated in the X (perpendicular to the common pump and probe polarization direction), Y (parallel to the polarization direction), and Z (axial) directions are found to be improved by 15 ± 6, 8 ± 8, and 21 ± 2% relative to those without a pupil filter. The standard deviation is obtained by the properly estimated background signal inevitable in the imaging. The experimental results, which are even better than our theoretical predictions, are discussed. We also demonstrate PT imaging of thick biological samples (cells from rabbit intestine and kidney) stained with hematoxylin and eosin dye with enhanced resolution.

2 INVERSE AND ANNULAR FILTER

First, we theoretically estimate the spatial resolution of the pump-probe signal using a vector diffraction model with three illumination modes produced using three different sets of a pupil and filters, including inverse-annular illumination. To investigate the differences in spatial resolution among the three modes, we used the following parameters for the three filters, as illustrated in Fig. 2.2.5.1(a): pupil diameterΦ_1, internal diameter of the annular filter, Φ_2, and pupil diameter of the objective lens Φ_3. The values selected for these parameters are listed in Table 2.2.5.1. The radial coordinate is represented in terms of numerical aperture (NA).

Inverse-annular illumination produces both circular and annular beams (blue area in Fig. 2.2.5.1 (a)). These beam areas determine the 3D spatial resolution in microscopy based on point-scanning systems [19, 20]. While Φ_3 depends on the focal length and NA of the objective lens, Φ_1 and Φ_2 can be set optimized by changing the beam-blocking area (gray area). In this study, we set $\Phi_1 = 3.1$ mm, $\Phi_2 = 6.3$ mm and used the objective lens parameter $\Phi_3 = 8.55$ mm. The vector diffraction

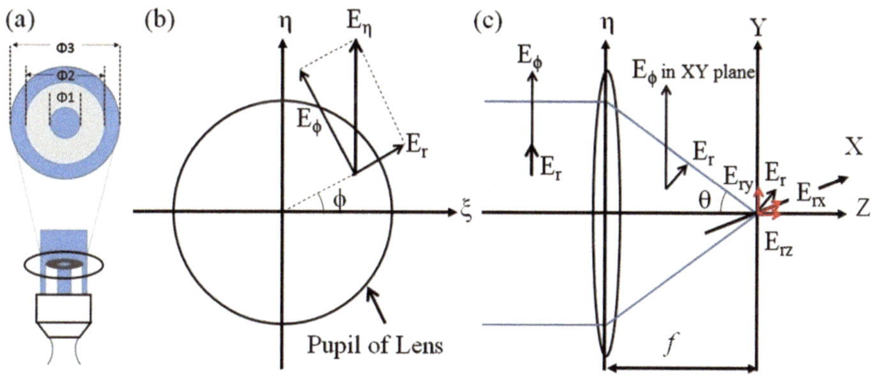

FIGURE 2.2.5.1 (a) Structure of inverse-annular filter before the objective pupil. The structures of circular, annular, and inverse-annular filters are described with different parameter sets (of Φ_1, Φ_2, Φ_3). This filter can be transmitted through the center and annular parts of the beams (blue-area). Diameters of the central and annular parts are expressed by Φ_1, and Φ_2, respectively. Φ_3, is the pupil diameter, which is determined by the objective lens. (b), (c) Electric field components in a lateral plane (b) and around a focal point (c). In this figure, a linear polarized laser is illuminated with polarization direction of the η-axis.

TABLE 2.2.5.1
Aperture Parameter Sets of the Filters for the Three Illumination modes

Illumination mode	Φ_1 [mm]	Φ_2 [mm]	Φ_3 [mm]
Circular	0	0	8.55
Annular	0	5.8	8.55
Inverse annular	3.1	6.3	8.55

theory used can determine the resolution with high accuracy in comparison with the scalar diffraction theory because the former includes polarization dependence in each direction. Both pump and probe fields in our microscope are linearly polarized parallel to Y-direction, so the intensity distribution on the focal plane is different between the parallel and perpendicular directions to the polarization [21, 22].

Fig. 2.2.5.1 (b) shows the electric field components in a lateral plane (XY). The illuminating laser is linearly polarized along the η-axis (or Y-axis), and the electric fields in the radial and declination directions in the microscopic region can be written as:

$$E_r = E_\eta \, sin\phi \tag{1}$$

$$E_\phi = E_\eta \, cos\phi, \tag{2}$$

where E_η is the electric field component of the laser along the η-axis. E_r and E_ϕ can be resolved into X, Y, and Z components, as illustrated in Fig. 2.2.5.1(c) and given as:

$$E_{rx} = E_r cos\theta \; cos\phi \tag{3}$$

$$E_{ry} = E_r cos\theta \; sin\phi \tag{4}$$

$$E_{rz} = E_r sin\theta \tag{5}$$

and

$$E\phi_{\,x} = -E\phi \; sin\phi \tag{6}$$

$$E\phi_{\,y} = E\phi \; cos\phi. \tag{7}$$

The components of the electric field amplitude in the X, Y, and Z directions P_x, P_y, and P_z in the microscopic region are obtained as follows after substituting Eqs. (1) and (2) into Eqs. (3)–(7) and

$$P_x = E_{rx} + E_{\phi\,x} = E_\eta \, (cos\theta \; sin\phi \; cos\phi - sin\phi \; cos\phi) \tag{8}$$

$$P_y = E_{ry} + E\phi_{\,y} = E_\eta \; cos\theta \; sin^2\phi - cos^2\phi \tag{9}$$

$$P_z = E_{rz} = E_\eta \; sin\theta \; sin\phi. \tag{10}$$

When an incident plane wave is converted into a convergent spherical wave, correction corresponding to the amplitude variation of the electric field $(\cos\theta)^{0.5}$ is required [23]. Electric field amplitudes of the X, Y, and Z-axis directions E_x (x, y, z), E_y (x, y, z), and E_z (x, y, z) including the correction term are determined by the Fourier transformation of Eqs. (8), (9), and (10) in a proper integration range. The electric field amplitude distributions are then described as follows

$$E_x(x,y,z) = C, \quad P_x(\theta,\phi)\sqrt{\cos\theta}\exp\{-ik(\xi x+\eta y+\zeta z)\}\,dS \tag{11}$$

$$E_x(x,y,z) = C, \quad P_y(\theta,\phi)\sqrt{\cos\theta}\exp\{-ik(\xi x+\eta y+\zeta z)\}\,dS \tag{12}$$

$$E_x(x,y,z) = C, \quad P_z(\theta,\phi)\sqrt{\cos\theta}\exp\{-ik(\xi x+\eta y+\zeta z)\}\,dS \tag{13}$$

Here, k is the wave number and is equal to $2\pi\, n/\lambda$, θ is the polar coordinate, and dS is the microscopic region on the convergence spherical plane given by $dS = \frac{d\xi d\eta}{\cos\theta} = \sin\theta d\theta d\phi$. The Cartesian coordinates representing the pupil plane, ξ and η, are given by

$$\xi = \sqrt{\xi^2+\eta^2}\cos\phi = \sin\theta\cos\phi \tag{14}$$

$$\xi = \sqrt{\xi^2+\eta^2}\cos\phi = \sin\theta\cos\phi \tag{15}$$

Substituting Eqs. (14) and (15) into Eqs. (11)–(13), the field amplitudes are obtained as follows

$$E_x(x,y,z) = C, \quad P_x\sqrt{\cos\theta}\exp\{-ik(x\sin\theta\cos\varphi + y\sin\theta\sin\varphi + z\cos\theta)\}\sin\theta d\theta d\varphi \tag{16}$$

$$E_x(x,y,z) = C, \quad P_y\sqrt{\cos\theta}\exp\{-ik(x\sin\theta\cos\varphi + y\sin\theta\sin\varphi + z\cos\theta)\}\sin\theta d\theta d\varphi \tag{17}$$

$$E_x(x,y,z) = C, \quad P_z\sqrt{\cos\theta}\exp\{-ik(x\sin\theta\cos\varphi + y\sin\theta\sin\varphi + z\cos\theta)\}\sin\theta d\theta d\varphi \tag{18}$$

Converting the integration variable θ to $r = n\sin\theta$ in the unit of numerical aperture (NA), Eqs. (16), (17) and (18) can be rewritten as follows:

$$E_x(x,y,z) = C, \quad A_x\exp\left\{-ik\left(rx\cos\varphi + ry\sin\varphi + z\left(1-(r/n)^2\right)^{\frac{1}{2}}\right)\right\}drd\varphi \tag{19}$$

$$E_x\left(x,y,z\right)=C, \quad A_y\exp\left\{-ik\left(rx\cos\varphi+ry\sin\varphi+z\left(1-(r/n)^2\right)^{\frac{1}{2}}\right)\right\}drd\varphi \tag{20}$$

$$E_x\left(x,y,z\right)=C, \quad A_z\exp\left\{-ik\left(rx\cos\varphi+ry\sin\varphi+z\left(1-(r/n)^2\right)^{\frac{1}{2}}\right)\right\}drd\varphi \tag{21}$$

With

$$A_x=\frac{r}{n}\left[\left\{1-\left(\frac{r}{n}\right)^2\right\}^{\frac{1}{4}}-\left\{1/\left(1-\left(\frac{r}{n}\right)^2\right)\right\}^{\frac{1}{4}}\sin\varphi\cos\varphi\right] \tag{22}$$

$$A_y=\frac{r}{n}\left[\left\{1/\left(1-\left(\frac{r}{n}\right)^2\right)\right\}^{\frac{1}{4}}\sin^2\varphi+\left\{1-\left(\frac{r}{n}\right)^2\right\}^{\frac{1}{4}}\cos^2\varphi\right] \tag{23}$$

$$A_z=\frac{r^3}{n}\left\{1/\left(1-\left(\frac{r}{n}\right)^2\right)^{\frac{1}{4}}\right\}\sin\varphi. \tag{24}$$

The light intensity distribution in the image volume, $I = |E_x(x, y, z)|^2 + |E_y(x, y, z)|^2 + |E_z(x, y, z)|^2$, is calculated by integration over the open aperture area represented by pitch coordinates of the NA, r, and declination angle Φ . The integration ranges in the radial direction are set as $0 < r < 0.95$ for no filter (circular illumination), $0.68 < r < 0.95$ for the annular filter and $0 < r < 0.34$ and $0.70 < r < 0.95$ for the inverse annular filter in the unit of numerical aperture NA. The integration range in the declination angle is $0 < \Phi < 2\pi$ for all calculation.

Calculated results of the resolution improvement in the pump-probe microscopy with circular illumination are shown in Figure 2.2.5.2, since signal intensity distribution of the pump-probe signal, $I_{sig}(x, y, z)$ is roughly speaking given by $I_{sig}(x, y, z) = I_{pu}(x, y, z) \times I_{pr}(x, y, z)$ [9]. The peripheral ring structures introduced by diffraction in both the pump and probe beams can be canceled out because the ring radii of the pump and probe are different, as the two beams are of different wavelength. This can be confirmed in Fig. 2.2.5.2, which shows resolution improvement of the pump-probe measurement by the similar calculation by the PSF. The improvements of the resolution in XY and XZ planes are 37–39%, respectively, as shown in Table 2.2.5.2 based on Fig. 2.2.5.2. Thus, in the case of PT imaging, inverse-annular illumination is even more advantageous.

The calculated results of intensity distribution are depicted in Figure 2.2.5.3, which shows distributions in the XY (a1, b1, c1) and XZ (a2, b2, c2) planes of the

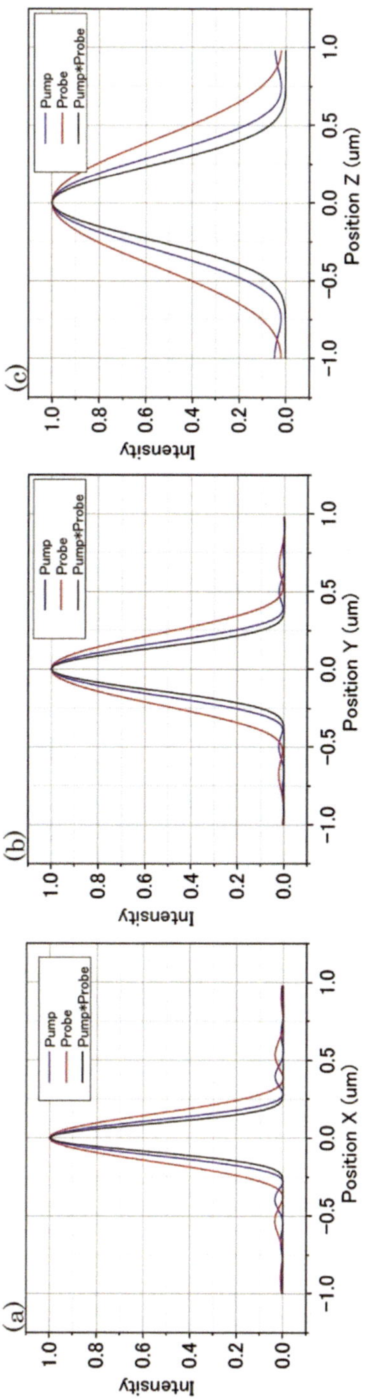

FIGURE 2.2.5.2 Calculated results of the intensity distribution to show the resolution improvement in pump-probe microscopy in X-axis (a), Y-axis (b) and Z-axis (c). The resolution of the intensity for the pump-probe microscopy (black lines) is better than the resolution for the pump (blue lines) and the probe (red lines) laser. In the calculation, the wavelengths of the pump and probe laser are 488 nm and 660 nm, respectively.

TABLE 2.2.5.2
FWHMs of Lateral lOntensity Distributions by the Pump Beam, the Probe Beam, and the Pump-probe Signal in Fig. 2.2.5.2; Improvement of Pump-Probe (PP) Results with Two Colors from the Cases of Single Color experiment[a]

Image direction	Pump (Blue) [nm]	Probe (Red) [nm]	Pump-Probe (PP) (Black) [nm]	Improvement PP /Pump	Improvement PP /Probe
X	234	315	197	16 %	37 %
Y	337	456	285	15 %	37 %
Z	635	863	527	17 %	39 %

[a]The laser wavelengths of pump and probe are 488 nm and 660 nm, respectively. The refractive index used in the calculation is set to be 1 (air).

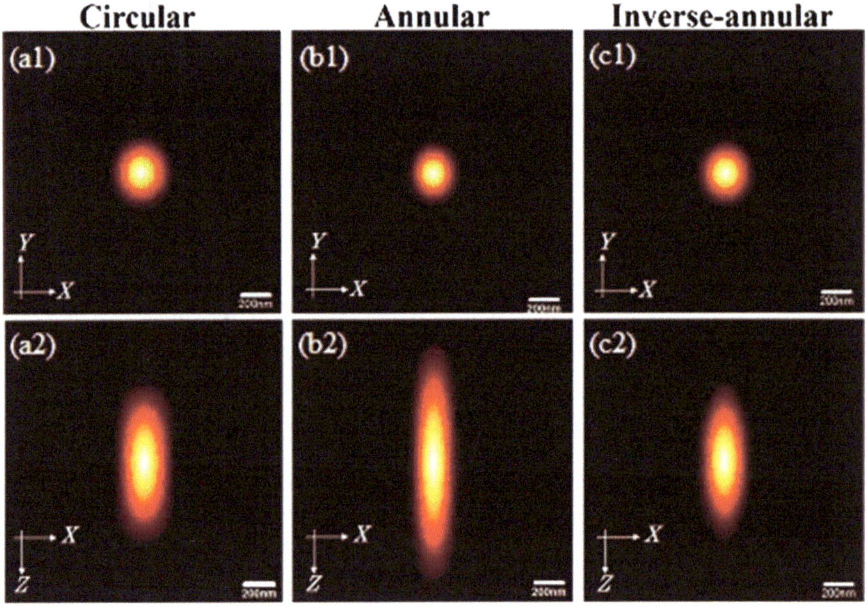

FIGURE 2.2.5.3 Intensity profile of images of a single GNP with 20-nm diameter near the focal point of the objective lens in the XY (a1, b1, and c1) and XZ (a2, b2, and c2) planes calculated for pump-probe experiment with different apertures. The wavelengths of pump and probe beams are 488 nm and 660 nm, respectively. Numerical aperture (NA) of the objective lens is 0.95.

pump-probe signal by circular illumination (a, without the filter, $\Phi_1 = \Phi_2 = 0$ mm), annular illumination (b, $\Phi_1 = 0$ mm, $\Phi_2 = 5.8$ mm), and inverse annular illumination (c). The polarization direction is along the Y-axis. Figure 2.2.5.4 shows cross-section profiles and Table 2.2.5.3 lists the calculated resolution of lateral (X- and Y-) directions and axial (Z-) direction in the case of the Y-polarization probe field in terms of the full width a half maximum (FWHM) values of the PSF. As shown in these figures and the table, the improvements of resolution with annular filter from the circular case in the lateral direction are 15% and 19%, respectively, in parallel (X-axis) and perpendicular (Y-axis) directions to the laser field (X-axis). However, the enhancement is with the expense of the lowering of the resolution in the axial direction by nearly 50% reduction indicating that the distribution is increased to 1.5 times in the spatial spread width of the image of a delta-function-like object. In contrast to the annular filter case, the improvements gained using the inverse-annular filter compared with the circular aperture case in the directions parallel and perpendicular to the laser field are 8% and 11%, respectively, and 6% improvement can be obtained even in the axial direction.

3 EXPERIMENTAL SETUP

Figure 2.2.5.5 shows a schematic illustration of our pump-probe microscopy system, which employs two laser diodes (LDs) as the pump and probe light sources. A 488-nm LD (Nichia, NDS4216) and a 660-nm LD (Mitsubishi, ML101U29) are used for the pump and probe beams, respectively. The two beams are modulated at frequencies of ω_1 (pump) and ω_2 (probe) using signal generators (SGs) and are introduced to beam collimators that contain spatial filters. After the beams are combined using a dichroic mirror, they irradiate the sample after passing through an objective lens (Olympus, UPLSAPO 40X2) with a NA of 0.95. Diameters of the pump and probe laser calculated by using $1.22 \times \lambda/NA$ are 626.7 nm and 847.6 nm, respectively. A condenser lens collects the transmitted probe beam from the sample and introduces the light to a detector. To remove the probe noise, we use an auto-balanced detector (Newfocus, Nirvana) containing two photodiodes; one is for the probe beam and the other is for a reference beam split off from the probe beam by a beam splitter (BS in Fig. 2.2.5.5) placed before a dichroic mirror[1] (DM in Fig. 2.2.5.5[10–15]).The intensity noise of these beams is canceled using this balance detection scheme. Thus, shot-noise limited[2] sensitivity can be achieved when the probe beam power is higher than ~10 mW on the photodiode surface [10]. A lock-in amplifier (Signal Recovery, 7270) detects the signal at the beat frequency (difference frequency) $|\omega_1 - \omega_2|$. In this study, we set modulation frequencies ω_1 and ω_2 to 1.000 MHz and 1.015 MHz, respectively, resulting in beat frequency of 15kHz. A frequency mixer[3] (Mini-circuit, ZAD-1) generates a $|\omega_1 - \omega_2|$ component for a reference signal of the lock-in amplifier. Closed loop piezo-actuators[4] (Thorlabs, MAX311D) scan the sample in the three axial directions. The time constant of the lock-in amplifier is 0.5 ms, and the dwell time is 1.0 ms in the gold nanoparticle experiments. The pixel number in the image of the GNP is 300×300 and hence the frame time of the image is 90 seconds and the motor driven stage step is 50 nm. The time constant of the lock-in amplifier and

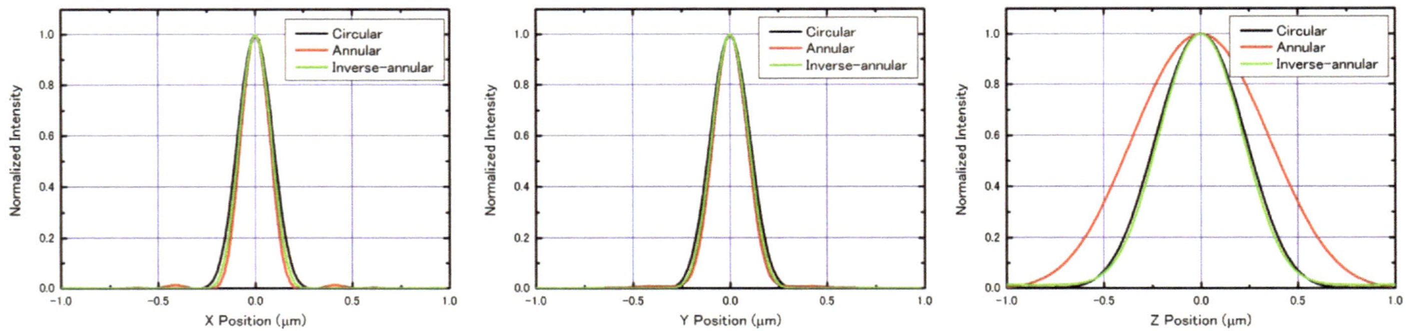

FIGURE 2.2.5.4 Cross-section profiles of the intensities of the pump-probe images shown in Fig.3 in X, Y, and Z directions. The width of the PSF in the X- and Y-axis direction is smaller, but the width in the Z-axis direction is much larger in annular illumination case (red lines) than the circular illumination case (black lines). The resolution is also improved in the inverse-annular illumination (green lines) in any direction.

TABLE 2.2.5.3
FWHMs of the Intensity Distribution of Images for Infinitely Small Particles by the Pump-Probe Calculation in Fig. 2.2.5.3.[a]

Imaging direction	Circular [nm]	Annular [nm]	Inverse-annular [nm]	Improvement (Annular filter)	Improvement (IA filter)
X-axis	208.0	167	186.1	19 %	11 %
Y-axis	232.9	198	214.9	15 %	8 %
Z-axis	527.4	793	494.4	-50 %	6 %

a The Laser wavelengths of pump and probe are 488 nm and 660 nm, respectively. The refractive index used in the calculation is set to be 1.5 (PVA film employed as a matrix material in the experiment of resolution evaluation using the gold nanoparticles).

the dwell time[5] are 0.5 ms and 1 ms, respectively, in the biological sample imaging. The pixel number is 800 × 800 and hence the frame time of the image is 640 seconds and the stage step is 18.75 nm.

To improve the 3D spatial resolution, an inverse-annular filter (IAF) with a planar donut structure is installed before the polarization beam splitter (PBS). The beam blocking area of this filter is an affixed metal washer (clip plate, facing ring) onto an optical window. A holder mounting the IAF is adjusted to produce a symmetric inverse-annular beam in the objective pupil plane.

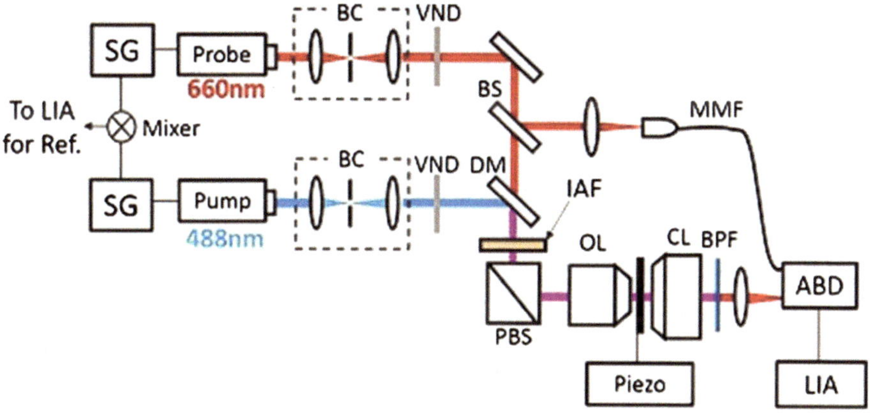

FIGURE 2.2.5.5 A scheme of pump-probe microscope system. SG: signal generator, Pump: 488-nm laser diode for pump light, Probe: 660-nm laser diode for probe light, BC: beam collimator, VND: variable neutral density filter, BS: probe beam splitter, DM: dichroic mirror, IAF: inverse-annular filter, PBS: polarized beam splitter, OL: objective lens, CL: condenser lens, BPF: band pass filter, ABD: auto-balanced detector, LIA: lock-in amplifier, MMF: multi-mode fiber.

4 RESULTS

A Evaluation of 3D Spatial Resolution by Imaging Gold Nanoparticle (GNP)

To evaluate the improvement in the 3D spatial resolution, we capture PT images of 20-nm gold nanoparticles (GNPs) dispersed in a polyvinyl alcohol (PVA) film. Although the GNPs have a large absorption cross section due to their plasmon resonance, the luminescence yield is extremely small [9]. Thus, the pump beam causes an increase in temperature around the GNPs due to the PT effect, which in turn results in a decrease in the local refractive index of the PVA at the modulation frequency ω_1. This index variation forming a thermal concave lens leads to a deflection of the probe beam and a time-dependent change in the probe beam intensity that is detected by the ABD. The beat signal at $|\omega_1-\omega_2|$ is also generated by the bilinear interaction of the pump and probe fields in the sample [12], which is detected using the lock-in scheme.

The sample is scanned in the focal plane (XY-plane, Figures 2.2.5.6 (a) and (c)) and optical axis plane (XZ-plane, Figs. 2.2.5.6(b) and 6(d)) with and without the IAF (i.e., inverse-annular and circular illumination), respectively. The pump and probe powers are respectively set to 3.8 mW and 0.3 mW for the inverse-annular illumination case and to 1.6 mW and 0.22 mW for the circular illumination case. The signal-to-noise ratios of the data of inverse-annular (black points) and circular (red points) illumination in Fig. 2.2.5.6 are 7.2 ± 2.8 and 11.5 ± 1.5 (e), 8.2 ± 3.5 and 12.8 ± 3.4 (f), 11.0 ± 6.3 and 6.5 ± 2.4 (g), respectively. The signal to background ratios of the data of inverse-annular (black dots) and circular (red dots) illumination mode in Fig. 2.2.5.6 are 12.1 ± 1.4 and 12.4 ± 1.1((e) along X), 10.8 ± 0.8 and 11.2 ± 0.9 ((f) along Y), 19.5 ± 1.8 and 11.9 ± 0.7 ((g) along Z), respectively. The reproducibility was checked by doing experiments for 5 to 10 GNPs depending on the illumination case. In each case the spatial dependence of the PT signal intensities is fitted with a Gaussian function using the least squares method. The data values in Fig. 2.2.5.6are summarized in Table 2.2.5.4. The values for experimental results ("Observed Circular" and "Observed Inverse annular" columns) in the Table 2.2.5.4 are mean values of statistical analysis of the FWHM values obtained by the Gaussian fit of the intensity distribution of several GNP images and the values after ± are the standard deviation of the statistical analysis. The spatial dependent intensities are averaged over the measured GNPs and the standard deviations of the FWHMs are calculated. The signal intensity of the inverse annular illumination case is substantially weaker than the circular illumination case; therefore, a larger number of GNPs was measured for the inverse annular illumination case. Table 2.2.5.4 clearly shows that the FWHM of the "inverse annular illumination" case is always smaller than the FWHM of the "circular" cases even when the standard deviation is taken into account. In the X- and Y-axis directions, FWHMs for the inverse-annular illumination are smaller than in the circular illumination. In addition, each FWHM value agrees well with its corresponding value calculated using vector diffraction theory, in Table 2.2.5.3. The resolution enhancements in the X- and Y-axis directions are

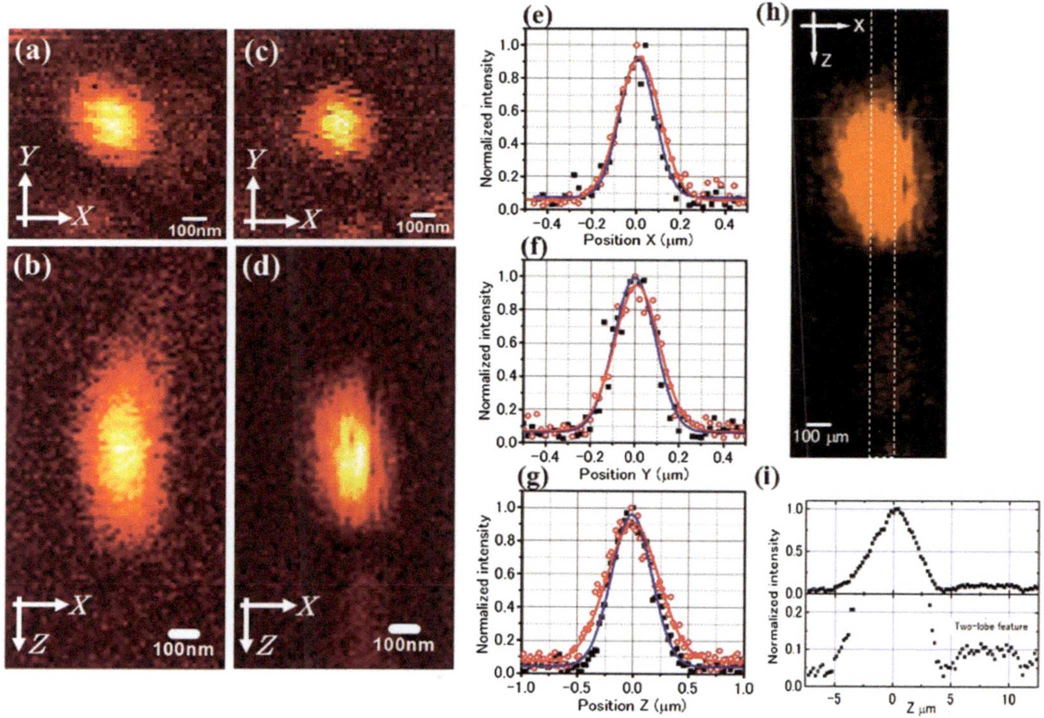

FIGURE 2.2.5.6 Photothermal images of a single GNP in PVA film and their intensity profiles. Circular illumination configuration in XY (a) and XZ-plane (b). Inverse-annular illumination configuration in XY (c) and XZ-plane (d). XZ-plane images are obtained directly by XZ scanning with piezo stage. Intensity cross-section profiles of circular (red open dots) and inverse annular illumination (black filled squares) in X, Y, and Z-axis directions are shown in (e), (f), and (g), respectively. Red and blue solid lines are Gaussian fitting curves in circular and inverse-annular illumination, respectively. Image with the two-lobe feature is shown in (h) and intensity along the axial direction integrated between two lines in (h) is shown in (i). The lower part shows the enlarged profile in the upper part.

TABLE 2.2.5.4
FWHMs of the Photothermal Intensity by the Pump-Probe Calculation and the Standard Deviations of the Gaussian Fitting to the Observed Intensity Distributions of GNP images in Fig. 2.2.5.6

Imaging direction	Calculated Circular [nm]	Observed Circular [nm]	Calculated Inversed-annular [nm]	Observed Inverse-annular [nm]	Improvement by calculation	Improvement by experiment
X-axis	208.0	219±6	186.1	187±8	11 %	15±6 %
Y-axis	232.9	239±10	214.9	219±11	8 %	8±8 %
Z-axis	527.4	550±10	494.4	430±8	6 %	21±2 %

evaluated to be 15 ± 6%, 8 ± 8%, and 21 ± 2%, respectively. Difference in resolution enhancement between the X- and Y-axis directions is due to the linear polarizations of the pump and probe beams being both along the Y-axis. In the Z-axis direction, the resolution obtained using inverse-annular illumination is also better than the conventional circular illumination. Surprisingly, the enhancements in resolution obtained experimentally are even better than those predicted theoretically, which is discussed in the following.

Selmke and coworkers [24] discussed the dependence of the image resolution on the relative position of the pump and probe focal points. They found that the apparent lateral resolution depends sensitively on the relative focal positions of the two beams. In some cases, the image along the axial direction had two lobes with a π-phase jump, resulting in an apparent reduction in the axial direction caused by a cancelation between the opposite signs of the two lobes due to the π-phase jump. Even though this effect may slightly distort the image, the location of the image of the signal-generating particle does not change much and therefore can be used in actual imaging when a precise axial intensity distribution of the target image is not required. By removing one of the lobes by shifting the relative position of the pump or probe focal plane, the two-lobe feature is detected as shown in Figs. 2.2.5.6 (h) and (i). This is because the relative contributions of the positive lobe (phase 0) and negative lobe (phase π) are changed by the probe position with respect to the pump peak position. Because of this effect, we observe minor changes smaller than 10% in the size of the image structure when we move the focal position in the axial direction as expected and observed previously in reference [24] and in our experiment [11.12].

B Biological Imaging

We also performed PT imaging of biological systems using a rabbit small intestine and kidney as samples. The resolution was evaluated for tissues stained with hematoxylin and eosin (HE). The thickness of the small intestine sample is about 10 μm. HE staining has been routinely used in path morphology and histopathology, e.g., in the diagnosis of tumor progression [25]. Since HE can be used to stain anionic and cationic structures (e.g., nuclei containing DNA and basic protein) [26], these structures can be distinguished using a bright-field microscope. While eosin has a fluorescence peak around 540 nm in water [27], hematoxylin is an essentially nonfluorescent pigment [25, 26], and the energy from the excitation laser is instead converted into heat, resulting in the photothermal effect. However we utilize both H and E as the source materials of the PT signal. Even in the case of eosin, it is much more advantageous to utilize the PT effect for imaging because of much higher robustness than fluorescence. Our pump-probe microscope can be used to capture PT images in both regions of a sample stained with eosin and hematoxylin at the same time.

Figures 2.2.5.7 (a) and (b) show the photothermal images of a small intestine taken for a rabbit intestine captured using the circular (a) and the inverse-annular (b) illumination in the focal (XY) plane. The pump and probe powers are 150 μW and 420 μW for (a), and 140 μW and 120 μW for (b), respectively. The signals originating from the elliptical structures in (a) and (b) are particularly well detected. It can

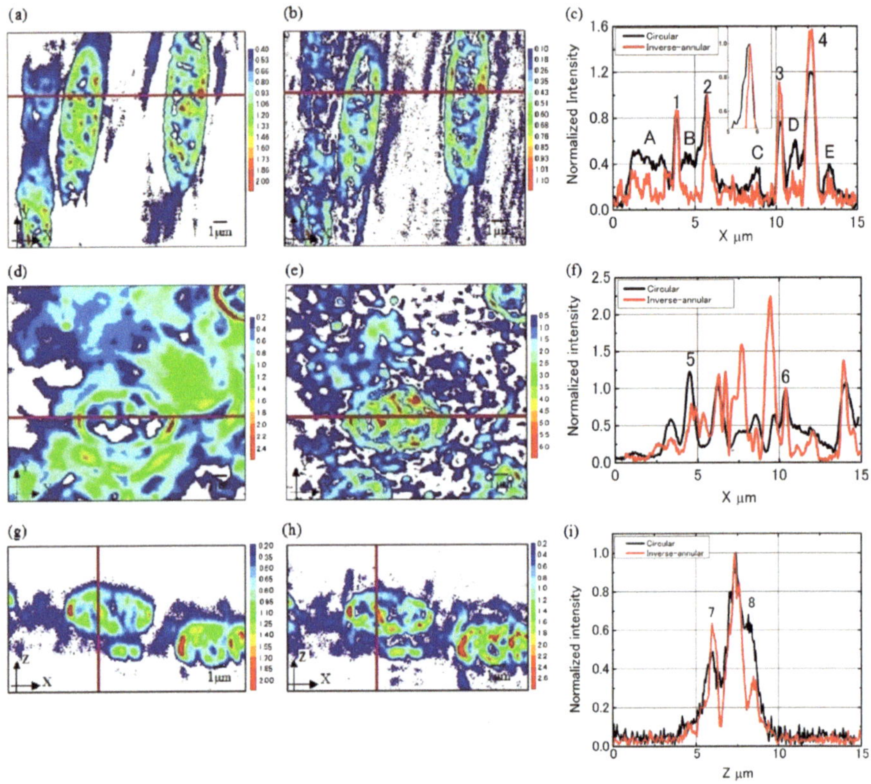

FIGURE 2.2.5.7 Images of several biological samples obtained without inverse annular filter (namely with circular illumination) and with inverse annular filter, and comparison in the lateral and axial resolution between the two illuminations. (a) and (b) Photothermal images of cells from rabbit small intestine for the circular and inverse-annular illumination in the focal (XY) plane. The pump and probe power are 140 μW and 120 μW for (a), 150 μW and 420 μW for (b), respectively. (c) The intensity profiles along the red lines in (a) and (b). The red and black solid lines are intensity profiles with the inverse-annular and circular illumination, respectively. Inset shows enlarged view of the peak No. 2. (d) and (e) Photothermal images of cells from rabbit kidney for the circular and inverse-annular illumination in the focal (XY) plane. The pump and probe power are 150 μW and 200 μW for (d), 700 μW and 650 μW for (e), respectively. (f) The intensity profiles along the black lines in (d) and (e). (g) and (h) Photothermal images of cells from rabbit kidney for the circular and inverse-annular illumination in optical axial (XZ) plane. The pump and probe power are 100 μW and 200 μW for (g), 780 μW and 260 μW for (h), respectively. (i) The intensity profiles along the red lines in (g) and (h). All scale bars are 1 μm. Most intense peak in (i) is not used for the resolution discussion because it suffers from the heavily neighboring peaks.

be concluded that these regions are cell nuclei, which are primarily stained with hematoxylin based on their shapes and sizes. To evaluate the lateral resolution of the two images obtained using the two illumination modes, we compare the intensity profiles [Fig. 2.2.5.7(c)] along the red lines in Figs. 2.2.5.7(a) and (b). The intensity peaks 1–4 in Fig. 2.2.5.7(c) are clearly separated for both illumination types. Peaks

A–E for circular illumination in Fig. 2.2.5.7(c) are missing for inverse-annular illumination, which can be explained as follows. They are considered to originate from signal produced in the domain located with longer distance than the z-direction-lobe size in the inverse-annular mode from the focal plane and thus do not appear when inverse-annular illumination is employed. This is proved by the fact that the absence of missing peaks remains when the locations of the focal planes of pump and probe are moved together with the same amount. In the experiment, moving the location of the focal plane is performed by moving the sample in the axial direction. In the GNP experiment, the FWHMs of the signal peak are 550 nm in the circular illumination, and 430 nm in the inverse-annular illumination in Table 2.2.5.4. Therefore, material structures located apart from the focal plane in the axial direction by 200 nm or more can show up by the circular illumination but not by inverse-annular illumination. This result indicates that the resolution is improved not only in the lateral directions but also in the axial direction. In the case of inverse-annular illumination, the signals produced in the vicinity of the focal plane are more efficiently detected, resulting in the observed improvement in the axial resolution over the case of circular illumination.

Figs. 2.2.5.7 (d) and (e) show the photothermal images of a rabbit kidney captured using the circular and the inverse-annular illumination in the focal, respectively, in the lateral (XY) plane. The pump and probe power are 150 μW and 200 μW for (d), and 700 μW and 650 μW for (e), respectively. The laser is set to the power level, which gives the best photothermal signal with high S/N and no detectable damage. Higher power could give higher signal intensity but suffer from damage during scanning the sample resulting in poor quality of the frame image. The transmittance of the inverse annular filter is 68%. As with the small intestine sample, the intensity profiles [Fig. 2.2.5.7(f)] shown by the red curve in the graph also indicate improvement in the lateral resolution. Figs. 2.2.5.7(g) and (h) show the PT images of a sample of rabbit kidney captured using circular and inverse-annular illumination, respectively, in the optical axis (XZ) plane. We have defined the Z-axis as the optical axis in Figs. 2.2.5.7(g) and (h). The pump and probe power are 100 μW and 200 μW for (g), and 780 μW and 260 μW for (h), respectively. The inner structures of the cell nuclei can be observed clearly with inverse-annular illumination compared to without it. The intensity profiles [Fig. 2.2.5.7(i)] along the black lines in Figs. 2.2.5.7 (g) and (h)] also indicate an improvement in the axial resolution. The data in Fig. 2.2.5.7 (c, f, XY plane) and (i, XZ plane), Table 2.2.5.5 and Table 2.2.5.6, respectively show the FWHM values and the error values in the fitting process of the Gaussian curves to the observed intensity distributions. All the FWHM values of the Gaussian curves are smaller for inverse-annular illumination than for circular illumination by 14–34%. The biological sample has thickness of about 10 μm, hence there is a possibility of the existence of neighboring structures on the planes above or below the focal plane causing the modification of the observing image of the target. This may be the reason for the variation of the enhancement depending on the position of target structure. Thus it is hard to remove completely the image deformation. In order to more strictly evaluate the resolution improvement, we analyzed the errors in the evaluation of resolution in terms of the values of FWHM of the intensity distribution of the image shown in Tables 2.2.5.5 and 2.2.5.6 To properly evaluate the effects

TABLE 2.2.5.5
FWHMs of the PT Intensity Peaks in XY-plane in Fig. 2.2.5.7[a]

Peak number	Circular [nm]	Inverse-annular [nm]	Difference of Width	$(\varepsilon_c^2 + \varepsilon_I^2)^{0.5}$	Improvement
1	344±54	271±32	73	62.8	21 %
2	478±63	320±35	158	72.1	33 %
3	399±19	263±17	136	25.5	34 %
4	735±61	585±27	150	66.7	20 %
5	458±4	341±67	117	67.1	26 %
6	314±15	271±8	43	17.0	14 %
Average					25(±8) %

[a]ε_c and ε_I: Error values of the peak intensity profiles using circular and inverse annular illumination, respectively.

TABLE 2.2.5.6
FWHMs of the Photothermal Intensity Peaks in XZ-plane in Fig. 2.2.5.7[a]

Peak number	Circular [nm]	Inverse-annular [nm]	Difference of Width	$(\varepsilon_c^2 + \varepsilon_I^2)^{0.5}$	Improvement
7	828±57	552±24	276	61.8	33 %
8	1004±39	774±40	230	55.9	23 %
Average					28 (±5)%

[a]ε_c and ε_I: Error values of the peak intensity profiles using circular and inverse annular illumination, respectively.

with and without the inverse-annular filter, we take the root of the sum of squared error values (corresponding to the widths of the two cases = $(\varepsilon_c^2 + \varepsilon_I^2)^{0.5}$). For peak 1, the root (542 + 322) is approximately 63, which is smaller than the difference of 73 between the cases of with and without shown in Table 2.2.5.5. In all other cases, the same inequality condition is satisfied as shown in Tables 2.2.5.5 and 2.2.5.6. Most intense peak in Fig. 2.2.5.7 (i) is not used for the resolution discussion because it suffers from overlap with the neighboring peaks and the stage step is 18.75 nm. The averages of the fractional increment of resolution by photothermal imaging by using inverse annular illumination are 25 ± 8% in the XY planes and 28 ± 5% in the XZ plane, respectively.

5 CONCLUSION

In conclusion, by simply adding an inverse-annular pupil filter to a conventional microscope configuration, the resolution of a pump-probe photothermal microscope has been significantly improved compared to that of a conventional microscope. The resolution in the X (perpendicular to the common pump and probe polarization direction), Y (parallel to the common pump and probe polarization direction), and Z (axial direction) directions has been enhanced by 15 ± 6, 8 ± 8, and 21 ± 2% relative to the case without the filter. The method has been applied to the photothermal imaging of thick biological samples (rabbit small intestine and kidney cells) stained with hematoxylin and eosin dye. The research described in this subsection is the products of cooperation of the following people: T. Kobayashi, K. Kawasumi, J. Miyazaki, and K. Nakata.

The above study was reported in the *Journal of Optics* and copyright is shared between the corresponding author (Takayoshi Kobayashi, TK) and AIP Advances. TK would like to express thanks to the AIP for permitting us to publish the paper properly revised to be suited as a section in the book *Photothermal and Raman Microscopy* (Taylor & Francis).

NOTES

1. dichroic mirror: a mirror with significantly different reflection or transmission properties at two different wavelengths, the regions of which are not so large in width. The specifications in some cases refer to frequently used laser lines, so that dichroic mirrors are often belonging to the category laser line optics.
2. shot-noise limited: is also called Poisson noise which is a type of noise which can be modeled by a Poisson process given by the following distribution function P(X=k) for stochastic variable X to be given by $P(X=k) = (\lambda^{k} e^{-\lambda})/k!$ with λbeing a parameter. In electronics shot noise originates from the discrete nature of electric charge. Shot noise also occurs in photon counting in optical devices, where shot noise is associated with the particle nature of light (photon picture).
3. frequency mixer: a mixer in electronics is an electrical circuit that creates new frequencies from two signals applied to the mixer, which produces new signals at the sum and difference of the original frequencies. Mixers are widely used to shift signals from one frequency range to another, a process known as heterodyning, for convenience in transmission or further signal processing such as superheterodyne receiver. Frequency mixers are also used for reference signal for lock-in detection as in the text.
4. closed loop piezo-actuator: is excellently suited for high precision positioning in the μm–nm range. But because of hysteresis the correlation between the applied voltage and the actuator's motion is not perfectly matched. There are some applications in practice where a high resolution of the motion is necessary, but the absolute positioning accuracy is not. The classic example is the problem of fiber positioning. The light of one fiber has to be coupled most efficiently into a second fiber, the knowledge of the absolute position of the fiber is not important. Another example is the case when it is possible to return after each positioning event (transaction) to the 0 voltage position, the hysteresis does not affect the event.
5. dwell time: In sample stage scanning mode, the amount of time that coupled pump/probe beams spend on a target.

REFERENCES

1. P. Samineni, A. deCruz, T. E. Villafana, W. S. Warren, and M. C. Fischer, "Pump-probe imaging of historical pigments used in paintings," *Opt. Lett.* **37**, 1310–1303 (2012).
2. B. A. Nechay, U. Siegner, M. Achermann, H. Bielefeldt, and U. Keller, "Femtosecond pump-probe near-field optical microscopy," *Rev. Sci. Instrum.* **70**, 2758–2764 (1999).

3. C. Y. Dong, C. Buehler, P. T. So, T. French, and E. Gratton, "Implementation of intensity-modulated laser diodes in time-resolved diodes in time-resolved, pump-probe fluorescence microscopy," *Appl. Opt.* **40**, 1109–1115 (2001).
4. T. E. Matthews, J. W. Wilson, S. Degan, M. J. Simpson, J. Y. Jin, J. Y. Zhang, and W. S. Warren, "In vivo and ex vivo epi-mode pump-probe imaging of melanin and microvasculature," *J. Biomed. Opt.* **2**, 1576–1583 (2011).
5. W. Min, S. Lu, S. Chong, R. Roy, G. R. Holtom, and X. S. Xie, "Imaging chromophores with undetectable fluorescence by stimulated emission microscopy," *Nature* **461**, 1105–1109 (2009).
6. L. Wei and W. Min, "Pump-probe optical microscopy for imaging nonfluorescent chromophores," *Anal. Bioanal. Chem.* **403**, 2197–2202 (2012).
7. L. Cognet, C. Tardin, D. Boyer, D. Choquet, P. Tamarat, and B. Lounis, "Single metallic nanoparticle imaging for protein detection in cells," *P. Natl. Acad. Sci. U.S.A.* **100**, 11350–11355 (2003).
8. P. Vermeulen, L. Cognet, and B. Lounis, "Photothermal microscopy: Optical detection of small absorbers in scattering environments," *J. Microsc.* **254**, 115–121 (2014).
9. S. Berciaud, D. Lasne, G. A. Blab, L. Cognet, and B. Lounis, "Photothermal heterodyne imaging of individual metallic nanoparticles: Theory versus experiment," *Phys. Rev. B* **73**, 045424 (2006).
10. J. Miyazaki, H. Tsurui, A. Hayashi-Takagi, H. Kasai, and T. Kobayashi, "Subdiffraction resolution pump probe microscopy with shot-noise limited sensitivity using laser diodes," *Opt. Express* **22**, 9024–9032 (2014).
11. J. Miyazaki, H. Tsurui, K. Kawasumi, and T. Kobayashi, "Optimal detection angle in sub-diffraction resolution photothermal microscopy: Application for high sensitivity imaging of biological tissues," *Opt. Express* **22**, 18833–18842 (2014).
12. J. Miyazaki, K. Kawasumi, and T. Kobayashi, "Resolution improvement in laser diodes-based pump probe microscopy with an annular pupil filter," *Opt. Lett.* **39**, 4219–4222 (2014).
13. J. Miyazaki, H. Tsurui, K. Kawasumi, and T. Kobayashi, "Simultaneous dual-wavelength imaging of nonfluorescent tissues with 3D subdiffraction photothermal microscopy," *Opt. Exp.* **23**(3), 3647–3656 (2015).
14. J. Miyazaki, H. Tsurui, K. Kawasumi, and T. Kobayashi, "Sensitivity enhancement of photothermal microscopy with radially segmented balanced detection," *Opt Lett.* **40**(4), 479–482 (2015).
15. J. Miyazaki, H. Tsurui, and T. Kobayashi, "Reduction of distortion in photothermal microscopy and its application to the high resolution three-dimensional imaging of nonfluorescent tissues," *Biomed. Opt. Exp.* **6**(9), 3217–3224 (2015).
16. B. Sick and B. Hech, "Orientational imaging of single molecules by annular illumination," *Phys. Rev. Lett.* **85**, 4482–4485 (2000).
17. J. Kim, D. C. Kim, and S. H. Back, "Demonstration of high lateral resolution in laser confocal microscopy using annular and radially polarized light," *Microsc. Res. Techniq.* **72**, 441–446 (2009).
18. K. Kitamura, K. Sakai, and S. Noda, "Sub-wavelength focal spot with long depth of focus generated by radially polarized, narrow-width annular beam," *Opt. Express* **18**, 4518–4525 (2010).
19. M. M. Corral, P. Andres, C. J. Z. Rodriguez, and M. Kowalczyk, "Three-dimensional superresolution by annular binary filters," *Opt. Commun.* **165**, 267–278 (1999).
20. M. M. Corral, P. Andres, J. O. Castaiieda, and G. Saavedra, "Tunable axial superresolution by annular binary filters. Application to confocal microscopy," *Opt. Commun.* **119**, 491–498 (1995).

21. B. Richards and E. Wolf, "Electromagnetic diffraction in optical systems. II. Structure of the image field in an aplanatic system," *Proc. Roy. Soc. London A* **253**, 358–379 (1959).
22. K. S. Youngworth and T. G. Brown, "Focusing of high numerical aperture cylindrical-vector beams," *Opt. Express* **7**, 77–87 (2000).
23. M. Shibuya, and H. Ooki, 回折と結像の 光学 (Asakura Publishing Co. Ltd, 2005), ISBN: 978-4-254-13731-6 C3342.
24. M. Selmke, M. Braun, and F. Cichos, "Photothermal single-particle microscopy: Detection of a nanolens," *Nano Lett.* **6**, 2741–2749 (2012).
25. M. W. Conklin, P. P. Provenzano, K. W. Eliceiri, R. Sullivan, and P. J. Keely, "Fluorescence lifetime imaging of endogenous fluorophores in histopathology sections reveals differences between normal and tumor epithelium in carcinoma in situ of the breast," *Cell Biochem. Biophys.* **53**, 145–157 (2009).
26. J. M. Apgar, A. Juarranz, J. Espada, A. Villanueva, M. Canete, and J. C. Stockert, "Fluorescence microscopy of rat embryo sections stained with haematoxylin–eosin and Masson's trichrome method," *J. Microsc.* **191**, 20–27 (1997).
27. H. F. de Carvalho and S. R. Taboga, "Fluorescence and confocal laser scanning microscopy imaging of elastic fibers in hematoxylin-eosin stained sections," *Histochem. Cell Biol.* **106**, 587–592 (1996).

2158-3226/2016/6(12)/125308/12, 6, 125308-1 (http://creativecommons.org/licenses/by/4.0/). [http://dx.doi.org/10.1063/1.4967948] AIP ADVANCES 6, 125308 (2016)

3.1 Sub-Diffraction Resolution Pump-probe Microscopy with Shot-Noise Limited Sensitivity Using Laser Diodes

1. INTRODUCTION

In pump-probe experiments, pump light promotes molecules to an allowed excited state, which in turn induces changes in probe transmissivity via excited-state absorption [1–3], ground state absorption bleaching [4], and stimulated emission [5–8]. This method has recently been used in optical microscopy to visualize the structure of biological tissues and to carry out functional analysis, such as the imaging of fluorophores that have short-lived excited states [1, 4, 5, 9, 10], the label-free biological imaging of red blood cells [2, 11], analysis of melamine pigments and melanoma diagnosis [12–14].

The relative change in the probe intensity in pump-probe microscopy is estimated to be no more than 10^{-7}–10^{-6} from the absorption cross section (typically 10^{-16}–10^{-15} cm^2) and focus spot size for a single molecule located in the focal volume of an objective lens [15]. Therefore, extremely stable lasers with special additional features are necessary to detect such weak signals, particularly as the laser noise interferes with their detection. The recent development of pump-probe microscopy has been carried out typically by using a mode-locked femtosecond pulse laser oscillator with high repetition rates and high-frequency lock-in detection schemes to avoid laser-intensity fluctuation, which is larger at lower frequencies. Xie and associates have performed stimulated emission microscopy by employing two optical parametric oscillators (OPOs) pumped by a high-intensity Ti:sapphire or Yb mode-locked laser to obtain wavelength-tunable and synchronous pump and probe pulse trains [5]. However, these mode-locked laser-based systems with OPOs are very expensive and complex.

In this study, using intensity modulated continuous wave laser diodes (LDs) with a balanced detection scheme [16, 17], we accomplished sub-diffraction resolution

DOI: 10.1201/9781003476030-10

pump-probe imaging with shot-noise limited sensitivity. We also demonstrated the use of this method for the biological imaging of mouse T cells and the axons of neurons in the cerebral cortex.

2. EXPERIMENT

2.1 Principle of Pump-probe Microscopy Using Intensity Modulated Lasers

The principle of pump-probe microscopy using an intensity modulated continuous wave laser is illustrated in Figure 3.1.1. The probe and pump-beam intensities were set to frequencies of ω_1 and ω_2, respectively. A beat signal at $|\omega_1 - \omega_2|$ is generated by the bilinear interaction of the pump field and the probe field in the sample. We used a lock-in technique to detect the signal, and the lock-in amplifier was referenced to the beat frequency. By changing the modulation frequencies (ω_1 and ω_2) while holding (or maintaining) $|\omega_1 - \omega_2|$ constant, this dual modulation scheme has the potential to acquire time-resolved information without a high-speed detector [7].

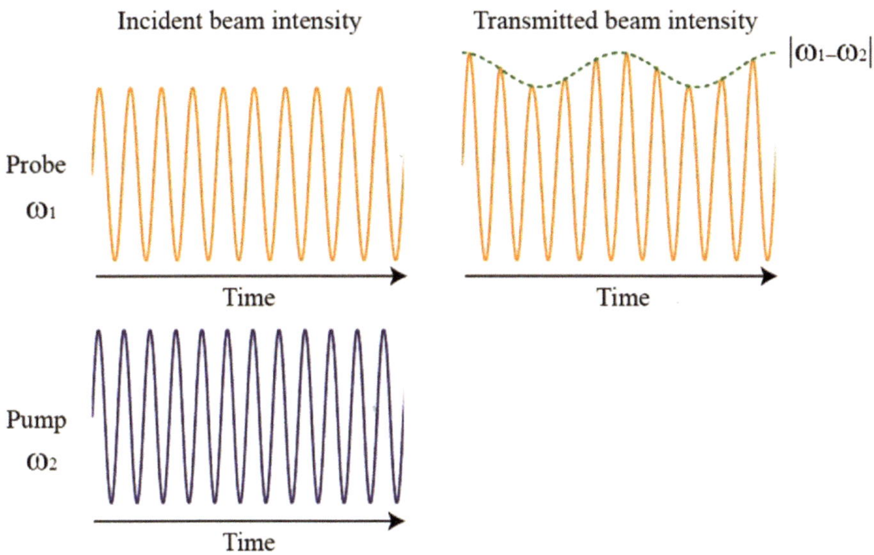

FIGURE 3.1.1 Principle of pump-probe microscopy using intensity modulated continuous wave laser. The probe and pump beam intensities are modulated at ω_1 and ω_2. A beat signal at $|\omega_1 - \omega_2|$ is generated by the bilinear interactions at sample position and detected using a lock-in amplifier. Copyright of this figure is partially held between the representative author (Takayoshi Kobayashi) and Optica (formally Optical Society of America).

2.2 Experimental System

Our pump-probe microscopy system, which uses simple LDs and a balance detector, is illustrated in Figure 3.1.2. Two laser diodes (Nichia NDS4216 and Mitsubishi ML101U29) were used for excitation at 488 nm and probing at 660 nm, respectively. Their output powers were 200 mW. A diode pumped solid state (DPSS) laser (Thorlabs, DJ525-40) with an output power of 40 mW was used as an alternative probe beam. The LDs and the DPSS were mounted on temperature-controlled laser-diode mounts (Thorlabs, TCLDM9 or LDM9T). Each beam was collimated through a spatial filter and the three beams were combined with dichroic beam-combining mirrors. All of the beams were linearly (vertically) polarized. A polarizing beam splitter (PBS) was used to direct the combined beam to the objective lens (Olympus: UPLSAPO 40X2 or UPLSAPO 100XO) with numerical apertures (NA) of 0.95 and 1.4. The beam size was adjusted to fill the back aperture of the objective lens. The sample position was raster-scanned using a three-axis positioning stage driven by piezo actuators (Thorlabs MAX311D). To evaluate the spatial resolution using a gold nanoparticle, we used a high-accuracy positioning stage (PI P-622.2) with a closed-loop resolution of 0.7 nm. A 1.4NA oil immersion condenser lens (Olympus U-AAC) was used to collect the transmitted light. A narrowband filter matching the laser line was placed in front of the detector to filter out both the excitation beam and the fluorescence from the sample, so that only the probe beam was detected.

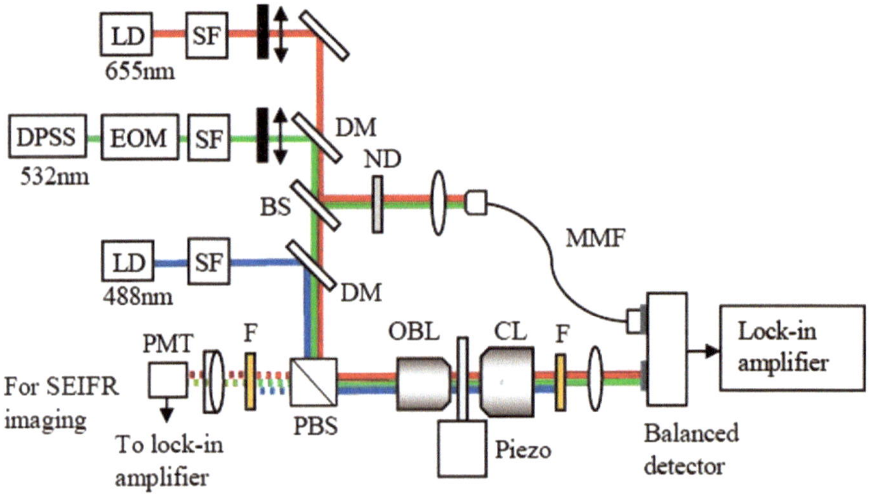

FIGURE 3.1.2 Experimental setup of pump-probe microscopy using laser diodes. LD: laser diode, DPSS: diode pumped solid state laser, EOM: electro optical modulator, SF: spatial filter, DM: dichroic mirror, BS: beam splitter, ND: neutral density filter, PBS: polarization beam splitter, OBL: objective lens, CL: condenser lens, F: band pass filter, PMT: photomultiplier tube, MMF: multimode fiber. Either the 660 nm LD or the 532 nm DPSS was used as a probe light source. Copyright of this figure is partially held between the representative author (Takayoshi Kobayashi) and Optica (formally Optical Society of America).

An auto-balanced photoreceiver (New Focus, Nirvana) was used to cancel out the intensity-fluctuation noise of the probe laser. A beam splitter placed in front of the focusing lens reflected the probe beam, which was then directed to the reference port of the balanced detector through a multimode fiber. The output of the balanced detector can be expressed as $V_{out} = G\gamma\ [p_{sig} - g\ p_{ref}]$, where G and g are the trans-impedance gain of the detector and the reference-channel gain, respectively, γ is the photodiode sensitivity and $p_{sig(ref)}$ is the beam power incident on the signal (reference) photodiode. A neutral density filter was used to set the reference-beam intensity to about twice the probe intensity for optimal common-mode noise rejection. We also set up a fluorescence detection scheme, wherein the fluorescence signal from the sample was detected by a photo-multiplier tube through the objective lens, a PBS and a multimode fiber with a core diameter of 100 μm.

A signal generator (Rigol DG4162) and an amplifier (Mini-circuit ZHL-6A +) were used to sinusoidally modulate the light intensity of each LD by adjusting the injection current from the kHz to the MHz range. The light intensity of the DPSS laser was externally modulated using an electro-optic modulator (EOM) (LINOS LM020). A lock-in amplifier (Signal Recovery 7270) was used to detect the pump-probe signal. The output of the balanced detector was connected to the lock-in amplifier through a low-pass filter (Mini-circuit LPFB0R3). The reference signal for the lock-in amplifier was delivered from the two signal generators through a frequency mixer (Mini-circuit ZAD-1) and a low-pass filter (Minicircuit LPF-B0R3). As it was crucial to avoid signal interference between the pump and the probe channels, we placed a 30-dB attenuator and an amplifier (Mini-circuit ZFL-500LN) between the signal generator and the frequency mixer. Furthermore, to reduce interference between the two channels due to electromagnetic (EM) waves, the LDs and the DPSS laser were placed in separate EM-shielding boxes.

2.3 Evaluation of the Noise Level and Signal-to-Noise Ratio

Using this setup, we first examined the noise level, signal intensity and signal-to-noise ratio (SNR) to ascertain the optimal probe-beam power. Figure 3.1.3(a) shows the noise level of the lock-in signal as a function of p_{sig} (660-nm LD). When p_{sig} was below ~10^{-5} W, circuit noise was dominant, at the level of 1.2 $10\times^{-7}$ ($V/\sqrt{Hz}$) (volt over square root of Hz). After canceling the laser-intensity fluctuation, we found the root mean square (RMS) of the noise of the lock-in signal to be close to the theoretical shot-noise level when $p_{sig} = 10^{-5}$–10^{-4} W.

Here, the shot-noise level was calculated using the equation $V_{shot} = G\sqrt{2q\gamma\, p_{sig}}$ ($V/\sqrt{Hz}$), where q is the elementary charge, $G = 1 \times 10^5$ (V/A), and $\gamma = 0.43$ (A/W). When p_{sig} was larger than 10^{-4} W, the noise level seemed to deviate from the shot-noise level due to the residual laser noise; however, the laser noise was substantially reduced, showing a decrease of about 26-dB.

Figure 3.1.3(b) shows the typical lock-in signal and the signal-to-noise ratio (SNR) as functions of the power of the probe beam power P_{probe} incident on a sample. The sample was a thin film polymer doped with CdSe/ZnS quantum dots, which emit

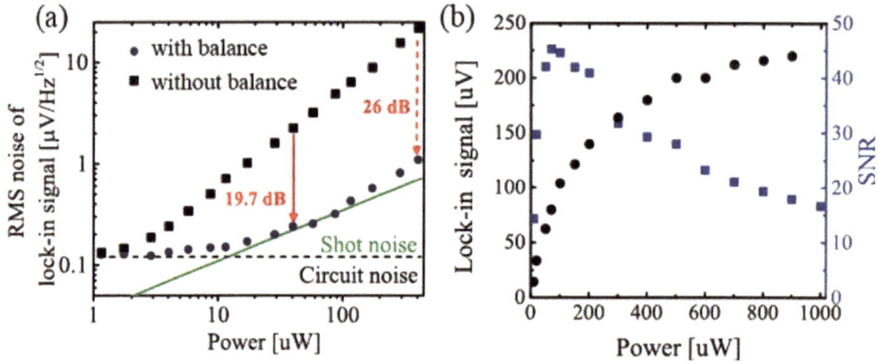

FIGURE 3.1.3 Evaluation of noise level, signal intensity and signal-to-noise ratio (a). Noise level of the lock-in signal as a function of laser beam power incident on the detector. Circles and squares denote, respectively, lock-in signals that do or do not involve a balanced detection scheme. Solid and broken lines are the shot-noise level and circuit noise level, respectively. The modulation frequency of the beams was 3.0 MHz and the lock-in frequency was set at 100 kHz. (b) Lock-in signal (circles) and the signal to noise ratio (SNR) (squares) as a function of probe beam power at the sample position. The pump beam power was set at 20 μW. The modulation frequencies of the probe and pump beams were 3.0 and 3.1 MHz, respectively. Copyright of this figure is partially held between the representative author (Takayoshi Kobayashi) and Optica (formally Optical Society of America).

fluorescence with a peak at 650 nm. We used 488-nm and 660-nm LDs for the pump and the probe, respectively. The probe-laser wavelength was found to overlap the fluorescence spectrum of the quantum dots, such that the main contributor to the pump-probe signal was considered to be the stimulated emission. The gain of signal intensity was found to saturate as P_{probe} increased up to ~100 μW. The maximum SNR was reached prior to signal saturation. Theoretical calculations using rate equations indicate that signal saturation occurs when k_{se} exceeds k_s in the limit of weak pump power. Here $k_{se} = \sigma_{se} p_{probe} /(h\nu_{probe} s)$ is the stimulated-emission rate; σ_{se} is the stimulated emission cross section; h is Planck's constant; and s is the beam waist at a focal point. k_s is the total decay rate of fluorescence radiative and nonradiative relaxations. The fluorescence lifetime of CdSe/ZnS was 13 ns, so the optimal value of p_{probe} that is close to saturation but not substantial saturation can be calculated as ~80 μW, by assuming $\sigma_{se} = 5 \times 10^{-16}$ cm^2 and s = 5 x 10^{-9} cm^2. This value is consistent with our experimental result, and at this laser power level, the SNR is mainly limited by shot noise. It is possible to reach this power level by LD, as the maximum output power is typically above 10 mW.

2.4 Enhancement of the Spatial Resolution

The spatial resolution obtained using a pump-probe microscope is better than that obtained using a conventional fluorescence intensity microscope, because the pump-probe signal is based on the nonlinear interaction between the two laser beams and the sample due to the fact that the signal intensity is proportional to the product of the two [18]. We evaluated the spatial resolution in our setup by measuring the 20-nm gold nanoparticles dispersed on a glass slide as shown in Figures 3.1.4(a) and (b). We used a 488-nm LD for the pump beam and a 532-nm DPSS for the probe beam. They were selected so that the wavelengths overlapped the absorption spectrum of the gold nanoparticles due to plasmon resonance with a peak at 524 nm. In this case, ground-state absorption bleaching and photothermal-induced light scattering may contribute to the measured signal [4, 19, 20]. To achieve the best resolution, we used a high numerical aperture (NA = 1.4) objective lens. Using the

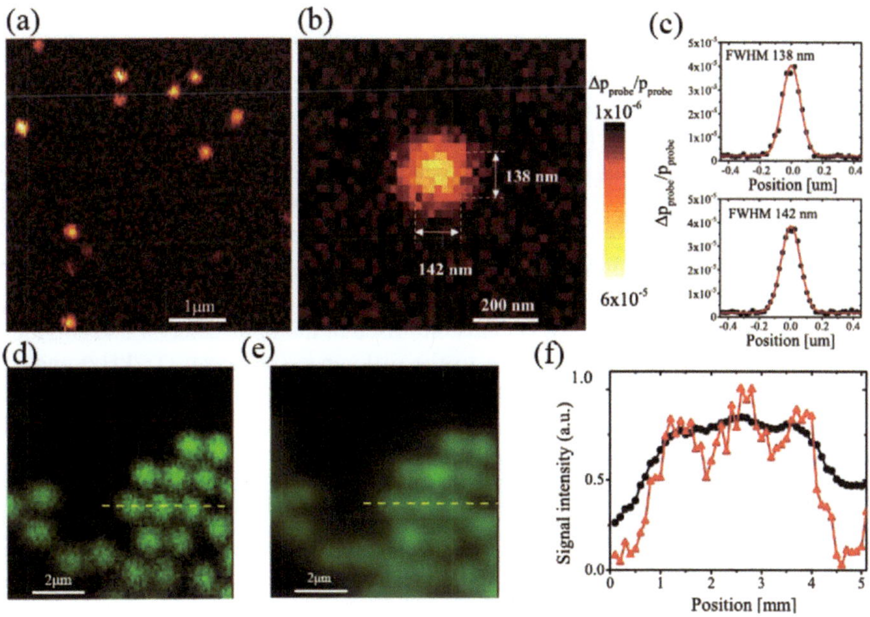

FIGURE 3.1.4 Evaluation of spatial resolution in pump-probe microscopy. (a), (b) pump-probe image of 20 nm gold nanoparticles dispersed on a glass slide. (c) Intensity profiles of the single nanoparticle in the vertical and horizontal direction. The average power of the probe and pump beam was 115 and 222 μW, respectively. The modulation frequencies of the pump and probe beams were 5.0 and 5.1 MHz, respectively. The time constant of the lock-in amplifier was 2 ms and the pixel dwell time was 6 ms. Global fitting of a single nanoparticle using the Gaussian results in full-width at half maximum of 138 nm and 142 nm in the vertical and transversal direction, respectively. Assuming the size of the nanoparticles to be 20 nm, deconvolution of the intensity profiles gives FWHMs of 136 nm and 140 nm for the effective PSF. (d), (e) Stimulated emission induced fluorescence-intensity reduction (SEIFIR) image and conventional fluorescence intensity image of 1 μm fluorescence beads. The excitation and emission stimulation beams were focused by an objective lens with NA = 0.95. The average power of the excitation and emission stimulation beam was 23 μW and 1.1 mW, respectively. (f) Intensity profiles along the broken lines in (d) (red) and (e) (black). Copyright of this figure is partially held between the representative author (Takayoshi Kobayashi) and Optica (formally Optical Society of America).

global fitting of a single particle with the Gaussian, the full width at half maximum (FWHM) values were calculated to be 142 ± 3 nm and 138 ± 3 nm in the transverse and vertical directions, respectively [Fig. 3.1.4(c)]. These values are substantially smaller than the calculated lateral FWHM value of the intensity distribution of a tightly focused probe beam waist under a diffraction-limited condition, w_{xy} ~0.51 λ/NA = 194 nm. The point-spreading function (PSF) produced by the pump-probe microscope is given by the product of the intensity distribution of the two beams at the focal point, and the lateral FWHM value is calculated as 132 nm, which is close to our experimental values. These results clearly demonstrate that sub-diffraction resolution imaging is achievable using our pump-probe microscopy, even for non-fluorescent species.

In the case of a fluorescent sample, it is possible to directly compare the conventional fluorescence-intensity image with the image created by the stimulated emission induced fluorescence-intensity reduction (we call this method SEIFIR) in epifluorescence detection [7, 8]. The images of 1 μm fluorescence beads (Invitrogen TetraSpeck microspheres) obtained by measuring SEIFIR and fluorescence intensity are shown in Figs 3.1.4(d) and 3.1.4 (e), respectively. We used the 488-nm LD and the 532-nm DPSS as light sources for excitation and emission stimulation, respectively. The modulation frequencies of the former and the latter were set to 20 kHz and 33 kHz, respectively. We measured the SEIFIR and conventional-fluorescence images simultaneously, using a system comprised of the same detector and two lock-in amplifiers with demodulation set to the difference frequency (13 kHz) and the fundamental excitation frequency (20 kHz), respectively. Figure 3.1.4(f) shows the intensity profiles along the broken lines in Figs 3.1.4(d) and (e). It is clear that SEIFIR microscopy improves spatial resolution.

2.5 Biological Imaging

We demonstrated pump-probe imaging using biological samples. The stimulated-emission images of mouse CD8 T cells[1] from the T-cell area of mesenteric lymph nodes,[2] with and without the scheme of balanced detection are shown in Figures 3.1.5(a) and (b), respectively. For preparing the sample, we removed mesenteric lymph nodes from adult mice (C57BL/6 from Sankyo-Labo Service Corporation Inc., Tokyo, Japan), embedded the nodes in an OCT compound (Sakura Seiki, Tokyo, Japan), immersed them in liquid nitrogen to freeze them and sliced them to 4-μm thickness. We stained the sections with an anti-CD8 antibody[3] (S3-6.7 from eBioscience, CA), labeled them with eFluor 650NC (from eBioscience) and observed them under pump-probe microscopy.

The antibody against CD8 was labeled with quantum dots, with the fluorescence peak at 650 nm. We were able to clearly visualize the cell structure using the balanced-detection scheme. Fig. 3.1.5(c) shows the intensity profiles along the broken lines in Figs 3.1.5(a) and (b). The relative background-fluctuation intensities in Figs 3.1.5(a) and (b) were 1.3 x 10 and 9.8 x 10, respectively.

The mesentery is an organ that attaches the intestines to the posterior abdominal wall and is formed by the double fold of peritoneum. It helps in storing fat and

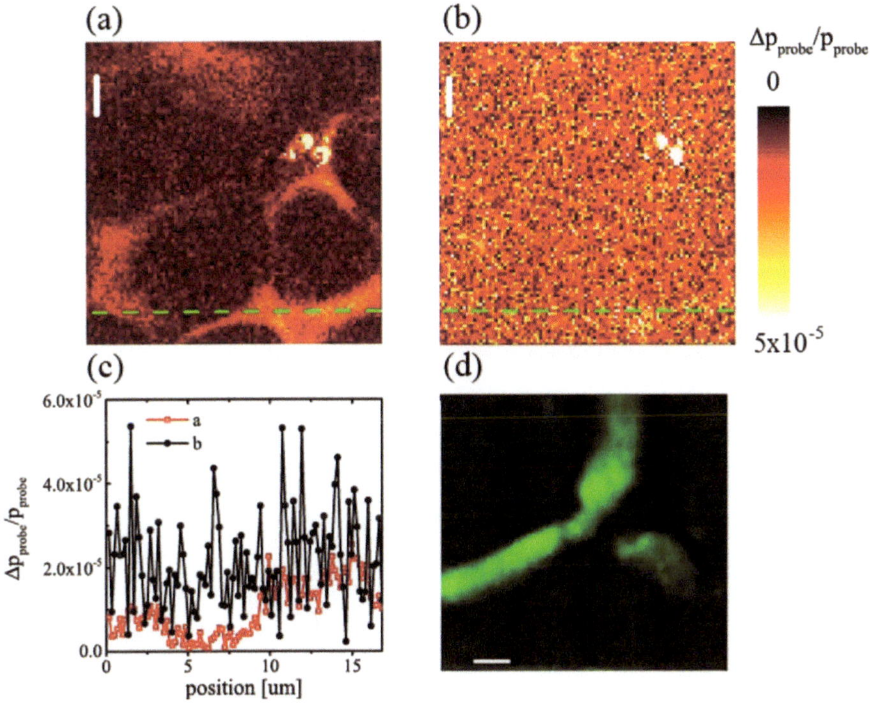

FIGURE 3.1.5 (a), (b). Stimulated emission images of a mouse T cell labeled with quantum dots with and without the scheme of the balanced detection. We used 488 nm LD for the pump and 660 nm LD for the probe. The average powers of the probe and pump beam were 210 and 1 mW, respectively. The modulation frequency of the pump and probe beams were 3 and 3.1 MHz, respectively. The time constant of the lock-in amplifier was 5 ms and the pixel dwell time was 8 ms. (c) Intensity profiles along the broken lines in (a) (red) and (b) (black). (d) Stimulated emission image of neuron-expressing yellow fluorescent proteins. The average powers of the probe and pump beam were 130 μW and 970 μW, respectively. The scale bars in (a), (b) and (d) indicate 2 μm. Copyright of this figure is partially held between the representative author (Takayoshi Kobayashi) and Optica (formally Optical Society of America).

allowing blood vessels, lymphatics, and nerves to supply the intestines, among other functions [1].

The mesocolon was thought to be a fragmented structure, with all named parts—the ascending, transverse, descending, and sigmoid mesocolons, the mesoappendix, and the mesorectum—separately terminating their insertion into the posterior abdominal wall. [2] However, in 2012, new microscopic and electron microscopic examinations showed the mesocolon to be a single structure derived from the duodenojejunal flexure and extending to the distal mesorectal layer. [2, 3] Thus, the mesentery is an internal organ. [4, 5]

Figs 3.1.5(a) and (b). The relative background-fluctuation intensities in Figs 3.1.5(a) and (b) were 1.3 x 10^{-6} and 9.8 x 10^{-6}, respectively.

Fig. 3.1.5(d) shows a stimulated-emission image of an axon of a neuron in the cerebral cortex, expressing yellow fluorescent proteins. For preparing the sample, we fixed adult transgenic mice (two months old, thy1-GFP H-line, Jackson Lab.) by cardiac perfusion with PBS and paraformaldehyde, cut 100-μm slices of the frontal cortex using a vibratome (Leica VT1200) and embedded the slices in UNICRYL resin (Funakoshi Ltd.). To induce stimulated emission, we used a 488-nm LD for the pump and a 532-nm DPSS for the probe, so that the absorption and fluorescence spectra overlapped the wavelengths of the pump beam and the probe beam, respectively. These results suggest that our method has potential for the study of biological nanostructures and the functional properties of material such as brain tissue.

3. DISCUSSION

In both the pulse and continuous wave laser systems, the ultimate SNR would be achieved when the pump beam intensity is enough to cause saturation and the probe beam intensity is close to saturation but not substantial saturation. Then imaging signal intensity is only limited by the shot noise of the probe beam. For the femtosecond pump-probe laser system, the detection limit of the relative intensity change of the probe beam is ~10^{-7} for 1-s time constant at the lock-in amplifier [5], and this value is comparable to our continuous wave system

For ground state depletion, signal is proportional to the ground state absorption cross section, which is usually comparable to the stimulated emission cross section because of microscopic reversibility. Furthermore, if the excited state absorption cross section is of the same order of magnitude or larger than that of the ground state, an even higher SNR signal can be expected. Because of the higher density of states in the higher electronic states, which can be the final states of the induced absorption signal, there is high possibility of intense induced absorption appearing in the available spectral range of the diode laser. Therefore, if it is possible to select the wavelength in some probe molecular systems appropriate for the probe beam, our method is expected to be applicable to excited-state absorption and ground state absorption bleaching in addition to stimulated emission all at the same time allowing the multi-color imaging.

During the sample scanning, the nonuniformity of the refractive-index and stationary absorption in the sample caused the probe beam to vary in intensity due to inevitable nonuniform concentration and various component species, which led to an imbalance in intensity between the signal and the reference path. In our experimental setup, this intensity imbalance was canceled out by the auto-gain controller with which the detector is equipped.

This controller automatically adjusts the gain so that (Please remove the three overbars) $g = \overline{p}_{sig} / \overline{p}_{ref}$ where $\overline{p}_{sig(ref)}$ is the low frequency component of the signal detected on the signal (reference) photodiode. To detect the lock-in signal while canceling both the laser noise and the intensity imbalance during the sample scanning, we selected a bandwidth for the gain controller with a cut-off frequency higher than the inverse of the pixel-dwell time and lower than the lock-in frequency. We measured an image of transparent 1 μm polystyrene beads with a dwell time of 6 ms and confirmed that the refractive-index variations in the sample did not disturbed the lock-in signal.

Finally, we discuss the effect of the thermal lens on the measured signal because the thermal lens microscope is based on the same pump-probe detection scheme. When the excitation beam is absorbed, the temperature of the sample increases, forming a refractive index distribution that changes the divergence of the transmitted probe beam. Thermal lensing may contribute to the measured lock-in signal if the characteristic length of heat transport during one modulation period is larger than the wavelength of probe light, i.e., $d = \sqrt{2D / \omega_{1,2}} > \lambda_{probe}$, where D is the thermal diffusivity (1.4 x 10^{-7} m^2/s in water). A typical characteristic of the thermal lens effect is that the signal intensity exhibits two peaks that correspond to focusing or defocusing when the sample position is moved in the axial direction [10, 21]. This becomes a drawback when the structures in the axial direction need to be resolved. To avoid thermal lens effect, we selected $\omega_{1,2}$ so that d was comparable to or smaller than λ_{probe}. Furthermore, we used a condenser lens with an NA that was larger or equal to the focusing lens. To examine thermal lens effect on the present setup, the signal intensity and phase dependences on the relative distance between the sample and the focal position of the probe beam were measured by moving the slides of thin film of quantum dots. We found that the amplitude of the lock-in signal exhibited only a single peak at the origin. Furthermore, the phase was nearly flat and did not exhibit the π shift at the origin. We also checked the frequency characteristics of the signals and found that the signal intensity was nearly independent of the pump and probe modulation frequency in the range from 0.5 to 5 MHz. Therefore, we concluded that the thermal lens had negligible effects on the measured signal.

4. CONCLUSION

In conclusion, we have demonstrated that sub-diffraction resolution pump-probe imaging with shot-noise limited sensitivity can be carried out using conventional laser diodes with a balanced-detection scheme. The cost of LD was 50 times lower than in the Ti:sapphire laser based system. Furthermore, LD is maintenance free and power consumption and space occupied by the instruments were reduced in comparison with the mode-locked laser-based system. Blue, red, and near-infrared laser diodes are readily available, as they are widely used in optical disks and in optical communication. In recent years, tremendous advances have been made in green-emitting (500–532 nm) InGaN-based LD technology, achieving an output power of more than 100 mW [22, 23]. Furthermore, green to yellow (543–570 nm) LDs with ~3 mW power can be used at room temperature [24]. By incorporating these LDs, our method becomes applicable to various types of probe materials including low molecular weight organic molecules, fluorescent proteins, and quantum dots. This technique has several applications for various types of induced transmission change, including excited-state absorption, ground state absorption bleaching and stimulated emission. Thus, this method offers new opportunities for studying the nanostructures and properties of biological tissues.

Contents described in this subsection are obtained by collaborative research activities among the following people: J. Miyazaki, H. Tsurui, A. Hayashi-Takagi, H. Kasai and T. Kobayashi [25].

NOTES

1. CD8 T cells: a type of T cell. T cells are one of the important types of white blood cells of the immune system and play a central role in the adaptive immune response.
2. mesenteric lymph nodes: The mesentery is an organ that attaches the intestines to the posterior abdominal wall and is formed by the double fold of peritoneum (membrane forming the lining of the abdominal cavity). It helps in storing fat and allowing blood vessels, lymphatics, and nerves to supply the intestines, among other functions. A mesenteric lymph node is a lymph node existing on the membrane.
3. anti-CD8 antibody: CD8 is a transmembrane glycoprotein expressed primarily on the surface of cytotoxic T cells. CD8a is also found on natural killer cells, cortical thymocytes, and dendritic cells.

REFERENCES

1. D. Fu, T. Ye, T. E. Matthews, G. Yurtsever, and W. S. Warren, "Two-color, two-photon, and excited-state absorption microscopy," *J. Biomed. Opt.* **12**(5), 054004 (2007).
2. D. Fu, T. Ye, T. E. Matthews, B. J. Chen, G. Yurtserver, and W. S. Warren, "High-resolution in vivo imaging of blood vessels without labeling," *Opt. Lett.* **32**(18), 2641–2643 (2007).
3. T. Chen, F. Lu, A. M. Streets, P. Fei, J. Quan, and Y. Huang, "Optical imaging of non-fluorescent nanodiamonds in live cells using transient absorption microscopy," *Nanoscale* **5**(11), 4701–4705 (2013).
4. S. S. Chong, W. Min, and X. S. Xie, "Ground-state depletion microscopy: Detection sensitivity of single molecule optical absorption at room temperature," *J. Phys. Chem. Lett.* **1**(23), 3316–3322 (2010).
5. W. Min, S. Lu, S. Chong, R. Roy, G. R. Holtom, and X. S. Xie, "Imaging chromophores with undetectable fluorescence by stimulated emission microscopy," *Nature* **461**(7267), 1105–1109 (2009).
6. T. Dellwig, P. Y. Lin, and F. J. Kao, "Long-distance fluorescence lifetime imaging using stimulated emission," *J. Biomed. Opt.* **17**(1), 011009 (2012).
7. C. Y. Dong, C. Buehler, P. T. So, T. French, and E. Gratton, "Implementation of intensity-modulated laser diodes in time-resolved, pump-probe fluorescence microscopy," *Appl. Opt.* **40**(7), 1109–1115 (2001).
8. C. Y. Dong, P. T. C. So, T. French, and E. Gratton, "Fluorescence lifetime imaging by asynchronous pumpprobe microscopy," *Biophys. J.* **69**(6), 2234–2242 (1995).
9. T. Chen, F. Lu, A. M. Streets, P. Fei, J. M. Quan, and Y. Y. Huang, "Optical imaging of non-fluorescent nanodiamonds in live cells using transient absorption microscopy," *Nanoscale* **5**(11), 4701–4705 (2013).
10. L. Tong, Y. Liu, B. D. Dolash, Y. Jung, M. N. Slipchenko, D. E. Bergstrom, and J. X. Cheng, "Label-free imaging of semiconducting and metallic carbon nanotubes in cells and mice using transient absorption microscopy," *Nat. Nanotechnol.* **7**(1), 56–61 (2011).
11. T. Ye, D. Fu, and W. S. Warren, "Nonlinear absorption microscopy," *Photochem. Photobiol.* **85**(3), 631–645 (2009).

12. D. Fu, T. Ye, T. E. Matthews, J. Grichnik, L. Hong, J. D. Simon, and W. S. Warren, "Probing skin pigmentation changes with transient absorption imaging of eumelanin and pheomelanin," *J. Biomed. Opt.* **13**(5), 054036 (2008).
13. T. E. Matthews, I. R. Piletic, M. A. Selim, M. J. Simpson, and W. S. Warren, "Pump-probe imaging differentiates melanoma from melanocytic nevi," *Sci. Transl. Med.* **3**(71), 71ra15 (2011).
14. M. J. Simpson, K. E. Glass, J. W. Wilson, P. R. Wilby, J. D. Simon, and W. S. Warren, "Pump-probe microscopic imaging of Jurassic-aged eumelanin," *J. Phys. Chem. Lett.* **4**(11), 1924–1927 (2013).
15. L. Wei and W. Min, "Pump-probe optical microscopy for imaging nonfluorescent chromophores," *Anal. Bioanal. Chem.* **403**(8), 2197–2202 (2012).
16. M. Celebrano, P. Kukura, A. Renn, and V. Sandoghdar, "Single-molecule imaging by optical absorption," *Nat. Photonics* **5**(2), 95–98 (2011).
17. P. Kukura, M. Celebrano, A. Renn, and V. Sandoghdar, "Single-molecule sensitivity in optical absorption at room temperature," *J. Phys. Chem. Lett.* **1**(23), 3323–3327 (2010).
18. J. R. Lakowicz, *Topics in Fluorescence Spectroscopy, Vol 5: Nonlinear and Two-Photon Induced Fluorescence*, Springer, 1997.
19. A. Gaiduk, M. Yorulmaz, P. V. Ruijgrok, and M. Orrit, "Room-temperature detection of a single molecule's absorption by photothermal contrast," *Science* **330**(6002), 353–356 (2010).
20. S. Berciaud, D. Lasne, G. A. Blab, L. Cognet, and B. Lounis, "Photothermal heterodyne imaging of individual metallic nanoparticles: Theory versus experiment," *Phys. Rev. B* **73**(4), 045424 (2006).
21. K. Uchiyama, A. Hibara, H. Kimura, T. Sawada, and T. Kitamori, "Thermal lens microscope," *Jpn. J. Appl. Phys.* **39**(Part 1, No. 9A), 5316–5322 (2000).
22. A. Avramescu, T. Lermer, J. Müller, C. Eichler, G. Bruederl, M. Sabathil, S. Lutgen, and U. Strauss, "True green laser diodes at 524 nm with 50 mW continuous wave output power on c-Plane GaN," *Appl. Phys. Express* **3**(6), 061003 (2010).
23. S. Takagi, Y. Enya, T. Kyono, M. Adachi, Y. Yoshizumi, T. Sumitomo, Y. Yamanaka, T. Kumano, S. Tokuyama, K. Sumiyoshi, N. Saga, M. Ueno, K. Katayama, T. Ikegami, T. Nakamura, K. Yanashima, H. Nakajima, K. Tasai, K. Naganuma, N. Fuutagawa, Y. Takiguchi, T. Hamaguchi, and M. Ikeda, "High-power (over 100 mW) green laser diodes on semipolar {2021} GaN substrates operating at wavelengths beyond 530 nm," *Appl. Phys. Express* **5**(8), 082102 (2012).
24. J. Kasai, R. Akimoto, T. Hasama, H. Ishikawa, S. Fujisaki, S. Tanaka, and S. Tsuji, "Green-to-yellow continuous-wave operation of BeZnCdSe quantum-well laser diodes at room temperature," *Appl. Phys. Express* **4**(8), 082102 (2011).
25. J. Miyazaki, H. Tsurui, A. Hayashi-Takagi, H. Kasai, and T. Kobayashi, "Sub-diffraction resolution pump-probe microscopy with shot-noise limited sensitivity using laser diodes," *Opt. Express* **22**(8), 009024 (2014).

3.2 Sensitivity Enhancement of Photothermal Microscopy with Radially Segmented Balanced Detection

1 INTRODUCTION

Photothermal microscopy is capable of visualizing single nonfluorescent nanoparticles and molecules using only their optical absorption [1–4]. Therefore it can be applied to nonfluorescent targets which are much more abundant in living systems except plant originated materials and exceptional luminescent insects. As nonfluorescent species suffer much less photodamage or photobleaching than fluorescent molecules, they are intrinsically useful for biological imaging. Recently, photothermal microscopy has been applied to imaging of the cell structure using gold nanoparticles [5–8], evaluation of the hydrodynamic properties of cytosol with photothermal absorption correlation spectroscopy [6, 9], and label-free and three-dimensional imaging of heme proteins with two photon excitation [10]. However, in photothermal microscopy, an integration time of 1 to 10 ms per pixel is typically needed to attain a sufficient signal-to-noise ratio (SNR) [11]. Thus, the image acquisition time is still slower than that achieved by conventional laser scanning fluorescence microscopy, with integration times as low as 1 to 10 μs per pixel.

In this subsection, we propose a novel detection scheme to improve the SNR on the basis of the characteristics of an angular-dependent photothermal signal. Typically, in photothermal microscopy, two laser beams for pumping and probing with different wavelengths are incident on the sample through a focusing lens [Figure 3.2.1(a)]. The pump beam increases the temperature ΔT around the focal point of the optical absorbing sample. This results in variations in the local refractive index and induces deflection of the probe beam. The variation of the refractive index (typically $\Delta n \sim 10^{-4}$ with ΔT 1 K) is detected by a lock-in detection scheme as a change in probe beam transmissivity.

DOI: 10.1201/9781003476030-11

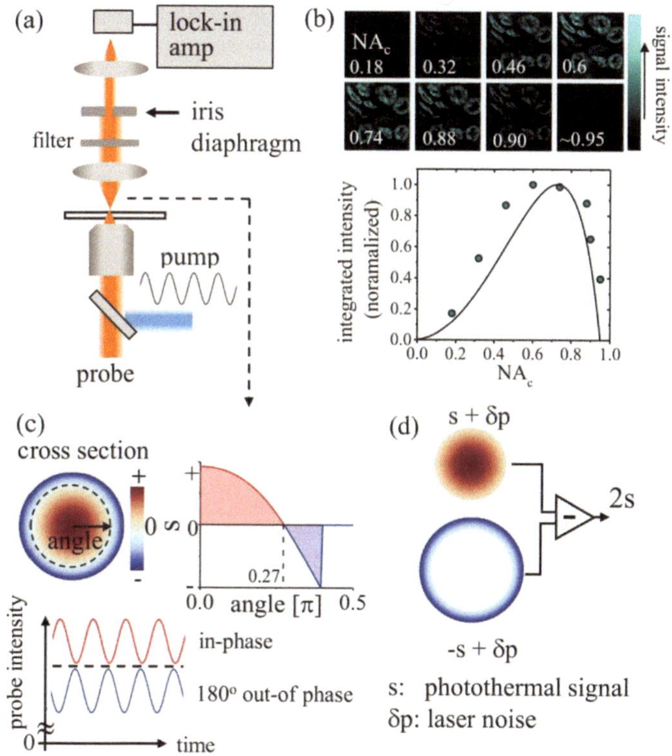

FIGURE 3.2.1 (a) Schematic illustration of a photothermal microscopic image with a conventional detection scheme. (b) Upper, photothermal images of a slice of mouse kidney stained with hematoxylin and eosin with various values of the numerical aperture of condenser (NAc); lower, integrated intensity as a function of the NAc [12]. The theoretical curve (solid curve) is calculated on the basis of the scattering theory [13]. (c) Angular dependence of the photothermal signal. Transmitted probe beam at a small (large) angle is modulated in phase (180° out of phase) with the modulating pump beam owing to the photothermal effect. The theoretical curve is calculated by assuming that the probe beam is incident on the sample through an objective lens with a converging angle of 0.4 πrad. (d) Conceptual scheme of the radially segmented balanced detection for improving signal intensity s and reducing intensity noise of the probe beam δp. The copyright of this figure is partially held between the representative author (Takayoshi Kobayashi) and Optica.

When the probe beam is detected in a forward direction, the photothermal signal mainly comes from the interference between the temporally modulated scattering field and the diverging transmitted probe field. In this case, the photothermal signal is most intense when the numerical aperture (NA) of the condenser is smaller than that of the focusing lens, as shown in Fig. 3.2.1(b) [12]. It should be noted that the signal intensity nearly vanishes when the entire transmitted probe beam is collected. This is because the signal is positive (in-phase with the modulating pump beam) for a transmitted probe beam at a small angle but negative (180° out of phase) for a beam at a large angle [Fig. 3.2.1(c)]. This characteristic behavior was partially explained on the basis of the scattering theory using the temporally modulated Yukawa-type potential [13] or the generalized Lorentz–Mie theory [14, 15].

In previous reports, a low-NA condenser or an iris diaphragm placed after the condenser was used to pass only a part of the probe beam at a small angle for maximization of the SNR [1, 9, 10, 13]. However, as the probe beam at a large angle is modulated out of phase with respect to that at a small angle, it is possible to improve signal intensity by up to two times if the transmitted probe beam is radially segmented into two and then detected by a balanced detector, as shown in Fig. 3.2.1(d).

It is important to note that this detection method also serves to reject the common mode noise of the probe beam. A balanced detector is commonly used to cancel the intensity noise of the probe beam in an absorption measurement [16]. In a typical setup, a part of the probe beam is separated before the sample to be detected as a reference beam. The signal and reference beams are detected separately and then input into a differential amplifier to cancel the common mode noise between the signal and reference beams. In laser-scanning pump-probe microscopy, because the non-uniformity of the refractive index and stationary absorption in the sample cause the probe beam to vary in intensity during the sample scan, an auto-balanced photodetector is used to compensate for the intensity imbalance between the signal and reference beams [17–21]. However, the scan speed is limited by the response time of the autobalancing loop, and thus this scheme is not applicable to fast imaging, as discussed below. The most important advantage of the present technique is that it is less affected by the intensity imbalance, because the transmitted probe beam is separated into two after the sample.

2 EXPERIMENTAL

In this study, a radially segmented balanced-detection scheme was implemented in a photothermal microscopy system using intensity-modulated laser diodes (LDs) (Figure 3.2.2) [21]. LDs of 450 nm and 660 nm were used for pumping and probing, respectively. Each beam was collimated through a single-mode fiber, and the two beams were combined with a dichroic mirror. The probe and pump beam intensities were set to frequencies of ω_1 and ω_2, respectively. A beat signal at $|\omega_1-\omega_2|$ was generated by the photothermal effect in the sample. A lock-in amplifier was referenced to the beat frequency. In this study, ω_1 and ω_2 were set at 1.0 and 1.1 MHz, respectively. This dual modulation scheme serves to reduce electric interference between the pump and probe channels in LD-based pump probe measurement [22]. The pump and probe beams were linearly (vertically) polarized. A polarizing beam splitter was used to direct the combined beam onto the objective lens (Olympus, UPLSAPO 40 × 2) with an NA of 0.95. The beam size was adjusted to fill the back aperture of the objective lens. The sample position was raster-scanned using a three-axis positioning stage driven by piezo actuators (Thorlabs, MAX311D). A condenser (Olympus, U-AAC) was used to collect the entire transmitted light. A narrowband filter matching the laser line was placed in front of the detector to filter out both the

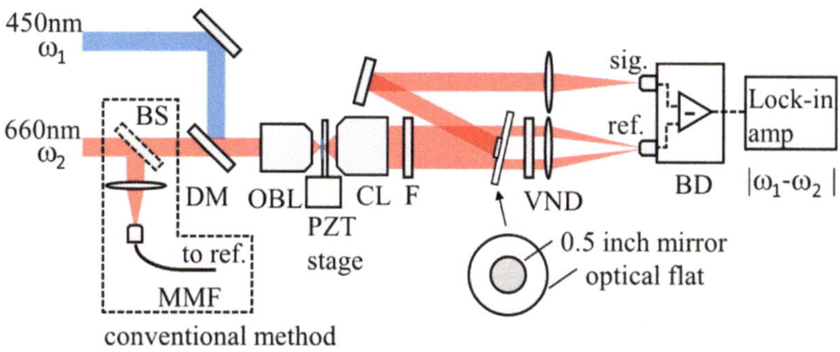

FIGURE 3.2.2 Experimental setup of laser diode-based photothermal microscopy with radially segmented balanced detection. DM: dichroic mirror; BS: beam splitter; OBL: objective lens; CL: condenser lens; F: band pass filter; VND: variable neutral-density filter; BD: balanced photodetector; MMF: multimode fiber. The copyright of this figure is partially held between the representative author (Takayoshi Kobayashi) and Optica.

excitation beam and the fluorescence from the sample, so that only the probe beam was directed to the detector.

To separate the inner portion of the collimated probe beam (inner beam) from the original beam, a mirror with a diameter of 0.5 inch (12.5 mm) was attached to an optical flat placed after the filter. An optical flat is an optical-grade piece of glass lapped and polished to be extremely flat on one or both sides, usually within a few tens of nanometers. They are used with a monochromatic light to determine the flatness (surface accuracy) of other surfaces (whether optical, metallic, ceramic, or otherwise), by means of wave interference. The position of the mirror with respect to the optical axis was adjusted so that the transmitted beam (outer beam) formed a symmetric annular shape. The diameter of the probe beam at the mirror was set at 17.5 mm, which corresponds to the split angle of 0.24 πrad (NA of 0.68). In this case, the intensity ratio of the inner portion to the outer portion was 0.7. The inner and outer beams were directed to the signal and reference ports of the balance photodetector (Newfocus Nirvana), respectively. This photodetector was used for either balanced or autobalanced detection by switching the operation mode. A variable neutral density filter was used to set the intensity of the outer beam to equal that of the inner beam for optimal common mode noise rejection.

The sample of mouse skin melanoma was prepared by culturing B16 melanoma cells RCB-1283 (Riken Bio Resource Centre) in RPMI-1640 supplemented with 10% fetal bovine serum, 100 units/ml penicillin, and 100 μg/ml streptomycin (Meiji Seika Kaisha Ltd.) at 37°C in a humidified atmosphere of 95% air and 5% carbon dioxide. A 50-μl suspension containing 0.5 million melanoma cells for each head was subcutaneously inoculated into female nude mice (6 weeks of age) as five aliquots on the dorsal sides of the mice from the base of the tail to the neck. Tissues containing inoculated cells were fixed with 4% paraformaldehyde for 3 days, embedded in melted paraffin, sliced at 15 to 20 μm, extended on glass slides, and enclosed with a coverslip.

3 RESULTS AND DISCUSSION

The spatial resolution of photothermal microscopy is dependent on the intensity profiles of the pump and probe beams and the refractive index profile at the focal point. When the spatial extent of the modulating refractive index is much smaller than the spot size of the probe beam, the resolution is determined by the product of the intensity profiles of the pump and probe beams; thus, the spatial resolution is better than that of the diffraction-limited optical microscope by approximately 30% for circular beams and about 50% for annular beams [23]. Furthermore, this nonlinear signal provides optical sectioning capability [12].

Photothermal imaging of mouse skin melanoma is demonstrated to evaluate the sensitivity and the resolution of the present method. Figures 3.2.3(a) and (b) show the bright field images of a slice of mouse melanoma observed through a conventional transmission optical microscope. Melanin granules seem to be present on the cell membrane. Fig. 3.2.3(c) shows a photothermal image measured without balanced detection (only the inner beam was incident on a photodiode), whereas Fig. 3.2.3(d) shows an image measured with radially segmented balanced detection. The area of observation is the same as that in Fig. 3.2.3(b). The time constant of the lock-in amplifier is 0.5 ms, and the pixel dwell time is 1 ms. The image size is 19.6 × 19.6 μm with 300 × 300 pixels. The pump and probe beam power incident on the sample are 0.1 mW and 0.25 mW, respectively. The distribution of the melanin granules is clearly visualized in Fig. 3.2.3(d). It can be clearly seen that the spatial resolution of the photothermal image is better than that of the bright-field image. The intensity noise of the LD is substantially reduced by the balanced detection, with a decrease of about 31 dB. We confirm that the noise level is close to the shot noise limit by 2–3 dB. The SNR is approximately 100 in Fig. 3.2.3(d).

The photothermal image measured by the radially segmented balanced detection was compared with those measured by the conventional balanced detection method [Figs. 3.2.3(e)–(h)]. To implement a conventional setup, a part of the probe beam was separated before the sample and used as a reference light instead of the transmitted outer beam (Fig. 3.2.3). The beam intensities incident on the sample were maintained at a constant level. We find that the signal intensity in the radially segmented balanced detection is about 1.7 times better than that in conventional detection [Fig. 3.2.3(g)]. Furthermore, in the case of the conventional balanced detection, a great deal of noise appears in an area with low transmissivity due to the intensity imbalance between the signal and the reference light [Figs. 3.2.3(e), (h)]. The transmissivity at the area marked by the arrows is estimated to be 70%. This intensity imbalance can be canceled by means of autobalanced detection [Figs. 3.2.3(f), (h)]. However, the pixel dwell time should be longer than the response time of the autobalancing loop (typically 10^{-4} to 10^{-2} s) to cancel the intensity imbalance during the sample scanning. In this regard, Xie's group recently conducted stimulated Raman scattering imaging using a fiber laser and homebuilt fast autobalanced detector with a lock-in frequency (modulation frequency of the pump beam ω_2) of 10 MHz and a cut-off frequency of the autobalancing loop of 500 kHz [17]. However, in photothermal microscopy, it is not possible to increase ω_2 up to 10 MHz, because the photothermal signal decreases

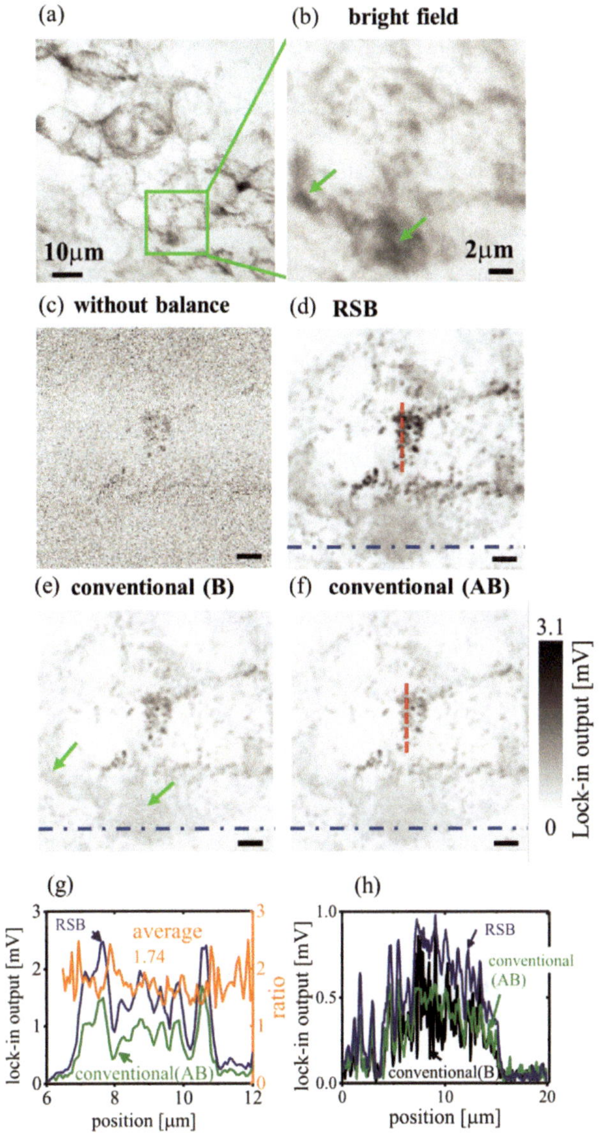

FIGURE 3.2.3 (a), (b) Bright field images of a slice of mouse melanoma observed by a conventional optical microscopy. Photothermal images of mouse melanoma slice (c) without balanced detection and with (d) radially segmented balanced detection (RSB), (e) conventional balanced detection (B), and (f) conventional autobalanced detection (AB). The arrows indicate the area in which a great deal of noise appears in (e) due to the imbalance between the signal and reference intensity. (g) Intensity profiles along the broken lines (red) in (d) and (f) and their ratio. (h) Intensity profiles along the dot-dashed lines (blue) in (d)–(f). Scale bar, 10 μm in (a) and 2 μm in (b)–(f). The pseudo color scheme represents signal intensity in (c)–(f). The copyright of this figure is partially held between the representative author (Takayoshi Kobayashi) and Optica.

by half at 1 to 2 MHz [1, 2, 11]. In contrast, in our detection technique, the laser noise is substantially reduced in the entire imaging area [Figs. 3.2.3(d), (h)] are based on a commonly used balanced detection method whose cut-off frequency is usually much higher than the ω_2 in photothermal microscopy (typically up to 1 MHz). In this case, the pixel dwell time is limited by ω_2. Thus the present method is potentially useful for high-speed imaging by the incorporation of a Galvano scanner into the system and placing the beam splitter of the transmitted probe beam on a plane conjugate to the pupil plane of the condenser lens.

In this study, photothermal signal is demodulated at the beat frequency $j\omega_1-\omega_2 j$ of 100 kHz to avoid the need to employ a high-frequency detector and high-frequency lock-in amplifier. However, to implement the high-speed imaging in the dual modulation scheme, it would be preferable to demodulate the signal at the sum frequency $j\omega_1\ \omega_2 j$ or increase ω_1 so that $j\omega_1-\omega_2 j>\omega_2$.

The background noise of the present method is smaller than that of the conventional setup by 37%, thus the SNR increases 2.3 times. The reduction of the background noise in the present scheme can be explained as follows: in the balanced detection, fluctuation in wavelength causes intensity noise when wavelength-sensitive optics are used in either reference or signal path. Furthermore, vignetting in either path maps beam pointing instability into intensity noise. In the conventional setup, use of the band pass filter and vignetting at the back aperture of the objective lens seem to cause the lack of noise suppression. In contrast, the proposed scheme is less affected by them, because the probe beam is separated after these optics.

4 CONCLUSION

In this subsection, a novel detection method is proposed for highly sensitive photothermal microscopic imaging. This method is based on the characteristics of an angular-dependent photothermal signal; a radially segmented balanced detection method is proposed for highly sensitive photothermal imaging [24, 25]. This method is based on the characteristics of an angular-dependent photothermal signal; it improves signal intensity by up to two times and rejects the intensity noise of the probe beam.

- We have performed the sub-diffraction resolution photothermal imaging of mouse melanoma and confirm that the signal intensity is enhanced 1.7 times compared with the conventional detection scheme. Furthermore, the intensity noise of the probe beam is rejected by about 31 dB even for a sample with non-uniformity of the refractive index and stationary absorption. This method is implemented by means of a commonly used balanced detector and is thus potentially useful for high-speed imaging. The present detection technique paves the way for high-speed photothermal imaging and will be useful for the study of melanosome transport in live cells [24, 25], because it would allow three-dimensional tracking of individual melanosome without labeling.

The work presented in this subsection is based on the collaborative work of the following people: J. Miyazaki, H. Tsurui, K. Kawasumi, and T. Kobayashi [25]. © 2015 Optical Society of America

http://dx.doi.org/10.1364/OL.40.000479

REFERENCES

1. S. Berciaud, D. Lasne, G. A. Blab, L. Cognet, and B. Lounis, *Phys. Rev. B* 73, 045424 (2006).
2. D. Boyer, P. Tamarat, A. Maali, B. Lounis, and M. Orrit, *Science* 297, 1160 (2002).
3. A. Gaiduk, M. Yorulmaz, P. V. Ruijgrok, and M. Orrit, *Science* 330, 353 (2010).
4. W.-S. Chang and S. Link, *J. Phys. Chem. Lett.* 3, 1393 (2012).
5. L. Cognet, C. Tardin, D. Boyer, D. Choquet, P. Tamarat, and B. Lounis, *Proc. Natl. Acad. Sci. USA* 100, 11350 (2003).
6. C. Leduc, J. M. Jung, R. P. Carney, F. Stellacci, and B. Lounis, *ACS Nano* 5, 2587 (2011).
7. C. Leduc, S. Si, J. Gautier, M. Soto-Ribeiro, B. Wehrle Haller, A. Gautreau, G. Giannone, L. Cognet, and B. Lounis, *Nano Lett.* 13, 1489 (2013).
8. D. Lasne, G. A. Blab, S. Berciaud, M. Heine, L. Groc, D. Choquet, L. Cognet, and B. Lounis, *Biophys. J.* 91, 4598 (2006).
9. V. Octeau, L. Cognet, L. Duchesne, D. Lasne, N. Schaeffer, D. G. Fernig, and B. Lounis, *ACS Nano* 3, 345 (2009).
10. S. Lu, W. Min, S. Chong, G. R. Holtom, and X. S. Xie, *Appl. Phys. Lett.* 96, 113701 (2010).
11. P. Vermeulen, L. Cognet, and B. Lounis, *J. Microsc.* 254, 115 (2014).
12. J. Miyazaki, H. Tsurui, K. Kawasumi, and T. Kobayashi, "Simultaneous dual color imaging of nonfluorescent tissues with 3D subdiffraction microscopy," *Opt. Express* 23, 3647 (2015).
13. J. Miyazaki, H. Tsurui, K. Kawasumi, and T. Kobayashi, *Opt. Express* 22, 18833 (2014).
14. M. Selmke and F. Cichos, *J. Opt. Soc. Am. A* 31, 2370 (2014).
15. M. Selmke, M. Braun, and F. Cichos, *Opt. Express* 20, 8055 (2012).
16. M. Celebrano, P. Kukura, A. Renn, and V. Sandoghdar, *Nat. Photonics* 5, 95 (2011).
17. C. W. Freudiger, W. Yang, G. R. Holtom, N. Peyghambarian, X. S. Xie, and K. Q. Kieu, *Nat. Photonics* 8, 153 (2014).
18. A. Gambetta, V. Kumar, G. Grancini, D. Polli, R. Ramponi, G. Cerullo, and M. Marangoni, *Opt. Lett.* 35, 226 (2010).
19. P. C. D. Hobbs, *Appl. Opt.* 36, 903 (1997).
20. K. Seto, Y. Okuda, E. Tokunaga, and T. Kobayashi, *J. Phys. D* 47, 345401 (2014).
21. J. Miyazaki, H. Tsurui, A. Hayashi-Takagi, H. Kasai, and T. Kobayashi, *Opt. Express* 22, 9024 (2014).
22. C. R. Hu, M. N. Slipchenko, P. Wang, P. Wang, J. D. Lin, G. Simpson, B. Hu, and J. X. Cheng, *Opt. Lett.* 38, 1479 (2013).
23. J. Miyazaki, K. Kawasumi, and T. Kobayashi, *Opt. Lett.* 39, 4219 (2014).
24. T. S. Kuroda and M. Fukuda, *Nat. Cell Biol.* 6, 1195 (2004).
25. J. Miyazaki, H. Tsurui, K. Kawasumi, and T. Kobayashi1, *Opt. Lett.* 40, 479 (2015). © 2015 Optical Society of America 480.

3.3 Fast 3D Visualization of Endogenous Brain Signals with High-sensitivity Laser Scanning Photothermal Microscopy

1 INTRODUCTION

Photothermal (PT) microscopy can visualize non-fluorescent chromophores with high sensitivity and high spatial resolution. Thus far, it has been applied to visualize the distribution of endogenous chromoproteins in biological specimens such as cytochromes in mitochondria [1–3], hemoglobin in blood vessels [4], and melanin pigments in skin cancer [5]. Furthermore, gold nanoparticles have been used to identify biomolecules in antibody labeling techniques applied in cellular imaging [6–8]. As gold nanoparticles are physiologically inert and do not suffer from photobleaching or photoblinking, they are highly useful in live cell imaging.

In PT imaging, a sample or a focused laser beam is scanned sequentially point-by-point and the pixel information is assembled into an image. Previously, PT imaging typically used a piezo stage to scan a sample, but stage scanning requires a longer time with integration times per pixel ranging from 1 to 10 ms. In this case, the acquisition time of an image was 4 to 40 minutes for 500 x 500 pixels. One of the advantages of PT microscopy is its 3D imaging capability like confocal microscopy [2, 9]; however, it takes too long to acquire a stack of images at different focal plane depths (a volume scan) using a stage scanning PT microscope. Photoacoustic microscopy (PAM) [10, 11] and PT optical coherent tomography (PTOCT) [12, 13] are powerful techniques for the 3D imaging of optical absorbers *in vivo*, but spatial resolution in PT microscopy is usually better than that of PAM and PTOCT by an order of magnitude. Besides 3D imaging, PT imaging at fast time scales has potential applications for live cell imaging and the tracking of biomolecular transport.

DOI: 10.1201/9781003476030-12

PT microscopy equipped with a scanning mirror is a straightforward approach for fast imaging and there have been several reports of laser scanning PT microscopy [4, 14], which allows image acquisition with an 80 μs pixel dwell time so that 500 x 500 pixel images including background imaging are acquired in 20 s.

To achieve fast imaging, it is also important to improve the signal-to-noise ratio (SNR) as it decreases with measurement time. The PT signal increases with the pump power, but lower pump power is preferable to avoid photo and/or thermal damage to the sample. Several techniques have been proposed to improve SNR in PT microscopy. As the PT signal is proportional to the change in refractive index, chromophores embedded in liquid crystal, which exhibits a large refractive index change with temperature, produce larger PT signals than those in water [15, 16]. This technique is especially useful for single-molecule absorption spectroscopy. In our previous study, a spatially segmented balanced detection (SBD) method was proposed to improve SNR in forward detection [17]. In SBD, the center and peripheral parts of the transmitted probe beam are separated and detected by a balanced detector. This method was incorporated in the stage scanning PT microscope and improved the PT signal by nearly two times, while also canceling the intensity noise of the probe beam to nearly the shot noise limit.

In this subsection, we report the development of a fast, high-sensitivity PT microscope by implementing the SBD scheme in a laser scanning microscope. For this purpose, a custom-made bifurcated fiber bundle was incorporated in a laser diode (LD)-based multimodal laser scanning microscope. Simultaneous PT and fluorescence (FL) imaging of gold nanoparticles and FL beads was performed to characterize the performance of the system. 3D visualization of endogenous mouse brain signals was also demonstrated.

2 EXPERIMENT

A laser scanning PT microscope equipped with Galvano mirrors was developed based on our LD-based multi-wavelength PT microscope (Figure 3.3.1) [5]. A 405-nm (Nichia NDV4B16) and a 520-nm (Osram PL-520-B1) LD were used for pumping, with their intensity modulated at different frequencies (f_1 and f_2). The two pump beams were combined using dichroic mirrors and collimated through polarization-maintaining single-mode fiber (PMSMF) and an off axis parabolic mirror (Thorlabs RC08FC-P01) for spatial-mode filtering. A 640-nm LD (Opnext HL63133DG) was used for probing, which was collimated through another PMSMF and a doublet lens. The offset between the pump and probe beams at the focus was controlled by adjusting the position of the doublet lens to reduce the distortion of the PT signal caused by the twin peaks corresponding to the focusing or defocusing of the probe beam [9]. The combined pump and probe beams were directed to an XY Galvano scanner (GSI VM500PLUS) and focused on a sample through an oil immersion objective lens (Olympus PLAPON60X) [numerical aperture (NA) = 1.4]. To acquire images in different focal planes (z-stack), the sample position was controlled in the axial direction using a positioning stage driven by a piezo actuator (MESS-TEK MA-140XLCa).

The SBD method was demonstrated to be useful for high-sensitivity PT imaging in forward detection as it improves signal intensity by nearly two times while also

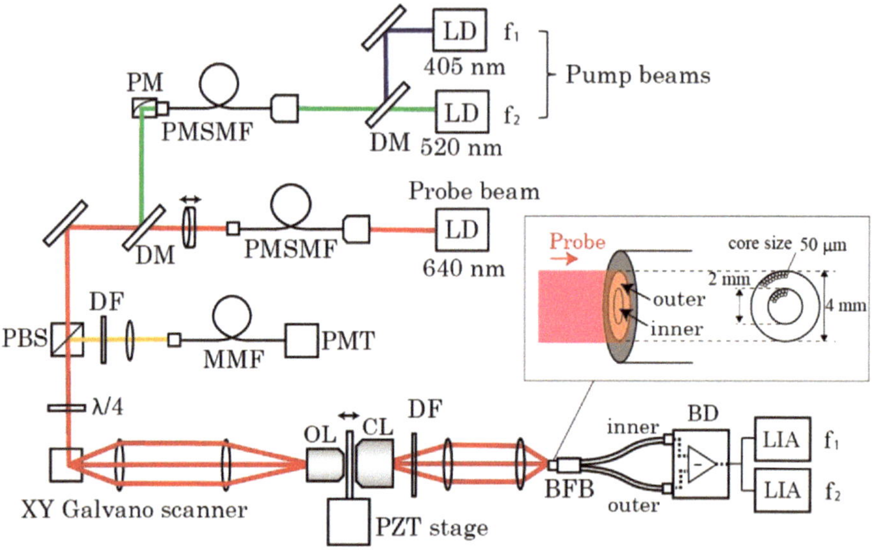

FIGURE 3.3.1 Schematic illustration of a photothermal microscope equipped with a Galvano scanner and spatially segmented balanced detection. LD: laser diode, PMSMF: polarization maintaining single-mode fiber, PM: parabolic mirror, DM: dichroic mirror, PBS: polarizing beam splitter, OL: objective lens, CL: condenser lens, DF: dielectric filter, BFB: bifurcated fiber bundle, BD: balanced detector, LIA: lock-in amplifier, MMF: multimode fiber, PMT: photomultiplier tube. Copyright of this figure is partially held between the representative author (Takayoshi Kobayashi) and Optica (formally Optical Society of America).

canceling the intensity noise of the probe beam [17]. In PT microscopy, the SNR is usually better in forward detection than in backward (epi-)detection [15, 18]. In our previous stage scanning PT microscope, a small mirror was used to separate the center and peripheral parts of the transmitted probe beam. However, it is impractical to incorporate this setup in a laser scanning microscope as it causes complexity in the optical setup. To implement the SBD method in a laser scanning microscope, a custom-made bifurcated fiber bundle with a core fiber size of 50 μm was used to separate the center and peripheral part of the transmitted probe beam. The diameters of the inner and outer circles were 2 and 4 mm, respectively. The incident port of the fiber bundle was placed on a plane conjugate to the pupil of the condenser lens. The center and peripheral parts of the beams separated by the fiber bundle were directed to a balanced photodetector (Newfocus 2307) to cancel the intensity fluctuation of the probe laser. The ratio of the laser powers between the separated two beams depends on the beam spot size at the incident port of the fiber. Thus, the focal lengths of the two lenses after the condenser lens were properly selected so that the laser powers of the separated two beams were nearly equal for optimal common mode noise rejection. PT signals (normal mode) at f_1 and f_2 were demodulated by two lock-in amplifiers (Signal recovery 7270). The output from the lock-in amplifiers was recorded by a multichannel data acquisition board (NI PCIe-6321), which was also used to control the Galvano scanner. The time constant of the lock-in amplifier t_c was set at 10 μs and the pixel dwell time was 20 μs. In this case, image acquisition time was about 6 s for 500 x 500 pixels. Modulation frequencies (ω_1 and ω_2) were typically set at 150 kHz and 250 kHz, respectively. We confirmed that the channel interference was negligible as the bandwidth of the lock-in amplifier $1/(2t_c)$ was sufficiently smaller than the difference in modulation frequencies $|f_1 - f_2|$.

A confocal FL detection scheme was also set up for simultaneous PT and FL imaging, in which the FL signal from the sample was detected by a photomultiplier through a multimode fiber.

To prepare a slice of mouse brain, we fixed adult transgenic mice (two months old, thy1GFP H-line, Jackson Lab.) by cardiac perfusion with a polarization beam splitter (PBS)[1] and paraformaldehyde[2], cut 100-μm slices of the frontal cortex[3] using a vibratome (Leica VT1200), and embedded the slices in UNICRYL resin (Funakoshi). To remove the glycogen, the slices of mouse brain were incubated with 10 mg/ml α-amylase[4] (Tokyo Chemical Industry) in PBS for one hour at 37°C. Then the slices were immunostained with anti-glycogen[5] antibody[6] (a gift from Dr. Otto Baba) to confirm the removal of glycogen. The slices were stained first with the anti-glycogen antibody (12 hours, dilution 1:10 at 4°C), followed by Alexa-Fluor-488[7] conjugated secondary antibody (Life technologies) staining (2 hours, dilution 1:1000 at 20°C).

3 RESULTS AND DISCUSSION

3.1 Evaluation of the Specially Segmented Balanced Detection Method

The PT image measured with SBD using the bifurcated fiber bundle was compared with that measured by the conventional detection scheme. Figure 3.3.2 shows the PT images of hematoxylin and eosin stained tissue with and without SBD. Here only a 520-nm beam was used for pumping. To obtain a conventional PT image without SBD, the iris diaphragm in the collimator lens was adjusted so that only the center part of the probe beam passed through to maximize the SNR [4, 19], and the transmitted beam was detected using a single photodiode. In contrast, in the SBD, the iris diaphragm of the collimator lens was fully opened and the entire transmitted beam was directed to the incident port of the bifurcated fiber bundle separating the center and peripheral area of the beam. It was confirmed that signal intensity in the SBD was about 1.8 times higher than with conventional detection. Furthermore, the intensity noise of the LD was reduced by 37%. Thus, the SNR was improved by a factor of 4.9. This indicates that image acquisition speed with SBD is 24 times faster than with the conventional scheme to obtain the same SNR using the same pump power, because the SNR is proportional to the square root of the measurement time. In the present setup, imaging speed was limited by the time constant of the lock-in amplifier (10 μs at minimum), which must be lower than the pixel dwell time. However, because the time response of the PT signal is in principle determined by the heat conductivity (~1 MHz at −6 dB point for tightly focused laser beam) [18, 20], it is possible to further decrease the pixel dwell time to a few microseconds by using a fast lock-in amplifier that works with a shorter time constant. To evaluate the effect

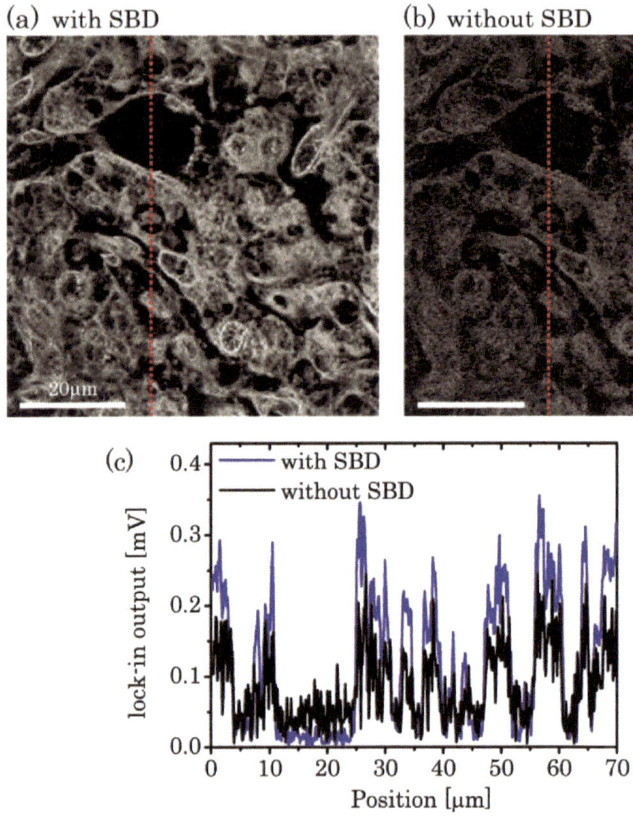

FIGURE 3.3.2 (a) Photothermal images of a hematoxylin and eosin stained rabbit ovary (a) with, and (b) without spatially segmented balanced detection (SBD). Pump and probe beam power incident on the sample were 8 μW and 4 mW, respectively. Image acquisition time was 6 s at 500 x 500 pixels. (c) Intensity profiles along the vertical lines in (a) (blue) and (b) (black). Copyright of this figure is partially held between the representative author (Takayoshi Kobayashi) and Optica (formally Optical Society of America).

of light scattering on the lock-in signal, we measured images without the pump beam and confirmed that light scattering is negligible (less than 5% of the PT signal).

3.2 Spatial Displacement between Photothermal and Fluorescence Images and Spatial Resolution

FL imaging is a well-established method for identifying specific biomolecules in a biological specimen, and the combination of PT and FL imaging can provide further information on the tissues because both the nonfluorescent and fluorescent molecules can be visualized at the same time. To characterize the performance of our imaging modality, the spatial displacement between the PT and FL images was examined by measuring 1 μm FL beads stained with Nile red [Figures. 3.3.3(a)–(d)]. Here, only the 520-nm beam was used for pumping. PT and FL signals were measured simultaneously during the beam scan. The central position of an individual bead was calculated by fitting them with the Gaussian, from which the magnitude of the displacement between the two images was estimated to be 0.18 μm in the lateral plane. PT and FL images of single bead were acquired by changing the sample position in the axial direction, and the displacement in the axial plane was estimated to be 0.22 μm. These displacements may be caused by misalignment of the confocal pinhole. Since the displacement vector in lateral direction is nearly uniform in the scanning area, it can be corrected by the pixel shift scheme. We confirmed that the signal from the FL beads was a result of the PT effect, as twin peaks appeared in the axial plane when the focal offset of the pump and probe beam was close to zero [9]. The ratio of the signal intensities between FL and PT images is different among individual beads due to a small displacement of image in the axial plane. This is the reason why some of the beads are seemingly presented in one of the imaging modes, though all beads appeared in the FL and PT modes.

The spatial resolution was evaluated by PT imaging of 20 nm gold nanoparticles dispersed in PVA on a glass slide [Figs 3.3.3(e)–(g)]. The average of the full width at half maximum (FWHM) values for the 61 measured spots was 149 ± 11 nm and 204 ± 10 nm with excitation at 405 nm and 520 nm, respectively. This is slightly larger than the diffraction-limited spot sizes of the pump beams (0.51λ/NA = 147 nm and 189 nm for λ = 405 nm and 520 nm). The point spread function (PSF) of PT microscopy is considered to be dependent on the intensity profiles of the pump and probe beams and the refractive index profile at the focal point, and the width of the PSF should be smaller than that of the diffraction-limited spot size of the pump and probe

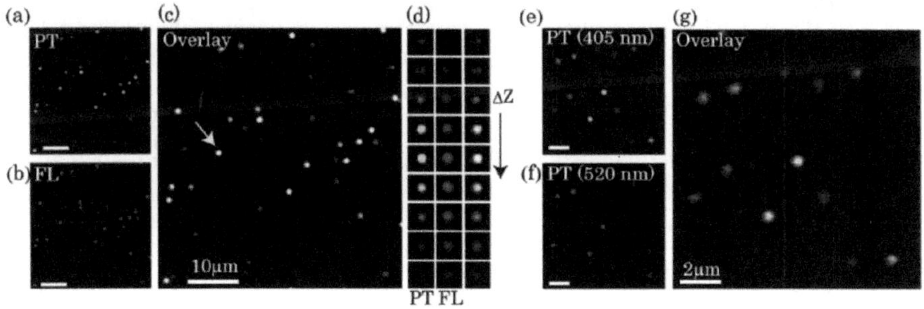

FIGURE 3.3.3 (a) Photothermal, (b) fluorescence, and (c) overlay images of 1 μm Nile red beads dispersed in PVA on a glass slide. Pump and probe beam power incident on the sample were 0.5 and 4 mW, respectively. Imaging area was 48 x 48 μm. Scale bars in (a)–(c): 10 μm. (d) Images of a single bead with a different focal plane for the bead designated by the arrow in (c). The step size was 0.27 μm. Simultaneous dual-wavelength photothermal images with excitation at (e) 405 nm and (f) 520 nm, and (c) overlay images. Pump beam power for both was 2 mW and probe beam power was 12 mW. Scale bars in (e)–(g): 2 μm. Copyright of this figure is partially held between the representative author (Takayoshi Kobayashi) and Optica (formally Optical Society of America).

beams [9]. We consider that wavefront distortion induced by optical elements may have caused degradation in the resolution and that further optimization of the optics should provide better resolution. From a stack of images at different focal planes, the FWHM for the axial plane was estimated to be 0.6 μm and 0.7 μm for the 405 nm and 520 nm beams, respectively.

Multi-wavelength PT imaging is invaluable for identifying the molecules observed in the PT image, as the magnitude of the PT signal is proportional to the absorption spectrum [2, 5]. To perform multi-wavelength PT imaging, it is important to investigate the spatial displacement between the images in the different channels, as chromatic aberration of the optical elements may cause distortion of the multi-wavelength images. PT images of gold nanoparticles were obtained simultaneously at the pumping wavelengths of 405 and 520 nm [Figs.3.3.3(e)–(g)]. The spatial displacement of the image between the two colors was estimated to be smaller than 50 nm in the lateral direction and 0.2 μm in the axial direction, which are about one-third and one-fourth of the PSF width, respectively.

3.3 Imaging of Endogenous Brain Signals

Various linear and nonlinear interactions of light with matter have been applied for label-free brain imaging, such as autofluorescence[8] [21], photoacoustic [11], spontaneous Raman scattering [22, 23], infrared absorption [22, 24], second harmonic generation [21, 25], third harmonic generation (THG) [26], coherent anti-Stokes Raman scattering [27], and stimulated Raman scattering [28]. PT microscopy is a powerful tool for visualizing endogenous signals in biological tissue, yet little brain imaging has been reported. We performed simultaneous PT and FL imaging of a fixed slice of transgenic mouse[9] brain expressing yellow fluorescent protein (YFP[10]). Figure 3.3.4 shows the results of simultaneous PT and confocal FL imaging of different layers of the mouse brain cortex. FL images visualize neurons expressing YFP at the dendrite terminals[11] [Fig. 3.3.4(a)], a dendrite [Fig. 3.3.4(b)], and cell body[12] [Fig. 3.3.4(c)], respectively. The PT images show distinct structures and organization. Red blood cells and vessels can be identified morphologically. They produce strong PT signals as they may contain heme proteins [Figs. 3.3.4(a) and 4(b)]. A Y-shaped junction of a vessel is also observed in Fig. 3.3.4(b). Although cardiac perfusion with PBS is conducted in the sample preparation, there is a possibility that red blood cells are fragmented or stacked in a vessel and result in the PT signal of a vessel.

The cell bodies of the neurons appear as dark shadows. It should be noted that fine structures are observed in all areas of the brain cortex. Figures 3.3.5(a) and (b) show close-up images around an axon[13] and the dendrites of neurons. Bright specks with diameters of around 1 μm appear surrounding the cell bodies. These are attributable to the lipofuscins[14] in lysosomes[15] as weak autofluorescence is detected [21]. Sponge-like structures of submicrometer size cover large areas of the tissue. The combined PT and FL images show that the axon and dendrites visualized in the FL image do not overlap with the fine structures in the PT images. This indicates that the contrasting agent of the sponge-like structure is outside the neuron. A stack of images in discrete z-steps clearly shows the depth profile of the shadow contrast of the cell

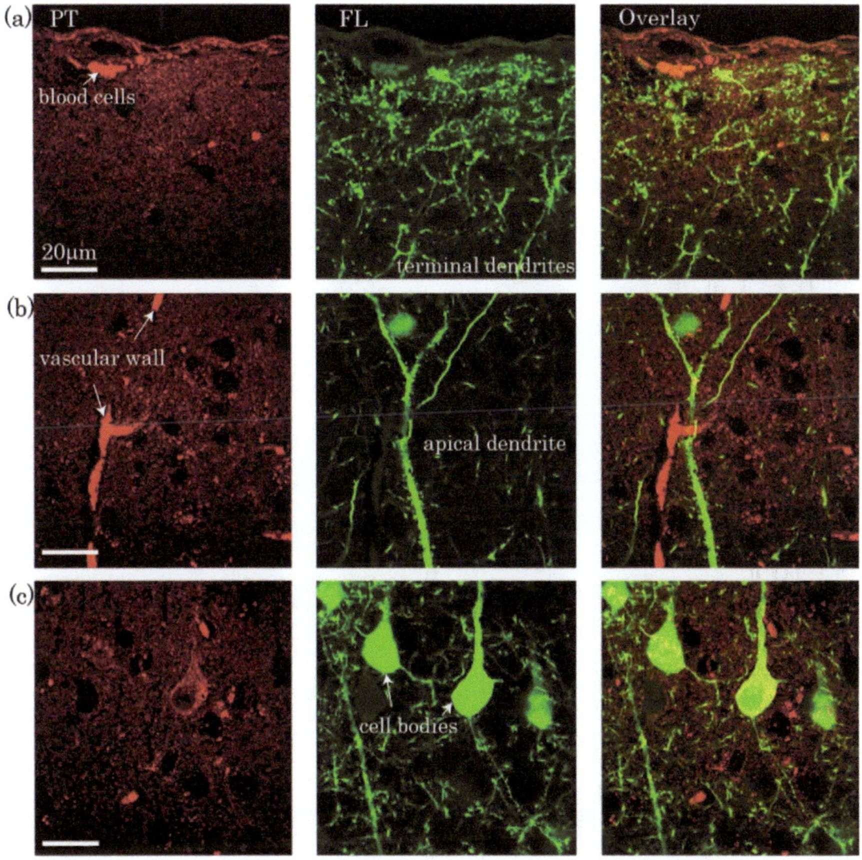

FIGURE 3.3.4 Simultaneous photothermal (PT) and fluorescence (FL) imaging of mouse brain cortex at (a) the surface, (b) layer II/III, and (c) layer IV. Left and center panels show the PT and FL images, respectively, and their overlay is shown in the right panel. The pump and probe beam power incident on the sample were 0.5 and 4 mW, respectively. Image acquisition time was 6 s at 500 x 500 pixels. Copyright of this figure is partially held between the representative author (Takayoshi Kobayashi) and Optica (formally Optical Society of America).

bodies and the fine structures. The 3D-imaging capability is also demonstrated [Fig. 3.3.5(c)] by the reconstructed 3D model of a volume scan.

To obtain further information on the fine structures observed in the PT image, simultaneous dual-wavelength imaging was performed using two pump beams at wavelengths of 405 and 520 nm [Fig. 3.3.5(d)]. From the dual-wavelength PT imaging of the sponge structures, the absorbance ratio between 520 and 405 nm is calculated to be about 0.3. The brain slice shows a faint yellow color. Thus, the light absorption of the brain tissue is predominantly attributed to the species contained in the sponge structures. In contrast, the intensity ratio of the bright specks around the dark shadows is ~0.7, which suggests that the bright specks and the sponge structures

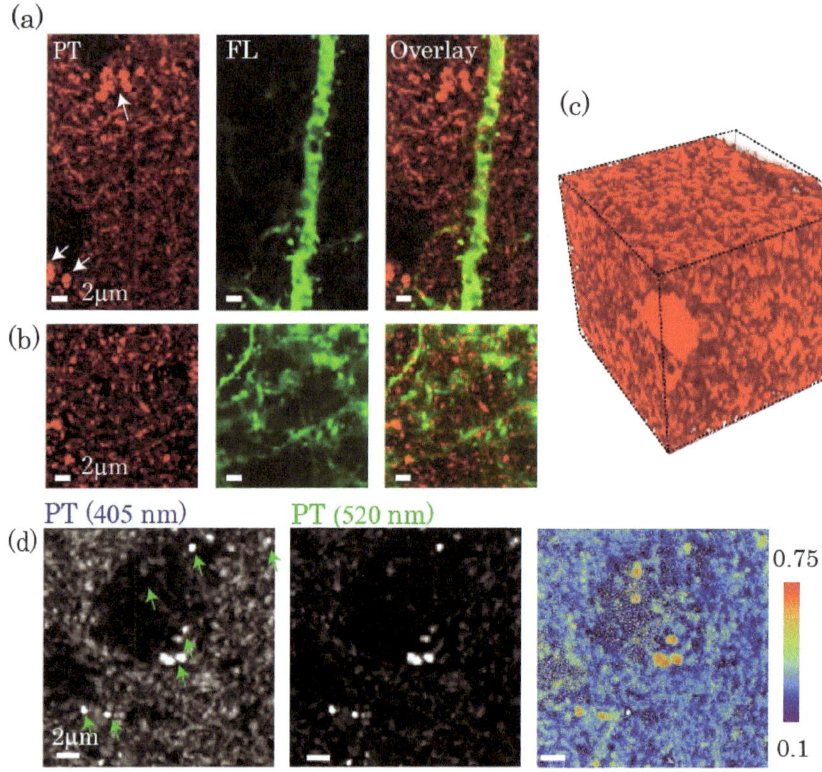

FIGURE 3.3.5 Close-up photothermal (PT) and fluorescence (FL) images of mouse brain cortex. PT and FL images around (a) a dendrite and (b) axons. Arrows shows bright specks localized around cell bodies, which are attributable to lipofuscins in lysosomes. (c) Reconstructed 3D model of PT signal in cortex. Image size is 23 x 23 x 21 μm with 400 x 400 x 80 voxels and acquisition time was about 9 minutes. A stack of 30 images of brain cortex was acquired by changing the sample position in the axial direction with a step size of 0.68 μm. Each image size is 70 x 70 μm with 600 x 600 pixels. (d) Acquisition times were 5 minutes, dual-wavelength PT images around a cell body with excitation at 405 nm (left) and 520 nm (center), and their ratio image (right). Copyright of this figure is partially held between the representative author (Takayoshi Kobayashi) and Optica (formally Optical Society of America).

are composed of different species. To exclude the possibility that transgenic expression of YFP affects the PT signal, we acquired PT images of a wild-type mouse brain and found identical fine sponge structures. This indicates that the PT signal originates mainly from the endogenous chromophores in the brain.

We presume that the fine sponge-like structures can be attributed to the lipids comprising 30–60% of the brain. Hydrophobic dyes such as lutein[16] and zeaxanthin[17] that can cross the blood-brain barrier[18] and/or lipid peroxides[19] may serve as

PT contrast agents. Lutein and zeaxanthin are carotenoid contained in grain, which are concentrated in the neural retina and brain neocortex [29, 30]. It is suggested that lipid has a steeper slope of the fitted line of refractive index *vs.* temperature (~1×10^{-3} K^{-1}) [31], and thus the chromophores in the lipids result in a significant PT signal. It is interesting to note that the PT image of the brain cortex is similar to the THG image [26]. Lipid bodies are a major source of contrast in THG microscopy, which is used to visualize lipid structures in the brain. Interestingly, however, unlike the THG signal, PT signal was not seen in neurons and myelin sheaths, but rather found in intracellular organelles of non-neuronal cells (Fig. 3.3.5). In this regard, in THG imaging, a partial phase match of third-order susceptibility at the focal produces the TH signal and the signal is most intense when a structure is half the size of the focal volume. This can result in the signal intensity distribution being different from the concentration distribution and may work as a kind of spatial filtering function to coarse grain the distribution, reducing the intensity of finer structure. This may be the reason why THG imaging provides more uniform distribution than the PT method [26]. On the other hand, since the contrast mechanisms are different between the THG and PT images, they may provide complemental information on the structure or composition of the lipid.

We also evaluated the possibility that the fine sponge-like structures represent glycogen granules, which are abundant in astrocytes. However, the PT signal did not change after treatment with amylase, which digests glycogen in the sample. Moreover, the PT signals rarely overlapped with glycogen immune-reactive granules.

Previous study showed that the PT signal of live cultured neurons originated from cytochrome c [32]. Cytochrome c is water solvable and the refractive index/temperature slope of water is 1×10^{-4} K^{-1}, while lipid has a steeper slope [31]. Therefore, the PT signal of lipid should be larger that of cytochrome c in imaging of brain slice even though they may have comparable optical absorption. However, there is still a possibility that cytochrome c contributes to the PT signal and careful study is need to this point.

Lipid metabolism in the brain is closely related to neurodegenerative illnesses such as Alzheimer's disease[20] [24, 33]. Hence, the ability to image both the neuron and the lipid structure at high spatial resolution will be a powerful technique for the study of such diseases.

4 CONCLUSION

The present subsection describes how fast, high-sensitivity photothermal (PT) microscopic imaging was conducted by incorporating the segment balanced detection (SBD) scheme into a laser scanning multi-wavelength and multimodal microscope [34].

It was further confirmed that SNR improves by a factor of 4.9; thus image acquisition time becomes 24 times faster than that of conventional detection with the same SNR and pump power. Furthermore, we have succeeded in realizing simultaneous bimodal (PT and confocal FL) imaging of transgenic mouse brain tissue expressing YFP with a pixel dwell time of 20 μs was demonstrated. Furthermore, the PT method detected endogenous non-fluorescent chromophores in brain tissue and various organizations such as blood cells, lipofuscin in lysosomes, and fine sponge structures, which are presumably attributable to lipids, all visualized with high 3D resolution. As this imaging modality is based on compact, low-cost laser diodes, it will be widely useful in the life and medical sciences, e.g., for lipid metabolism in the brain and related neurodegenerative diseases [34]. The research presented in this subsection is based on collaboration of the following people: J. Miyazaki, T. Iida, S. Tanaka, A. Hayashi-Takagi, H. Kasai, S. Okabe and T. Kobayashi [34].

NOTES

1. polarization beam splitter (PBS): a splitter of a light beam into a transmitted and a reflected beam. It is a crucial part of many optical experimental and measurement systems, such as interferometers, also widespread application in fiber optic telecommunications. In its most common form it is made from two triangular transparent glass prisms.
2. paraformaldehyde: the smallest polyoxymethylene, the polymerized of formaldehyde with a typical degree of polymerization of n = 8–100 units. $HO[(CH2)O]_{8\text{-}100}H$.
3. frontal cortex: the largest of the four major lobes of the brain in mammals, and is located at the front of each cerebral (brain) hemisphere (in front of the parietal lobe (a major sensory processing hub for the brain) and the temporal lobe (one of the four major lobes of the cerebral cortex in the brain of mammals)).
4. α-amylase: an enzyme that hydrolyses α bonds of large, α-linked polysaccharides, such as starch and glycogen, yielding shorter chains thereof, dextrins, and maltose.
5. glycogen: a multibranched polysaccharide of glucose that serves as a form of energy storage in animals, fungi, and bacteria. It is the main storage form of glucose in the human body.
6. antibody: a large, Y-shaped protein used by the immune system to identify and neutralize foreign objects such as pathogenic bacteria and viruses. The antibody recognizes a unique molecule of the pathogen, called an antigen.
7. Alexa-Fluor-488: a fluorescent compound with an excitation peak at 499 nm and an emission peak at 520 nm. There are several other fluorescent dyes in the family group of Alexa such as Alexa-555 and Alexa-647.
8. autofluorescence: autofluorescence is the natural emission of light by biological structures such as mitochondria and lysosomes when they have absorbed light, and is used to distinguish the light originating from artificially added fluorescent markers (fluorophores). The most commonly observed autofluorescencing molecules are NADPH (Nicotinamide adenine dinucleotide phosphate), a cofactor used in anabolic reactions, such as the Calvin cycle and lipid and nucleic acid syntheses.
9. transgenic mouse: a genome altered or genetically engineered mouse commonly used for research or as animal models of human diseases and are also used for research on genes.
10. YFP: Yellow fluorescent protein is a genetic mutant of green fluorescent protein (GFP) originally derived from the jellyfish *Aequorea victoria*. Its excitation peak is 513 nm and its emission peak is 527 nm. Like GFP, YFP is a useful cell in molecular biology because the excitation and emission peaks of YFP are distinguishable from GFP which allows for the study of multiple processes/proteins in the same experiment.

11. dendrite terminal: a branched protoplasmic extension of a nerve cell that propagates the electrochemical stimulation received from other neural cells to the cell body[12], or soma, of the neuron from which the dendrites project. Electrical stimulation is transmitted onto dendrites by upstream neurons (usually via their axons:[13]) via synapses which are located at various points throughout the dendritic tree.
12. cell body: often used to refer to neurons, it can also refer to other cell types as well, including astrocytes, oligodendrocytes, and microglia. There are many different specialized types of neurons, and their sizes vary from as small as about 5 μm to over 10 mm for some of the smallest and largest neurons of invertebrates, respectively.
13. axon: a long, slender projection of a nerve cell, or neuron, in vertebrates, that typically conducts electrical impulses known as action potentials away from the nerve cell body. The function of the axon is to transmit information to different neurons, muscles, and glands. In certain sensory neurons (pseudounipolar neurons), such as those for touch and warmth, the axons are called afferent nerve fibers and the electrical impulse travels along these from the periphery to the cell body and from the cell body to the spinal cord along another branch of the same axon.
14. lipofuscin: fine yellow-brown pigment granules composed of lipid-containing residues of lysosomal15 digestion. It is considered to be one of the aging or "wear-and-tear" pigments, found in the liver, kidney, heart muscle, retina, adrenals, nerve cells, and ganglion cells.
15. lysosome: a membrane-bound organelle found in many animal cells. They are spherical vesicles that contain hydrolytic enzymes that can break down many kinds of biomolecules.
16. lutein: synthesized only by plants, and like other xanthophylls is found in high quantities in green leafy vegetables such as spinach, kale, and yellow carrots. In green plants, xanthophylls act to modulate light energy and serve as non-photochemical quenching agents to deal with triplet chlorophyll, an excited form of chlorophyll which is overproduced at very high light levels during photosynthesis. Animals obtain lutein by ingesting plants. In the human retina, lutein is absorbed from blood specifically into the macula lutea, although its precise role in the body is unknown. Lutein is also found in egg yolks and animal fats. Lutein is isomeric with zeaxanthin, differing only in the placement of one double bond. Lutein and zeaxanthin can be interconverted in the body through an intermediate called meso-zeaxanthin.
17. zeaxanthin: one of the most common carotenoids used in the xanthophyll cycle in nature. Synthesized pigment in plants and some micro-organisms, ~~it~~This carotenoid dye is the that gives characteristic color to paprika (made from bell peppers), corn, saffron (spice derived from the flower of *Crocus sativus*, known as "saffron crocus"), goji (wolfberries), and many other plants and microbes.
18. blood-brain barrier: a highly selective semipermeable border of endothelial cells that regulate the transfer of solutes and chemicals between the circulatory system and the central nervous system, thus protecting the brain from harmful substances in the blood. This system allows the passive diffusive passage of some small hydrophobic molecules such as O_2, CO_2 and hormones as well as selective and active transport of various nutrients, ions, organic anions, and macromolecules such as glucose (sugar with formula $C_6H_{12}O_6$) and amino acids (organic compounds containing both amino and carboxylic acid functional groups) that are crucial to neural function.
19. lipid peroxide: lipid peroxidation is the process of conversion of lipids to peroxide and hydroperoxide derivatives. These derivatives, known as lipid peroxides or lipid oxidation products (LOPs), are susceptible to further reactions that are relevant to DNA and

protein modification, radiation damage, and aging. Lipid peroxidation mainly applies to unsaturated fats, especially polyunsaturated fats such as those derived from linoleic acid.

20. Alzheimer's disease: a neurodegenerative disease that usually starts slowly and progressively worsens, and is the cause of 60–70% of cases of dementia (decline in cognitive abilities that impacts a person's ability to perform everyday activities).

REFERENCES

1. E. Tamaki, K. Sato, M. Tokeshi, K. Sato, M. Aihara, and T. Kitamori, "Single-cell analysis by a scanning thermal lens microscope with a microchip: direct monitoring of cytochrome c distribution during apoptosis process," *Anal. Chem.* **74**(7), 1560–1564 (2002).
2. D. A. Nedosekin, E. I. Galanzha, S. Ayyadevara, R. J. Shmookler Reis, and V. P. Zharov, "Photothermal confocal spectromicroscopy of multiple cellular chromophores and fluorophores," *Biophys. J.* **102**(3), 672–681 (2012).
3. A. V. Brusnichkin, D. A. Nedosekin, E. I. Galanzha, Y. A. Vladimirov, E. F. Shevtsova, M. A. Proskurnin, and V. P. Zharov, "Ultrasensitive label-free photothermal imaging, spectral identification, and quantification of cytochrome c in mitochondria, live cells, and solutions," *J. Biophotonics* **3**(12), 791–806 (2010).
4. S. Lu, W. Min, S. Chong, G. R. Holtom, and X. S. Xie, "Label-free imaging of heme proteins with two-photon excited photothermal lens microscopy," *Appl. Phys. Lett.* **96**(11), 113701 (2010).
5. J. Miyazaki, H. Tsurui, K. Kawasumi, and T. Kobayashi, "Simultaneous dual-wavelength imaging of nonfluorescent tissues with 3D subdiffraction photothermal microscopy," *Opt. Express* **23**(3), 3647–3656 (2015).
6. C. Leduc, S. Si, J. Gautier, M. Soto-Ribeiro, B. Wehrle-Haller, A. Gautreau, G. Giannone, L. Cognet, and B. Lounis, "A highly specific gold nanoprobe for live-cell single-molecule imaging," *Nano Lett.* **13**(4), 1489–1494 (2013).
7. C. Leduc, J. M. Jung, R. P. Carney, F. Stellacci, and B. Lounis, "Direct investigation of intracellular presence of gold nanoparticles via photothermal heterodyne imaging," *ACS Nano* **5**(4), 2587–2592 (2011).
8. L. Cognet, C. Tardin, D. Boyer, D. Choquet, P. Tamarat, and B. Lounis, "Single metallic nanoparticle imaging for protein detection in cells," *Proc. Natl. Acad. Sci. U.S.A.* **100**(20), 11350–11355 (2003).
9. J. Miyazaki, H. Tsurui, and T. Kobayashi, "Reduction of distortion in photothermal microscopy and its application to the high-resolution three-dimensional imaging of nonfluorescent tissues," *Biomed. Opt. Express* **6**(9), 3217–3224 (2015).
10. A. P. Jathoul, J. Laufer, O. Ogunlade, B. Treeby, B. Cox, E. Zhang, P. Johnson, A. R. Pizzey, B. Philip, T. Marafioti, M. F. Lythgoe, R. B. Pedley, M. A. Pule, and P. Beard, "Deep in vivo photoacoustic imaging of mammalian tissues using a tyrosinase-based genetic reporter," *Nat. Photonics* **9**, 239–246 (2015).
11. J. Yao, L. Wang, J. M. Yang, K. I. Maslov, T. T. Wong, L. Li, C. H. Huang, J. Zou, and L. V. Wang, "Highspeed label-free functional photoacoustic microscopy of mouse brain in action," *Nat. Methods* **12**(5), 407–410 (2015).
12. C. Pache, N. L. Bocchio, A. Bouwens, M. Villiger, C. Berclaz, J. Goulley, M. I. Gibson, C. Santschi, and T. Lasser, "Fast three-dimensional imaging of gold nanoparticles in living cells with photothermal optical lock-in optical coherence microscopy," *Opt. Express* **20**(19), 21385–21399 (2012).

13. J. M. Tucker-Schwartz, T. A. Meyer, C. A. Patil, C. L. Duvall, and M. C. Skala, "In vivo photothermal optical coherence tomography of gold nanorod contrast agents," *Biomed. Opt. Express* **3**(11), 2881–2895 (2012).
14. S. Arunkarthick, M. M. Bijeesh, G. K. Varier, M. Kowshik, and P. Nandakumar, "Laser scanning photothermal microscopy: fast detection and imaging of gold nanoparticles," *J. Microsc.* **256**(2), 111–116 (2014).
15. W.-S. Chang and S. Link, "Enhancing the sensitivity of single-particle photothermal imaging with thermotropic liquid crystals," *J. Phys. Chem. Lett.* **3**(10), 1393–1399 (2012).
16. A. N. G. Parra-Vasquez, L. Oudjedi, L. Cognet, and B. Lounis, "Nanoscale thermotropic phase transitions enhancing photothermal microscopy signals," *J. Phys. Chem. Lett.* **3**(10), 1400–1403 (2012).
17. J. Miyazaki, H. Tsurui, K. Kawasumi, and T. Kobayashi, "Sensitivity enhancement of photothermal microscopy with radially segmented balanced detection," *Opt. Lett.* **40**(4), 479–482 (2015).
18. S. Berciaud, D. Lasne, G. A. Blab, L. Cognet, and B. Lounis, "Photothermal heterodyne imaging of individual metallic nanoparticles: theory versus experiment," *Phys. Rev. B* **73**(4), 045424 (2006).
19. P. Vermeulen, L. Cognet, and B. Lounis, "Photothermal microscopy: optical detection of small absorbers in scattering environments," *J. Microsc.* **254**(3), 115–121 (2014).
20. D. Boyer, P. Tamarat, A. Maali, B. Lounis, and M. Orrit, "Photothermal imaging of nanometer-sized metal particles among scatterers," *Science* **297**(5584), 1160–1163 (2002).
21. A. C. Kwan, K. Duff, G. K. Gouras, and W. W. Webb, "Optical visualization of Alzheimer's pathology via multiphoton-excited intrinsic fluorescence and second harmonic generation," *Opt. Express* **17**(5), 3679–3689 (2009).
22. C. Krafft, M. Kirsch, C. Beleites, G. Schackert, and R. Salzer, "Methodology for fiber-optic Raman mapping and FTIR imaging of metastases in mouse brains," *Anal. Bioanal. Chem.* **389**(4), 1133–1142 (2007).
23. C. Krafft, "Bioanalytical applications of Raman spectroscopy," *Anal. Bioanal. Chem.* **378**(1), 60–62 (2004).
24. N. Benseny-Cases, O. Klementieva, M. Cotte, I. Ferrer, and J. Cladera, "Microspectroscopy (μFTIR) reveals Co-localization of lipid oxidation and amyloid plaques in human alzheimer disease brains," *Anal. Chem.* **86**(24), 12047–12054 (2014).
25. D. A. Dombeck, K. A. Kasischke, H. D. Vishwasrao, M. Ingelsson, B. T. Hyman, and W. W. Webb, "Uniform polarity microtubule assemblies imaged in native brain tissue by second-harmonic generation microscopy," *Proc. Natl. Acad. Sci. U.S.A.* **100**(12), 7081–7086 (2003).
26. S. Witte, A. Negrean, J. C. Lodder, C. P. de Kock, G. Testa Silva, H. D. Mansvelder, and M. Louise Groot, "Label-free live brain imaging and targeted patching with third-harmonic generation microscopy," *Proc. Natl. Acad. Sci. U.S.A.* **108**(15), 5970–5975 (2011).
27. C. L. Evans, X. Xu, S. Kesari, X. S. Xie, S. T. Wong, and G. S. Young, "Chemically-selective imaging of brain structures with CARS microscopy," *Opt. Express* **15**(19), 12076–12087 (2007).
28. C. W. Freudiger, W. Min, B. G. Saar, S. Lu, G. R. Holtom, C. He, J. C. Tsai, J. X. Kang, and X. S. Xie, "Label free biomedical imaging with high sensitivity by stimulated Raman scattering microscopy," *Science* **322**(5909), 1857–1861 (2008).
29. B. R. Hammond, Jr. and L. M. Renzi, "Carotenoids," *Adv. Nutr.* **4**(4), 474–476 (2013).
30. E. J. Johnson, "A possible role for lutein and zeaxanthin in cognitive function in the elderly," *Am. J. Clin. Nutr.* **96**(5), 1161S–1165S (2012).

31. F. J. van der Meer, D. J. Faber, I. Cilesiz, M. J. van Gemert, and T. G. van Leeuwen, "Temperature-dependent optical properties of individual vascular wall components measured by optical coherence tomography," *J. Biomed. Opt.* **11**(4), 041120 (2006).
32. D. Lasne, G. A. Blab, F. De Giorgi, F. Ichas, B. Lounis, and L. Cognet, "Label-free optical imaging of mitochondria in live cells," *Opt. Express* **15**(21), 14184–14193 (2007).
33. J. P. Liu, Y. Tang, S. Zhou, B. H. Toh, C. McLean, and H. Li, "Cholesterol involvement in the pathogenesis of neurodegenerative diseases," *Mol. Cell. Neurosci.* **43**(1), 33–42 (2010).
34. J. Miyazaki, T. Iida, S. Tanaka, A. Hayashi-Takagi, H. Kasai, S. Okabe and T. Kobayashi, "Fast 3D visualization of endogenous brain signals with high-sensitivity laser scanning photothermal microscopy," *Biomed. Opt. Express* **6**(1), 22088 (2016).

4.1.1 Numerical Calibration of the Spatial Overlap for Subtraction Microscopy

1 INTRODUCTION

Superresolution microscopic techniques, such as stimulated emission depletion (STED) microscopy [1], structured illumination microscopy (SIM) [2], photoactive localization microscopy (PALM) [3], stochastic optical reconstruction microscopy (STORM) [4], and others [5–10], have achieved great success in breaking the diffraction barrier of optical microscopy with lateral resolution beyond 100 nm or even 20 nm [11]. Meanwhile, subtraction microscopy [12–15] developed recently attracts interest for providing a simpler way to obtain superresolution by mathematically processing a group of fluorescence images. The concept is to subtract the acquired two fluorescence images of the sample that are excited with a single wavelength but two different spot shapes. The produced image gets an enhanced resolution compared with the initial confocal images. Because of the problems of negative signals and over-subtraction in such microscopes, several progressive ideas have also been proposed. To reduce the nonlinearity brought by negative sidebands, excitation lights with radial and azimuthal polarizations and with extended solid and hollow spot shapes have been put forward [16, 17]. Further to achieve appropriate subtraction, the concept of a subtraction threshold has also been provided [18]. However, inevitable displacements of the two excitation light spots could exist in the experiment. It may result from imprecise experimental adjustments or inaccurate localization of the excitation spot centers during data processing. In STED or pump-probe microscopy, the effects of this imperfect overlap are small or even negligible due to the quenching characteristic in STED or insensitivity of the product profile of pump-probe spots. In contrast, in subtraction microscopy, the resolution-enhanced images are derived from the direct subtraction of two fluorescent images. The displacements may induce the reduction of resolution and intensity of the signal, or even distortions in the subtracted images. In these cases, the effects of the displacements on the subtracted results have to be found out. However, there is neither such

DOI: 10.1201/9781003476030-13

a report on explaining these effects nor a model to realize analysis on the quality of subtracted images. A fast, precise approach for inspecting the overlap of spot centers or digitally calibrating the acquired fluorescence images is also desired for reliable subtraction microscopes. In this subsection a common model to evaluate the quality of subtracted images is established based on vector beam diffraction theory (VBDT). Parameters of peak intensity, negative sidebands, or peak position shift are numerically described. With this model, the effects of imperfect overlaps in lateral and axial directions on the resolution, signal intensity, and optimum subtraction threshold are performed. It quantitatively clarifies the importance of the overlap to improve the performance of subtraction microscopes. With the calculations, lateral displacement of 0.1λ for the excitation light spots is found to have about 3% deterioration of image resolution and 10% loss of signal peak intensity in the case of sample size of 0.1λ, focused numerical aperture of 1.4, and subtraction factor of unity. The peak position shift versus subtraction factors can be used for conveniently and simply calibrating the overlap. The two spots aligned with an axial displacement of 0.3λ can suppress the negative sidebands by 20% compared with complete overlap while the corresponding resolution and signal intensity loss are 10% and 15%, respectively. It provides a strategy to reduce the nonlinearities brought by negative sidebands without introducing special optics. Besides, the simulated results with different sample sizes, numerical aperture of the objective lens, and polarization combination for excitations are also investigated and discussed for calibration of the spatial overlaps.

2 PRINCIPLE AND NUMERICAL MODELS

The schematics of the spatial displacements in a subtraction microscope and their effects on the subtracted results are shown in Figure 4.1.1.1. When the two excitation spots have lateral displacement, illustrated in Fig. 4.1.1.1(a), the subtracted images could generate inconspicuous or notable deformations as in the images in Fig. 4.1.1.1(b) and (c). Fig. 4.1.1.1(d) shows the horizontal cross sectioning profiles of the images, in which asymmetrical distortion can be observed. Figs. 4.1.1.1(e)–(h) show the results with displaced excitation spots in the axial plane. Although the shapes of the subtracted spots are not deformed, the resolution degradation in subtracted images can be clearly seen.

To quantitatively analyze the effects of the spatial displacements on the quality of the images and to investigate the appropriate parameters for inspecting the overlap in the subtraction microscope, a numerical calculation is performed based on the following conditions. The location of point spread function (PFS) of the excitation beam is parallely shifted in space rather than tilted light incidence for simulating the displacements and simplifying the calculation.

The sample is assumed to be thin and have a fluorophore concentration distribution of Gaussian shape. The confocal detection and the fluorescence Stokes shift are not taken into account, to avoid the complexity introduced by the inhomogeneous fluorescence collection efficiency and various detection PSFs. The optical collection efficiencies and the photoelectric conversion efficiencies for the two excitation light

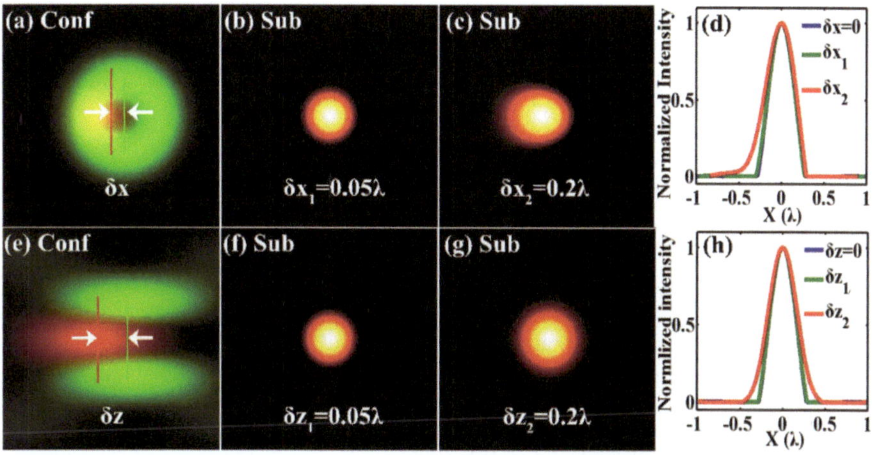

FIGURE 4.1.1.1 Schematics of spatial displacements in subtraction microscope. (a)–(d) Excitation spots with lateral misalignment of δx, subtracted images with factor of 0.3 while misaligned values are $\delta x_1 = 0.05\lambda$ and $\delta x_2 = 0.2\lambda$, and the cross sectioning profiles. (e)–(f) Excitation spots with axial misalignment of δz, subtracted images with factor of 0.3 while misaligned values are $\delta z_1 = 0.05\lambda$ and $\delta z_2 = 0.2\lambda$, and their cross sectioning profiles. Copyright of this figure is shared between the representative author (Takayoshi Kobayashi) and Optica (formally Optical Society of America)

spots are considered to be the same. Then, the theoretical model can be written with a group of mathematical formulas. The focused excitation point spread function can be obtained from VBDT [19–23]:

$$E_{x,y,z}(x,y,z) = \frac{iC}{\lambda}\int_0^{2\pi}\int_0^{\arcsin(NA/n)} A_{AMP}\cdot A_{\text{Phase}}\cdot A_L\begin{pmatrix}P_x\\P_y\\P_z\end{pmatrix} \sin\theta\cdot e^{-ik\left\{z\cos\theta+\sqrt{x^2+y^2}\sin\theta\cos\left[\varphi-\arctan(y/x)\right]\right\}}d\theta d\varphi \quad (1)$$

$$I_{G,D}(x,y,0) = \left|E_x\right|^2 + \left|E_y\right|^2 + \left|E_z\right|^2 \quad (2)$$

Here C is the normalization constant; λ is the wavelength of incident light; NA is the numerical aperture of the objective lens; n is the refractive index of the contact medium between the objective and the cover glass for the sample, the value of which is 1.518 for the immersion oil; k is the wave number; A_{AMP} is the amplitude factor; A_{phase} is the phase factor. The set values of A_{AMP}, A_{phase} and the polarization unit vector matrix for I_G and I_D are shown in Table 4.1.1.1. The I_G and I_D with radial and azimuthal polarizations are only used in the simulations in Section 3.5.

A_L is the vector weight matrix of the aplanatic objective lens, calculated as follows:

$$A_L = \sqrt{\cos\theta}\begin{pmatrix} 1+(\cos\theta-1)\cos^2\varphi & (\cos\theta-1)\cos\varphi\sin\varphi & -\sin\theta\cos\varphi \\ (\cos\theta-1)\cos\varphi\sin\varphi & 1+(\cos\theta-1)\sin^2\varphi & -\sin\theta\sin\varphi \\ \sin\theta\cos\varphi & \sin\theta\sin\varphi & \cos\theta \end{pmatrix} \tag{3}$$

The sample has a preset full width at half maximum (FWHM) value of r, express as:

$$Ob(x,y,r) = C_0 e^{-(4\ln 2)\frac{x^2+y^2}{r^2}} \tag{4}$$

where C_0 is the intensity normalization constant.

The subtracted images are calculated as

$$\begin{aligned} I_{RAW}(x,y,z,\delta x,\delta z,r,\gamma) &= I_G(x+\delta x/2, y, z+\delta z/2) * Ob(x,y,r) \\ &\quad - \gamma \cdot I_D(x-\delta x/2, y, z-\delta z/2) * Ob(x,y,r) \\ I(x,y,z,\delta x,\delta z,r,\gamma) \\ = \begin{cases} I_{RAW}(x,y,z,\delta x,\delta z,r,\gamma) & I_{RAW}(x,y,z,\delta x,\delta z,r,\gamma) \geq 0 \\ 0 & I_{RAW}(x,y,z,\delta x,\delta z,r,\gamma) < 0 \end{cases} \end{aligned} \tag{5}$$

where δ x and δ z are the displacements in the X component (or lateral) and Z component (or axial) in the Cartesian coordinates, γ is the subtraction factor, ∗ is the convolution. After the convolution and subtraction, the generated negative sidebands are removed and the final image $I(x, y, z, \delta x, \delta z, r, \gamma)$ is obtained.

Additionally, we give definitions of other parameters to quantitatively model the quality of subtracted images. The evolution of peak intensities, or Ipeak, against subtraction factors is used to evaluate the loss of signal intensities. The negative sidebands, which may bring nonlinearity into the subtraction microscope, are also defined by the following:

$$I_{\text{peak}} = \frac{\text{MAX}\left[I(x,y,z,\delta x,\delta z,r,\gamma)\right]}{\text{MAX}\left[I(x,y,z,\delta x,\delta z,r,0)\right]} \tag{6}$$

$$\left|I_{Neg}\right| = \begin{cases} 0 & \text{MIN}\left[I_{RAW}(x,y,z,\delta x,\delta z,r,\gamma)\right] \geq 0 \\ \dfrac{\left|MIN\left[I_{RAW}(x,y,z,\delta x,\delta z,r,\gamma)\right]\right|}{MAX\left[I(x,y,z,\delta x,\delta z,r,\gamma)\right]} & \text{MIN}\left[I_{RAW}(x,y,z,\delta x,\delta z,r,\gamma)\right] < 0 \end{cases} \tag{7}$$

The subtraction threshold line, or $r_0 - \gamma_0$ line, describing an appropriate subtraction, is defined as the resolved FWHM value is equal to the preset FWHM value. The threshold peak intensity factor or I_P factor, defined as the normalized peak intensity

TABLE 4.1.1.1
Set values of I_C and I_D

I_{PSF}	Polarization	A_{AMP}	A_{phase}	$[P_x\ P_y\ P_z]$
I_G	Right Circular	1	1	$[1\ -i\ 0]/\sqrt{2}$
I_D	Left Circular	1	$e^{i\varphi}$	$[1\ i\ 0]/\sqrt{2}$
I_G	Radial	1	1	$\left[\cos\varphi \quad \sin\varphi \quad 0\right]$
I_D	Azimuthal	1	1	$\left[-\sin\varphi \quad \cos\varphi \quad 0\right]$

subtracted with the factor at the threshold, describes the threshold signal intensity with proper subtraction. They are expressed as:

$$FWHM_{Ob}(x,y,r_0) = FWHM_I(x,y,z,\delta\ x,\delta\ z,r_0,\gamma_0) \tag{8}$$

$$I_P = \frac{MAX\left[I\left(x,y,z,\delta x,\delta z,r_0,\gamma_0\right)\right]}{MAX\left[I\left(x,y,z,\delta x,\delta z,r_0,0\right)\right]} \tag{9}$$

The peak position shifted values in the lateral plane are defined as the relative shifted values of peak positions versus subtraction factors according to the initial position without subtraction.

$$\left(\Delta x,\Delta y\right) = (X,Y)_{MAX\left[I\left(x,y,z,\delta x,\delta y,r,\gamma\right)\right]} - \left(X_0,Y_0\right)_{MAX\left[I\left(x,y,z,\delta x,\delta y,r,0\right)\right]} \tag{10}$$

With these defined parameters, the FWHM resolution, signal intensity, maximum values of negative signals after subtraction, and the subtraction thresholds can be investigated and evaluated with different spatial displacements as well as the sample size, *NA* of objective lens and the polarization of incident beams.

3 CALCULATION RESULTS

3.1 Lateral Displacement

Firstly, effects of lateral position shifts with the amount of $\delta\ x$=0, 0.05λ, 0.1λ, and 0.2λ are analyzed. As are shown in Figure 4.1.1.2(a) and 2(b), when the cross sectioning lines of excitation spots are shifted from each other with $\delta\ x$, the subtracted lines with factors of γ=0.3 and of γ= 1 exhibit resolution degradation, signal intensity decrease, and peak position shifts. Calculating with the defined formulas, the evolutions of the FWHM values, normalized peak intensities, and maximum absolute values of the negative signal are plotted against the subtraction factors shown in Fig. 4.1.1.2(c)–(e) for different amounts of displacement. It can be found that, if the displacement value is smaller than 0.1λ, the FWHM is about 3% larger than the case

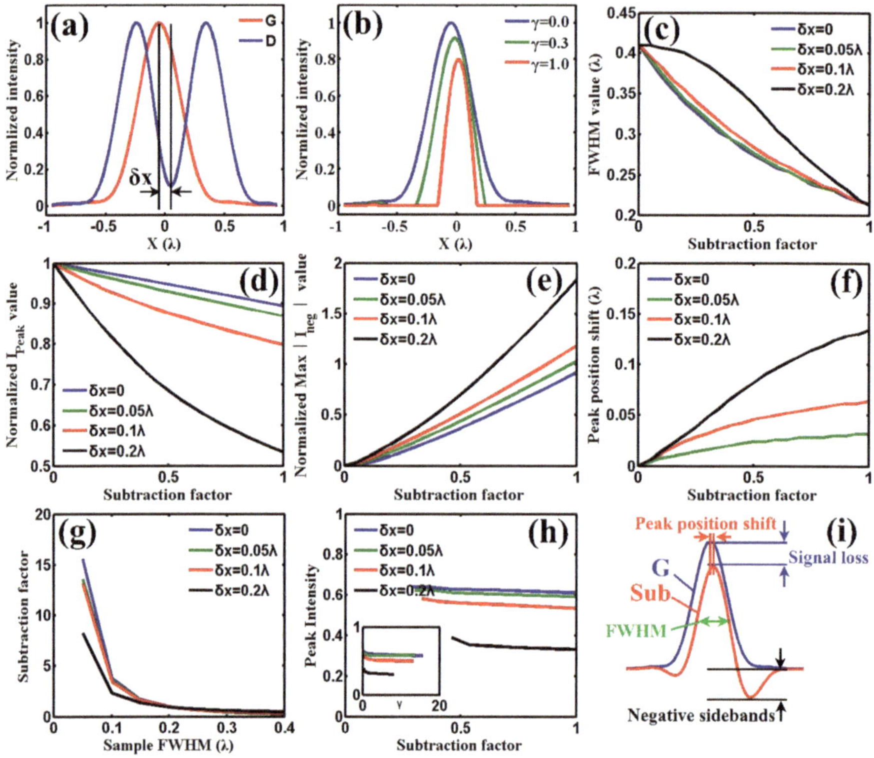

FIGURE 4.1.1.2 (a) Cross sectioning lines of lateral shifted excitation spots. (b) Subtracted results with γ= 0, 0.3, 1. (c)–(f) Evolution of subtracted FWHM values, normalized peak intensities, maximum values of negative sidebands and the subtracted peak position shifts versus the subtraction factors with excitation displacements of $\delta x = 0, 0.05\lambda, 0.1\lambda, 0.2\lambda$. (g)–(h) Subtraction threshold lines and the threshold peak intensity factors with excitation displacements of $\delta x = 0, 0.05\lambda, 0.1\lambda, 0.2\lambda$. (i) Schematic of key parameters in the subtracted microscopes. The preset sample size is $r = 0.1\lambda$. Copyright of this figure is partially held between the representative author (Takayoshi Kobayashi) and Optica (formally Optical Society of America)

of perfect overlap; the relative peak intensity decreases quickly with increasing misalignment higher than 0.1λ; the absolute values of the negative signal after subtraction increase with subtraction factors and with δx. The results indicate that although slight lateral displacement within 0.1λ does not degrade the FWHM values more than 3%, the loss of signal intensity by 10% and the increased negative sideband by 20% may deteriorate the image quality significantly. In this case, it is necessary to inspect whether or not there is offset between the two fluorescence images or excitation spots.

Here the peak position shift is found to be effective parameters. As is shown in Fig. 4.1.1.2(f), when the misalignment exists, the subtracted peak positions are shifted with the increasing subtraction factors. The larger the slope of this peak position shift against subtraction factors, the larger is the misalignment. It is also more convenient and rapid to realize the algorithm of subtracting the two fluorescence images and locating the position of the peak rather than locating the centers of excitation spots [12], where fitting the donut spot is much so difficult and requires accuracy. In addition, the fluorescence peak intensity or the value of negative sidebands can be a secondary parameter used for calibrating the misalignments even though it is less convenient to apply. Shifting the confocal images, subtracting them with the same factor and then checking if the peak intensity in the subtracted image is increased or if the value of negative sidebands is decreased is essential. The signal intensity is maximal or the value of negative sidebands is minimal when the overlap is optimized.

There is a third strategy, i.e., combining the peak intensity curves with the subtraction threshold lines to inspect the overlap of the excitation spots. The concept can be simply introduced as follows. Get the subtraction factor value when the peak intensity after subtraction drops to 0.6, corresponding to the proper subtraction without displacement [18]. Find the sample size (or FWHM value) from the subtraction threshold line with this factor. Compare the FWHM value obtained in the subtracted image with this predicted value. If they cannot be matched, there is substantial misalignment existing. Optimize the displacement and repeat the same procedure to check. However, this method is not recommended because it is too time consuming and because of the error which may be induced in the process of matching the FWHM values. Moreover, the ellipticity of the spot shape after subtraction can also be used to analyze the misalignments. Due to its lower effectiveness and possibly incorrect ellipticity values caused by the same amounts of displacements in the two axes, it is not discussed here further.

The subtraction threshold lines and threshold peak intensity factors with separate amounts of displacement are shown in Fig. 4.1.1.2(g) and 2(h). They describe the obtained FWHM values in subtracted images equal to the preset FWHM values of the sample. To resolve the image of a sample with small sizes of 0.05λ, a subtraction factor as large as 15 is needed. It can be found that the larger displacement can have a smaller subtraction threshold factor and smaller threshold peak intensity. It indicates that displaced fluorescence images would suffer from signal degradation and are vulnerable to over-subtraction for reaching appropriate subtraction. The parameters of FWHM, peak position shift, signal loss and negative sidebands are illustrated in Fig. 4.1.1.2(i).

3.2 Axial Displacement

Similarly, the simulations with axial shifts of δz can also be realized, shown in Figure 4.1.1.3. After subtracting the cross sectioning lines of excitation PSFs, the obtained peak intensities are reduced without peak-position shifts as shown in Figs. 4.1.1.3(a) and (b). Fig. 4.1.1.3(c) shows the FWHM value becomes pretty large for position shifts larger than 0.3λ. It is interesting to see that the peak intensity after subtraction decreases more slowly and the absolute negative values are reduced by about 20% in the cases of displaced excitations, as displayed in Figs. 4.1.1.3(d) and

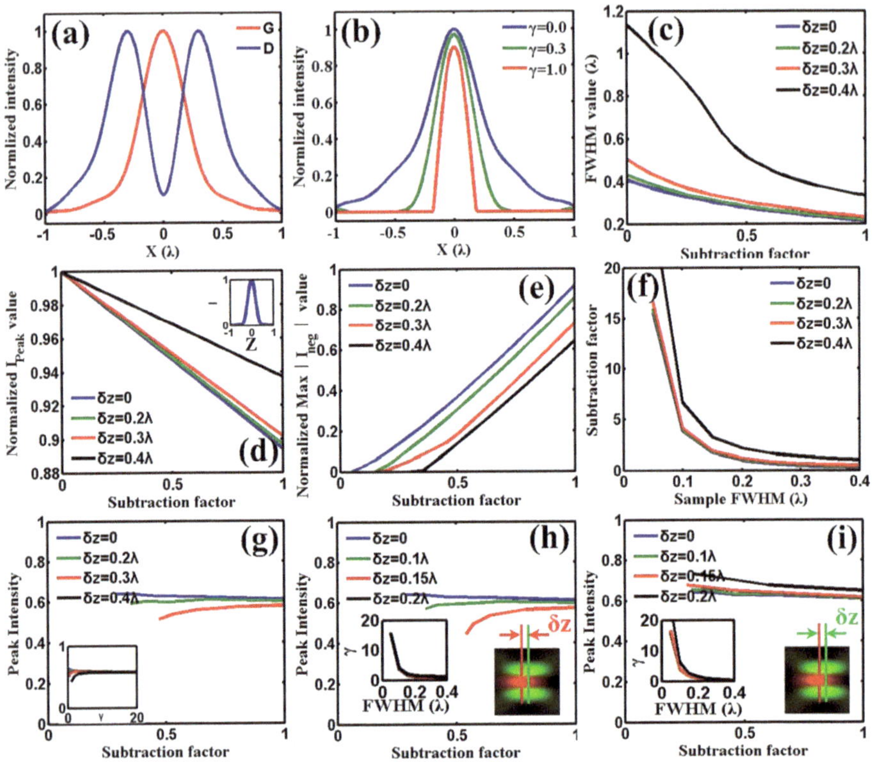

FIGURE 4.1.1.3 (a) Intensities along the axial shifted excitation spots. (b) Subtraction results of them with $\gamma = 0, 0.3, 1$. (c)–(e) Evolution of subtracted FWHM values, normalized peak intensities and maximum values of negative signal versus the subtraction factors with excitation displacements of $\delta z = 0, 0.2\lambda, 0.3\lambda, 0.4\lambda$. (f)–(g) Subtraction threshold lines and the peak intensity factors with excitation displacements. (h)–(i) Subtraction threshold lines and the peak intensity factors with only I_G or I_D shifted from central focal plane. The preset sample size is $r = 0.1\lambda$. Copyright of this figure is shared between the representative author (Takayoshi Kobayashi) and Optica (formally Optical Society of America)

3(e). It means that a small axial position shift can reduce the nonlinearity brought by negative sidebands at the cost of a small amount of reduction in resolving power and signal intensity. The peak intensity seems to be apparently increased with displacements. However, because the focused spot is shifted from the center of the focal plane, the signal loss in subtracted images contains two parts. One is from the signal decrease in subtraction; the other is from the signal decrease due to the focal spot shifting from the center. Therefore, the actual signal after subtraction is the multiplication of the I_{peak} values in Fig. 4.1.1.3(d) with the focused PSF beam profile along the axial direction, shown with a blue curve in the top right figure in Fig. 4.1.1.3(d). The subtraction threshold lines and the threshold peak intensity factors are shown in Figs. 4.1.1.3(f) and (g), respectively. The corresponding subtraction threshold factors are larger and the peak intensity factors are smaller in the cases of axial position shifts. It indicates that subtraction results with axial displacements are relatively insensitive to over-subtraction while the signal intensity reduction cannot yet be avoided at the subtraction threshold. The subtraction threshold lines and peak intensity factors for displacements with Gaussian spot only and donut spot only shift from the central focal plane are shown in Figs. 4.1.1.3(h) and (i). The only Gaussian-spot shift can readily lead to over-subtraction rather than the case of only donut-spot shift. However, further calculations find that the only donut-spot shift generates larger negative signals than the case of only Gaussian-spot shift. It indicates a trade-off between tolerance of over-subtraction and production of negative side bands.

3.3 Sample Size

Simulating with a larger sample size of $r = 0.3\lambda$, the subtracted FWHM, peak intensities, and maximum values of negative signal are not sensitive to the amount of displacement as illustrated in Figures 4.1.1.4(a)–(c). With lateral displacement of $\delta x = 0.1\lambda$, the resolution is degraded by 3% and the peak intensity is reduced by 6%. Compared with former results obtained with sample size of $r = 0.1\lambda$, it demonstrates that the sample with a larger size is less sensitive to the negative effects of lateral misalignments. In contrast, the peak position shifts are still effective and sensitive to investigate the good overlap, the curves of which with different amount of lateral displacement are distinguishable and shown in Fig. 4.1.1.4(d). The subtracted results with axial displacements are displayed in Figs. 4.1.1.4(e)–(g). The maximum negative values can be significantly reduced by almost 50% according to axial displacement of 0.3λ. It indicates that the reduction of the negative sidebands with axially shifted excitations is more effective for the sample with larger sizes.

3.4 Numerical Aperture

Results of simulation with an objective of $NA = 0.9$ describe that the loose focus in subtraction microscopy permits a higher robustness against the displacements, shown in Figure 4.1.1.5. Compared with the values with a perfect overlap, the FWHM values after subtraction vary within 6% even with lateral displacement of 0.2λ or with axial displacement of 0.5λ as depicted in Figs. 4.1.1.5(a) and (e). Figs. 4.1.1.5(b) and (f) show that the subtracted peak intensities corresponding to the above cases are 10% weaker than or almost the same (effects of axial PSF excluded) as the cases of perfect overlaps, respectively. The related maximum values of the negative signal after subtraction in Figs. 4.1.1.5(c) and (g) result in a 30% increase or have almost no change. The peak position shift is still sensitive to the subtraction factors with lateral shifts, as shown in Fig. 4.1.1.5(d). Therefore it is proved to be an effective parameter to calibrate the misalignment. In Figs. 4.1.1.5(h) and (j), the subtraction threshold lines are similar to the ones without misalignments. It means the required subtraction factors are approximately the same in both cases with and without misalignments. However, the misalignment in the lateral plane leads to the peak intensity loss at subtraction threshold, while in the axial plane it almost does not change as shown in Figs. 4.1.1.5(i) and (k).

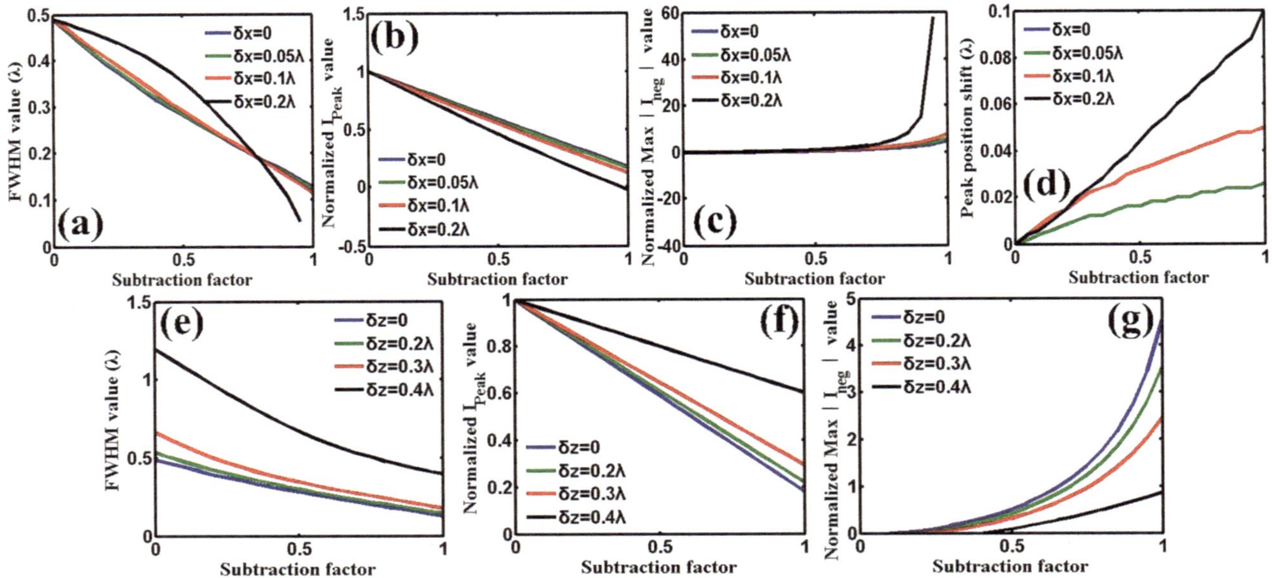

FIGURE 4.1.1.4 Simulation results with preset sample size of $r = 0.3\lambda$. (a)–(d) FWHM values, normalized peak intensities, normalized absolute values of maximum negative signals and peak position shifts versus subtraction factors with lateral displacements of $\delta\ x = 0, 0.05\lambda, 0.1\lambda, 0.2\lambda$. (e)–(g) FWHM values, normalized peak intensities and normalized absolute values of maximum negative signals versus subtraction factors with axial displacements of $\delta\ z = 0, 0.2\lambda, 0.3\lambda, 0.4\lambda$. (h)–(k) Corresponding subtraction threshold lines and peak intensity factors with lateral and axial displacements. Copyright of this figure is shared between the representative author (Takayoshi Kobayashi) and Optica (formally Optical Society of America)

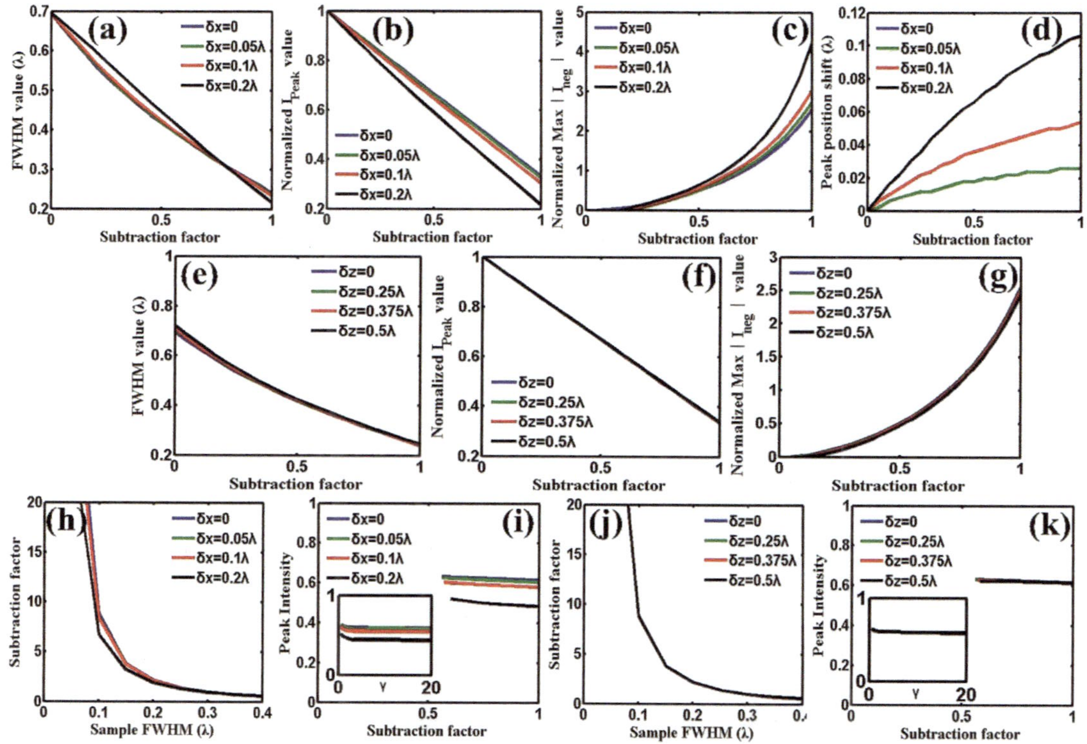

FIGURE 4.1.1.5 Results of simulation with objective NA =0.9. (a)–(d) FWHM, peak intensities, maximum negative values and peak position shifts versus the subtraction factors with lateral displacements. (e)–(g) FWHM, peak intensities, maximum negative value versus the subtraction factors with axial displacements. (h)–(k) Subtraction threshold lines and peak intensity factors with lateral and axial displacements, respectively. The preset sample size is $r = 0.4\lambda$. Copyright of this figure is shared between the representative author (Takayoshi Kobayashi) and Optica (formally Optical Society of America)

3.5. Polarization

Simulated results with polarization combination of radial and azimuthal polarizations and focused objective of NA = 1.4 are shown in Figure 4.1.1.6. Similar to the results in Figs. 4.1.1.2 and 4.1.1.3, spatial misalignment of $\delta x = 0.1\lambda$ or $\delta z = 0.3\lambda$ is acceptable for the subtracted FWHM value and the corresponding peak intensity decreases are within 10% or almost the same (effects of axial PSF excluded) compared with the results without misalignments, as shown in Figs. 4.1.1.6(a), (b), (e) and (f). However, the absolute negative values are increased by 50% or decreased by 10% compared with the results without displacements, displayed in Figs. 4.1.1.6(c) and (g). Fig. 4.1.1.6(d) describes the peak intensity shifts versus the subtraction factors with different misalignments in lateral direction. It can be seen that it is also useful to estimate the misalignments. The subtraction threshold lines in Figs. 4.1.1.6(h)–(k) show similar rules with former polarizations. However, the subtracted line shapes versus the subtraction factors are not as straight as the lines in Fig. 4.1.1.2(h) and Fig. 4.1.1.3(g). The FWHM curve with axial displacement of $\delta z = 0.4\lambda$ is also not smooth. It is due to appeared fringes of excitation PSFs with a large axial displacement. Furthermore, a large shift in axial direction leads to asymmetric spot deformation after subtraction. It results in the slight increase of FWHM for $\delta x = 0.2\lambda$ in Fig. 4.1.1.6(a).

3.6 A Simulation with 2D Excitation Spot Shifts

Four groups of subtracted images with corresponding 2D evolution of peak position shifts and the peak intensity value of beads sample are simulated and shown in Figure 4.1.1.7. The wavelengths of the light sources are both 488 nm. The objective lens is an oil-immersion type with NA = 1.4. The sample is a combination of beads with four sizes of 200 nm, 150 nm, 100 nm, and 50 nm. The pixel sizes in the images are 2 nm. The illumination light with Gaussian spot shape is spatially position-shifted with different amounts in the lateral plane. The first groups results with position shifts δx = 12 nm and δy = 20 nm are shown in Figs. 4.1.1.7(a)–(c). From the evolution of the peak position shifts, the directions of the displacement can be estimated while the amount of the displacement can be compared. Both of the positive shift values in X and Y directions indicate that the I_G has relative positive shifts to I_D images in both X and Y directions. Meanwhile, the shifted amount in Y direction increases more rapidly than that in the X direction. It indicates that the I_G images have larger shifted value in Y direction than in X direction. In contrast, the shifts cannot be directly resolved from the subtracted images with this small amount of displacements. Similarly, the trends of position shifts in the second and third groups also agree with the preset displacement values of δx = 40 nm, δy = -24 nm, and δx=-52 nm, δy = -56 nm as shown in Figs. 4.1.1.7(d)–(i). Compared with non-displaced results shown in Figs. 4.1.1.7(j)–(l), it can be seen that the images of beads in the third groups are distorted after subtraction. The observed shifted directions agree with the directions analyzed from Figs. 4.1.1.7(g) and (h). From the peak position shifts of the former three groups of displacements, it can also be found that

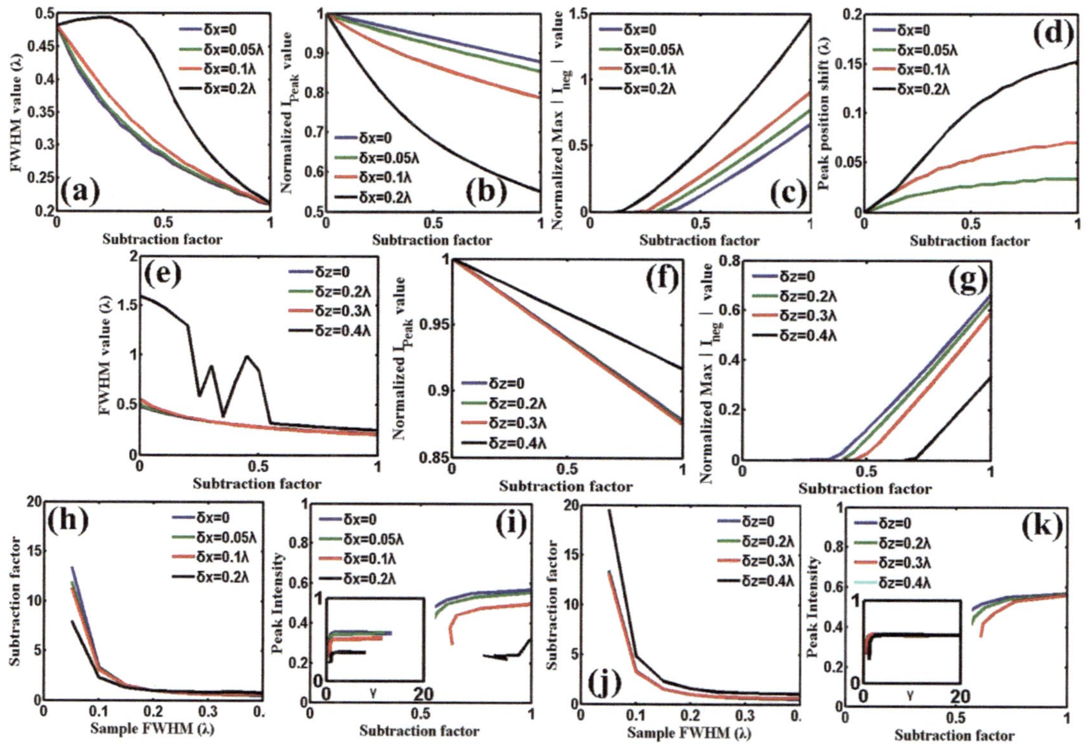

FIGURE 4.1.1.6 Simulation results with radial and azimuthal polarizations. (a)–(d) FWHM, peak intensities, maximum negative values and peak position shifts versus the subtraction factors with lateral displacements. (e)–(g) FWHM, peak intensities, maximum negative value versus the subtraction factors with axial displacements. (h)–(k) Subtraction threshold lines and peak intensity factors with lateral and axial displacements, respectively. Sample size is $r = 0.1\lambda$. Copyright of this figure is shared between the representative author (Takayoshi Kobayashi) and Optica (formally Optical Society of America)

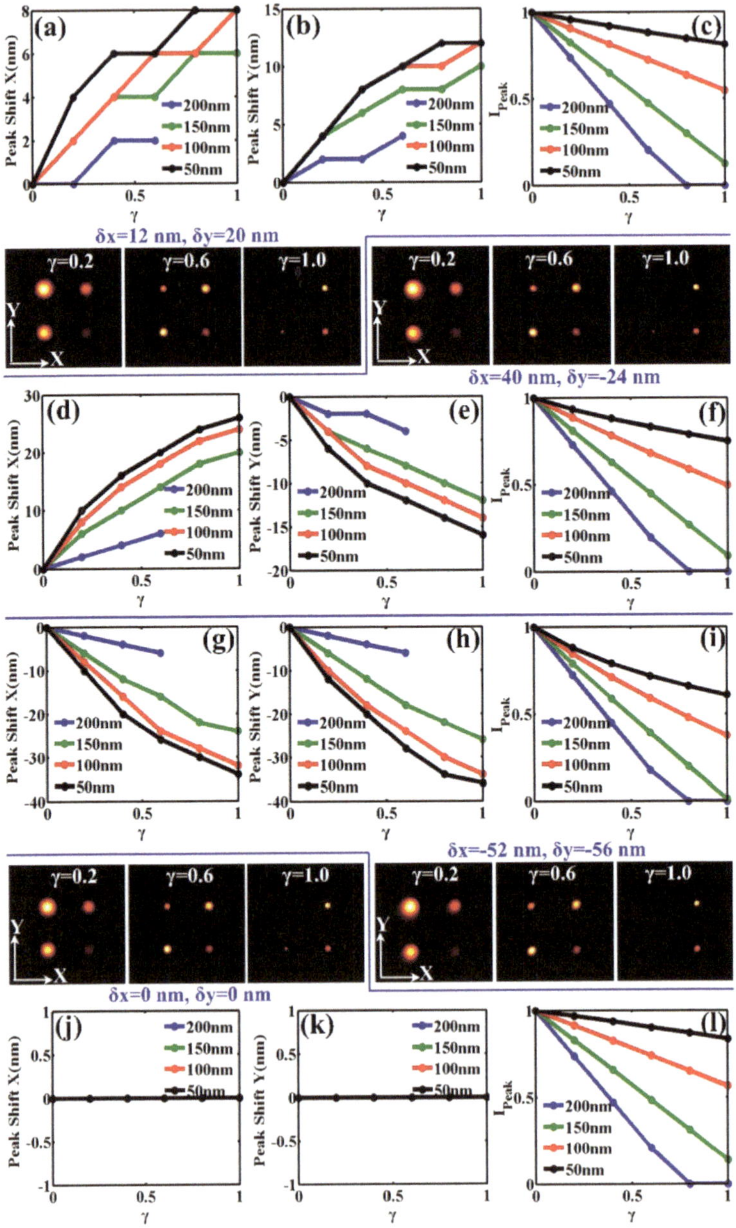

FIGURE 4.1.1.7 Simulations with 2D position shifts with beads of FWHM = 200, 150, 100 and 50 nm (from left to right, top to bottom in the subtracted images). (a)–(c) Evolution of peak position shifts in *X*, *Y* direction and the peak intensity values with I_G displaced with $\delta\ x = 12$ nm and $\delta\ y = 20$ nm. Similarly, (d)–(e) Results with $\delta\ x = 40$ nm and $\delta\ y = -24$ nm. (g)–(i) Results with $\delta\ x = -52$ nm and $\delta\ y = -56$ nm. (j)–(l) Results without misalignment. Copyright of this figure is shared between the representative author (Takayoshi Kobayashi) and Optica (formally Optical Society of America)

the larger sample sizes correspond to a smaller position shift. It can be a potential parameter to compare the relative sample sizes.

4 DISCUSSION

The calculation in this chapter describes the numerical calibration method of spatial overlap in the subtraction microscope. When there is misalignment in the lateral plane, it reduces the resolving power and signal intensity, and also introduces larger negative values which bring in nonlinearity. By checking the evolution of peak position shifts versus subtraction factors, the misalignment can be evaluated. The overlap without misalignment gets no peak position shifts. What is more, the algorithm of checking the peak of the image is easier to realize than it is of fitting the donut spot; it can be easily integrated in the data acquisition software, such as LabVIEW. Then the acquired data can be analyzed in real time for optimizing the alignments and improving the efficiency. If the misalignments appear in the axial plane, the resolution becomes worse and the peak intensity also drops. However, it is not necessary to optimize the axial overlap to reach perfect overlap because the slight misalignment can reduce the negative sidebands while it does not seriously affect the resolution and the signal intensity.

The calculations in this chapter do not take the confocal detection and Stokes-shifts of the fluorophores into consideration. In the case of a small confocal pinhole for detection, the fluorescence emitted from the displaced excitation spot could deviate from the confocal detection position and give rise to a low collection efficiency or even image distortion. Therefore, a more accurate result can be obtained with specific detection parameters and fluorescent labels included. In addition, the misalignment is simulated with the simple shifting of the two focal spots. In the experiment, the misalignments may be caused by unintentionally tilted illumination. That is, the incident light is not perpendicular to the focal plane.

In this case, inhomogeneous detection could generate aberrations in the fluorescence images. Generally, if the amount of tilt is too large, the obtained images are highly distorted. For the experimental conditions of tiny tilt, which generates slight position shift, the excitation spots can be considered to be homogeneous and parallel position shifted. The noise effect is not included in the present simulation. Usually, the effect of the background noise and the fluctuation of the fluorescence intensity can be ignored if the signal noise ratio (SNR) of the microscope is higher than ten in the experiments. For subtraction microscopes, the raw images are taken with common confocal microscopes. It may be easy to obtain common confocal images with SNR larger than ten. If the quality of the images is really not good, e.g., the fluorophore has weak emission efficiency or the detector is very noisy, in this case, fitting the curves to get the peak can also be used for calibration. Furthermore, the excitation saturation and biological samples with complex structures are not discussed here temporally. Meanwhile, the resolution of the subtraction microscope, relating with the implemented subtraction factors and sample sizes, is still the main problem of the subtraction microscope. In this chapter, the smallest sample we used for simulation has a preset FHWM of 0.05λ. With a corresponding subtraction factor of around 15, the resolved FWHM reaches 0.05λ. However, if this factor of 15 is used, the samples with sizes larger than 0.05λ are over-subtracted. Because of the rapid progress in imaging technology, concepts to solve the problem of size-dependent and factor-dependent subtraction may hopefully be realized in the near future.

5 CONCLUSION

In summary, a common model to analyze the quality of subtracted images for subtraction microscopy is established based on vector beam diffraction theory [24]. With this model, the effects of lateral and axial displacements on the resolution, signal intensity, nonlinearity, subtraction threshold, and peak intensity factor are investigated quantitatively combining with the sample sizes, numerical aperture of the objective lens, and the polarization combination for excitations. The lateral misalignments of the excitation spots should be inspected and calibrated for they can reduce the image resolution and signal intensity and increase the nonlinearity. The threshold subtraction factors and peak intensity factors also decrease with the displacement, indicating easier over-subtractions and inevitable signal loss to reach the appropriate subtraction. The peak position shift after subtraction versus the subtraction factors is found to be convenient and sensitive to check the amounts and the directions of misalignments. It can also be a potential parameter for estimating the relative sample sizes. The larger sample size has a smaller amount of peak position shift. Dependances of peak intensity after subtraction and the value of negative sidebands on the subtraction factors can be secondary effective parameters to inspect the overlap. The signal intensity is maximal or the negative sideband intensity is minimal when the overlap is optimized. The contents of this subsection are the based on the collaboration of N. Wang and T. Kobayashi [24].

REFERENCES

1. S. W. Hell and J. Wichmann, "Breaking the diffraction resolution limit by stimulated emission: stimulated emission-depletion fluorescence microscopy," *Opt. Lett.* **19**(11), 780–782 (1994).
2. M. G. L. Gustafsson, "Surpassing the lateral resolution limit by a factor of two using structured illumination microscopy," *J. Microscopy* **198**, 82–87 (2000).
3. E. Betzig, G. H. Patterson, R. Sougrat, O. W. Lindwasser, S. Olenych, J. S. Bonifacino, M. W. Davidson, J.L. Schwartz, and H. F. Hess, "Imaging intracellular fluorescent proteins at nanometer resolution," *Science* **313**, 1642–1645 (2006).
4. M. J. Rust, M. Bates, and X. W. Zhuang, "Sub-diffraction-limit imaging by stochastic optical reconstruction microscopy (STORM)," *Nat. Meth.* **3**, 793–796 (2006).
5. M. P. Gordon, T. Ha, and P. R. Selvin, "Single-molecule high-resolution imaging with photobleaching," *Proc. Natl. Acad. Sci. USA* **101**(17), 6462–6465 (2004).
6. A. Sharonov and R. M. Hochstrasser, "Wide-field sub-diffraction imaging by accumulated binding of diffusing probes," *Proc. Natl. Acad. Sci. USA* **103**(50), 18911–18916 (2006).
7. T. Dertinger, R. Colyer, G. Iyer, S. Weiss, and J. Enderlein, "Fast, background-free, 3D super-resolution optical fluctuation imaging (SOFI)," *Proc. Natl. Acad. Sci. USA* **106**(52), 22287–22292 (2009).

8. J. Yao, L. Wang, C. Li, C. Zhang, and L. V. Wang, "Photoimprint photoacoustic microscopy for three-dimensional label-free subdiffraction imaging," *Phys. Rev. Lett.* **112**, 014302 (2014).
9. J. Miyazaki, H. Tsurui, A. Hayashi-Takagi, H. Kasai, and T. Kobayashi, "Sub-diffraction resolution pump-probe microscopy with shot-noise limited sensitivity using laser diodes," *Opt. Express* **22**(8), 9024–9032 (2014).
10. J. Miyazaki, K. Kawasumi, and T. Kobayashi, "Resolution improvement in laser diode-based pump-probe microscopy with an annular pupil filter," *Opt. Lett.* **39**(14), 4219–4222 (2014).
11. E. Rittweger, K. Y. Han, S. E. Irvine, C. Eggeling, and S. W. Hell, "STED microscopy reveals crystal colour centres with nanometric resolution," *Nat. Photonics* **3**, 144–147 (2009).
12. C. F. Kuang, S. Li, W. Liu, X. Hao, Z. T. Gu, Y. F. Wang, J. H. Ge, H. F. Li, and X. Liu, "Breaking the diffraction barrier using fluorescence emission difference microscopy," *Sci. Rep.* **3**, 1441 (2013).
13. H. Dehez, M. Piche, and Y. D. Koninck, "Resolution and contrast enhancement in laser scanning microscopy using dark beam imaging," *Opt. Express* **21**(13), 15912–15925 (2013).
14. S. Segawa, Y. Kozawa, and S. Sato, "Resolution enhancement of confocal microscopy by subtraction method with vector beams," *Opt. Lett.* **39**(11), 3118–3121 (2014).
15. N. Wang, J. Miyazaki, J. He, K. Seto, and T. Kobayashi, "Sub-diffraction-limit imaging using mode multiplexing," *Opt. Commun.* **343**, 28–33 (2015).
16. S. Segawa, Y. Kozawa, and S. Sato, "Demonstration of subtraction imaging in confocal microscopy with vector beams," *Opt. Lett.* **39**(15), 4529–4532 (2014).
17. S. You, C. Kuang, Z. Rong, and X. Liu, "Eliminating deformations in fluorescence emission difference microscopy," *Opt. Express* **22**(21), 26375–26385 (2014).
18. N. Wang and T. Kobayashi, "Numerical study of the subtraction threshold for fluorescence difference microscopy," *Opt. Express* **22**(23), 28819–28830 (2014).
19. B. Richards and E. Wolf, "Electromagnetic diffraction in optical systems. II. structure of the image field in an aplanatic system," *Proc. R. Soc. Lond. A* **253**, 358–379 (1959).
20. D. Ganic, X. Gan, and M. Gu, "Focusing of doughnut laser beams by a high numerical-aperture objective in free space," *Opt. Express* **11**(21), 2747–2752 (2003).
21. R. Dorn, S. Quabis, and G. Leuchs, "Sharper focus for a radially polarized light beam," *Phys. Rev. Lett.* **91**, 233901 (2003).
22. G. M. Lerman and U. Levy, "Effect of radial polarization and apodization on spot size under tight focusing conditions," *Opt. Express* **16**(7), 4567–4581 (2008).
23. S. Deng, L. Liu, Y. Cheng, R. Li, and Z. Xu, "Investigation of the influence of the aberration induced by a plane interface on STED microscopy," *Opt. Express* **17**(3), 1714–1725 (2009).
24. N. Wang and T. Kobayashi, "Numerical calibration of the spatial overlap for subtraction microscopy," *Opt. Express.* **23**(10), 13411–13422 (2015).

4.1.2 Polarization Modulation for Fluorescence Emission Difference Microscopy

1 INTRODUCTION

Various types of superresolution techniques have achieved great success in improving the resolving power of optical far-field microscopes during the last 20 years, such as stimulated emission depletion (STED) microscopy [1], structured illumination microscopy (SIM) [2], photoactive localization microscopy (PALM) [3], stochastic optical reconstruction microscopy (STORM) [4], and others [5–10]. Recently, mathematically calibrated microscopes, such as fluorescence emission difference (FED) microscopy [11], scanning laser mode microscopy (SLAM) [12], and others [13, 14], have attracted interest for being able to provide a simple approach to break the diffraction barrier. The concept is to subtract two fluorescence images captured by two excitation lasers with the same wavelength but with different spatial intensity distributions, e.g., solid and donut shapes. Because the quality of subtracted images strongly depends on the spot shapes of the two excitation lasers and the implemented subtraction factors, the possible problems of nonlinearities and over-process resulting from the subtraction could be generated. To reduce the distortion and negative sidebands brought by improper illumination, excitation beams with radial and azimuthal polarizations and with extended solid and hollow spot shapes have been proposed [15, 16]. Further to achieve appropriate subtraction, the concept of a subtraction threshold has also been constructed [17]. However, the imaging quality can be easily deteriorated by imperfect polarizations in coherent optical microscopes [18–21]. Especially for fluorescence subtraction microscopy, the resolution enhancement is obtained from direct subtraction of fluorescence images, in which the quality of the polarization for excitation should be considered significant. However, there has been no report on the detailed study of the effects of the imperfect polarization on the subtracted parameters in such microscopes until now. In this chapter, a numerical model is built based on the vector beam diffraction theory (VBDT) and it is applied to investigate the importance of the polarizations in fluorescence

DOI: 10.1201/9781003476030-14

difference microscopy. The resolution, peak intensity, and further parameters in subtracted images are quantitatively evaluated by altering the polarization described by the Jones matrices. It is found that when the wave plate for polarization conversion deviates from normal incidence by 10°, the resolution deterioration is about 5%; peak intensity decreases within 3%; the corresponding amplitude of negative sidebands is reduced by about 5%. The deviation from appropriate polarization conditions lead to easier over-subtraction and the threshold peak intensity decreases for sample sizes of smaller than 0.1λ. The subtraction factor of 0.6 is found to eliminate the spot deformation after subtraction which was caused by the deviation of the polarizations. Simulation results with combined radially and azimuthally polarized spots for excitation are also given.

2 PRINCIPLE AND NUMERICAL MODELS

The schematic of polarization conversion in the subtraction microscope is depicted in Figure 4.1.2.1(a). To avoid the interference between each other, two orthogonally linear-polarized excitation beams are combined with a polarization beam splitter and converted into desired polarization combinations with corresponding wave

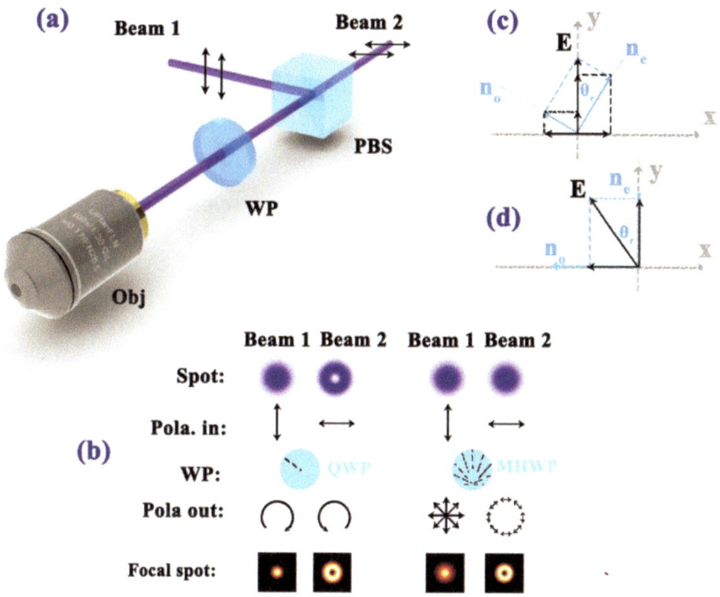

FIGURE 4.1.2.1 (a) Schematic of polarization conversion for excitation beams in subtraction microscopy. (b) Two groups of polarization combinations for excitation. (c) Electric field components with rotated wave plate. (d) Electric field components with fixed wave plate. PBS: polarization beam splitter; WP: wave plate; Obj: objective lens; QWP: quarter wave plate; MHWP: mosaic half wave plate. Copyright of this figure is shared between the representative author (Takayoshi Kobayashi) and Optica (formally Optical Society of America).

plates. Two combinations of spots and polarizations are usually used for excitation, as are shown in Fig. 4.1.2.1(b). The first group is the combination of a solid-shape spot and a donut-shape spot generated with a vortex phase plate. Then they are converted into right-handed and left-handed circular polarization after passing through a quarter-wave plate (QWP). The second group is the combination of radially and azimuthally polarized spots converted from two solid-shape spots using a mosaic half wave plate (MHWP). With each of the polarization combinations, the solid and donut focal spots for excitation can be obtained.

The mathematical formula of subtracted image can be expressed as [17, 22]:

$$I_{RAW}(x,y,z,\theta_r,r,\gamma) = I_S(x,y,z,\theta_r)*Ob(x,y,r) - \gamma \cdot I_D(x,y,z,\theta_r)*Ob(x,y,r)$$
$$I(x,y,z,\theta_r,r,\gamma) = \begin{cases} I_{RAW}(x,y,z,\theta_r,r,\gamma) & I_{RAW}(x,y,z,\theta_r,r,\gamma) \geq 0 \\ 0 & I_{RAW}(x,y,z,\theta_r,r,\gamma) < 0 \end{cases} \tag{1}$$

where γ is the subtraction factor, $*$ is the convolution, I_S and I_D are the focused solid and donut spots, the point spread functions (PSFs) of which can be derived from VBDT [23–25]. *Ob* is the sample. They are expressed as:

$$E_{x,y,z}(x,y,z,\theta_r) = \frac{iC}{\lambda}\int_0^{2\pi}\int_0^{\arcsin(NA/n)} A_{AMP}\cdot A_{\text{Phase}}\cdot A_L \begin{bmatrix} P_x(\theta_r) \\ P_y(\theta_r) \\ P_z(\theta_r) \end{bmatrix} \sin\theta \cdot e^{-ik\left\{z\cos\theta + \sqrt{x^2+y^2}\sin\theta\cos\left[\varphi - \arctan(y/x)\right]\right\}} d\theta d\varphi \tag{2}$$

$$I_{S,D}(x,y,z,\theta_r) = \left|E_x\right|^2 + \left|E_y\right|^2 + \left|E_z\right|^2 \tag{3}$$

$$Ob(x,y,r) = C_0 e^{-(4\ln 2)\frac{x^2+y^2}{r^2}} \tag{4}$$

Here C and C_0 are the normalization constants; λ is the wavelength of incident light; NA is the numerical aperture of the objective lens; n is the refractive index of the contact medium between the objective and the cover glass of the sample, the value of which is 1.518 for the immersion oil; k is the wave number; A_{AMP} is the amplitude factor; A_{phase} is the phase factor; r is the preset full width at half maximum (FWHM) of the sample envelope. The $A_{AMP} = 1$ means the plane wave is used to simulate the intensity distribution of the incident excitation lasers; the $A_{phase} = e^{i\phi}$ is the phase distribution of the vortex phase plate. The set values of A_{AMP} and A_{phase} for I_S and I_D are shown in Table 4.1.2.1.

A_L is the vector weight matrix of the aplanatic objective lens, calculated as follows:

$$A_L = \sqrt{\cos\theta}\begin{bmatrix} 1+(\cos\theta-1)\cos^2\varphi & (\cos\theta-1)\cos\varphi\sin\varphi & -\sin\theta\cos\varphi \\ (\cos\theta-1)\cos\varphi\sin\varphi & 1+(\cos\theta-1)\sin^2\varphi & -\sin\theta\sin\varphi \\ \sin\theta\cos\varphi & \sin\theta\sin\varphi & \cos\theta \end{bmatrix} \tag{5}$$

TABLE 4.1.2.1
Parameter Sets for I_S and I_D

	Group 1			Group 2		
I_{PSF}	Polarization	A_{AMP}	A_{PHASE}	Polarization	A_{AMP}	A_{PHASE}
I_S	Right Circular	1	1	Radial	1	1
I_D	Left Circular	1	$e^{i\varphi}$	Azimuthal	1	1

[Px, Py, Pz] is the unit vector matrix of the polarization. According to the Jones Calculus [26, 27], the electric field of s- or p-polarized light has components along the tilted crystal axis and then they are projected to the horizontal and vertical axes when rotating the wave plate. The phase delay between the two birefringent optical axes is $\delta = kd(n_o - n_e)$, where d is the thickness of the wave plate; n_o and n_e are the refractive indices of ordinary and extraordinary optical axes, respectively. For a QWP of negative crystal, δ is $\pi/2$. From Fig. 4.1.2.1(c), Jones matrices of the horizontal and vertical electric field components for s- and p-polarized beams after a negative-type QWP can be written as:

$$s-p:\begin{bmatrix}P_x\\P_y\\P_z\end{bmatrix}=\begin{bmatrix}(1-i)\sin\theta_r\cos\theta_r\\\cos^2\theta_r+i\sin^2\theta_r\\0\end{bmatrix},p-p:\begin{bmatrix}P_x\\P_y\\P_z\end{bmatrix}=\begin{bmatrix}\sin^2\theta_r+i\cos^2\theta_r\\(1-i)\sin\theta_r\cos\theta_r\\0\end{bmatrix} \tag{6}$$

where θ_r is the angle between the electric field of the s-polarized beam and the fast axis. Then the intensity distributions of focused spots with different angles of the QWP can be calculated. Because the long and short axes of the focused spots are always along the fast and slow axes of the wave plates, it is inconvenient to analyze the resolutions along two perpendicular axes when the focused spots are tilted. Fixing the wave plate and rotating the polarization can be a calculation strategy to solve this problem. Then the long and short axes of the focused spots can be fixed to the horizontal or vertical axes. As is shown in Fig. 4.1.2.1(d), the Jones matrices used to calculate the sectioning lines can be written as:

$$s-p:\begin{bmatrix}P_x\\P_y\\P_z\end{bmatrix}=\begin{bmatrix}-i\sin\theta_r\\\cos\theta_r\\0\end{bmatrix},p-p:\begin{bmatrix}P_x\\P_y\\P_z\end{bmatrix}=\begin{bmatrix}\cos\theta_r\\-i\sin\theta_r\\0\end{bmatrix} \tag{7}$$

For the radially or azimuthally polarized beams generated from MHWP, the Jones matrices are

[28]:

$$s-p:\begin{bmatrix}P_x\\P_y\\P_z\end{bmatrix}=\begin{bmatrix}\cos(\varphi-\theta_r)\\\sin(\varphi-\theta_r)\\0\end{bmatrix},p-p:\begin{bmatrix}P_x\\P_y\\P_z\end{bmatrix}=\begin{bmatrix}\sin(\varphi-\theta_r)\\-\cos(\varphi-\theta_r)\\0\end{bmatrix} \tag{8}$$

where ϕ is the azimuthal angle. Then the focal PSFs and subtracted spots related with the angles θ_r for different polarizations can be obtained. In order to quantitatively analyze the quality of the image, the peak intensity and amplitude of negative sidebands after subtraction are defined as:

$$I_{\text{peak}} = \frac{\text{MAX}\left[I\left(x, y, z, \theta_r, r, \gamma\right)\right]}{\text{MAX}\left[I\left(x, y, z, \theta_r, r, 0\right)\right]} \tag{9}$$

$$\left|I_{Neg}\right| = \begin{cases} 0 & \text{MIN}\left[I_{RAW}\left(x, y, z, \theta_r, r, \gamma\right)\right] \geq 0 \\ \dfrac{\left|MIN\left[I_{RAW}\left(x, y, z, \theta_r, r, \gamma\right)\right]\right|}{MAX\left[I\left(x, y, z, \theta_r, r, \gamma\right)\right]} & \text{MIN}\left[I_{RAW}\left(x, y, z, \theta_r, r, \gamma\right)\right] < 0 \end{cases} \tag{10}$$

The subtraction threshold condition, which describe the appropriate subtraction [17], and normalized peak intensity at threshold under the condition, are expressed by Eqs. (11) and (12), respectively.

$$FWHM_{Ob\left(x,y,r_0\right)} = FWHM_{I\left(x,y,z,\theta_r,r_0,\gamma_0\right)} \tag{11}$$

$$I_P = \frac{MAX\left[I\left(x, y, z, \theta_r, r_0, \gamma_0\right)\right]}{MAX\left[I\left(x, y, z, \theta_r, r_0, 0\right)\right]} \tag{12}$$

Then the effects of the polarization on the FWHM resolution, ellipticity, peak intensity, negative sidebands, subtraction threshold, and peak intensity after subtraction can be carried out.

3 SIMULATION RESULTS

3.1 Linear and Circular Polarizations

The calculated fluorescence and subtracted spots with polarizations of different circularity are shown in Figure 4.1.2.2. The preset sample size is $r = 0.1\lambda$. From Fig. 4.1.2.2(a), it can be seen that the shape of the focused spot varies with different polarization status. The focused solid spot has elliptical shape when the incident light is linearly or elliptically polarized. Because of the phase distribution, the center of the donut spot is round and has zero intensity only if θ_r is −45°, where the polarization is left-handed circular. This is the appropriate angle for illumination in a subtraction microscope. As is shown in Fig. 4.1.2.2(b), the subtracted images with $\theta_r = -45°$keep the same circularity while images with this angle ±10°have varying circularities after subtraction. Figs. 4.1.2.2(c)–(e) display the FWHM values of the cross sectioning profiles along the long and short axes of subtracted focal spots and corresponding ellipticities with factors of $\gamma = 0, 0.3, 0.6, 1.0$. It can be seen that the FWHM values of subtracted spots are fluctuating by about 0.1λ when rotating the

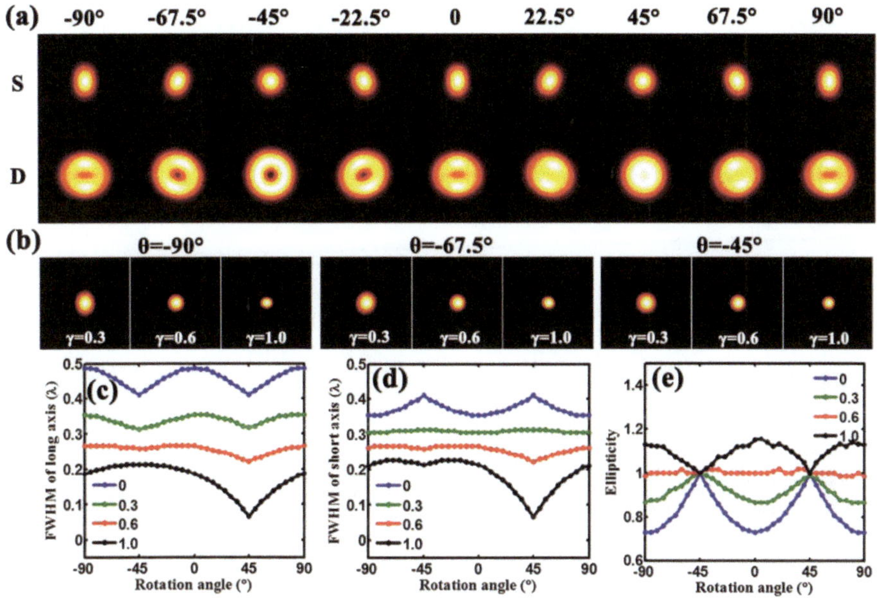

FIGURE 4.1.2.2 (a) Convoluted solid and donut fluorescence images with different rotation angles of the QWP. The sample has size of r = 0.1λ. (b) Subtracted images with factors of γ = 0.3, the long and short axes, and ellipticity with different subtraction factors of 0.6, 1.0 for θr =–90°, –67.5°, –45°. (c)–(e) FWHM values of sectioning profiles along γ = 0, 0.3, 0.6, 1.0 versus different rotation angles (θ_r). Copyright of this figure is shared between the representative author (Takayoshi Kobayashi) and Optica (formally Optical Society of America).

QWP by π; the FWHM values change within 5% even when the angles of the QWP deviate from the right angle for 10°; the ellipticities of the spots after subtraction with the two critical angles of –45° and 45° are both unity; with subtraction factors of larger than 0.6, the long and short axes of the subtracted spots are reversed. The ellipticity here is defined as the ratio of FWHM values on the primary short and long axes of the focused spot. What is more, it is interesting to find that the factor of 0.6 can almost eliminate the errors of ellipticity induced by the misalignments of the polarization. It also indicates that when the factor of 0.6 is implemented in the subtraction scheme, the problem of polarization aberration cannot be intuitively found out from the subtracted images.

The evolution of peak intensity, amplitude of negative sidebands and the subtraction threshold are shown in Figure 4.1.2.3. From Fig. 4.1.2.3(a), it can be seen that the subtracted peak intensities are maximal when $\theta_r = -45°$ even with different subtraction factors. With $\theta_r = 45°$, the center of donut spot is not completely dark, so the peak intensity drops significantly after the convolution with samples and the following subtraction. Fig. 4.1.2.3(b) describes that the decreased amount of peak intensity after subtraction is about 2.5% if the angle of QWP is displaced with 10° and a subtraction factor of 1. Thus, the problem of signal loss due to the imperfect

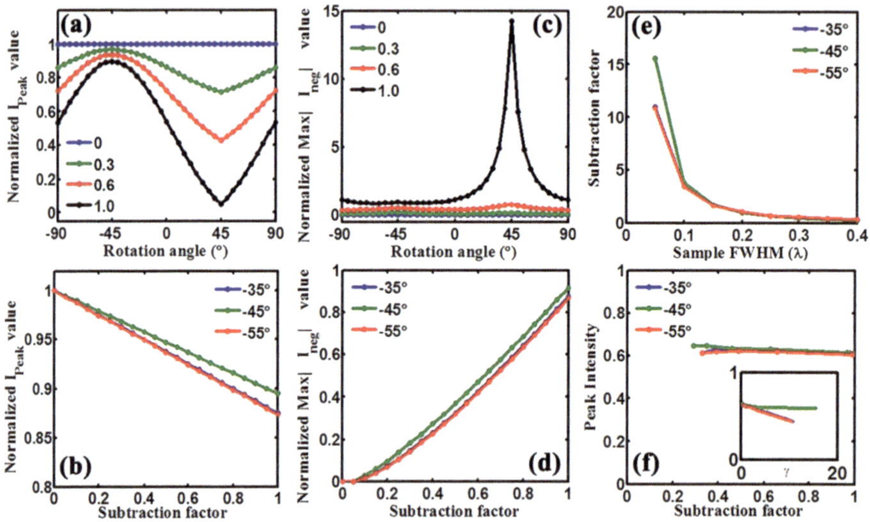

FIGURE 4.1.2.3 (a) Peak intensity and (c) Negative sidebands versus rotation angles of QWP with subtraction factors of $\gamma = 0, 0.3, 0.6, 1.0$. (b) Peak intensity and (d) Negative sidebands versus subtraction factors with rotation angles of $\theta_r = -35°, -45°, -55°$. (e) Subtraction threshold line. (f) Threshold peak intensity curves. Copyright of this figure is shared between the representative author (Takayoshi Kobayashi) and Optica (formally Optical Society of America).

polarization for excitation within 10° is not serious. Besides, the negative sidebands are generated inevitably after subtraction. However, from Fig. 4.1.2.3(c), the values of negative sidebands introduced are similar in a wide range from −90° to 0°. The deviation of the polarization may reduce the negative sidebands by 5% as is shown in Fig. 4.1.2.3(d). It describes that the amplitude of negative sidebands after subtraction is insensitive to the ellipticity of the polarization, as well. Similarly, the calculations on the resolution, peak intensity, and negative sidebands with different sample sizes have been carried out. The curves give similar evolutions to the present ones, which will not be illustrated in further detail. For a sample size of 0.4λ, the variation of FWHM values of subtracted spots is by about 0.1λ versus the rotation angles of the QWP. The peak intensities or negative sidebands are affected by less than 1% with 10° deviation from the right angle. The subtraction threshold lines are shown in Fig. 4.1.2.3(e). It can be seen that the threshold subtraction factors for θ_r = -45° are a little larger than the cases of −35° and −55° with sample sizes of smaller than 0.1λ. It means the polarization-deviated excitation is vulnerable to over-subtraction and it mainly acts on the fine structures in the sample. Meanwhile, from Fig. 4.1.2.3(f), it can be seen that the threshold peak intensity factors slightly drop when increasing the subtraction factors. It means that the deviation of polarization will not introduce significant signal loss at the subtraction threshold for sample sizes larger than 0.1λ.

3.2 Radial and Azimuthal Polarizations

The calculation results with radially and azimuthally polarized excitation spots are shown in Figure 4.1.2.4. The focused PSFs with different rotation angles of MHWP are shown in Fig. 4.1.2.4(a). Because the evolutions of the subtracted spot shapes appear periodically, only the angles between −45° and 45° are analyzed here. From Fig. 4.1.2.4(b), it can be seen that the subtracted spot shapes are not deformed with displaced polarization angles while the resolutions are a little worse than the case of 0° for MHWP. Figs. 4.1.2.4(c)–(e) depict the FWHM values, peak intensities, and the values of negative sidebands when changing the angles of the MHWP. It can be

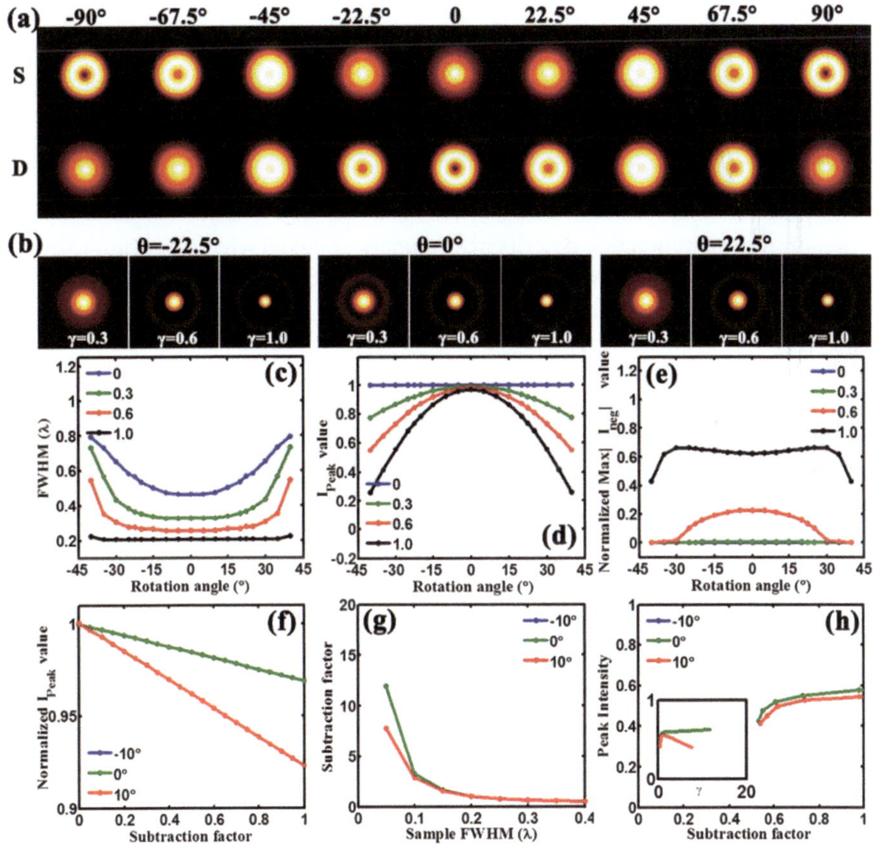

FIGURE 4.1.2.4 Simulation results with radial and azimuthal polarization. (a) Excitation spots versus different angles of wave plate. (b) Subtracted images with factors of 0.3, 0.6 and 1.0 for the rotation angles of -22.5°, 0°, 22.5°. Copyright of this figure is shared between the representative author (Takayoshi Kobayashi) and Optica (formally Optical Society of America).

found that when the angle of MHWP is 0° and the subtraction factor stays the same, the resolved FWHM values are smaller than those with other angles. Meanwhile, the peak intensity is also maximal at the angle of 0°. It is possible to use these two parameters to calibrate the angles of the used MHWP. When rotating the plate, the angle with maximal peak intensity and minimal FWHM values corresponds to the appropriate angle for MHWP in a subtraction microscope scheme. The amplitude of negative sidebands is insensitive to the rotation angles of the plate, as is shown in Fig. 4.1.2.4(e). From Fig. 4.1.2.4(f), it can be seen that the peak intensity decrease is about 5% when the angle is deviated with 10° and a subtraction factor of 1. The subtraction threshold lines and the threshold peak intensity lines are shown in Figs. 4.1.2.4(g) and (h), respectively. Similar to the results in the case of circular polarization, the threshold lines show that the right angle of 0° can stand the over-subtraction better than the incorrect polarizations. Meanwhile, the threshold peak intensity factors are a little smaller for the displaced angles. It means that with the appropriate subtraction, the peak intensity is slightly decreased, especially for the sample sizes of smaller than 0.1λ.

4 CONCLUSION

In summary, this subsection describes the numerical model which was built to evaluate the effects of polarization for a subtraction microscope based on the vector beam diffraction theory. By modulating the Jones matrices, the effects of different polarization status on the parameters of resolution, ellipticity, peak intensity, negative sidebands, and subtraction threshold in fluorescence emission difference microscopy are quantitatively investigated. The resolution deterioration, signal loss, and intensity reduction of negative sidebands are all within 5% for both of the polarization combinations for excitation even when the angles of wave plates are deviating from the normal incidence by 10°.

The corresponding threshold peak intensity curve finds that the signal loss at appropriate subtraction is negligibly small for sample sizes of larger than 0.1λ. The rotating angle of polarizer being 45° must be critically avoided, since it can generate serious signal intensity loss and resolution deterioration. The subtraction factor of 0.6 is found to be a specially behaving parameter for the case when the polarization is modified with a quarter wave plate. With this factor, the ellipticity is always unity even when the excitation beams are linearly or elliptically polarized. To calibrate the polarization, ellipticity variation, or peak intensity after subtraction can be used. Furthermore, the simulation here is carried out based on the model in ideal conditions. However, calculations in this chapter can basically describe and evaluate the effects of polarization on the parameters for fluorescence subtraction microscopy even though the confocal detection parameter or noise effect is not included in the model. The improved model including them is to be presented in future work. Sensitive fluorescence detection with high signal-to-noise ratio is necessary for experimental demonstration of the present calculation.

The research presented in this subsection was performed collaboratively by N. Wang and T. Kobayashi [29].

REFERENCES

1. S. W. Hell and J. Wichmann, "Breaking the diffraction resolution limit by stimulated emission: stimulated emission-depletion fluorescence microscopy," *Opt. Lett.* **19**(11), 780–782 (1994).
2. M. G. L. Gustafsson, "Surpassing the lateral resolution limit by a factor of two using structured illumination microscopy," *J. Microscopy* **198**, 82–87 (2000).
3. E. Betzig, G. H. Patterson, R. Sougrat, O. W. Lindwasser, S. Olenych, J. S. Bonifacino, M. W. Davidson, J.L. Schwartz, and H. F. Hess, "Imaging intracellular fluorescent proteins at nanometer resolution," *Science* **313**, 1642–1645 (2006).
4. M. J. Rust, M. Bates, and X. W. Zhuang, "Sub-diffraction-limit imaging by stochastic optical reconstruction microscopy (STORM)," *Nat. Meth.* **3**, 793–796 (2006).
5. M. P. Gordon, T. Ha, and P. R. Selvin, "Single-molecule high-resolution imaging with photobleaching," *P. Natl. Acad. Sci. USA* **101**(17), 6462–6465 (2004).
6. A. Sharonov and R. M. Hochstrasser, "Wide-field sub-diffraction imaging by accumulated binding of diffusing probes," *P. Natl. Acad. Sci. USA* **103**(50), 18911–18916 (2006).
7. K. Fujita, M. Kobayashi, S. Kawano, M. Yamanaka, and S. Kawata, "High-resolution confocal microscopy by saturated excitation of fluorescence," Phys. Rev. Lett. **99**, 228105 (2007).
8. T. Dertinger, R. Colyer, G. Iyer, S. Weiss, and J. Enderlein, "Fast, background-free, 3D super-resolution optical fluctuation imaging (SOFI)," *P. Natl. Acad. Sci. USA* **106**(52), 22287–22292 (2009).
9. J. Yao, L. Wang, C. Li, C. Zhang, and L. V. Wang, "Photo imprint photoacoustic microscopy for three-dimensional label-free subdiffraction imaging," *Phys. Rev. Lett.* **112**, 014302 (2014).
10. J. Miyazaki, H. Tsurui, A. Hayashi-Takagi, H. Kasai, and T. Kobayashi, "Sub-diffraction resolution pump-probe microscopy with shot-noise limited sensitivity using laser diodes," *Opt. Express* **22**(8), 9024–9032 (2014).
11. C. F. Kuang, S. Li, W. Liu, X. Hao, Z. T. Gu, Y. F. Wang, J. H. Ge, H. F. Li, and X. Liu, "Breaking the diffraction barrier using fluorescence emission difference microscopy," *Sci. Rep.* **3**(1441) (2013).
12. H. Dehez, M. Piche, and Y. D. Koninck, "Resolution and contrast enhancement in laser scanning microscopy using dark beam imaging," *Opt. Express* **21**(13), 15912–15925 (2013).
13. S. Segawa, Y. Kozawa, and S. Sato, "Resolution enhancement of confocal microscopy by subtraction method with vector beams," *Opt. Lett.* **39**(11), 3118–3121 (2014).
14. N. Wang, J. Miyazaki, J. He, K. Seto, and T. Kobayashi, "Sub-diffraction-limit imaging using mode multiplexing," *Opt. Commun.* **343**, 28–33 (2015).
15. S. Segawa, Y. Kozawa, and S. Sato, "Demonstration of subtraction imaging in confocal microscopy with vector beams," *Opt. Lett.* **39**(15), 4529–4532 (2014).
16. S. You, C. Kuang, Z. Rong, and X. Liu, "Eliminating deformations in fluorescence emission difference microscopy," *Opt. Express* **22**(21), 26375–26385 (2014).
17. N. Wang and T. Kobayashi, "Numerical study of the subtraction threshold for fluorescence difference microscopy," *Opt. Express* **22**(23), 28819–28830 (2014).
18. X. Hao, C. Kuang, T. Wang, and X. Liu, "Effects of polarization on the de-excitation dark focal spot in STED microscopy," *J. Opt.* **12**, 115707 (2010).
19. C. Yuan, G. Situ, G. Pedrini, J. Ma, and W. Osten, "Resolution improvement in digital holography by angular and polarization multiplexing," *Appl. Opt.* **50**(7), B6–B11 (2011).

20. S. Galiani, B. Harke, G. Vicidomini, G. Lignani, F. Benfenati, A. Diaspro, and P. Bianchini, "Strategies to maximize the performance of a STED microscope," *Opt. Express* **20**(7), 7362–7374 (2012).
21. Y. Chen, H. Lin, M. J. Ludford-Menting, A. H. Clayton, M. Gu, and S. M. Russell, "Polarization of excitation light influences molecule counting in single-molecule localization microscopy," *Histochem. Cell Biol.* **143**(1), 11–19 (2015).
22. N. Wang and T. Kobayashi, "Numerical calibration of the spatial overlap for subtraction microscopy," *Opt. Express.* accepted.
23. B. Richards and E. Wolf, "Electromagnetic diffraction in optical systems. II. structure of the image field in an aplanatic system," *Proc. R. Soc. Lond. A* **253**, 358–379 (1959).
24. D. Ganic, X. Gan, and M. Gu, "Focusing of doughnut laser beams by a high numerical-aperture objective in free space," *Opt. Express* **11**(21), 2747–2752 (2003).
25. G. M. Lerman and U. Levy, "Effect of radial polarization and apodization on spot size under tight focusing conditions," *Opt. Express* **16**(7), 4567–4581 (2008).
26. R. C. Jones, "A new calculus for the treatment of optical systems," *J. Opt. Soc. Am.* **31**(7), 488–493 (1941).
27. W. H. Mcmaster, "Matrix representation of polarization," *Rev. Mod. Phys.* **33**(1), 8–28 (1961).
28. M. Stalder and M. Schadt, "Linearly polarized light with axial symmetry generated by liquid-crystal polarization converter," *Opt. Lett.* **21**(23), 1948–1950 (1996).
29. N. Wang and T. Kobayashi, Polarization modulation for fluorescence emission difference microscopy" *Opt. Exp.* **23**(10), 13704–13712 (2015).

4.1.3 Subtraction Threshold for an Isotropic Fluorescence Emission Difference Microscope

1 INTRODUCTION

With various types of superresolution microscopes, such as stimulated emission depletion microscopy [1], structured illumination microscopy [2], photoactive localization microscopy [3], stochastic optical reconstruction microscopy [4], and others [5–13], the diffraction-limit is not regarded as an unbreakable barrier for obtaining resolution beyond tens of nanometers in the lateral plane in fluorescence microscopy. To enhance the resolving power in three dimensions (3D), several methods have also been proposed and realized [14–18]. Recently, subtraction microscopy [19–22] is introduced for providing a simple way to enhance the spatial resolution. With an annular 0-π phase plate instead of a vortex phase plate (VPP) for generating a negative excitation spot, 3D resolution enhancement can be achieved in the fluorescence imaging, referred to as isotropic fluorescence emission difference (iFED) microscopy [23, 24]. The method is to subtract the 3D negative fluorescence image from the 3D solid fluorescence image, in which the excitation beams have separate donut and solid spot shapes. The obtained image gets an enhanced resolution compared with the common 3D fluorescence image. However, the factors for subtracting the images in this microscope are empirically used at present. A specific factor is required to get the appropriate subtraction in both lateral and axial planes, while to our knowledge it has not been studied. Meanwhile, the basic methodology to evaluate the lateral subtraction threshold for a VPP-based subtraction microscope [25] has been established and the effects of the parameters in the microscope have also been discussed [26, 27]. Then it is possible to further develop the model into 3D and obtain the subtraction threshold for this 0-π phase plate based iFED. Based on vector diffraction

DOI: 10.1201/9781003476030-15

theory, a 3D calculation model is introduced in this chapter and the enhancement can be achieved in the fluorescence imaging, referred to as isotropic fluorescence emission difference (iFED) microscopy [23, 24]. The method is to subtract the 3D negative fluorescence image from the 3D solid fluorescence image, in which the excitation beams have separate donut and solid spot shapes. The obtained image gets an enhanced resolution compared with the common 3D fluorescence image. However, the factors for subtracting the images in this microscope are empirically used at present. A specific factor is required to get the appropriate subtraction in both lateral and axial planes while to our knowledge it has not been studied. Meanwhile, the basic methodology to evaluate the lateral subtraction threshold for a VPP-based subtraction microscope [25] has been established and the effects of the parameters in the microscope have also been discussed [26, 27]. Then it is possible to further develop the model into 3D and obtain the subtraction threshold for this 0-π phase plate based iFED. Based on vector diffraction theory, a 3D calculation model is introduced in this chapter and the subtraction threshold is numerically investigated for the iFED. It finds that the factors in lateral and axial planes in the microscope are approximately equal when appropriate subtraction is achieved. The threshold peak intensity factors after subtraction in lateral and axial planes are both around 0.6. They prove that the microscope has isotropic resolution at 3D subtraction threshold level. The effects of a 0-π phase plate with different radius ratios for negative spot generation on the subtracted parameters are also investigated.

2 PRINCIPLES AND MODELS

The 0-π phase plate based iFED can be carried out with the layout illustrated in Figure 4.1.3.1(a). To pursue the subtraction threshold of this microscope, a 3D numerical model is developed and introduced firstly. The final subtracted images of the microscope can be expressed by (4.1.3.1a).

$$\begin{aligned} I_{\mathrm{RAW}}(x,y,z,R,r,\gamma) &= I_{\mathrm{S}}(x,y,z,R)*\mathrm{Ob}(x,y,z,r) \\ &\quad -\gamma\cdot I_{\mathrm{N}}(x,y,z,R)*\mathrm{Ob}(x,y,z,r) \end{aligned} \tag{4.1.3.1a}$$

$$I(x,y,z,R,r,\gamma) = \begin{cases} I_{\mathrm{RAW}}(x,y,z,R,r,\gamma) & I_{\mathrm{RAW}}(x,y,z,R,r,\gamma)\geqslant 0 \\ 0 & I_{\mathrm{RAW}}(x,y,z,R,r,\gamma)<0 \end{cases}$$

where I_S and I_N are focal point spread functions (PSFs) excited with solid and negative spot beams, expressed by (4.1.3.2) shown below [28–30]; Ob is the three-dimensional distribution function of the fluorophores in the sample, the preset value of full width at half maximum (FWHM) of which are both r in lateral and axial plane, being shown in (4.1.3.3); γ is the subtraction factor, * is the convolution. After the convolution and subtraction, the negative signals are removed.

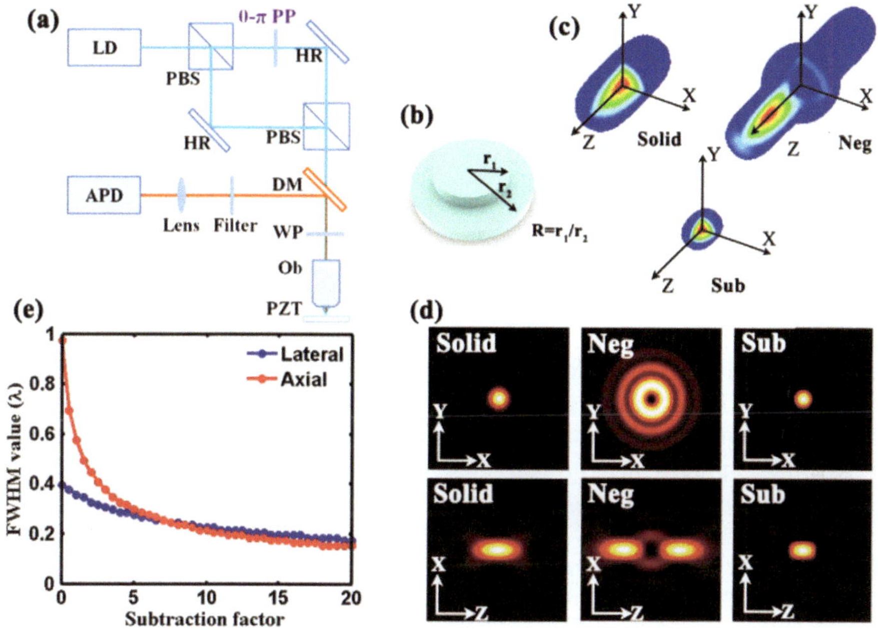

FIGURE 4.1.3.1 (a) A schematic for iFED microscope. LD: laser diode; PBS: polarization beam splitter; HR: high reflection mirror; PP: phase plate; DM: dichroic mirror; WP: quarter wave plate; Ob: objective lens; PZT: sample mounted on piezoelectric driven stage; APD: photon counting avalanche photodiode. (b) A schematic of 0-π phase plate. (c) 3D focal PSFs of solid, negative, and subtraction spots. (d) Solid, negative, and subtraction PSFs in the sectioning lateral (XY) and axial (XZ) plane. (e) FWHM values of subtracted PSFs in the lateral and axial direction. Here R = 0.6 is set for 0-π annular phase plate. Copyright of this figure is partially held between the representative author (Takayoshi Kobayashi) and IOP Science.

$$E_{x,y,z}\left(x,y,z,R\right) = \frac{\mathrm{i}}{\lambda}\int_0^{2\pi}\int_0^{\arcsin(NA/n)} A_{\mathrm{AMP}}\cdot A_{\mathrm{PHASE}}\left(R\right)\cdot A_{\mathrm{L}}\begin{bmatrix}P_x\\P_y\\P_z\end{bmatrix} \cdot\sin\theta\cdot \mathrm{e}^{-\mathrm{i}k\left\{z\cos\theta+\sqrt{x^2+y^2}\sin\theta\cos\left[\varphi-\arctan\left(y/x\right)\right]\right\}}\,\mathrm{d}\theta\mathrm{d}\varphi \tag{4.1.3.2a}$$

$$I_{\mathrm{S,N}}\left(x,y,z,R\right)=C\left(\left|E_x\right|^2+\left|E_y\right|^2+\left|E_z\right|^2\right), \tag{4.1.3.2b}$$

$$\mathrm{Ob}\left(x,y,z,r\right)=C_0\mathrm{e}^{-(4\ln 2)\frac{x^2+y^2+z^2}{r^2}}. \tag{4.1.3.3}$$

TABLE 4.1.3.1
Parameter sets for I_S and I_N

I_{PSF}	Polarization	A_{AMP}	A_{PHASE}
I_S	Right Circular	1	1
I_N	Left Circular	1	$\begin{cases} 1 & \theta \geq R\sin^{-1}\left(NA\ n^{-1}\right) \\ e^{i\pi} & \theta < R\sin^{-1}\left(NA\ n^{-1}\right) \end{cases}$

Here C and C_0 are the intensity normalization constants; λ is the wavelength of incident light; NA is the numerical aperture of the objective lens with value of 1.4; n is the refractive index of the immersion oil with value of 1.518; k is the wave number; A_{AMP} is the amplitude factor; A_{PHASE} is the phase factor. The unity amplitude factor, $A_{AMP} = 1$, means the plane wave is used to simulate the intensity distribution of the incident excitation lasers; the A_{PHASE} is the phase distribution of the 0-π phase plate. The set values of A_{AMP} and A_{PHASE} for I_S and I_N are shown in Table 4.1.3.1. The A_L is the vector weight matrix of the aplanatic objective lens; [Px, Py, Pz] is the unit vector matrix of the polarization [25]; R is the radius ratio of the inner and outer plate of the 0-π phase plate illustrated in Figure 4.1.3.1(b) and expressed by $R = r_1/r_2$ in which r_1 and r_2 are the radii of the inner and outer plates, respectively.

To achieve appropriate subtraction, a subtraction threshold is defined to select the subtraction factors. The criterion that the obtained FWHM value in subtracted images is equal to the preset sample FWHM is used to evaluate the threshold. It can be expressed as (4.1.3.4). The peak intensity of the subtracted fluorescence image for a sample at the subtraction threshold is expressed by (4.1.3.5).

$$I_P = \frac{\mathrm{MAX}\left[I\left(x,y,z,R,r_0,\gamma_0\right)\right]}{\mathrm{MAX}\left[I\left(x,y,z,R,r_0,0\right)\right]}. \tag{4.1.3.5}$$

Meanwhile, the amplitude of negative sidebands generated from the subtraction can also be analyzed with the (4.1.3.6) depending the positive (and zero) case or negative value case of MIN[I_{RAW} as follows.

$$\left|I_{\mathrm{Neg}}\right| = \begin{cases} 0 \\ \mathrm{MIN}\left[I_{\mathrm{RAW}}\left(x,y,z,R,r,\gamma\right)\right] \geqslant 0 \\ \dfrac{\left|\mathrm{MIN}\left[I_{\mathrm{RAW}}\left(x,y,z,R,r,\gamma\right)\right]\right|}{\mathrm{MAX}\left[I\left(x,y,z,R,r,\gamma\right)\right]} \\ \mathrm{MIN}\left[I_{\mathrm{RAW}}\left(x,y,z,R,r,\gamma\right)\right] < 0 \end{cases} \tag{4.1.3.6}$$

With these equations, the subtracted images versus annular phase plate with different R can be obtained. Then the threshold subtraction factor, peak intensities at threshold and the negative sidebands can also be investigated.

3 SIMULATION RESULTS

3.1 Spatial Resolution with Focal PSFs

The intensity normalized focal PSFs of solid, negative excitation spots, and subtracted spots are shown in Figures 4.1.3.1(c) and (d). The solid spot is the focal spot with conventional fluorescence microscopy. The negative spot is obtained with a phase modulated incident beam using an annular 0-π phase plate. The ratio of the two radii for the plate is R = 0.6 (it will be discussed in section 3.2). With this plate, a 3D donut spot can be produced at the focus. Subtracting with different factors, the lateral and axial FWHM values attained from PSFs only can be obtained and shown in Figure 4.1.3.1(e). It can be seen that the resolving power in the two axes approaches nearly equal values when the subtraction factor is larger than five. Copyright of this figure is partially held between the representative author (Takayoshi Kobayashi) and IOP Science.

3.2 Radius Ratio of 0-π Annular Phase Plate

The resolution enhancement of the microscope is determined by the spot shape of the focused negative beam while the spot shape of the negative beam relies on the radius ratio of the annular phase plate. As is shown in Figure 4.1.3.1(b), the inner plate has a radius of r_1 and a phase shift of π relative to the substrate, the radius of which is r_2. In the focused beam, their radius ratio of $R = r/ r_{1\ 2}$ determines the primary phase distributions of the incident beam which are further applied to the focal numerical apertures of the incident light. The R = 1 means that full NA of the focused light is π phase shifted. Figure 4.1.3.2(a) gives the focused spots with $R = 0.5$, 0.6, 0.7 in an axial plane of negative excitation spots. The cross sectioning profiles of them along the z direction are shown in Figure 4.1.3.2(b). It can be seen that only a specific value of $R = 0.6$ gives complete dark centers. The axial and lateral FWHM values of the subtracted spots with different ratios are shown in Figures 4.1.3.2(c) and (d). It can be seen that the radius ratio of the phase plate affects the resolution more effectively in axial plane than in lateral plane. When the radius ratio deviates from $R = 0.6$, the FWHM values of subtracted spots become rather small. It is mainly caused by the non-zero dark center of negative beams. It can produce over-processing after subtraction which leads to signal loss and possible imaging distortions.

3.3 Subtraction Threshold

The previous two sections investigate the direct subtraction of PSFs and the effects of the radius ratios R of the phase plate on the excitation spot shapes. However, the direct subtraction of focal PSFs cannot describe the subtraction microscope sufficiently [25] due to the removal of negative sidebands. Infinite resolution could be

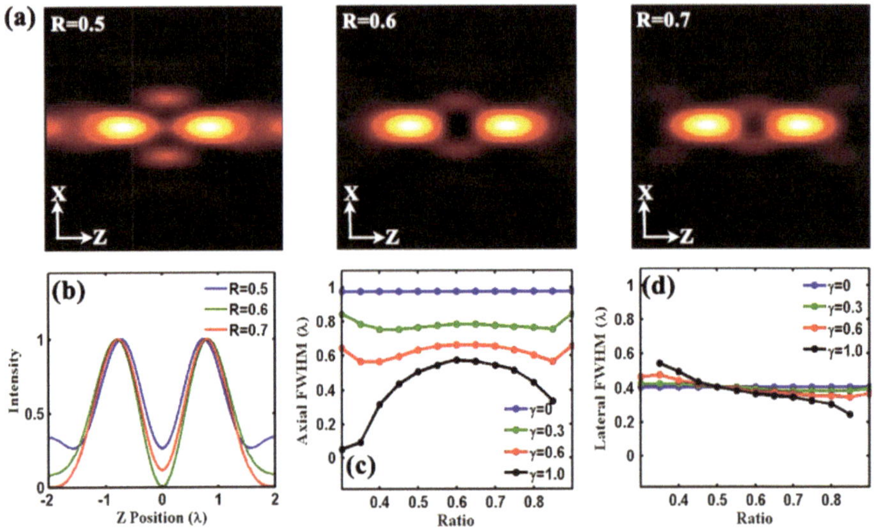

FIGURE 4.1.3.2 Radius ratio of 0-π phase plate and related focal PSFs. (a) Focal PSFs of negative spots with radius ratios of R = 0.5, 0.6, 0.7. (b) Cross sectioning lines in axial direction. (c)–(d) Subtracted FHWM values in axial and lateral planes with factors of γ= 0, 0.3, 0.6, 1.0. Copyright of this figure is partially held between the representative author (Takayoshi Kobayashi) and IOP Science.

obtained without taking the sample into account in the simulation. With (1) and R = 0.6 for the phase plate, the focal PSFs convolute with sample of different sizes and then they are subtracted with different factors. The finally obtained imaging resolution can compare with the preset sample size to derive the suitable subtraction. The resolved FWHMs in lateral and axial planes are shown in Figures 4.1.3.3(a) and (b), respectively. The black dotted lines, which describe the resolved FWHM values equal to the preset FWHM values of the samples, give the subtraction threshold in the lateral and axial directions. These two lines are picked up and shown together in Figure 4.1.3.3(c). It can be seen that the appropriate subtraction factors for different sample sizes are different; the required subtraction factors in lateral and axial planes for the same sample size are similar to each other. It provides evidence that the resolving power of subtraction microscope using 3D negative excitation spot is almost isotropic at the subtraction threshold. Figure 4.1.3.3(d) describes the subtracted peak intensity I_P at the threshold status. It can be concluded that the subtraction factor reaches the subtraction threshold when the peak intensity is about 0.6 of the beginning, which is also insensitive to the sample size. It can be a criterion to inspect whether the appropriate subtractions are achieved or not.

3.4 Validation of Subtraction Threshold

A numerical simulation is carried out to verify the obtained subtraction threshold. The excitation light sources of the two beams have the same wavelength at 488 nm.

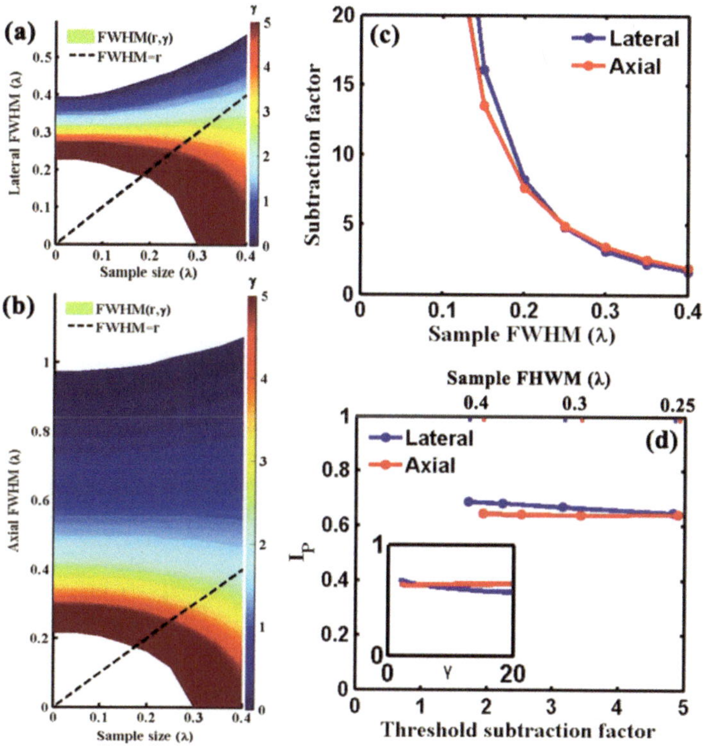

FIGURE 4.1.3.3 Subtraction threshold for isotropic fluorescence difference microscope. (a)–(b) FWHM(r, γ) in lateral and axial planes. (c)–(d) Threshold subtraction factors and subtracted threshold peak intensities in lateral and axial planes.

The focused numerical aperture is 1.4. The selected radius ratio of the 0-π phase plate is $R = 0.6$. Beads with Gaussian shape and FWHM values of 50, 100, 150, and 200 nm in 3D are used for imaging, illustrated in Figure 4.1.3.4(a). By using the 3D convolution between the focused excitation PSFs and the beads, the solid and negative fluorescence images are obtained, as shown in Figures 4.1.3.4(b) and (c). All of the excitation PSFs and the beads are digitally sampled with the step size of 20 nm in three directions with the volume of 27 μm3. Subtracting with different subtraction factors, resolution enhanced 3D images of the sample can be obtained. As are shown in Figure 4.1.3.4(d), the subtracted images of the four beads with factors of $\gamma = 1.5, 2.0, 2.5$ display the evolution of the resolved beads of 50, 100, 150, 200 nm in rows. The values of lateral and axial FWHMs are shown in Figures 4.1.3.4(e) and (f). Usually, it cannot be a priori known that the resolved structures are appropriate or over-subtracted. According to the obtained subtraction threshold in Fig. 4.1.3.3, it is possible to say that when the peak intensities of the beads after subtraction drop below about 0.65 of the initial values, it is probably over subtracted. From Figure 4.1.3.4(g), it can be seen that when $\gamma = 1.5, 2.0, 2.5$, the subtracted peak intensities of the 200

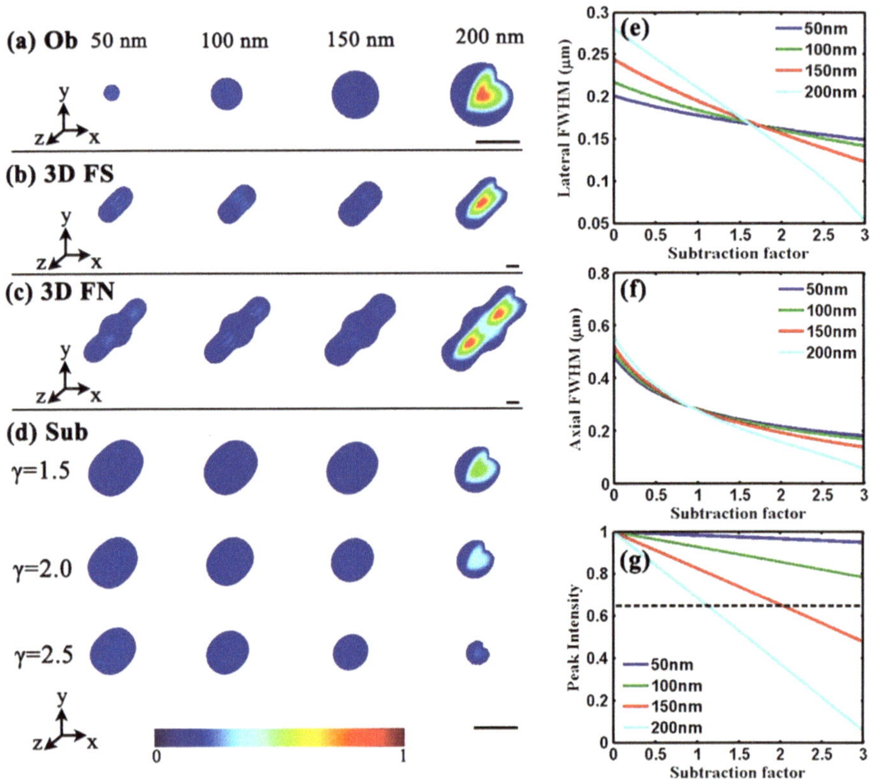

FIGURE 4.1.3.4 (a) Simulated 3D beads with sizes of 50, 100, 150, 200 nm as the sample. (b) 3D solid fluorescence images of beads (50, 100, 150, 200 nm). (c) 3D negative fluorescence images of the beads. (d) Subtracted images with factors of γ= 1.5, 2.0, 2.5. (e)–(g) Lateral resolution, axial resolution and peak intensity after subtraction of beads with different subtraction factors. Black scale bars shown by short horizontal lines at the right bottom of each figure sets of (a), (b), (c), and (d): 400 nm. Copy right of this figure is partially held between the representative author (Takayoshi Kobayashi) and IOP Science.

nm bead drop below 0.65 of its initial value; meanwhile, the FWHM values in lateral and axial planes of this bead subtracting with γ = 1.5, 2.0, 2.5 are 176 nm, 141 nm, 103 nm and 210 nm, 157 nm, 109 nm, which are almost all over-subtracted. Similarly, the 150 nm bead is considered to approach over-subtracted when γ = 2 due to that the peak intensity drops below around 0.65. Its FWHM values in lateral and axial directions are 156 and 193 nm, respectively. The same conclusion can also be established by analyzing the results of 50 and 100 nm beads. Therefore, the subtraction threshold value obtained here can be commonly used in the 3D iFED microscopy.

3.5 Peak Intensity and Negative Sideband

The subtraction threshold factors and peak intensities at threshold with different ratios of R in lateral (dotted lines) and axial (solid lines) planes are shown in Figures

4.1.3.5(a) and (b). It can be seen that although the threshold factors with R = 0.65 have a better agreement in lateral and axial planes, the peak intensities at threshold I_P decrease with increased subtraction factors. It means that, when appropriate subtraction is reached, the signal loss is substantially large for R = 0.65 especially with large subtraction factors, corresponding to smaller sample sizes. It is not convenient either to judge whether the subtraction threshold has been reached or not by evaluating the peak intensity after subtraction with this radius ratio R. That is, the subtraction threshold with R = 0.65 shows better isotropic characteristic while problems of signal loss and over-subtraction with it are more serious. Therefore, R = 0.6 for the phase plate is still favorable for the iFED microscope. For a specific sample size of 0.1λ, with increasing subtraction factors, the peak intensity substantially decrease with radius ratios of phase plate being unequal 0.6, as is shown in Figure 4.1.3.5(c). Figure 4.1.3.5(d) depicts the negative sideband signals in lateral and axial planes with different values of R. It can be seen that the amplitude of negative values with R = 0.6 is smaller than the others and the amplitude of negative sidebands in the lateral plane is smaller than that in the axial plane. Thus the subtraction in the axial plane needs to be used carefully due to large negative sidebands which may produce distortions in the imaging.

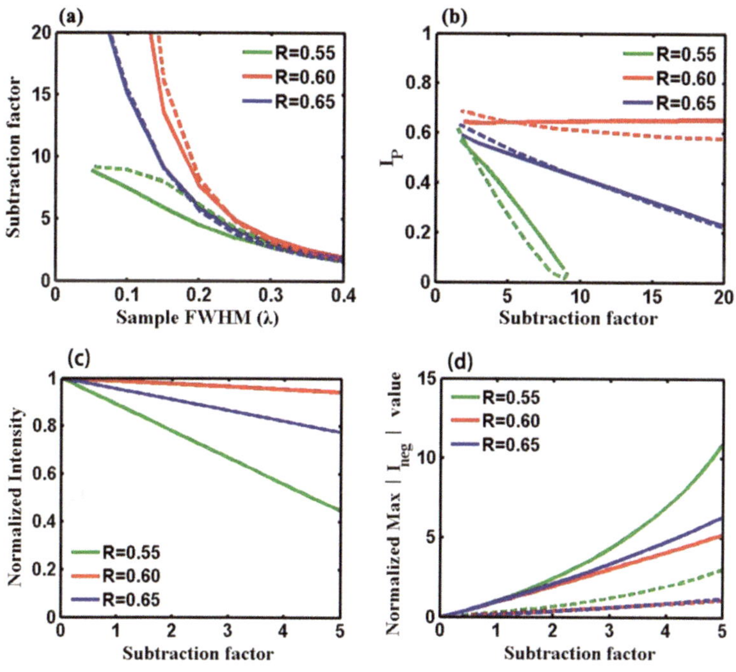

FIGURE 4.1.3.5 (a) Threshold subtraction factors (b) subtracted threshold peak intensities (c) normalized peak intensities after subtraction (d) amplitude of negative sidebands in lateral (dotted lines) and axial (solid lines) planes with R = 0.55, 0.60, 0.65. Copy right of this figure is partially held between the representative author (Takayoshi Kobayashi) and IOP Science.

DISCUSSION

From the calculation, the values of threshold subtraction factors in the 0-π phase plate based 3D isotropic subtraction microscope here are quite different from the values [25] in the VPP based fluorescence difference microscope, in which a 2D donut spot is generated for excitation. The threshold factors here can be larger than five for sample size of 0.2λ, while in the case of factors larger than one, over-subtraction can readily take place in the VPP-type microscope. It is because the energy of negative excitation spot is mainly distributed along the axial direction and off the focus with the 0-π phase plate-based microscope. Then the effective subtraction contribution in the axial direction is enhanced while resolution enhancement in the lateral direction is less effective. Therefore, the required excitation power of negative beam in the 3D isotropic subtraction microscope would be much higher than the case of VPP-type microscope to reach the same lateral resolution enhancement. Nevertheless, it is still convenient to realize the appropriate resolution enhancement with the subtraction threshold in both of the two types of microscopes. It is also interesting to find that the threshold peak intensities after subtraction of the two types of subtraction microscopes are quite similar. Because the threshold subtraction factors are sample-size dependent, the sample-size insensitive I_P factor is an effective way to apply the subtraction threshold. Then the subtractions based on multi-peaks processing are expected for either VPP-type or 0-π type isotropic subtraction microscopes. Using this dynamic subtraction algorithm, it is expected to implement the subtraction microscope to uncover complex structures of samples including biological samples with enhanced resolution. To carry out the experiment, the special 0-π phase plate [18] is required for generating the 3D donut spot while it is not commercially available. Collaborations to make the phase plate or supplies can be found, the numerically investigated subtraction threshold can be demonstrated with real images and employed to the prospective applications. Moreover, the noise problem affects many superresolution imaging techniques. In subtraction microscopy, the resolution enhancement is based on the subtraction of two confocal images. With the help of a sensitive photon counting detector, the SNR of the confocal microscope is usually better than ten. In this case, the noise effect is small and can be ignored in this type of microscope. If the detection system has substantive noise or the fluorescence is very weak, a curve fitting method can be used to obtain a better SNR in the image. Then the subtraction method can be applied to the images. Therefore, the current method can be applied and is robust in the real subtraction microscopy.

CONCLUSION

Isotropic fluorescence emission difference microscopy proposed recently provides a simple method to enhance the spatial resolution in three-dimensions (3D) for fluorescence imaging. However, the subtraction threshold to achieve the conditions for appropriately resolving the sample in 3D has not been studied. Thus the subtraction factors used in this type of microscopes are still experientially chosen. Based on vector diffraction theory and a 3D numerical model developed here, the subtraction threshold is numerically investigated for the isotropic fluorescence subtraction microscopy [31]. The subtraction factors and peak intensities at the threshold are obtained and compared both in lateral and axial planes for achieving the most appropriate subtraction and inspecting the isotropic characteristic. In other words, in the present subsection, the subtraction threshold for an isotropic subtraction microscope with an annular 0-π phase plate for generating a negative excitation spot is numerically investigated based on vector diffraction theory. The appropriate subtractions in both lateral and axial planes require nearly the same subtraction factors [31]. Furthermore, the annular phase plate with different radius ratios of inner and outer area are investigated along with their effects on the resolution, threshold subtraction factors, threshold peak intensities and amplitude of negative sidebands [31]. The simulations here could be useful for investigating and improving the performance of isotropic subtraction microscopes. Research described in this subsection was collaboratively studied by N. Wang and T. Kobayashi [31].

REFERENCES

1. Hell S W and Wichmann J. 1994. Breaking the diffraction resolution limit by stimulated emission: stimulated emission-depletion fluorescence microscopy. *Opt. Lett.* 19 780–2.
2. Gustafsson M G L. 2000. Surpassing the lateral resolution limit by a factor of two using structured illumination microscopy. *J. Microsc.* 198 82–7.
3. Betzig E, Patterson G H, Sougrat R, Lindwasser O W, Olenych S, Bonifacino J S, Davidson M W, Schwartz J L and Hess H F. 2006. Imaging intracellular fluorescent proteins at nanometer resolution. *Science* 313 1642–5.
4. Rust M J, Bates M and Zhuang X W. 2006. Sub-diffraction-limit imaging by stochastic optical reconstruction microscopy (STORM). *Nat. Methods* 3 793–6.
5. Gordon M P, Ha T and Selvin P R. 2004. Single-molecule high resolution imaging with photobleaching. *Proc. Natl Acad. Sci. USA* 101 6462–5.
6. Sharonov A and Hochstrasser R M. 2006. Wide-field subdiffraction imaging by accumulated binding of diffusing probes. *Proc. Natl Acad. Sci. USA* 103 18911–6.
7. Fujita K, Kobayashi M, Kawano S, Yamanaka M and Kawata S. 2007. High-resolution confocal microscopy by saturated excitation of fluorescence. *Phys. Rev. Lett.* 99 228105.
8. Dertinger T, Colyer R, Iyer G, Weiss S and Enderlein J. 2009. Fast, background-free, 3D super-resolution optical fluctuation imaging (SOFI). *Proc. Natl Acad. Sci. USA* 106 22287–92.
9. Yao J, Wang L, Li C, Zhang C and Wang L V. 2014. Photoimprint photoacoustic microscopy for three-dimensional label-free sub diffraction imaging. *Phys. Rev. Lett.* 112 014302.
10. Miyazaki J, Tsurui H, Hayashi-Takagi A, Kasai H and Kobayashi T. 2014. Sub-diffraction resolution pump–probe microscopy with shot-noise limited sensitivity using laser diodes. *Opt. Express* 22 9024–32.
11. Yew E Y S and Sheppard C J R. 2007. Tight focusing of radially polarized Gaussian and Bessel–Gauss beams. *Opt. Lett.* 32 3417–9.
12. Wei J and Yan H. 2014. Strong nonlinear saturation absorption induced optical pinhole channel and super-resolution effects: a multi-layer system model. *Opt. Lett.* 39 6387–90.
13. He J, Miyazaki J, Wang N, Tsurui H and Kobayashi T. 2015. Label-free imaging of melanoma with nonlinear photothermal microscopy. *Opt. Lett.* 40 1141–4.

14. Vaziri A, Tang J, Shroff H and Shank C V. 2008. Multilayer three-dimensional super resolution imaging of thick biological samples. *Proc. Natl Acad. Sci. USA* 105 20221–6.
15. Han K Y, Willig K I, Rittweger E, Jelezko F, Eggeling C, and Hell S W. 2009. Three-dimensional stimulated emission depletion microscopy of nitrogen-vacancy centers in diamond using continuous-wave light. *Nano Lett.* 9 3323–9.
16. Fiolka R, Shao L, Rego E H, Davidson M W and Gustafsson M G L. 2012. Time-lapse two-color 3D imaging of live cells with doubled resolution using structured illumination. *Proc. Natl. Acad. Sci. USA* 109 5311–5.
17. Jia S, Vaughan J C and Zhuang X. 2014. Isotropic three dimensional super-resolution imaging with a self-bending point spread function. *Nat. Photonics* 8 302–6.
18. Iketaki Y, Kumagai H, Jahn K and Bokor N. 2015. Creation of a three-dimensional spherical fluorescence spot for superresolution microscopy using a two-color annular hybrid wave plate. *Opt. Lett.* 40 1057–60.
19. Kuang C F, Li S, Liu W, Hao X, Gu Z T, Wang Y F, Ge J H, Li H F and Liu X. 2013. Breaking the diffraction barrier using fluorescence emission difference microscopy. *Sci. Rep.* 3 1441.
20. Dehez H, Piche M and Koninck Y D. 2013. Resolution and contrast enhancement in laser scanning microscopy using dark beam imaging. *Opt. Express* 21 15912–25.
21. Segawa S, Kozawa Y and Sato S. 2014. Resolution enhancement of confocal microscopy by subtraction method with vector beams. *Opt. Lett.* 39 3118–21.
22. Wang N, Miyazaki J, He J, Seto K and Kobayashi K. 2015. Subdiffraction-limit imaging using mode multiplexing. *Opt. Commun.* 343 28–33.
23. You S, Kuang C, Rong Z, Liu X and Ding Z. 2014. Isotropic superresolution imaging for fluorescence emission difference microscopy. *Appl. Opt.* 53 7838–44.
24. You S, Kuang C, Li S, Liu X and Ding Z. 2015. Three-dimensional super-resolution imaging for fluorescence emission difference microscopy. *AIP Adv.* 5 084901.
25. Wang N and Kobayashi T. 2014. Numerical study of the subtraction threshold for fluorescence difference microscopy. *Opt. Express* 22 28819–30.
26. Wang N and Kobayashi T. 2015. Polarization modulation for fluorescence emission difference microscopy. *Opt. Express* 23 13704–12.
27. Wang N and Kobayashi T. 2015. Numerical calibration of the spatial overlap for subtraction microscopy. *Opt. Express* 23 13410–22.
28. Richards B and Wolf E. 1959. Electromagnetic diffraction in optical systems. II. Structure of the image field in an aplanatic system. *Proc. R. Soc. A* 253 358–79.
29. Ganic D, Gan X and Gu M. 2003. Focusing of doughnut laser beams by a high numerical-aperture objective in free space. *Opt. Express* 11 2747–52.
30. Hao X, Kuang C, Wang T and Liu X. 2010. Effects of polarization on the de-excitation dark focal spot in STED microscopy. *J. Opt.* 12 115707.
31. Wang N and Kobayashi T. 2015. Subtraction threshold for an isotropic fluorescence emission difference microscope. *J. Opt.* 17 125302. doi:10.1088/2040-8978/17/12/125302

4.2 Noise Cancelation with Phase-detection Technique for Pump-probe Measurement and Application to stimulated Raman Imaging

1 INTRODUCTION

Various types of pump-probe microscopies such as stimulated Raman scattering (SRS) [1–11], stimulated emission [12–15], and photothermal [15–20] microscopies have been investigated to achieve superresolution with high sensitivity and high-speed imaging. In pump-probe microscopy, a pump beam is irradiated on a sample as a stimulus, and a probe beam is irradiated to detect the stimulation effects. The signal is obtained as the intensity modulation of the probe beam induced by the stimulation on the background of the beam. While the optical source noise is multiplicative to the signal and noise is dominated by the shot noise in the cases of observation of light-emitting effects such as fluorescence and spontaneous Raman, the intensity noise of the probe beam is also additively included in pump-probe measurements, as it is not factored by the modulation depth of the signal. For example, 1×10^{-4} of the intensity fluctuation of the source deteriorates the signal-to-noise ratio (SNR) by 1×10^{-4} through the multiplicative effect, while it degrades the SNR to 1 for 1×10^{-4} of the modulation depth through the additive effect. Hence, the stability of the background signal of the probe beam is critical in pump-probe microscopy. Lock-in detection is frequently applied to reject the intensity noise of the probe beam as much as possible [21–23]. In this detection scheme, the pump intensity is modulated at a specific frequency, and the intensity modulation of the probe beam around this frequency is detected. The intensity noise apart from the frequency is rejected, thereby enhancing the SNR. Additionally, a substantial contribution from noise components at some frequencies is rejected by setting the lock-in frequency at

DOI: 10.1201/9781003476030-16

an appropriate level to avoid these frequencies. For example, a contribution from the acoustic noise at a low frequency is rejected by a high-frequency lock-in detection in an MHz region. However, noises independent of frequency such as the thermal noise of a load resistor for a photodetector, shot noise of the light source, and white intensity noise of the light source are not rejected by the lock-in detection at any lock-in frequency. The contribution of the thermal noise is reduced with a resonant circuit as a load impedance with a combination of the high impedance and response speed [24]. However, the intensity noise of the light source is irrelevant to the thermal noise, and thus, the intensity noise cannot be reduced by the resonator. In this case, the use of a narrow bandwidth efficiently rejects any noise from nearby frequencies. The bandwidth of frequency measurement is determined by the time constant, and the bandwidth is made narrower by using a longer time constant. Although a narrow bandwidth enhances the SNR inversely proportionally to its bandwidth, it decreases the response speed. Therefore, a lock-in detection scheme involves a trade-off between the SNR and response speed.

To overcome this trade-off problem, we present the following method for the detection of the phase modulation in the detected signal, rather than the intensity modulation, of which physical quantity is common to the intensity noise. We split an intensity modulated probe beam (e.g., pulsed beam) with a frequency of ω_c into two beams. One beam overlaps with the pump beam, the other is delayed by a quarter of the intensity modulation cycle. The phase of the former beam coincides with that of the pump beam, forming an "in-phase beam". The phase of the latter is displaced by 90°, forming a "quadrature beam". When the in-phase and pump beams are transmitted through a sample, the intensity of the in-phase beam is modulated. The in-phase beam is then detected after being superposed on the quadrature beam. By capturing this superposed beam as a "carrier" of the phase-amplitude modulation, the instantaneous amplitude is determined by the root of the square sum of the intensities of the in-phase and quadrature beams, while the instantaneous phase is the ratio between the two beam intensities. The instantaneous phase of the modulated carrier is represented as $\omega_c t+\theta+\theta_m(t)$, where θ and $\theta_m(t)$ are the initial phase and phase deviation, respectively. The intensity modulation of only the in-phase beam causes a modulation of the ratio between the in-phase and quadrature beams, and hence, this intensity modulation creates the phase deviation, $\theta_m(t)$ [Figure 4.2.1(a)]. In this scheme, because the in-phase and quadrature beams are generated from the same light source, the instantaneous amplitudes of the in-phase and quadrature beams are common. Thus, the intensity ratio or phase deviation is not affected by the variation of the instantaneous amplitude of the carrier induced by intensity noise of the light source [Fig. 4.2.1(b)]. Therefore, through the phase detection of the carrier beam, the intensity noise of the source is canceled, while the signal of the sample is detected independently of the intensity noise.

Because this noise cancelation does not limit the band, the aforementioned trade-off is principally removed. In realizing this scheme, the crosstalk between the phase deviation and instantaneous amplitude of the carrier in the phase detection must be removed to prevent the intensity noise from being detected as phase noise. To fulfill this requirement, we introduce a method for detecting the phase deviation independently of the instantaneous amplitude.

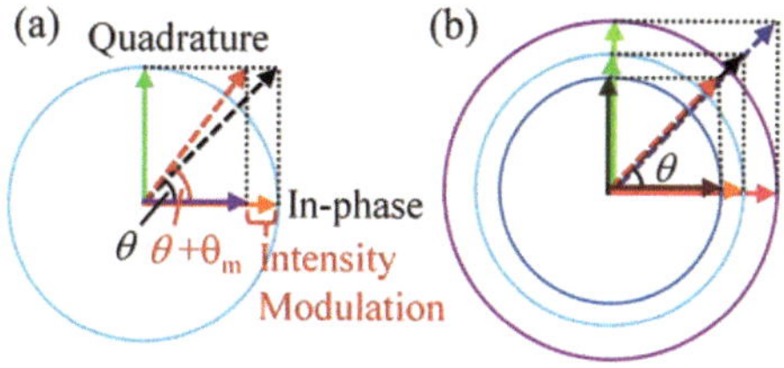

FIGURE 4.2.1 Principle of the phase-detected pump-probe measurement displayed in a complex plane. (a) The orange, purple, and green arrows indicate the in-phase beam without intensity modulation by the sample, the in-phase beam with intensity modulation, and the quadrature beam, respectively. The black and red dotted arrows indicate the carrier beam (superposition of the in-phase and quadrature beams) with and without the intensity modulation by the sample being excited by the pump. θ and θ_m indicate the initial phase and phase deviation of the carrier, respectively. (b) The effect of a light source intensity change. The radii of the circles indicate the intensity of the light source or the instantaneous amplitude of the carrier. The copyright of this figure is shared between the representative author (Takayoshi Kobayashi) and Optica.

In this study, we apply the present method to SRS microscopy. Pulsed beams with higher and lower photon energies are used in the SRS experiment. When the difference in their energy coincides with a vibrational energy of a molecule, a photon of the higher energy is annihilated and the counterpart is created in the SRS process. The typical modulation depth of the beam intensity in this process has been reported to be in the range of 10^{-3} to 10^{-5} in microscopy [2, 4, 9, 10]. By sweeping the beams or translating a sample, the intensity modulation signal of the beams provides an image of the molecular vibration of the sample without labeling [2, 3]. In this study, a titanium-sapphire laser oscillator generates the beam of the lower photon energy. The beam of the higher photon energy is prepared in the form of a white pulsed beam by supercontinuum generation with the same light source and a photonic crystal fiber (PCF) [25]. In PCFs, supercontinuum generation is possible even with the low peak power of the input pulse since the tight confinement of the input beam in a core surrounded by air voids results in a large amount of nonlinearity. In the present chapter, the beam of the lower photon energy serves as the pump beam, and the white beam serves as the probe (or carrier) beam. In the present SRS microscopy scheme, the pump and probe beams from the common light source render the synchronization of plural sources unnecessary. Furthermore, we readily conduct spectral imaging through the spectral dispersion of the white probe beam [26, 27]. However, the intensity noise of the white beam with the PCF is substantially high [28–32], causing a serious problem when the PCFs are applied to high-speed pump-probe microscopy. This cannot be solved simply by using high-frequency lock-in detection since this noise is nearly white, even in a megahertz frequency region [29–32]. Therefore, we demonstrate the feasibility of applying the present method to pump-probe imaging with the use of a continuum and extending the capability of light source choices through noise rejection.

2 OPTICS

Figure 4.2.2 illustrates the optics system for an SRS microscope with phase-sensitive detection. The optics system is similar to that in our previous works [26, 33]. A titanium-sapphire laser (Mira 900P, COHERENT) generates the 76.3 MHz optical pulse repetition. The pulse duration is 2.5 ps, and the center wavelength is set at 802 nm. This beam is split into probe and pump arms. In the probe arm, the polarization angle is adjusted by a half-wave plate, and then coupled into a 1-m length PCF (NL-2.3-790-2, THORLABS) with a zero-dispersion wavelength (ZDW) of 790 nm by a microscope objective lens (×40, NA = 0.65) to generate a white-light continuum. The spectral bandwidth is preferred to be a requisite minimum at certain conversion efficiency into the white light to maximize spectral power density. The input power tends to be more distributed in shorter wavelength components with a PCF of a shorter ZDW at a certain length of the PCF [25]. Hence, a PCF with a ZDW

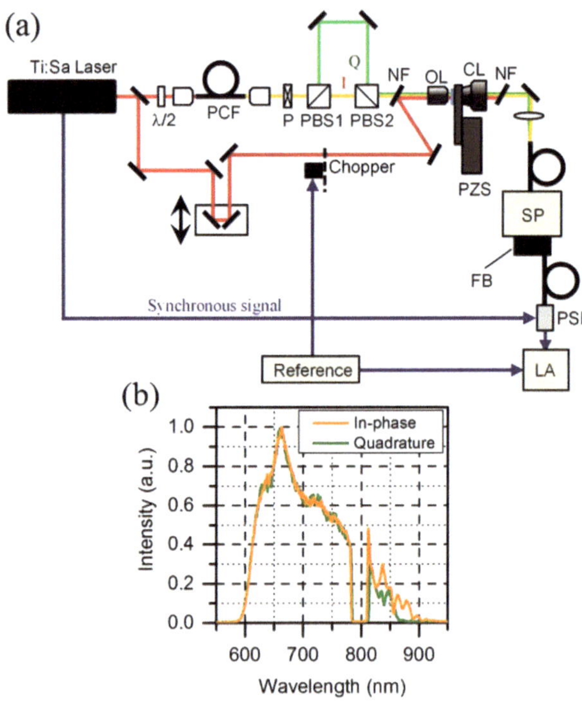

FIGURE 4.2.2 Optics system of a stimulated Raman microscope with (a) present phase-sensitive detection, and (b) spectra of in-phase and quadrature beams. (a) The red, orange, green, and blue lines indicate 802-nm, in-phase, quadrature pulse beams, and electrical signal flows, respectively. λ /2, half-wave plate; PCF, photonic crystal fiber; P, GlanThompson polarizer; PBSs, polarization beam splitters; I, in-phase beam; Q, quadrature beam; NF, notch filter; OL, objective lens; CL, condenser lens; PZS, piezo stage; SP, spectrometer; FB, fiber bundle; PSD, present phase-sensitive detector; and LA, lock-in amplifier. The copyright of this figure is shared between the representative author (Takayoshi Kobayashi) and Optica.

of 790 nm, which is closer to the input wavelength than the 750 nm in the previous work [33], is selected. In addition, the length is increased to 1 m from 30 cm in the previous work [33] to enhance conversion efficiency. The spectral power density is increased by ~10 times higher than that in the previous work. The efficiency of the white-light generation also depends on the polarization angle of the input pulse due to small birefringence in the PCF. Hence, the polarization angle is adjusted to maximize the efficiency.

A microscope objective lens (magnification power = ×40, NA = 0.65) collimates the generated white light, and a Glan-Thompson polarizer adjusts and fixes the polarization angle. A polarization beam splitter (PBS) divides the white pulse beam into the in-phase and quadrature arms. The splitting ratio is adjusted by the polarization angle determined by the polarizer. Fixation of the polarization angle is critically important for the cancelation of the optical noise in the white-light source. The polarization angle of the white pulse beam fluctuates from pulse to pulse. The polarization angle must be fixed before the beam splitter to reduce the fluctuation in the splitting ratio. Otherwise, this fluctuation would result in the phase fluctuation of the carrier. In the quadrature arm, the beam is delayed for a quarter cycle of the pulse repetition by a 98-cm additional optical path, and prepared as the quadrature beam. The in-phase and quadrature beams are spatially overlapped and aligned collinearly with each other on another PBS, and function as the carrier beam. A notch filter (NF808-34, THORLABS) rejects the spectral component near the 802-nm input beam. The spectra of the in-phase and quadrature beams cover 600–780 nm [Fig. 4.2.2(b)]. This spectral bandwidth of ~4200 cm−1 is enough for most of the molecular vibrations used in SRS microscopy.

In the pump arm, the pulse beam is guided through the delay stage to adjust the timing so that it temporally overlaps with only the in-phase pulse beam. The pump beam is chopped at 4.48 kHz, and then spatially overlapped and aligned collinearly with the carrier beam on the notch filter. The beams are sent to a microscope and focused on a sample by a microscope objective (×10, NA 0.25 or ×40, NA 0.85). In SRS imaging, the sample is scanned by piezo actuators on a stage (NanoMax300, THORLABS). The beams transmitted through the sample are collected and collimated through a condenser lens. A notch filter (NF808-34, THORLABS) rejects the spectral component of the pump beam at the level of 55 dB. The carrier beam is guided into a spectrometer (SpectraPro-300i, ACTON) by a piece of optical fiber. The focal length and F number of the spectrometer are 30 cm and 4, respectively. The groove density on the grating, blaze wavelength, and line dispersion are 300 groves/mm, 500 nm, and 10.8 nm/mm, respectively. The spectrum of the carrier is imaged on the entrance facet of a 128 × 16 fiber bundle (core/clad: 100/110 μm). The bundle splits the spectrum into 128 components, and 16 fibers of the same spectral component are bundled. The bundle of 16 fibers guides a spectral component into a developed phase-sensitive detector (PSD), which is described in detail in the following section. The optical pulse from the laser oscillator is detected with a fast photodiode to generate the synchronous signal for the phase-sensitive detection. The output signal from the PSD is lock-in detected at the chopping frequency.

3 COMPOSITION OF PHASE-SENSITIVE DETECTOR

The circuit block diagram of the PSD is shown in Figure 4.2.3(a). The in-phase and quadrature beams are detected by an avalanche photodiode [referred to as APD in Fig. 4.2.3(a)]. The photocurrent signals of the in-phase and quadrature beams are multiplied with the APD. The multiplication is needed to detect low intensity beams and suppress noise effects caused by the subsequent circuit. Then the photocurrent around the carrier frequency is converted into a voltage signal by a parallel resonant circuit (RES). This voltage signal is buffered and amplified by a preamplifier (PA) with a high input impedance. This configuration contributes to the suppression of the thermal noise contribution owing to a high load impedance of the RES while maintaining the required response speed [24]. The voltage signal is multiplied by the synchronous signal of the light source with a multiplier (ML), to obtain an error for a feedback control of a phase deviation of the synchronous signal.

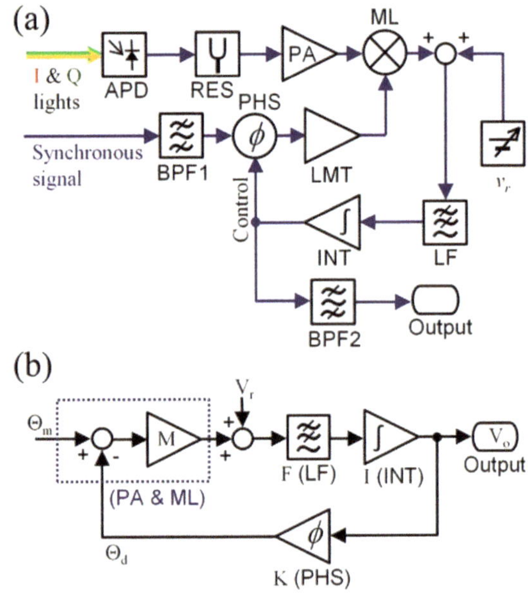

FIGURE 4.2.3 (a) Circuit block diagram of the present phase-sensitive detector. (b) Block diagram of the phase feedback control. APD, avalanche photodiode; RES, resonator of carrier frequency; PA, preamplifier; ML, multiplier; BPFs, bandpass filters; PHS, phase shifter; LMT, amplitude limiter; v_r, bias voltage source for offset; LF, loop filter; INT, integrator; Θ_m, Θ_d, V_r, and V_o: Laplace transformations of the phase deviation of the carrier signal [$\theta_m t$], the phase deviation of the synchronous signal ($\theta_d t$), the offset voltage (v_r), and the output signal ($v_o t$), respectively; M: response function of the combined gain of the multiplier and pre-amplifier; F, I, and K: response functions of the loop filter, integrator, and phase shifter, respectively. The copyright of this figure is shared between the representative author (Takayoshi Kobayashi) and Optica.

The synchronous signal from the light source is filtered by a bandpass filter (BPF1) to obtain the fundamental component. A phase deviation is applied on this signal in a phase shifter (PHS). The phase deviation is controlled by a feedback signal based on the above-mentioned error signal to track the deviation on the optical signal. The amplitude is fixed by a limiter[1] (LMT) for the sake of simplicity of the design, and sent to the ML. An offset (v_r) is added to the output (the error signal) from the ML to adjust the phase difference between the optical signal and the synchronous signal. This v_r contains an unintentional offset caused by the electrical components. The unintentional offset can be compensated for by an intentional offset. This signal is integrated by an integrator[2] (INT) through a loop-filter[3] (LF), such as a low-pass filter.[4] Then the integrated signal controls the phase deviation of the synchronous signal to track the phase deviation of the optical signal. The LF compensates for an inappropriate removal of a high-frequency component in the error signal by the INT. Therefore, it enables a high-speed feedback. In this configuration, the control signal is proportional to the phase deviation with an offset of the phase difference. This signal is transmitted through a band- pass filter (BPF2), and observed with the lock-in amplifier. Owing to the lock-in detection of the output through the BPF2, dynamic range is enhanced by eliminating a direct cur- rent offset and unwanted noise in the high-frequency range.[5]

We use an APD of S5343, HAMAMATSU as a photodetector. Then, the photocurrent is converted to a voltage signal by the RES. This signal is buffered with a non-inverting amplifier (functions as the PA) based on an operation amplifier (LMH6624) of ×5.3 voltage gain. This voltage signal is amplified by 20 dB with a transistor (2SC3356)-based amplifier and input into a multiplier device (AD835) (functions as the ML).

In the arm of the synchronous signal, it is amplified by 20 dB by a narrow-band amplifier based on a field-effect transistor (FET) (2SK241). Its bandwidth is 10 MHz and it functions as the BPF1. Then the signal is sent to a voltage controlled phase shifter, or all-pass filter (functions as the PHS) based on an operation amplifier (LMH6609) and variable capacitance diode (BB148). The phase deviation of the synchronous signal is controlled by a feedback bias voltage signal on the variable-capacitance diode. The synchronous signal is then converted into a square wave by a digital comparator based on ADCMP600 to maintain the amplitude at constant (i.e., LMT), and input into the ML.

The v_r is adjusted with the applied voltage on the summing input terminal of the AD835. The LF is a third-order Bessel type low-pass filter of 3 MHz cutoff frequency based on the LMH6609. The INT is an integrator circuit of 0.16 μs integration time based on the LMH6609. The integrated signal for the feedback control is divaricated and filtered through a bandpass filter (functions as the BPF2) and output. The BPF2 is composed of a second-order Butterworth type high-pass filter of 400 Hz cutoff frequency and a second-order Butterworth type low-pass filter of 100 kHz cutoff frequency with 20 dB gain.

Note, the present configuration is similar to the one in our previous report [26] to observe the difference signal between the probe beam and a reference beam from the same light source of the probe. In [26], the optical signal is multiplied by the synchronous signal, and its phase deviation is feedback controlled, based on the low-frequency component in the multiplied signal. However, the present and previous ones are different in the following way. The output positions in the diagrams and the purposes of the feedbacks are different from each other. In the previous work, the output is obtained just after the ML through a low-pass filter to obtain the difference signal of the probe and reference beams, while it is the feedback signal itself in the present circuit that observes the phase deviation of the synchronous signal, which tracks the deviation of the optical signal. The different function of the feedback creates a difference in the response speed. In the previous work, the function of the feedback is compensation for the difference in the transmissions for the probe and reference beams, and thus, the feedback response frequency must be set lower than the modulation frequency of the pump beam so as not to cancel the signal. Hence it is set at 65 Hz for 4.48 kHz modulation frequency. In the present work, it must be set higher than the modulation frequency to track the phase deviation on the carrier. As mentioned later, the response frequency is set at 32 kHz.

4 OPERATION PRINCIPLE

The working principle is based on our previous work [26]. In the previous work, the difference between the probe and reference signals (corresponding to the in-phase and quadrature signals, respectively) is observed. The gains for the signals are determined by the phase deviation of the synchronous signal. Then the phase deviation is feedback controlled to cancel the difference with a response speed slow enough not to respond to the intensity modulation of the probe signal by the sample. The background signal, including the intensity noise but excluding the modulation signal, is canceled in the difference signal. In the present PSD, the feedback scheme similar to that in [26] is operated, but the response is faster to also cancel the intensity modulation of the in-phase signal by the sample. In this configuration, the phase deviation of the synchronous signal is controlled so that its instantaneous phase is always perpendicular to that of the optical signal (carrier). The signal for control of the phase deviation, but neither the carrier nor synchronous signal, is observed. The modulation of this control signal is proportional to the phase modulation of the carrier, as the control signal swings to force the instantaneous phase of the synchronous signal to track that of the carrier signal at a 90° interval.

Because the phase of the quadrature beam is delayed by 90° against the in-phase beam, the fundamental component of the quadrature beam is expressed as a sine wave when that of the in-phase beam is expressed as a cosine wave. The splitting ratio of the white pulse beam into the in-phase and quadrature beams is expressed as a:b'+(a+b'= 1). Thus, the in-phase signal is given as

$$a\{1+\mathrm{m}(t)\}\,\mathrm{A}(t)\cos(\omega_c t), \tag{1}$$

and the quadrature signal is

$$b'\mathrm{A}\left(t-\frac{T}{4}\right)\sin(\omega_c t). \tag{2}$$

Here, A(t) is the instantaneous amplitude and is proportional to the intensity of the light source, including amplitude noise;

m(t) is the amplitude modulation of the signal on the in-phase beam; ω is the carrier frequency; and t is the carrier cycle.

When the in-phase and quadrature beams are the pulse beams, their optical currents from the photodetector are proportional to the following terms, respectively:

$$a \sum_{n=-\infty}^{0} A_n \{1+\mathrm{m}(nT)\}\delta(t-nT) \tag{3}$$

$$b' \sum_{n=-\infty}^{0} A_n \delta\left(t-nT-\frac{T}{4}\right). \tag{4}$$

Here, A_n is the pulse intensity of the nth pulse and δ (t) is the Dirac delta function. The Laplace transformations of these currents are

$$a \sum_{n=-\infty}^{0} A_n \{1+\mathrm{m}(nT)\}\exp(-snT) \tag{5}$$

and

$$b' \sum_{n=-\infty}^{0} A_n \exp\left\{-s\left(n+\frac{1}{4}\right)T\right\}. \tag{6}$$

These optical currents are converted into voltage signals by the thermal noise effect less the RES in the present system.

One might be concerned about the phase change caused by the RES. We describe the effect of the RES as follows. The responses of an inductance L, capacitance C, and resistance R are sL, $1/(sC)$, and R, respectively, and hence the response of the RES is

$$\left(\frac{1}{sL}+sC+\frac{1}{R}\right)^{-1} = \frac{L\omega_0^2 s}{(s+\zeta\omega_0)^2+(1-\zeta^2)\omega_0^2}. \tag{7}$$

where ω_0 and ζ are the resonant frequency and damping factor, and determined by the element values, as follows:

$$\omega_0^2 = \frac{1}{LC}, \text{ and } 2\zeta = \frac{1}{Q} = \frac{L\omega_0}{R} \tag{8}$$

This R contains a contribution of unintentional energy losses in the inductor and capacitor. From Eqs. (5) and (7), the voltage signal of the in-phase beam in the time domain is

$$\begin{aligned} &a\frac{L\omega_0^2}{\sqrt{1-\zeta^2}}\Sigma_{n=-\infty}^{0} A_n \{1+\mathrm{m}(nT)\}\exp\{-\zeta\omega_0(t-nT)\} \\ &\times\cos\left\{\sqrt{1-\zeta^2}\,\omega_0(t-nT)+\beta\right\}, \end{aligned} \tag{9}$$

where

$$\beta = \tan^{-1} \frac{\zeta}{\sqrt{1-\zeta^2}}. \tag{10}$$

Here, by adjusting the resonant frequency as follows,

$$\sqrt{1-\zeta^2}\,\omega_0 = \omega_c, \tag{11}$$

the signal is written as

$$a\left[\frac{L\omega_0^2}{\sqrt{1-\zeta^2}} \sum_{n=-\infty}^{0} A_n \left\{1+\mathrm{m}(nT)\right\} \exp\left\{-\zeta\omega_0 (t-nT)\right\}\right] \times \cos(\omega_c t + \beta). \tag{12}$$

The ζ should be set at a value large enough to observe high-speed phenomena. In other words, the change of m(t) is slower than the damping speed. Hence, m(t) within the time constant of the resonator, and Eq. (12) is approximated in the same form as Eq. (1), where β is set at 0 by adjusting the initial phase of the synchronous signal:

$$a\left\{1+\mathrm{m}(t)\right\} \mathrm{A}(t) \cos(\omega_c t) \tag{13}$$

$$\text{where} \quad \mathrm{A}(t) = \left[\frac{L\omega_0^2}{\sqrt{1-\zeta^2}} \sum_{n=-\infty}^{0} A_n \exp\left\{-\zeta\omega_0 (t-nT)\right\}\right]. \tag{14}$$

The voltage signal of the quadrature beam is similarly obtained as

$$b' \exp\left(\frac{\zeta\omega_0 T}{4}\right) \mathrm{A}(t) \sin(\omega_c t). \tag{15}$$

Therefore, the instantaneous amplitude of the quadrature beam [Eq. (2)] is proportional to that of the in-phase beam, as follows:

$$\mathrm{A}\left(t - \frac{T}{4}\right) = \exp\left(\frac{\zeta\omega_0 T}{4}\right) \mathrm{A}(t). \tag{16}$$

Note that the instantaneous amplitude is not that of the light source itself but the convolution of it by the resonator. However, this effect does not destroy the correlation between the instantaneous amplitudes of the in-phase and quadrature beams [Eq. (16)], and it allows the following noise cancelation. Hereafter, the quadrature signal is written as follows, for simplicity,

$$b\,\mathrm{A}(t) \sin(\omega_c t), \tag{17}$$

where the damping effect is contained in the optical splitting ratio:

$$b'\exp\left(\frac{\zeta\omega_0 T}{4}\right) = b. \tag{18}$$

The superposition of the in-phase and quadrature signals [Eqs. (13) and (17)],

$$\mathrm{A}(t)\left[a^2\{1+\mathrm{m}(t)\}^2+b^2\right]^{1/2}\sin\left\{\omega_c t+\theta_\mathrm{m}(t)\right\}, \tag{19}$$

is captured as a phase-amplitude-modulated wave and input to the ML. Here, the phase deviation is determined by the ratio between the instantaneous amplitudes of the in-phase and quadrature signals, as follows:

$$\theta_\mathrm{m}(t) = \tan^{-1}\left[\frac{a}{b}\left\{1+\mathrm{m}(t)\right\}\right]. \tag{20}$$

Note that this phase deviation is dependent only on m(*t*) and independent of A(*t*). The instantaneous amplitude, including the amplitude noise of the light source at any frequency, is canceled. When m(t) $\ll$ 1,

$$\theta_\mathrm{m}(t) \cong \frac{ab}{a^2+b^2}\,\mathrm{m}(t)+\tan^{-1}\left(\frac{a}{b}\right). \tag{21}$$

The size of phase modulation is approximately proportional to the amplitude modulation only of the in-phase signal.

The synchronous signal around the fundamental component of amplitude D and phase deviation $\theta_\mathrm{d}t$ is expressed as

$$D\cos\left\{\omega_c t+\theta_\mathrm{d}(t)\right\}. \tag{22}$$

Here, we assume that $\theta_\mathrm{d}(t)$ is designed to be controlled by the output of the INT, $\mathrm{v}_\mathrm{o}(t)$ as mentioned later, with the following proportionality relation with constant *k*:

$$\theta_\mathrm{d}(t) = k\mathrm{v}_\mathrm{o}(t). \tag{23}$$

The output from the ML is the product of the carrier and synchronous signals:

$$\begin{aligned}&\frac{1}{2}\mathrm{A}(t)D\left[a^2\{1+\mathrm{m}(t)\}^2+b^2\right]^{1/2}\\&\times\left[\sin\left\{\theta_\mathrm{m}(t)-\theta_\mathrm{d}(t)\right\}+\sin\left\{2\omega_c t+\theta_\mathrm{m}(t)+\theta_\mathrm{d}(t)\right\}+\mathrm{HF}(t)\right].\end{aligned} \tag{24}$$

Here, HF(*t*) refers to the higher frequency components originating in the harmonic components in the square wave by the digital limiter. We then add the offset value of v_r. The LF rejects the high-frequency components containing that at $2\omega_c$. Hence, we focus on the first term in Eq. (24) with v_r:

$$\frac{1}{2}\mathrm{A}(t)D\left[a^2\{1+\mathrm{m}(t)\}^2+b^2\right]^{1/2}\times\sin\{\theta_\mathrm{m}(t)-\theta_\mathrm{d}(t)\}+v_r. \tag{25}$$

When $\theta_m t(t) - \theta_d(t) \ll 1$, this term is approximated as

$$\frac{1}{2}\mathrm{A}(t)D\left[a^2\{1+\mathrm{m}(t)\}^2+b^2\right]^{1/2}\times\{\theta_\mathrm{m}(t)-\theta_\mathrm{d}(t)\}+v_r. \tag{26}$$

This signal is integrated, and the integrated signal controls $\theta_d(t)$. At the same time, this integrated signal is transmitted through the BPF2 to reject offset and electrical noise. When the observing frequency (ω) is sufficiently lower than $2\omega_c$, the cutoff frequency of the INT can be set to be low enough against $2\omega_c$, and the LF is not required because the high-frequency components are properly rejected by the INT. In other words, the rejection of the high-frequency components by this LF allows the cutoff frequency of the INT to be set at a higher frequency than that without the LF. This enhances the response speed of the feedback loop, and high-speed observation becomes possible.

The present scheme is considered to be a feedback system of $\theta_d(t)$ that tracks $\theta_m(t)$, as described in the block diagram in Fig. 4.2.3(b). Based on Eq. (26), the photodetector and the ML are captured as the combination of a subtracter for $\theta_m(t)$ and $\theta_d(t)$, and an amplifier of the gain M(*t*) in this scheme.

Here,

$$\mathrm{M}(t)=\frac{1}{2}\mathrm{A}(t)D\left[a^2\{1+\mathrm{m}(t)\}^2+b^2\right]^{1/2}. \tag{27}$$

The response function of this amplifier is defined as M(*s*). V_r (the Laplace transformation of v_r) is then added to the signal. The signal is filtered by the LF of response function

F integration times and integrated. The response function of this INT with an integration time *T* is expressed as

$$\mathrm{I}(s)=\frac{1}{sT}. \tag{28}$$

This integrated signal controls the PHS of response function ($K = k$) and provides the output. When M(t) is assumed to be time invariant, i.e., A(*t*) is constant and m(*t*) is ignored, this feedback system is considered to be a linear time-invariant system. Hence, $\Theta_d(s)$ is given as follows:

$$\Theta_d(s)=K\mathrm{I}(s)\mathrm{F}(s)\left[M\{\Theta_m(s)-\Theta_d(s)\}+V_r\right], \tag{29}$$

where ($\theta\, m(s)$ and $\theta\, d(s)$ are the Laplace transformations of $\theta\, m(t)$ and $\theta\, d(t)$, respectively. Based on Eq. (29), $\Theta_d(s)$ is calculated as follows:

$$\Theta_d(s)=\frac{1}{\dfrac{sT}{k\,\mathrm{F}(s)M}+1}\Theta_m(s)+\frac{1}{\dfrac{sT}{k\,\mathrm{F}(s)}+M}V_r. \tag{30}$$

Therefore, the final value of θ_d is calculated as follows:

$$\theta_d(\infty) = \lim_{s\to 0}\{s\Theta_d(s)\} = \theta_m(\infty) + \frac{1}{M}v_r. \tag{31}$$

In Eq. (31), v_r functions as the reference value for the difference between $\theta_m t$ and $\theta_d t$, and when v_r 0, $\theta_d t$ is controlled so that the synchronous signal is perpendicular to the carrier signal. Equation (30) indicates that, when the frequency ω in the observed signal is too high, the system is unable to respond, and results in a large displacement between $\theta_m t$ and $\theta_d t$. This situation is restated as a shortage of the response speed to detect high-speed phenomena. When the feedback gain is set at a large value (e.g., by setting T as a small value), this effect is compensated and the cutoff frequency of the system can be increased. However, only increasing the gain results in an insufficient rejection of the $2\omega_c$ and HF(t) components in Eq. (24). This problem is solved by the LF and an adequate rejection of only the high-frequency components. The high cutoff frequency of the feedback is permitted with the high gain. The phase retardation of the feedback loop containing the LF must not exceed 180° at the cutoff frequency; otherwise, the positive feedback would cause the system to oscillate. As a rule of thumb, a maximum compromised retardation between stability and response speed can be found in the range of 120–135°.

For the present system, $D = 1$ and the maximum value of A is 1. Then, the maximum gain of the ML is 1/2, where $a \to 1$ or $b \to 1$. The open-loop response function at the largest gain (the most unstable condition) is

$$K\mathrm{I}(s)\mathrm{F}(s)M = k\frac{1}{sT}\,\mathrm{F}(s)\frac{1}{2}. \tag{32}$$

The LF is designed as a third-order Bessel-type low-pass filter with a 3 MHz cutoff frequency. Its response function (Fs) is calculated as follows:

$$\mathrm{F}(s) = 1.11 \times 10^{23} \times (s^3 + 1.15 \times 10^8 s^2 + 5.65 \times 10^{15} s + 1.11 \times 10^{23})^{-1}: \tag{33}$$

The integration time T is set at 0.16 μs and the response of the PHS k =65.7 mrad/V. The cutoff frequency is calculated as 32 kHz, and the phase retardation is 92°. The high-frequency component of $2\omega_c$ at 152.6 MHz is suppressed by -185 dB.

Mt is not time invariant because of the fluctuation of At caused by noise from the light source. Therefore, the requirement of detecting the phase deviation without the optical noise is fulfilled by eliminating the contribution of Mt. Without the time invariance assumption, Eq. (29) is replaced by the following:

$$\Theta_d(s) = K\mathrm{I}(s)\mathrm{F}(s)\times\left[\mathrm{M}(s)\otimes\{\Theta_m(s) - \Theta_d(s)\} + V_r\right], \tag{34}$$

where $\otimes$ is the convolution operator. When the deconvolution by M is expressed by ($1/\mathrm{M}_\otimes$), Eq. (34) is deduced as follows:

$$(1/\mathrm{M}\otimes)\left\{\frac{sT\Theta_d(s)}{k\,\mathrm{F}(s)}\right\} + \Theta_d(s) = \Theta_m(s) + (1/\mathrm{M}\otimes)V_r. \tag{35}$$

Here, when the response speed is high enough (i.e., $|kF(j\omega / T)| \gg \omega$), the first term in the left side of Eq. (35) is ignored:

$$\Theta_d(s) = \Theta_m(s) + (1/\mathrm{M}\otimes)V_r. \tag{36}$$

Based on Eq. (36), when $v_r = 0$, the effect of M(t) is rejected. $\theta_d(t)$ is deduced as follows:

$$\theta_d(t) = \theta_m(t) \text{ or } v_o(t) = \frac{\theta_m(t)}{k}. \tag{37}$$

Regardless of the values of At, a, and b, the synchronous signal is perpendicular to the carrier signal. Therefore, the optical noise is canceled, but the output v_0t is proportional to the carrier phase deviation at any optical splitting ratio. Therefore, to achieve the maximum noise cancelation with this scheme, it is important to control the instantaneous phase of the synchronous signal perpendicular to that of the carrier signal.

Finally, we mention the case for a light source of intensity modulated continuous wave (CW). From Eqs. (1) and (2), the phase deviation with the CW light source is

$$\theta_m(t) = \tan^{-1}\left[\frac{a\,\tilde{A}(t)}{b'\tilde{A}\left(t - \frac{T}{4}\right)}\{1 + m(t)\}\right]. \tag{38}$$

Therefore, the phase noise caused by the light source originates in an uncorrelated nature between At and A(t - T/4).

We define the spectra of these terms as $\tilde{A}_I(\omega)$ and $\tilde{A}_Q(\omega)$, respectively. Here,

$$\begin{aligned}\tilde{A}_Q(\omega) &= \exp\left(-j\omega\frac{T}{4}\right)\tilde{A}_I(\omega) \\ &= \tilde{A}_I(\omega) + \left\{\exp\left(-j\omega\frac{T}{4}\right) - 1\right\}\tilde{A}_I(\omega).\end{aligned} \tag{39}$$

The second term in the right side explains the difference in the correlated noise component with the noise component in the in-phase beam, and an addition of the uncorrelated component perpendicular to that of the in-phase beam. The noise not to be canceled is the incoherent summation of this noise at respective frequencies. The power spectrum of this term is

$$\left|\exp\left(-j\omega\frac{T}{4}\right) - 1\right|^2 \left|\tilde{A}_I(\omega)\right|^2. \tag{40}$$

Thus the observed noise with the band of ω_{ob} is

$$\left[\int_{-\omega_{ob}/2}^{\omega_{ob}/2}\left|\exp\left(-j\omega\frac{T}{4}\right)-1\right|^{2}\left|\tilde{A}_{I}(\omega)\right|^{2}d\omega\right]^{\frac{1}{2}}. \tag{41}$$

Here, when the noise is assumed to be white,

$$\left|\tilde{A}_{I}(\omega)\right|^{2}=N^{2}, \tag{42}$$

the noise is calculated as follows:

$$\begin{aligned}&N\left[\int_{-\omega_{ob}/2}^{\omega_{ob}/2}\left|\exp\left(-j\omega\frac{T}{4}\right)-1\right|^{2}d\omega\right]^{\frac{1}{2}}\\&=N\left[2\omega_{ob}-\frac{16}{T}\sin\left(\frac{T}{8}\omega_{ob}\right)\right]^{\frac{1}{2}}.\end{aligned} \tag{43}$$

This result is approximated as

$$N\left[2\omega_{ob}-\frac{16}{T}\sin\left(\frac{T}{8}\omega_{ob}\right)\right]^{\frac{1}{2}}\cong\frac{\pi\omega_{ob}}{16\sqrt{3}\omega_{c}}N\sqrt{\omega_{ob}}, \tag{44}$$

under the assumption of $\omega_c \gg \omega_{ob}$. This result indicates that the factor to the noise without our method $(N(\omega_{ob})^{1/2})$ is $\pi\omega_{ob}/16(3\omega_c)^{1/2}$, and the noise component in the quadrature signal uncorrelated to the in-phase signal is increased proportionally to the bandwidth; the noise cancelation performance is degraded proportionally to the bandwidth. However, a substantial noise cancelation is still possible with a considerable bandwidth. As an example, when the band is set as $\omega_c/10$, this factor is ~0.045, and ~27 dB cancellation is still ω_{ob} expected.

5 EXPERIMENT

A Signal Gain Dependence on Optical Input Power and Splitting Ratio

We investigated the dependence of the output signal intensity of the PSD on the optical input power at the fixed splitting ratio of a = b. We adjusted the optical power of the carrier beam using a variable neutral density filter just after the polarizer. The power of the pump beam was fixed at 74 mW, and the maximum power of the carrier beam was 24 mW on the sample. The gain of the APD was fixed. The carrier and pump beams were focused into the cyclohexane in a 5-mm-length cell by a ×10 NA = 0.25 microscope objective lens. We observed the SRS signal of a CH stretching mode of cyclohexane at 2,854 cm^{-1}. We set the time constant of the lock-in amplifier at 1 s.

We then investigated the dependence of the signal intensity on the splitting ratio into the in-phase and quadrature beams at a fixed optical power. We set the APD gain as large as possible within less than saturation of the circuit. The polarizer adjusted the splitting ratio.

B Noise Cancellation Performance on Optical Splitting Ratio

We observed noise spectra of the 802-nm pulsed beam from the laser oscillator and the white beam. We input the beams into a photodiode module (PDA36A-EC, THORLABS) and observed spectra of the output with a spectrum analyzer (DSA815, RIGOL). All wavelength components of the white beam except the 802-nm component were input. The band frequency of the photodiode module ranges from DC to 10 MHz. The bandwidth was set at 300 Hz for the observation of noise spectra. The optical power of the 802-nm pulse from the oscillator was adjusted by a neutral density filter in such a way that the average output power from the photodiode module was equal to that of the white beam.

We evaluated the output noise as the intensity-modulation equivalent value at various optical splitting ratios. First, we measured the intensity modulation of the cyclohexane SRS using the present circuit to obtain the factor between the intensity modulation of the in-phase beam and the output intensity of the phase modulation of the carrier. To observe the intensity modulation of the SRS, we blocked the quadrature beam, fixed the phase of the synchronous signal, and lock-in detected the signal just after the LF with the in-phase and chopped pump beams. In this case, the ML worked only as a down converter, and the output at the LF was proportional to the power of the in-phase beam. Then the signal intensity of the in-phase beam was measured with only the chopped in-phase beam directly inputted. The modulation equivalent intensity of the SRS signal was measured through normalization according to the signal intensity of the directly chopped in-phase beam. Next, the feedback of the phase was set to be active, and the SRS signal intensity was measured on the PSD with the quadrature and chopped pump beams. We determined the modulation depth of the in-phase beam according to the SRS and PSD-detected signal intensity, and obtained the factor between them. We set the time constant of the lock-in detection at 1 s in all of these measurements.

The intensity noise of the in-pulse beam was measured by inputting only a non-chopped in-phase beam and fixing the phase of the synchronous signal according to the preceding method. We set the time constant at 10 ms and measured the noise signals 140 times. We divided the root mean square (RMS) value of the 140 noise signals by 10 to convert it into the unit bandwidth value, and normalized it according to the in-phase beam intensity to calculate the equivalent noise density of the intensity modulation.

The phase noise of the carrier observed on the present PSD was evaluated with the phase feedback by inputting the in-phase and quadrature beams without the pump beam and without chopping. We set the time constant at 10 ms and measured the noise signals 140 times. We divided the RMS of the 140 values by 10, and then converted it into the intensity modulation equivalent noise density according to the preceding protocol. We repeated the protocol at various splitting ratios, and evaluated the dependency of the noise cancellation on the ratio.

C Application to SRS Microscopy

We prepared the sample with 4 μm polystyrene (PS) beads dispersed in a polyvinyl alcohol (PVA) film on a glass slide. We captured the SRS image of the 3,054 cm^{-1} Raman band of PS with a ×40 NA 0.85 objective lens. The pixel size was set at 0.25 μm, and the image size was 10 × 10 μm (41 × 41 pixels). The powers of the 802 nm pump, in-phase, and quadrature beams were 43, 10, and 10 mW, respectively, at the sample surface. First, we captured the image by modulating the amplitude of the in-phase beam. As detailed in the experiment in Section 5.B, we blocked the quadrature beam, fixed the phase of the synchronous signal, and lock-in detected the signal at just after the LF. Second, we fed the phase back and captured the image through the present PSD with the in-phase and quadrature beams. In both cases, the time constant of the lock-in detection and the dwelling time were set at 100 and 200 ms, respectively.

6 RESULTS AND DISCUSSION

A Signal Gain Dependence on Optical Input Power and Splitting Ratio

Red plots in Fig. 4.2.4(a) show the dependence of the SRS signal intensity on the optical power of the white beam source through the present PSD. The optical input power is the sum of the power of the in-phase and quadrature beams at the APD. The intensity modulation depth of the in-phase beam by SRS is 1.7×10^{-3}. The signal intensity is constant at ~30 mV and independent of the optical power from 1.5 to 8 μW, as expected in the phase detection excluding the multiplicative source-intensity effect.

Figure 4.2.4(b) explains the dependence of the signal intensity on the optical unbalance between the in-phase and quadrature beams. The optical unbalance is defined by $(I- Q)/(I + Q)$ with the optical powers of the in-phase (I) and quadrature (Q) beams. As in Eq. (21), the gain of $\theta_{m}t$ for mt depends on the optical splitting ratio, and the gain depends on the optical unbalance as in the following term:

$$\frac{1-x^2}{2\left(1+x^2\right)};\ x:=\frac{I-Q}{I+Q} \tag{45}$$

The signal should be proportional to the term in Eq. (45) when using the phase-detected pump-probe method. The green line in Fig. 4.2.4(b) indicates the fitted curve with the term in Eq. (45) on the experimental data (red plots). In that fitting, only the amplitude is varied. The curve thoroughly explains the signal intensity dependence, as expected in successful phase detection.

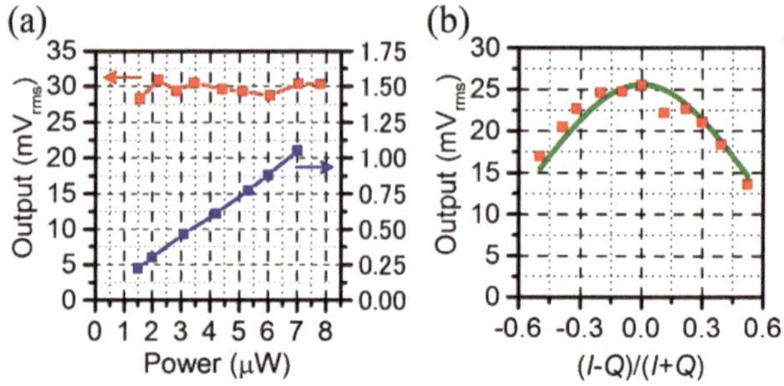

FIGURE 4.2.4 (a) Dependence of signal intensity of a cyclohexane SRS signal on the optical input power. The red and blue plots indicate the phase detection scheme and the auto-balanced detection scheme in [26], respectively. The input power is the sum of those of the in-phase(probe) and quadrature (reference) beams measured at the APD. (b) Dependence of signal intensity on the intensity unbalance of the in-phase (I) and quadrature (Q) beams. The red plots indicate experimental values, and the green line indicates the theoretical value of the phase-detected pump-probe method. The Copy right of this figure is partially held between the representative author (Takayoshi Kobayashi) and Optica.

B Noise Cancelation Performance on Optical Splitting Ratio

Figure 4.2.5(a) shows noise spectra of the beams in a 0–100 kHz range, and 5(b) shows the same in a 0–10 MHz range. The average output powers of the beams are 4.6 dBm. The noise power of the oscillator overlaps (the red plots) that of the electrical noise of the detection system (the blue plots). The intensity of the beam from the oscillator is stable and the noise is below the detection limit. The noise spectrum of the white beam (the green plots) is observed above the detection limit from DC to 10 MHz, and is independent of frequency, or white. This white noise originates in an amplification of the shot noise of the laser oscillator [31]. Thus, high-frequency lock-in detection to avoid acoustic noise is not effective to attain high SNR for the white beam with PCFs. This fact indicates that an active noise rejection around the observation frequency is required to achieve high SNR.

The intensity noise of the white beam source is 1.1×10^{-4} Hz$^{-1/2}$ [the pink line in Fig. 4.2.5(c)]. In Fig. 4.2.5(c), the noise observed through the developed PSD is converted into the intensity modulation equivalent value of the in-phase beam and plotted as red dots. The optical noise is canceled by ~26 dB (20 times) to $\sim 5 \times 10^{-6}$ Hz$^{-1/2}$. The results of the experiments described in Sections 5.A and 5.B clearly explain that the phase modulation of the carrier, rather than the intensity modulation of the carrier beam, is detected independently of the intensity and/or intensity noise of the carrier. The present scheme demonstrates a successful phase-sensitive detection of signal and intensity noise cancelation.

Certain balanced detection methods [26, 33–35] offer other ways of cancelation of the optical intensity noise on a probe beam. In these methods, a reference beam is prepared from the same beam source as that of the probe beam, and the detected

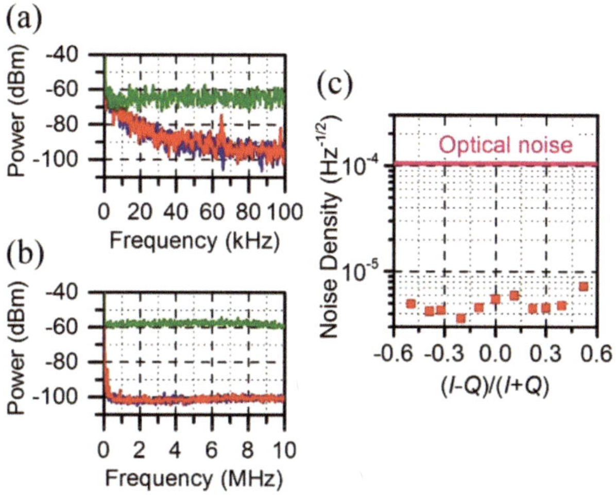

FIGURE 4.2.5 (a) and (b) Noise spectra of 802-nm pulsed beam from the laser oscillator and white beam source. The blue, red, and green plots indicate electrical noise power of the observation system, noise power of the pulsed beam from the oscillator, and that of the white beam source. (a) The frequency range is 0–100 kHz. (b) The range is 0–10 MHz. (c) Dependence of observed noise on the intensity balance of the I and Q beams, converted to the intensity modulation equivalent noise density. The pink line indicates the intensity noise on the in-phase beam measured at (I −Q) /(I+Q) = 0. The red plots indicate the observed noise through the present PSD. Copyright of this figure is shared between the representative author (Takayoshi Kobayashi) and Optica.

signal of the reference beam is subtracted from that of the probe beam. The intensity of the pulsed probe beam after measuring the sample is

$$\begin{aligned} &\sum_{n=-\infty}^{0} A_n \left\{1+\mathrm{m}\left(nt\right)\right\}\delta\left(t-nT\right) \\ &= \sum_{n=-\infty}^{0} A_n \ \mathrm{m}\left(nt\right)\delta\left(t-nT\right)+\sum_{n=-\infty}^{0} A_n\delta\left(t-nT\right). \end{aligned} \tag{46}$$

The multiplicative and additive noise components are included in A_n of the first and second terms of the right member, respectively. The balanced detection is intended to cancel only the additive component by the subtraction. The signal intensity versus the probe and reference beam power with the balanced detection of [26] is shown as blue plots in Fig. 4.2.4(a). The intensity is proportional to the beam power, which demonstrates that the multiplicative effect to the signal is not canceled. The noise cancelation efficiency is maximized when the signal intensities of the probe and reference beams are equated. When the intensity balance is displaced, the intensity noise is insufficiently or excessively subtracted, and the noise persists depending on the amount of the unbalance [26, 33]. In the present phase detection, the multiplicative effect of A_n is canceled, as is shown as the independence of the beam power in

Fig. 4.2.4(a). The additive noise is also canceled since it is proportional to the multiplicative one [Eq. (46)]. Additionally, because of the cancelation of the multiplicative effect, the noise-rejection performance is principally independent of the intensity balance, as shown in Fig. 4.2.5(c). Through this feature, even when the optical balance is displaced during a measurement, the maximized noise cancelation is always maintained. This is especially advantageous for imaging measurement applications, which we describe in detail later. This feature also helps make parallel noise cancelations at plural wavelengths feasible [26, 33]. The optical splitting ratios at different wavelengths are mismatched because of the dependence of the performance of practical optics such as beam splitters on wavelengths. As a result of this mismatch, when the optical splitting ratio is set so that maximum noise cancelation is attained at a specific wavelength, the noise cancelation performance on the other wavelengths is degraded without the ratios being compensated. In the present method, the performance is independent of the ratio, and the parallel measurement is realized readily with a number of the present PSDs. One might think a combination of the spectrometer and a fast CMOS camera operating in a ~ MHz readout frequency [36] can be applied as an array of the photodetectors and MLs for this purpose. However, the readout timing (or phase deviation) must be controlled independently for respective pixels in the present method. This operation is difficult with the available CMOS cameras. Thus we use the fiber bundle to split the spectrum and input into the PSDs in the present work.

This feature, even when the optical balance is displaced during a measurement, means that the maximized noise cancelation is always maintained. This is especially advantageous for imaging measurement applications, which we describe in detail later. This feature also helps make parallel noise cancelations at plural wavelengths feasible [26, 33]. The optical splitting ratios at different wavelengths are mismatched because of the dependence of the performance of practical optics such as beam splitters on wavelengths. As a result of this mismatch, when the optical splitting ratio is set so that maximum noise cancelation is attained at a specific wavelength, the noise cancelation performance on the other wavelengths is degraded without the ratios being compensated. In the present method, the performance is independent of the ratio, and the parallel measurement is realized readily with a number of the present PSDs. One might think a combination of the spectrometer and a fast CMOS camera operating in a ~ MHz readout frequency [36] can be applied as an array of the photodetectors and MLs for this purpose. However, the readout timing (or phase deviation) must be controlled independently for respective pixels in the present method. This operation is difficult with the available CMOS cameras. Thus we use the fiber bundle to split the spectrum and input into the PSDs in the present work.

As in usual balanced detection schemes, the noise floor of the present method is determined by the fluctuation of the splitting ratio at the beam splitter. This fluctuation originates in the quantum nature of photons (shot noise). The observed shot-noise current (σ_L) is calculated as follows:

$$\sigma_L = \sqrt{2qI_L F}; \; I_L = SP, \tag{47}$$

where q, I_L, F, S, and P are the elemental charge, average optical current, bandwidth, photosensitivity, and average optical power, respectively. When larger amounts of photons by ε are distributed into the in-phase arm, the amount of those distributed into the quadrature arm is less distributed by ε. The phase deviation is determined by the ratio of the in-phase and quadrature signals. Thus, the ratio of the in-phase and quadrature beam powers is

$$\frac{I+\varepsilon}{Q-\varepsilon} \cong \frac{I}{Q}\left(1+\frac{\varepsilon}{I}+\frac{\varepsilon}{Q}\right), \tag{48}$$

when $\varepsilon \ll I$ and Q. Therefore, when the optical splitting ratio is equal ($I = Q$), the shot noise is double that of the in-phase beam.

In the experiment, the optical power of the in-phase beam is 3.5 μW at the APD. The S of the APD is 0.42 A/W, and the excessive noise caused by multiplication is negligible because of the small multiplication factor (~2 times). Using these parameters, the shot noise is calculated as 9.3×10^{-7} $Hz^{-1/2}$.

The noise caused by the circuit measured without an optical input is 7.6×10^{-7} $Hz^{-1/2}$. The experimental value is ~5×10^{-6} $Hz^{-1/2}$ and larger by ~5 times. The noise floor of the present equipment is probably determined by the phase noise of the synchronous signal caused by the circuit, since the floor is upraised when the optical signal is input and the phase difference between the optical and synchronous signals is defined. The simple digital comparator converts the sinusoidal wave signal to a square wave by amplitude clipping. The timing when the wave signal crosses the threshold voltage of the comparator is disturbed by the electrical noise on the wave and/or power supply. This process converts the electrical noise in the whole frequency range into phase noise of the synchronous signal. In the near future, we will replace the comparator with an analog auto-gain control circuit as the LMT. Let us compare with our previous work [26] on the auto-balanced scheme with the similar configuration. In [26], the difference is obtained by flipping the sign of either the probe or reference signals through multiplication by the synchronous signal. The gains for the signals are adjusted with the phase deviation. The noise floor in [26] is almost the same value of ~5×10^{-6} $Hz^{-1/2}$ with similar probe-beam power of 3 μW. This result is explained by the fluctuation of the gains caused by the phase noise, since the circuit involving the synchronous signal is almost the same. Therefore, the same technical factor determines the noise floor for both schemes although the different physical quantities are observed.

In the present PSD, a common photodetector measures the in-phase and quadrature beams. This scheme helps make measurement feasible via a common spectrometer for the two beams [26]. The noise in the white continuum on one wavelength is different from that on a different wavelength. Hence, if the beams are detected by different photodetectors, two different spectrometers are required for spectroscopy. The center wavelength and spectral bandwidth must be equated in the two; otherwise, the correlation between the intensities of the two beams is disturbed, resulting in the phase noise of the carrier. Meeting this requirement requires a complex system. However, in the present study, we satisfy this requirement in a simple and robust system using the common spectrometer.

C Application to SRS Microscopy

Figure 4.2.6(a) shows an SRS image obtained by the intensity modulation on the in-phase beam. The modulation equivalent RMS value on the 144 pixels surrounded by the orange square in Fig. 4.2.6(a) is calculated as 3.1×10^{-4}. The modulation equivalent noise density is 1.0×10^{-4} $Hz^{-1/2}$ from this value when the origin of these signals is assumed as noise. This value is almost the same as that of the noise density of the in-phase beam, and it means that these signals originate from the intensity noise of the white beam source. In this image, the intensity noise of the in-phase beam is dominant, and no meaningful structure is captured at this time constant. Figure 4.2.6(b) shows an image through the present PSD. The signal intensity is converted into the intensity modulation depth equivalent value of the in-phase beam, and the relative color scale is adjusted as in Fig. 4.2.6(a). The fluctuation of the signal is quite small in the film region compared with that in Fig. 4.2.6(a) because of noise suppression. A discoid signal distribution with a 4 μm diameter corresponding to the PS bead is captured with the modulation equivalent value of $\sim 1.5 \times 10^{-4}$. The RMS value on the 144 pixels surrounded by the square is 2.6×10^{-5}. The noise density is 8.4×10^{-6} $Hz^{-1/2}$ on the assumption that all contributions to these signals are the intensity noise of the source. This value is near the noise floor of $\sim 5 \times 10^{-6}$ $Hz^{-1/2}$ determined by the present circuit configuration, and the noise cancelation to the noise floor is almost achieved in the imaging application. The reason for this larger value may be the offset caused by the cross phase modulation effect [2] on the PVA film. In fact the noise density is obtained as 4.6×10^{-6} $Hz^{-1/2}$ without the pump beam. Therefore, the intensity noise of the light source is canceled at least to the circuit noise. The detection limit for the present system is determined by the circuit noise.

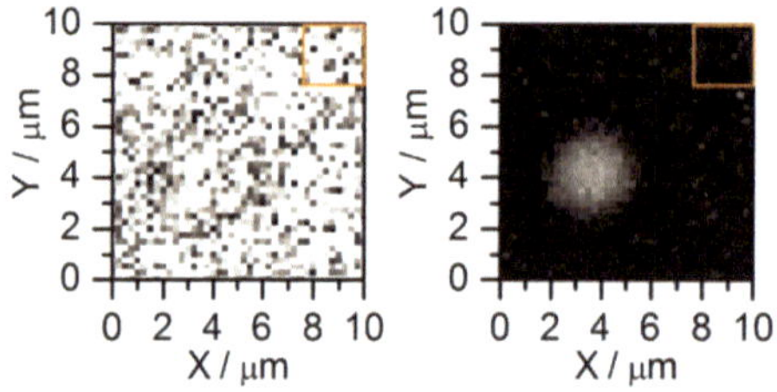

FIGURE 4.2.6 SRS images on a 3054-cm^{-1} Raman band of a 4-μm polystyrene bead embedded in a polyvinyl alcohol film. The orange squares indicate 144 pixels for noise evaluation. The images are constructed on (a) the intensity modulation depth of the in-phase beam and (b) the phase modulation of the carrier. The signal intensity is converted into the equivalent value of the intensity modulation, and the relative color scale is unified with (a). The copyright of this figure is shared between the representative author (Takayoshi Kobayashi) and Optica.

These results confirm the feasibility of applying the method to SRS imaging, with noise cancelation and no suppression of the SRS signal.

The high-speed imaging could not be demonstrated on the present system used at the time when the experimental study described in this subsection was performed. It takes 14.5 hours to capture an image of 512 x 512 pixels with the 200 ms dwelling time. The most important factor preventing the high-speed imaging in this result is the low SNR, which requires the long time constant. The noise is composed of that caused in the circuit and the optical shot noise, and the former one determines the noise floor in the result. The most dominant factor for the circuit noise is the phase noise of the synchronous signal caused in the present amplitude limitation, and this noise is eliminated by modifying the circuit configuration, as described in sub-sub-section 6.B. The shot noise subsequently limits the SNR. The contribution of the shot noise is hindered with a higher power probe beam. The circuit noise composed of the thermal noise and that by the active devices finally limits the SNR. The contributions of these effects are reduced with an enhanced dynamic range of the circuit to accept the high power probe light without deteriorating the shot-noise-limited SNR. One of the useful ways is changing the PA with a low noise FET and the active multiplication IC (AD835) into a passive mixer. Then, the observation frequency should be increased with high-frequency modulation of the pump beam and high-frequency lock-in detection. The feedback response speed should be also increased with a shorter time constant of the LF and INT. For instance, the noise density is calculated as 5.5×10^{-8} $Hz^{-1/2}$ with a power of 1 mW of the in-phase beam. Therefore, SNR = 10 for an SRS signal of 10^{-4} requires a time constant of 30.5 μs, and thus, 61 μs dwelling time is allowed. It takes 16 s for an image of 512 × 512 pixels.

Here we compare the present method with the usual balanced detection in terms of the factor determining the maximum dwelling speed. When balanced detection is applied in imaging, the optical intensity ratio between the probe and reference beams is disturbed because of the dependence of transmittance on the sample position. To compensate for this effect, a feedback scheme known as auto-balance [13, 26, 34, 37] automatically controls the gain of the reference and/or probe beam. It is preferable for the response speed to be faster than the dwelling speed in the imaging; otherwise, the auto-balance would not track the transmittance change, resulting in less-than-maximum noise cancelation. The response speed also must be slower than the intensity modulation speed of the signal on the probe beam; otherwise the signal would be eliminated by the feedback. In other words, the signal modulation and auto-balance feedback scheme doubly limit the dwelling speed. As an example, the response frequency of the auto-balance must be 10 times lower than the modulation frequency to observe 90% of the full modulation depth of the signal. The dwelling speed should be half the response speed to compensate for half the intensity unbalance effect change. Thus, the maximum dwelling speed is determined to be 20 times slower than the modulation speed. Therefore, to perform the high-speed imaging, the modulation frequency is set at a much higher frequency than the dwelling frequency for high feedback cutoff frequency. In a high-speed imaging of 4 μs dwelling time [37], the modulation and cutoff frequencies are set at 10 MHz and 500 kHz, respectively. In the present method, unbalance compensation is not required for the noise cancelation. Although the feedback scheme is used in the present PSD, the response speed must be higher than the modulation speed of the signal to track the phase deviation. In the present system, the feedback frequency must be 10 times higher than the modulation frequency to detect 90% of the full modulation depth. The dwelling speed is limited by the modulation frequency, as in the lock-in detection scheme.

In addition, having a common circuit in the present PSD for the in-phase and quadrature signals simplifies the high-speed imaging process. In the usual balanced detection, the probe and reference signals are involved in different circuit paths. To apply this scheme to high-speed imaging, the timing difference between inputting the signals into a subtraction circuit should be precisely regulated to maintain the intensity correlation. Hence, the difference in delay and the frequency dependence (dispersion) between the paths is non-negligible. This factor depends on the individual characteristics of the elements, circuit pattern, temperature, and so on, requiring dispersion compensation [37] for individual equipment. In the present PSD, the common path of the signals makes the dispersion automatically common, rendering dispersion compensation unnecessary.

Additionally, let us mention a parallel measurement on a number of wavelengths. According to Refs. [10, 11, 38], ~100 mW average input power with a picosecond pulse is possible for a biological sample. Hence, the parallel measurement on ~100 channels is not impossible for the above-mentioned condition with ~100 mW white light of narrower spectral bandwidth than that in the present work. The quadrature beam can bypass the microscope to avoid damage on the sample. Additionally, such high a power supercontinuum light source is already realized [39, 40] and commercially available [41, 42].

We believe the high-speed imaging will be realized on the present scheme through satisfying the above-mentioned requirements because of the amplitude noise cancelation in the phase deviation at any frequencies [Eq. (20)] and observation of the deviation independently of the instantaneous amplitude [Eq. (37)].

7 CONCLUSION

In this pump-probe microscopy study, we demonstrate a scheme for detecting a signal as the phase modulation on a periodically intensity-modulated carrier beam to reject intensity noise on the probe beam [43]. Contrary to the lock-in detection, the scheme does not resort to band limitation. The source beam is split into in-phase and quadrature beams, which are then overlapped to create the carrier beam. The instantaneous phases of the in-phase and pump beams are made equal, and that of the quadrature beam is made perpendicular to that of the pump beam through an additional optical path. The intensity modulation on the in-phase beam, which is caused by sample measurement, is converted into the phase modulation on the carrier beam. Through this process, the intensity noise on the light source is not converted to the phase modulation, but is canceled, and the SNR is enhanced. This scheme principally enables high-speed measurement because it does not limit the band. The intensity noise is canceled as the multiplicative effect in the present scheme rather than the additive effect to be canceled in the usual balanced detection. Therefore, the maximum noise cancelation is maintained independently of the intensity unbalance between the in-phase and quadrature beams. This feature contributes to the robustness against the intensity unbalance between the beams caused in the measurement, as well the auto-balanced scheme [43].

Phase detection without crosstalk with the instantaneous amplitude of the carrier signal is of vital importance when taking this phase-detected pump-probe measurement. Therefore, we develop the PSD to detect the phase deviation of the carrier signal independently of its instantaneous amplitude. In this detection system, we compare the instantaneous phases of the carrier signal and the synchronous signal of the light source by multiplication, and control the phase deviation of the synchronous signal to maintain the phase difference at 90°. We observe that the phase control signal, rather than the carrier or synchronous signals, detects the phase modulation. Maintaining the 90° phase difference is the key to rejecting the intensity noise.

In the detection system presented here in this subsection, a common photodetector detects the in-phase and quadrature beams. This configuration permits the use of a common spectrometer for the beams. The common spectrometer guarantees that the beams at the photodetector will have the same center wavelength and spectral bandwidth. Thus, the correlation of the intensity fluctuations is maintained for the beams, even when the scheme is applied to white light, in which case the intensity fluctuation differs depending on wavelength. Parallel measurements at different wavelengths are readily taken with many PSDs. In addition, the in-phase and quadrature signals are involved in the common circuit. This configuration is free of the dispersion difference between the circuits for the signals, and does not destroy the intensity correlation between them. These features help to keep the system simple and robust for high-speed imaging applications, as delicate dispersion compensation is not required. Although the principal noise floor is double that of the shot noise of the in-phase beam, the present noise cancelation cannot achieve such a performance, probably because of the circuit configuration. We believe that the shot-noise-limited performance is achieved through a modification of the circuit [43].

We apply the present method to SRS microscopy to demonstrate its feasibility [43]. In this microscopy, the white pulsed beam from a piece of the PCF functions as the carrier beam, and the SRS image is captured on a Raman band through the spectrometer. The present method cancels the significant noise on the white beam and dramatically modifies the SNR of the image. With many PSDs, multi-color or spectral imaging is readily performed because of its high suitability for spectroscopy. The present method allows for multiple simple measurements in highly sensitive and high-speed microscopy by expanding the light source choices.

In summary, intensity noise on a probe beam is a serious obstacle to highly sensitive and high-speed pump-probe microscopy. In this report, a reference beam of the probe is prepared and delayed. The intensity modulation by the sample is measured as the phase modulation of the superposition of detected electrical signals of the probe and reference beams, and the intensity noise is canceled. We evaluate performance of the noise cancelation using the supercontinuum light from a piece of photonic crystal fiber, and find that the noise is canceled by ~26 dB. We then apply the method to a stimulated Raman microscope. This method contributes to highly sensitive and high-speed pump-probe imaging with various light sources.

The research presented in this subsection is the output of the collaboration of the following people: K. Seto, T. Tsukada, Y. Okuda, E. Tokunaga, and T. Kobayashi [43]. http://dx.doi.org/10.1364/JOSAA.32.000809

NOTES

1. Limiter (LMT): In electronics, a limiter is a circuit that allows signals below a specified input power or level to pass unaffected while attenuating (lowering) the peaks of stronger signals that exceed this threshold. Limiting is a type of dynamic range compression. Clipping is an extreme version of limiting.
2. Integrator (INT): The operational amplifier integrator is an electronic integration circuit. Based on the operational amplifier (op-amp), it performs the mathematical operation of integration with respect to time; that is, its output voltage is proportional to the input voltage integrated over time. The integrator circuit is mostly used in analog computers, analog-to-digital converters, and wave-shaping circuits. A common wave-shaping use is as a charge amplifier and they are usually constructed using an operational amplifier though they can use high gain discrete transistor configurations.
3. Loop-filter (LF): a term used in the design of Phase-Locked Loops (PLLs). It is a low-pass filter that smooths the output of the charge pump and determines the cutoff frequency (Fc). The loop bandwidth affects the phase tracking speed and phase noise performance
4. Low-pass filter: a low-pass filter that smooths the output of the charge pump and determines the cutoff frequency (Fc).
5. Bandpass filter: (BPF): A bandpass filter or bandpass filter (BPF) is a device that passes frequencies within a certain range and rejects (attenuates) frequencies outside that range. It can be a combined system of high-frequency transmitting (or attenuating) filter and low-frequency transmitting (or attenuating) filter. A bandpass filter system allows through components in a specified band of frequencies, called its passband but blocks components with frequencies above or below this band. The idea can be used in various fields such as electronics, optical signals, and sound.

REFERENCES

1. E. Ploetz, S. Laimgruber, S. Berner, W. Zinth, and P. Gilch, "Femtosecond stimulated Raman microscopy," *Appl. Phys. B* 87, 389–393 (2007).
2. C. W. Freudiger, W. Min, B. G. Saar, S. Lu, G. R. Holtom, C. He, J. C. Tsai, J. X. Kang, and X. S. Xie, "Label-free biomedical imaging with high sensitivity by stimulated Raman scattering microscopy," *Science* 322, 1857–1861 (2008).
3. P. Nandakumar, A. Kovalev, and A. Volkmer, "Vibrational imaging based on stimulated Raman scattering microscopy," *New J. Phys.* 11, 033026 (2009).
4. A. Gambetta, V. Kumar, G. Grancini, D. Polli, R. Ramponi, G. Cerullo, and M. Marangoni, "Fiber-format stimulated-Raman scattering microscopy from a single laser oscillator," *Opt. Lett.* 35, 226–228 (2010).
5. Y. Ozeki and K. Itoh, "Stimulated Raman scattering microscopy for live-cell imaging with high contrast and high sensitivity," *Laser Phys.* 20, 1114–1118 (2010).

6. H. T. Beier, G. D. Noojin, and B. A. Rockwell, "Stimulated Raman scattering using a single femtosecond oscillator with flexibility for imaging and spectral applications," *Opt. Express* 19, 18885–18892 (2011).
7. D. Fu, F.-K. Lu, X. Zhang, C. Freudiger, D. R. Pernik, G. Holtom, and X. S. Xie, "Quantitative chemical imaging with multiplex stimulated Raman scattering microscopy," *J. Am. Chem. Soc.* 134, 3623–3626 (2012).
8. F.-K. Lu, M. Ji, D. Fu, X. Ni, C. W. Freudiger, G. Holtom, and X. S. Xie, "Multicolor stimulated Raman scattering microscopy," *Mol. Phys.* 110, 1927–1932 (2012).
9. E. Ploetz, B. Marx, T. Klein, R. Huber, and P. Gilch, "A 75 MHz light source for femtosecond stimulated Raman microscopy," *Opt. Express* 17, 18612–18620 (2009).
10. B. G. Saar, C. W. Freudiger, J. Reichman, C. M. Stanley, G. R. Holtom, and X. S. Xie, "Video-rate molecular imaging in vivo with stimulated Raman scattering," *Science* 330, 1368–1370 (2010).
11. Y. Ozeki, W. Umemura, Y. Otsuka, S. Satoh, H. Hashimoto, K. Sumimura, N. Nishizawa, K. Fukui, and K. Itoh, "High-speed molecular spectral imaging of tissue with stimulated Raman scattering," *Nat. Photonics* 6, 845–851 (2012).
12. W. Min, S. Lu, S. Chong, R. Roy, G. R. Holtom, and X. S. Xie, "Imaging chromophores with undetectable fluorescence by stimulated emission microscopy," *Nature* 461, 1105–1109 (2009).
13. J. Miyazaki, H. Tsurui, A. Hayashi-Takagi, H. Kasai, and T. Kobayashi, "Sub-diffraction resolution pump-probe microscopy with shot-noise limited sensitivity using laser diodes," *Opt. Express* 22, 9024–9032 (2014).
14. W. Min, "Label-free optical imaging of nonfluorescent molecules by stimulated radiation," *Curr. Opin. Chem. Biol.* 15, 831–837 (2011).
15. J. Miyazaki, K. Kawasumi, and T. Kobayashi, "Resolution improvement in laser diode-based pump-probe microscopy with an annular pupil filter," *Opt. Lett.* 39, 4219–4222 (2014).
16. E. Tamaki, K. Sato, M. Tokeshi, K. Sato, M. Aihara, and T. Kitamori, "Single-cell analysis by a scanning thermal lens microscope with a microchip: direct monitoring of cytochrome c distribution during apoptosis process," *Anal. Chem.* 74, 1560- 1564 (2002).
17. D. A. Nedosekin, E. I. Galanzha, E. Dervishi, A. S. Biris, and V. P. Zharov, "Super-resolution nonlinear photothermal microscopy," *Small* 10, 135–142 (2014).
18. A. Gaiduk, P. V. Ruijgrok, M. Yorulmaz, and M. Orrit, "Detection limits in photothermal microscopy," *Chem. Sci.* 1, 343–350 (2010).
19. S. Arunkarthick, M. M. Bijeesh, G. K. Varier, M. Kowshik, and P. Nandakumar, "Laser scanning photothermal microscopy: fast detection and imaging of gold nanoparticles," *J. Microsc.* 256, 111–116 (2014).
20. J. Miyazaki, H. Tsurui, K. Kawasumi, and T. Kobayashi, "Optimal detection angle in sub-diffraction resolution photothermal microscopy: application for high sensitivity imaging of biological tissues," *Opt. Express* 22, 18833–18842 (2014).
21. Y. Ozeki, Y. Kitagawa, K. Sumimura, N. Nishizawa, W. Umemura, S. Kajiyama, K. Fukui, and K. Itoh, "Stimulated Raman scattering microscope with shot noise limited sensitivity using subharmonically synchronized laser pulses," *Opt. Express* 18, 13708–13719 (2010).
22. A. Owyoung and E. D. Jones, "Stimulated Raman spectroscopy using low-power cw lasers," *Opt. Lett.* 1, 152–154 (1977).
23. A. Owyoung, "Sensitivity limitations for CW stimulated Raman spectroscopy," *Opt. Commun.* 22, 323–328 (1977).

24. M. N. Slipchenko, R. A. Oglesbee, D. Zhang, W. Wu, and J.-X. Cheng, "Heterodyne detected nonlinear optical imaging in a lock-in free manner," *J. Biophotonics* 5, 801–807 (2012).
25. J. M. Dudley, G. Genty, and S. Coen, "Supercontinuum generation in photonic crystal fiber," *Rev. Mod. Phys.* 78, 1135–1184 (2006).
26. K. Seto, Y. Okuda, E. Tokunaga, and T. Kobayashi, "Multiplex stimulated Raman imaging with white probe-light from a photonic-crystal fibre and with multi-wavelength balanced detection," J. Phys. D 47, 345401 (2014).
27. K. Seto, Y. Okuda, E. Tokunaga, and T. Kobayashi, "Development of a multiplex stimulated Raman microscope for spectral imaging through multi-channel lock-in detection," *Rev. Sci. Instrum.* 84, 083705 (2013).
28. U. Møller, S. T. Sørensen, C. Jakobsen, J. Johansen, P. M. Moselund, C. L. Thomsen, and O. Bang, "Power dependence of supercontinuum noise in uniform and tapered PCFs," *Opt. Express* 20, 2851–2857 (2012).
29. J. N. Ames, S. Ghosh, R. S. Windeler, A. L. Gaeta, and S. T. Cundiff, "Excess noise generation during spectral broadening in a microstructured fiber," *Appl. Phys. B* 77, 279–284 (2003).
30. N. R. Newbury, B. R. Washburn, K. L. Corwin, and R. S. Windeler, "Noise amplification during supercontinuum generation in microstructure fiber," *Opt. Lett.* 28, 944–946 (2003).
31. K. L. Corwin, N. R. Newbury, J. M. Dudley, S. Coen, S. A. Diddams, K. Weber, and R. S. Windeler, "Fundamental noise limitations to supercontinuum generation in microstructure fiber," *Phys. Rev. Lett.* 90, 113904 (2003).
32. K. L. Corwin, N. R. Newbury, J. M. Dudley, S. Coen, S. A. Diddams, B. R. Washburn, K. Weber, and R. S. Windeler, "Fundamental amplitude noise limitations to supercontinuum spectra generated in a microstructured fiber," *Appl. Phys. B* 77, 269–277 (2003).
33. K. Seto, T. Tsukada, Y. Okuda, E. Tokunaga, and T. Kobayashi, "Development of a balanced detector with biased synchronous detection and application to near shot noise limited noise cancelling of supercontinuum pulse light," *Rev. Sci. Instrum.* 85, 023702 (2014).
34. P. C. D. Hobbs, "Ultrasensitive laser measurements without tears," *Appl. Opt.* 36, 903–920 (1997).
35. K. Nose, Y. Ozeki, T. Kishi, K. Sumimura, N. Nishizawa, K. Fukui, Y. Kanematsu, and K. Itoh, "Sensitivity enhancement of fiber laser-based stimulated Raman scattering microscopy by collinear balanced detection technique," *Opt. Express* 20, 13958–13965 (2012).
36. W. Rock, M. Bonn, and S. H. Parekh, "Near shot-noise limited hyperspectral stimulated Raman scattering spectroscopy using low energy lasers and a fast CMOS array," *Opt. Express* 21, 15113–15120 (2013).
37. C. W. Freudiger, W. Yang, G. R. Holtom, N. Peyghambarian, X. S. Xie, and K. Q. Kieu, "Stimulated Raman scattering microscopy with a robust fibre laser source," *Nat. Photonics* 8, 153–159 (2014).
38. R. Kawakami, K. Sawada, A. Sato, T. Hibi, Y. Kozawa, S. Sato, H. Yokoyama, and T. Nemoto, "Visualizing hippocampal neurons with in vivo two-photon microscopy using a 1030 nm picosecond pulse laser," *Sci. Rep.* 3, 1014 (2013).
39. M. Seefeldt, A. Heuer, and R. Menzel, "Compact white-light source with an average output power of 2.4 W and 900 nm spectral bandwidth," *Opt. Commun.* 216, 199–202 (2003).
40. T. Schreiber, J. Limpert, H. Zellmer, A. Tünnermann, and K. P. Hansen, "High average power supercontinuum generation in photonic crystal fibers," *Opt. Commun.* 228, 71–78 (2003).

41. NKT Photonics, "Super EXTREME Supercontinuum White Light Laser Sources," http://www.nktphotonics.com/ superkextreme?cid=7856.
42. Leukos, "Innovative Optical Systems," http://www.leukossystems.com/spip.php ?rubrique63.
43. K. Seto, T. Tsukada, Y. Okuda, E. Tokunaga, and T. Kobayashi, "Noise cancellation with phase-detection technique for pump-probe measurement and application to stimulated Raman imaging," *JOSA A Vol.* 32(5), 809–821 (2015).

4.3 Simultaneous Dual-wavelength Imaging of Nonfluorescent Tissues with 3D Subdiffraction Photothermal Microscopy

1 INTRODUCTION

Photothermal microscopy is an efficient method for observing molecules or nanomaterials using only their optical absorption [1–16]. Typically, two laser beams (pump and probe beams) are incident on the sample through a focusing lens. The pump beam increases the temperature around the optical absorbing sample in the focal spot range. This results in local refractive index variations (typically ~10^{-5}–10^{-3} with $\Delta T = 0.1–10\ K$) [6, 7] and induces deflection of the probe beam. Variation in the refractive index is detected as a change in probe beam transmissivity by a lock-in detection scheme with a modulated pump beam. This technique has been applied for biological imaging, such as in imaging biological tissues using gold nanoparticles [1, 10, 12], evaluating the hydrodynamic properties of cytosol with photothermal absorption correlation spectroscopy [3–5] and label-free and three-dimensional imaging of heme proteins with two-photon excitation [9]. However, previous photothermal imaging has been single color only, and multi-wavelength imaging with multiplex biomolecule labeling is needed to visualize a variety of cellular components in high specificity.

Multi-wavelength microscopic imaging is essential to visualize a variety of nanoscale cellular components with high specificity and high spatial resolution. However, previous techniques are based on fluorescence, and thus cannot visualize nonfluorescent species, which suffer much less from photodamage or photobleaching and hence are intrinsically useful in a wider range of optical microscopy. Here,

 DOI: 10.1201/9781003476030-17

we show that simultaneous multi-wavelength imaging of nonfluorescent species can be achieved with the use of a photothermal microscope

The photothermal signal is proportional to the absorption cross-section of the sample material [6, 7]. Thus, when the sample contains different species, the measured signal is given by the linear sum of the signals from the different species. If we know the number of species in the sample and their absorption spectra, the abundance of species may be determined from spectral images (images measured with different pump wavelengths) through the spectral unmixing procedure. For unknown spectra or materials, spectral images can be processed by blind unmixing based on principle or independent component analysis as with stimulated Raman scattering microscopy [17].

In the previous report, photothermal spectral images were acquired by using a wavelength-tunable optical parametric oscillator (OPO) as a pump beam with switching the laser wavelength and repeating scan [8]. However, this scheme would cause an error in performing spectral unmixing because of mechanical drift and photobleaching of the tissue. On the other hand, it is too costly to implement simultaneous multi-wavelength excitation using multiple OPOs. Thus, an OPO-based schematic is not suited for multi-wavelength imaging. Recently, we implemented photothermal (pump-probe) microscopy using intensity modulated laser diodes (LDs) in which shot-noise-limited sensitivity was achieved using a balanced detection with a dual modulation scheme [18, 19]. Major advances have been made recently in high power blue to yellow emitting LDs which make this scheme readily applicable to simultaneous multi-wavelength imaging by incorporating LDs with different wavelengths.

In the study presented in this subsection, simultaneous dual-wavelength sub-diffraction imaging of biological tissues stained with hematoxylin and eosin (H&E) is performed using two pump beams of LDs modulated at different frequencies. Highly reliable unmixed imagies reflecting the distribution of each species is obtained. Three-dimensional label-free imaging of mouse melanoma tissue section is also presented to demonstrate the effectiveness of the enhanced spatial resolution. Our technique can be implemented using cost-effective and compact laser diodes and is applicable for various types of both fluorescent and nonfluorescent tissues.

2 EXPERIMENTAL SETUP

2.1 Experimental System and Principle of Simultaneous Dual-wavelength Imaging

A 488-nm LD (Nichia NDS4216) and 532-nm diode pumped solid-state (DPSS) laser (Thorlabs, DJ525-40) were used for pumping, and a 660-nm LD (Mitsubishi ML101U29) was used for probing [Figure 4.3.1(a)]. Each beam was collimated through a polarization maintaining single mode (SM) fiber for spatial mode filtering. Three beams were combined using dichroic mirrors. All beams were linearly polarized in the vertical plane. The combined beams were directed to the objective lens [40X, numerical aperture (NA) = 0.95]. The sample position was raster scanned using a three axis positioning stage driven by piezo actuators (Thorlabs, MAX311D).

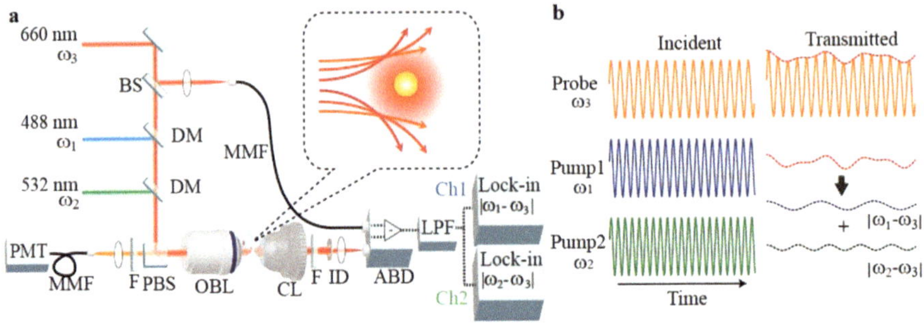

FIGURE 4.3.1 Experimental setup and principle of dual-wavelength photothermal microscopy using intensity-modulated lasers. (a) Schematic of the experimental setup. DM: dichroic mirror, BS: beam splitter, PBS: polarization beam splitter, OBL: objective lens. CL: condenser lens, F: band-pass filter, ID: iris diaphragm, PMT: photomultiplier tube, MMF: multimode fiber, ABD: auto-balanced detector, LPF: low-pass filter. (b) Principle of dual-wavelength photothermal microscopy using intensity-modulated continuous-wave lasers. The pump-beam intensities are modulated at ω_1 and ω_2, and the probe beam intensity is modulated at ω_3. Beat signals are generated by the photothermal effect at the sample position, and they are detected using two lock-in amplifiers referenced at the beat frequencies. Copyright of this figure is partially held between the representative author (Takayoshi Kobayashi) and Optica.

A condenser lens (Olympus, U-AAC) was used to collect the transmitted light. A narrowband filter matching the laser line was placed in front of the detector to filter out both the pump beam and fluorescence from the sample so that only the probe beam was detected. An iris diaphragm is used to adjust effective NA of the condenser lens for optimizing signal-to-noise ratio. An auto-balanced photodetector (Newfocus, Nirvana) was used to cancel the intensity fluctuation of the probe laser and achieve nearly shot-noise-limited sensitivity. A signal generator (Rigol DG4162) and an amplifier (Mini-circuit ZHL6A +) were used to sinusoidally modulate the light intensity of the LD by controling the injection current. The light intensity of the DPSS laser was externally modulated using an electro-optical modulator (LINOS LM020). To reduce the relative displacement between the images in the two channels, all beams were set to be collinear within the inter-beam angle of the order 10^{-5} rad. Furthermore, the beam divergence angle was adjusted by changing the position of the doublet lens after the SM fiber so that the relative difference in divergence among the three beams was within the order of 10^{-5} rad. We also set up a confocal fluorescence detection scheme, wherein the fluorescence signal from the sample was detected by a photo-multiplier through a multimode fiber of 0.6 Airy units at a fluorescence wavelength of 600 nm.

Simultaneous dual-wavelength photothermal imaging is demonstrated using two pump beams modulated at different frequencies (ω_1–ω_2) [Figure 4.3.1(b)], with the probe beam modulated at ω_3. Photothermal signals are generated at beat frequencies of $|\omega_1-\omega_3|$ and $|\omega_2-\omega_3|$ by the bilinear interaction of the pump and probe fields at the sample. Each frequency component is demodulated using a lock-in amplifier referenced at the beat frequency. This dual-modulation scheme is capable of characterizing the microsecond to picosecond relaxation dynamics of sample species without using a high-frequency (wideband) detector and lock-in amplifier, whose principle and technique have been detailed in the previous paper [19]. This scheme also serves to eliminate the contribution of residual fluorescence that passes through the filter. To avoid channel interference, the beat frequency difference ($|\omega_1-\omega_3|-|\omega_2-\omega_3|$) is set (typically at 30 kHz) to be sufficiently larger than the inverse of the time constant of the lock-in amplifiers (1–2 kHz).

2.2 Sample Preparation

In skin melanoma imaging, B16 melanoma[1] cells RCB-1283 (Riken Bio-Resource Centre) were cultured in RPMI (Rowell Park Memorial Institute)-1640 supplemented with 10% FBS[2], 100 units/ml penicillin and 100 μg/ml streptomycin (Meiji Seika Kaisha Ltd.) at 37°C in a humidified atmosphere of 95% air and 5% CO_2. A 50-μl suspension containing 0.5 million melanoma cells for each head was subcutaneously inoculated into female nude mice (6 weeks of age) as five aliquots on the dorsal sides from the base of the tail to the neck. Tissues containing inoculated cells were fixed with 4% paraformaldehyde for 3 days, embedded in melted paraffin, sliced at 15–20 μm, extended on glass slides, and enclosed with a coverslip. H&E stained biological samples were purchased from Artec Co Ltd.

3 RESULTS

3.1 Spatial Resolution and Optical Sectioning Capability

The spatial resolution of photothermal microscopy is considered to be dependent on the intensity profiles of the pump and probe beams and the refractive index profile at the focal point [2]. When the spatial extent of the modulating refractive index is much smaller than the spot size of the probe beam, the resolution (point spread function) is determined by the product of the intensity profiles of the pump and probe beams; thus, the spatial resolution is better than that of the diffraction-limited optical microscope [20]. Furthermore, this nonlinear signal provides optical sectioning capability as with confocal fluorescence and two-photon excitation microscopy [21].

The spatial resolution and displacement between the images in the two channels are examined by measuring single gold nanoparticles in the lateral and axial planes [Figure 4.3.2(a)–2(e)]. The full width at half maximum value of the measured spot in the lateral plane is down to 234 nm, which is 34% smaller than the focal spot size of the probe beam ($0.51\lambda/NA$ = 354 nm). Furthermore, the axial spot size is down to 510 nm, which is 35% smaller than the diffraction-limited axial size of the probe beam (689 nm). The displacements between the two channels are 29 nm in the lateral plane and 48 nm in the axial plane, which are within ~10% of the focal spot size. Use of an annular pupil filter further improves the lateral-plane spatial resolution, reaching nearly half of the diffraction-limited spot size; however, it reduces spatial resolution in the axial direction [20].

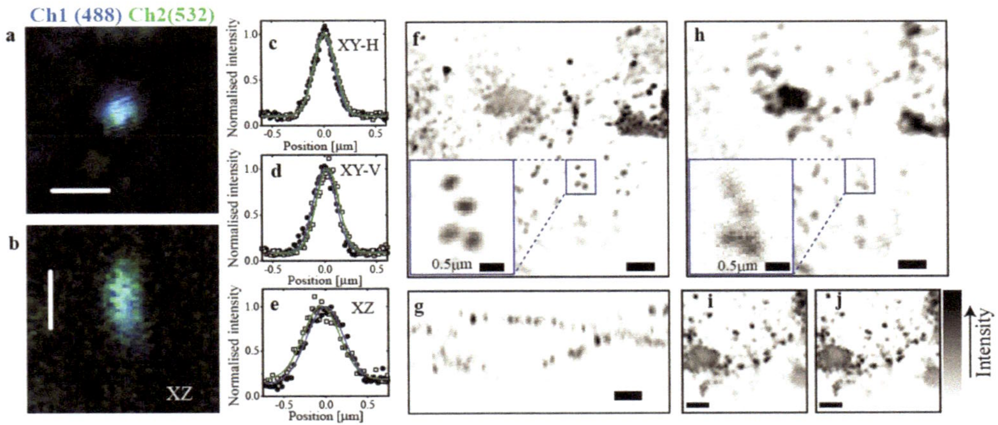

FIGURE 4.3.2 Photothermal images of signal gold nanoparticle and label-free 3D imaging of mouse melanoma tissue section. (a, b) Combined images of a 20-nm gold nanoparticle embedded in polyvinyl alcohol with excitation at 488 nm (blue) and 532 nm (green). The wavelengths of the two pump beams overlapped the absorption spectrum of the gold nanoparticles due to plasmon resonance with a peak at 524 nm. The 488-nm and 532-nm pump beams were modulated at 1.07 and 1.1 MHz, respectively, and the probe beam was modulated at 1.0 MHz. The time constant of the lock-in amplifier was 1 ms, and the pixel dwell time was 2 ms. Scale bars, 500 nm. (c, d) Intensity profiles of the single nanoparticle with excitation at 488 nm (filled circles) and 532 nm (open squares) in the horizontal and vertical directions, respectively. For the 488-nm pump beam, fitting of a single particle with the Gaussian (solid curves) gives the full-width at half-maximum (FWHM) values of 234 nm and 249 nm in the horizontal and vertical directions, respectively. For the 532-nm pump beam, the FWHM values are 256 nm and 257 nm in the horizontal and vertical directions, respectively. (e) Intensity profiles along the axial direction. FWHM values are 525 nm and 510 nm for the 488-nm and 532-nm pump beams, respectively. (f, g) Photothermal images of a slice of mouse melanoma in the lateral and axial planes, respectively. Time constant of the lock-in amplifier was 0.5 ms, and pixel dwell time was 1 ms. The pump and probe beam powers incident on the sample were 50 and 150 μW, respectively. A set of 30 images of a slice of mouse melanoma is subsequently acquired by changing the sample position in the axial direction with a step size of 0.28 μm (see Media 1). Each image size is 19.6 x 19.6 μm at 200 x 200 pixels. (h) Bright field image of slice of mouse melanoma in the same area as in (f). (i,j) Photothermal images in the simultaneous measurement with excitation at (i) 488 nm and (j) 532 nm . Scale bars, 2 μm in (f-j) and 0.5 μm in the inset of (f, h). Copyright of this figure is partially held between the representative author (Takayoshi Kobayashi) and Optica.

To demonstrate the effectiveness of the spatial resolution enhancement in photothermal microscopy, we performed label-free three-dimensional imaging of mouse melanoma [Figs. 4.3.2(f) and 2(g), https://figshare.com/articles/media/Media_1_Simultaneous_dual-wavelength_imaging_of_nonfluorescent_tissues_with_3D_subdiffraction_photothermal_microscopy/4949045?file=8316899]. 488-nm and 660-nm LDs are used for pump and probe beams, respectively. It can be clearly seen that the spatial resolution of photothermal image is better than that of bright-field image [Fig. 4.3.2(h)], which is measured by transmission-mode nearly confocal microscopy (Airy units 1.8) with the same probe beam. The distribution of melanin granules is clearly visualized in both planes in photothermal microscopy. Such three-dimensional structural information would be useful to differentiate melanoma from melanocytic nevi for diagnostic purposes. There was minimal damage to the sample after measurements.

Since the absorption spectrum of melanin is nearly flat in visible range, identical photothermal images are acquired in the simultaneous dual-wavelength measurement of mouse melanoma tissue section [Figs. 4.3.2(i) and 2(j)]. Furthermore, by measuring images with blocking either pump beam, we confirm that interference between the two channels is negligibly small (less than 1% with respect to signal intensity). These facts ensure the multispectral ability of the present scheme. The advantage of the present multi-wavelength method is explained later by showing the results of two species.

3.2 Simultaneous Dual-Wavelength Imaging

We demonstrated dual-wavelength photothermal imaging of a biological tissue slice stained with H&E, a popular staining method widely used in histology[3] and histopathology[4]. Hematoxylin is nonfluorescent, whereas eosin is fluorescent with a quantum yield of 10–20% [22]. Although the absorption spectra of hematoxylin and eosin overlap [Figure 4.3.3(a)], their intensities depend on the wavelength: hematoxylin absorbance at 532 nm is 1.8 times larger than that at 488 nm, whereas eosin excitation efficiency at 532 nm is 2.3 times larger than that at 488 nm. This difference allows us to unmix the two channel images to determine the distribution of each species. Here, the absorption spectrum of eosin is calculated from the excitation spectrum of eosin contained in the specimen. The magnitude of the excitation spectrum is related to the molar absorption coefficient by assuming that fluorescence quantum yield does not depend on the excitation wavelength because of Kasha's rule and that the area of absorption spectrum of eosin in the specimen is nearly equal to that in aqueous solution.

Figures 4.3.3(b) and 3(c) show photothermal images of a part of a rabbit oocyte with excitation at 488 nm and 532 nm, respectively. Both pump beam powers were set to be sufficiently weak (40 μW) to avoid saturated excitation. Figure 4.3.3(d) shows a blight field image of the tissue in the same area as that in Figs. 4.3.3(b) and (c). The same objective lens and a doublet lens are used to produce the image on a CCD with a white LED backlight. Linear spectral unmixing was then performed [Fig. 4.3.3(e)–(g)], in which pseudo-colored blue and red denote the abundance of hematoxylin and eosin, respectively. In the linear spectral unmixing process, the absorption spectra of hematoxylin and eosin [Fig. 4.3.3(a)] are used as reference spectra. The residual spatial displacement between the two channels is not calibrated because it is smaller than the 100-nm pixel step size. The figure clearly shows that a granule cell is stained with hematoxylin while the zona pellucida and cytoplasm are stained with eosin. We also measured sectional (XZ) images [Fig. 4.3.3(h)–(j)] and found that tissue morphology is clearly visualized in the depth direction.

As fluorescence comes only from eosin, it is helpful to detect fluorescence to verify the results of the spectral unmixing. Figure 4.3.3(k) shows the combined image of confocal fluorescence (green) and photothermal (red) signals. Here, only the 488-nm beam was used for pumping. Photothermal and fluorescence signals are simultaneously measured during the sample scan (multimodal imaging). The scanning area is the same as that in Figs. 4.3.3(b) and (c). We confirm that intense fluorescence is detected from the zona pellucida and cytoplasm, while nearly uniform and weak fluorescence is detected from the granule cell, consistent with the images of Figs. 4.3.3(e)–(g). The position of the maximum of the fluorescence signal and that of the

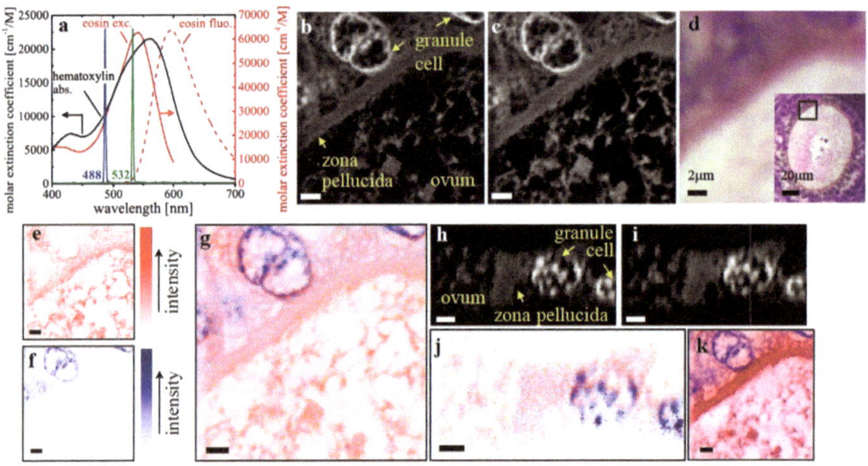

FIGURE 4.3.3 Dual-wavelength photothermal images of hematoxylin and eosin (H&E) stained biological tissues. (a) Absorption (red solid curve) and fluorescence (red dashed curve) spectra of eosin and absorption spectrum of hematoxylin (black solid curve). Laser wavelengths of the two pump beams are also shown. (b) and (c) Photothermal images of H&E-stained rabbit ovary slice with excitation at 488 nm and 532 nm, respectively. Probe beam power was 300 µW. Image size was 19.6 x 19.6 mm with 300 x 300 pixels. The time constant of the lock-in amplifier was 0.5 ms, and the pixel dwell time was 1 ms. Image acquisition time was ~120 s. (d) Bright field image of the tissue in the same area as that in (b,c). The inset is the entire picture of the oocyte. (e, f) Unmixed images reflecting the distribution of eosin (red) and hematoxylin (blue), respectively. (g), Dual-wavelength image produced by combining images e and f. (h, i) Cross-sectional images of a slice of rabbit ovary with excitation at 488 nm and 532 nm, respectively. Image size was 19.6 x 9.8 mm with 200 x 100 pixels, image acquisition time was ~30 s. (j) Unmixed image reflecting the distribution of eosin and hematoxylin. (k), Combined image of photothermal (blue) and fluorescence (red) signals. Scale bars, 2 µm. Copyright of this figure is partially held between the representative author (Takayoshi Kobayashi) and Optica.

photothermal signal seems to be slightly shifted by ~0.5 μm in the depth direction [23].

The degree of cross-talk between the unmixed images in Fig. 4.3.3 is evaluated by calculating a factor

$$\gamma_{\mathrm{he}} = \iint S_{\mathrm{h}}(x,y) / S_{\mathrm{e}}(x,y)\,dxdy, \tag{4.3.1}$$

where $S_{h(e)}(x,y) = s_{h(e)}(x,y)/s_{h(e)}^{max}$ with $s_{h(e)}(x,y)$ the intensity at the position (x, y) and $s_{h(e)}^{max}$ the maximum intensity in the unmixed image for hematoxylin (eosin). Integration is performed over the zona pellucida and cytoplasm domains by assuming that they are stained only by eosin, which results in $\gamma_{h/e}$ of ~6%.

3.3 Collection Angle and Modulation Frequency Dependence

To ensure that the observed signal is not affected by other pump-probe signals such as stimulated emission or excited state absorption, we examined the relation between the collection angle of the transmitted probe beam and signal intensity. When the photothermal signal is detected in the forward (transmission) configuration, the measured signal arises from the interference between the scattered field due to photothermal effect and the transmitted diverting probe field [2, 14]. In this case, the signal is most intense when the numerical aperture of the condenser lens (NA_c) is smaller than that of the focusing lens. We found that the signal is most intense at NA_c ~0.6, and the experimental result fairly agrees with the curve calculated based on the scattering theory with modulated Yukawa potential [14] (Figure 4.3.4). This characteristic behavior suggests that the measured signal is attributable to the photothermal effect.

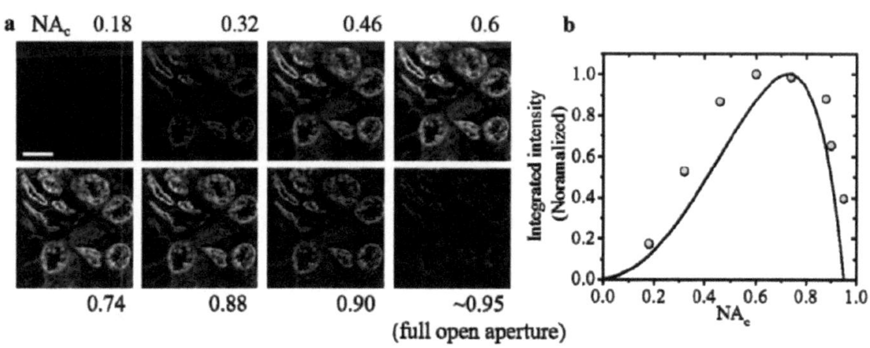

FIGURE 4.3.4 Relation between the signal intensity and collection angle of the probe beam. (a) Images of a slice of rabbit kidney stained with hematoxylin and eosin with various values of the numerical aperture of the condenser lens system (NA_c). Scale bar, 5 μm. (b) Integrated intensity as a function of the NA_c. The theoretical curve (solid curve) is calculated using Eq. (15) in Ref [14] with r_c = 0.2 μm. Copyright of this figure is partially held between the representative author (Takayoshi Kobayashi) and Optica.

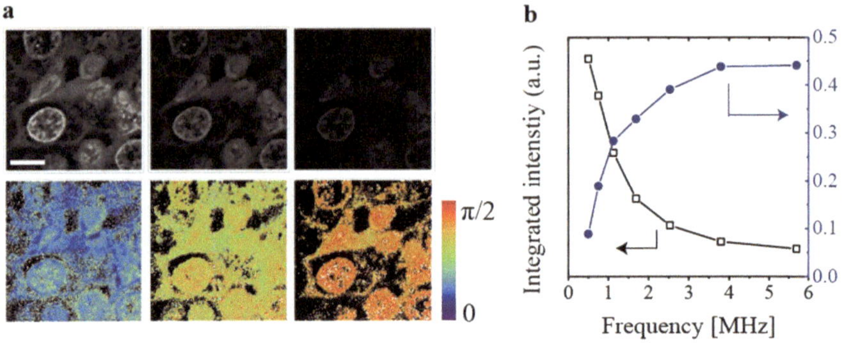

FIGURE 4.3.5 Modulation frequency dependence in photothermal microscopy. (a) Intensity (upper) and phase (lower) images of a slice of rabbit kidney stained with hematoxylin and eosin with excitation at 488 nm. Modulation frequencies of the pump beam are 0.5, 1.13, and 2.53 MHz (from left to right), respectively. Scale bars, 5 μm. (b) Integrated intensity and averaged phase as a function of modulation frequency of the pump beam. Copyright of this figure is partially held between the representative author (Takayoshi Kobayashi) and Optica.

Furthermore, the frequency response of the signal is measured by synchronously varying the modulation frequencies of the pump and probe beam while holding the frequency difference at 15 kHz. In photothermal microscopy, signal intensity decreases when the spatial extent of the modulating refractive index is smaller than the focal spot size, and the cutoff frequency is thus estimated to be several MHz assuming a thermal diffusivity of ~0.1 mm^2/s. We verified that the signal intensity (magnitude) decreased by half at ~1 MHz and the phase approaches $\pi/2$ as the modulation frequencies increase (Figure 4.3.5). The $\pi/2$ out of phase component comes from the retardation effect in heat conduction [2, 15]. Thus, the phase images offer the possibility of analyzing thermal diffusivity in tissues and would be useful in studying the effects of heat transport in biological tissues.

4 DISCUSSION

Although both fluorescent and nonfluorescent materials can be detected in photothermal microscopy because they release heat through vibrational relaxation, nonfluorescent molecules are more suited for (multi-wavelength) imaging because fluorescent molecules are easily damaged or quenched through a triplet state as they have longer excited state lifetime compared with a short-lived nonfluorescent molecule. In this regard, eosin has a quantum yield for triplet state formation of about 70% [22], and an increase in pump power easily induces saturated excitation. These saturation effects would cause an error in performing linear spectral unmixing. Furthermore, the light to heat conversion efficiency of nonfluorescent material is usually better than that of fluorescent material. In the measurement of the H&E stained sample, we confirmed that there was minimal damage to a part of tissue stained by hematoxylin, while eosin is slightly bleached after several measurements.

The simultaneous measurement with linear unmixing is considered to be much less affected by photobleaching compared with the frame-by-frame measurement. This can be explained as follows. The photothermal signal for the ith pump channel S_i is given by the linear sum of the signals from the different species

$$S_i \propto \Sigma_j^2 I_i \sigma_j (\lambda_i) c_{j'}, \tag{4.3.2}$$

where I_i is the intensity of the ith pump beam (i = 1, 2), $\sigma_j(\lambda_i)$ is the absorption cross section of the jth species at the wavelength λ_i, and c_j is the concentration of the jth species. To take into account the effect of photobleaching, average concentration $c_{j'}$ is used for time varying c_j during the measurement. This does not give serious errors because of the weak degradation of eosin (< 20%). In the simultaneous measurement, linear unmixing can be performed using two values (S_1 and S_2) with the same $c_{j'}$ since it is the same for the two pumping processes at λ_1 and λ_2. In contrast, in the frame by frame measurement, $c_{j'}$ changes with repeated scan and is different from channel to channel. This would cause an error in unmixing calculation. Therefore, the simultaneous measurement is crucial for multi-wavelength imaging.

In the experimental setup, two pump beams were collimated with different SM fibers to avoid the effect of residual chromatic aberration of a doublet lens placed after the SM fiber on the focal position. However, all beams could be collimated through a single SM fiber without chromatic aberration if a reflective collimator is employed. This could dramatically simplify the system and eliminate the need to independently adjust the laser beams.

Finally, we consider the detection limit of the photothermal microscopy. The magnitude of the modulating refractive index at the focal volume is approximated by

$$\Delta n \sim \frac{N\sigma P}{sfC\rho r_c^3}\frac{\partial n}{\partial T}, \tag{4.3.3}$$

where N is the number of molecules in the focal volume, σ (~10^{-16} cm^2) is the absorption cross-section of dye molecules, P (~100 μW) is the pump beam power, s (~10^{-9} cm^2) is the beam waist area, f (~1 MHz) is the modulation frequency, C [~1 J/(gK)] and ρ (~1 g/cm^3) are heat capacity and density of the medium, respectively, r_c is the spatial extent of the modulating refractive index (r_c^3 ~10^{-14} cm^3), and $\partial n/\partial T$ (~10^{-14} $_K^{-1}$) is refractive index variations by temperature increase. Assuming that the relative change in probe beam transmissivity and Δn is of the same order of magnitude, and the noise level of the probe beam is nearly shot-noise limit ($\delta P'/P'$ ~1×10^{-5} when the time constant of the lock-in amplifier is 0.5 ms), the detection limit (number of molecules when SNR is unity) is calculated to be N' ~10^2-10^3. This is consistent with the experimental results of a 20 nm-sized gold nanoparticle, whose absorption cross section is ~10^{-12} cm^2 and SNR is ~10^2. Since σP is wavelength dependent, the detection limit in the dual-wavelength photothermal microscopy is determined by a smaller value of σP in either of the channels.

5 CONCLUSION

Simultaneous dual-wavelength imaging has been demonstrated by simply incorporating LDs with different wavelengths.

Multi-wavelength microscopic imaging is essential to visualize a variety of nanoscale cellular components with high specificity and high spatial resolution. However, previous techniques are based on fluorescence, and thus cannot visualize nonfluorescent species, which have suffered much less from photodamage or photobleaching and hence are intrinsically useful in wider range of optical microscopy. Here, we show that simultaneous multi-wavelength imaging of nonfluorescent species can be achieved with the use of a photothermal microscope.

The advantages of photothermal microscopy implemented with this scheme described in this subsection are (1) sub-diffraction resolution by ~30% with optical sectioning capability, (2) applicable to both nonfluorescent materials and fluorescent systems even at the same time, (3) multicolor and multimodality, (4) dual-wavelength sub-diffraction imaging of biological tissues stained with hematoxylin and eosin is demonstrated, (5) three-dimensional label-free imaging of mouse melanoma tissue section is also presented to demonstrate the effectiveness of the enhanced spatial resolution. Our technique can be implemented using cost-effective and compact laser diodes and is applicable for various types of both fluorescent and nonfluorescent tissues. Thus, this method would be widely useful in biological and medical imaging [24].

The research presented in this subsection was performed by the collaborative team of J. Miyazaki, H. Tsurui, K. Kawasumi, and T. Kobayashi [24].

NOTES

1. B16 melanoma: a murine tumor cell line used for research as a model for human skin cancers. B16 cells are useful models for the study of metastasis and solid tumor formation, and were one of the first effective murine tools for metastasis research. Metastasis is a pathogenic agent's spread from an initial or primary site to a different or secondary site within the host's body; the term is typically used when referring to metastasis by a cancerous tumor [22]. The newly pathological sites, then, are called metastases (mets). It is generally distinguished from cancer invasion, which is the direct extension and penetration by cancer cells into neighboring tissues.
2. FBS: Fetal bovine serum (FBS) is a byproduct of harvesting cattle for the meatpacking industry. It is used as a supplement to basal growth medium in cell culture applications to promote cell growth and division12. FBS is frequently added to cell culture media because of its high content of growth factors and other nutrients that meet the metabolic needs of eukaryotic cells.
3. histology: Histology is known to be microscopic anatomy or microanatomy, which is the branch of biology that studies the microscopic anatomy of biological tissues and it is the microscopic counterpart to gross anatomy, which looks at larger structures visible without a microscope.
4. histopathology: Histopathology refers to the microscopic examination of tissue to study the manifestations of disease. Specifically, in clinical medicine, histopathology refers to the examination of a biopsy or surgical specimen by a pathologist, after the specimen has been processed and histological sections have been placed onto glass slides.

REFERENCES

1. L. Cognet, C. Tardin, D. Boyer, D. Choquet, P. Tamarat, and B. Lounis, "Single metallic nanoparticle imaging for protein detection in cells," *Proc. Natl. Acad. Sci. U.S.A.* **100**(20), 11350–11355 (2003).
2. S. Berciaud, D. Lasne, G. A. Blab, L. Cognet, and B. Lounis, "Photothermal heterodyne imaging of individual metallic nanoparticles: theory versus experiment," *Phys. Rev.* B **73**(4), 045424 (2006).
3. V. Octeau, L. Cognet, L. Duchesne, D. Lasne, N. Schaeffer, D. G. Fernig, and B. Lounis, "Photothermal absorption correlation spectroscopy," *ACS Nano* **3**(2), 345–350 (2009).
4. R. Radünz, D. Rings, K. Kroy, and F. Cichos, "Hot Brownian particles and photothermal correlation spectroscopy," *J. Phys. Chem. A* **113**(9), 1674–1677 (2009).
5. P. M. R. Paulo, A. Gaiduk, F. Kulzer, S. F. G. Krens, H. P. Spaink, T. Schmidt, and M. Orrit, "Photothermal correlation spectroscopy of gold nanoparticles in solution," *J. Phys. Chem. C* **113**(27), 11451–11457 (2009).
6. A. Gaiduk, P. V. Ruijgrok, M. Yorulmaz, and M. Orrit, "Detection limits in photothermal microscopy," *Chem. Sci.* **1**(3), 343–350 (2010).
7. A. Gaiduk, M. Yorulmaz, P. V. Ruijgrok, and M. Orrit, "Room-temperature detection of a single molecule's absorption by photothermal contrast," *Science* **330**(6002), 353–356 (2010).
8. D. A. Nedosekin, E. V. Shashkov, E. I. Galanzha, L. Hennings, and V. P. Zharov, "Photothermal multispectral image cytometry for quantitative histology of nanoparticles and micrometastasis in intact, stained and selectively burned tissues," *Cytometry A* **77A**(11), 1049–1058 (2010).
9. S. Lu, W. Min, S. Chong, G. R. Holtom, and X. S. Xie, "Label-free imaging of heme proteins with two-photon excited photothermal lens microscopy," *Appl. Phys. Lett.* **96**(11), 113701 (2010).
10. C. Leduc, J. M. Jung, R. P. Carney, F. Stellacci, and B. Lounis, "Direct investigation of intracellular presence of gold nanoparticles via photothermal heterodyne imaging," *ACS Nano* **5**(4), 2587–2592 (2011).
11. W.-S. Chang and S. Link, "Enhancing the sensitivity of single-particle photothermal imaging with thermotropic liquid crystals," *J. Phys. Chem. Lett.* **3**(10), 1393–1399 (2012).
12. C. Leduc, S. Si, J. Gautier, M. Soto-Ribeiro, B. Wehrle-Haller, A. Gautreau, G. Giannone, L. Cognet, and B. Lounis, "A highly specific gold nanoprobe for live-cell single-molecule imaging," *Nano Lett.* **13**(4), 1489–1494 (2013).
13. M. Selmke and F. Cichos, "Photothermal single particle Rutherford scattering microscopy," *Phys. Rev. Lett.* **110**(10), 103901 (2013).
14. J. Miyazaki, H. Tsurui, K. Kawasumi, and T. Kobayashi, "Optimal detection angle in sub-diffraction resolution photothermal microscopy: application for high sensitivity imaging of biological tissues," *Opt. Express* **22**(16), 18833–18842 (2014).
15. M. Selmke, A. Heber, M. Braun, and F. Cichos, "Photothermal single particle microscopy using a single laser beam," *Appl. Phys. Lett.* **105**(1), 013511 (2014).
16. P. Vermeulen, L. Cognet, and B. Lounis, "Photothermal microscopy: optical detection of small absorbers in scattering environments," *J. Microsc.* **254**(3), 115–121 (2014).
17. Y. Ozeki, W. Umemura, Y. Otsuka, S. Satoh, H. Hashimoto, K. Sumimura, N. Nishizawa, K. Fukui, and K. Itoh, "High-speed molecular spectral imaging of tissue with stimulated Raman scattering," *Nat. Photonics* **6**(12), 845– 850 (2012).
18. J. Miyazaki, H. Tsurui, A. Hayashi-Takagi, H. Kasai, and T. Kobayashi, "Sub-diffraction resolution pump-probe microscopy with shot-noise limited sensitivity using laser diodes," *Opt. Express* **22**(8), 9024–9032 (2014).

19. J. Miyazaki, K. Kawasumi, and T. Kobayashi, "Frequency domain approach for time-resolved pump-probe microscopy using intensity modulated laser diodes," *Rev. Sci. Instrum.* **85**(9), 093703 (2014).
20. J. Miyazaki, K. Kawasumi, and T. Kobayashi, "Resolution improvement in laser diode-based pump-probe microscopy with an annular pupil filter," *Opt. Lett.* **39**(14), 4219–4222 (2014).
21. W. Denk, J. H. Strickler, and W. W. Webb, "Two-photon laser scanning fluorescence microscopy," *Science* **248**(4951), 73–76 (1990).
22. A. Penzkofer, A. Beidoun, and M. Daiber, "Intersystem-crossing and excited-state absorption in eosin-Y solutions determined by picosecond double pulse transient absorption-measurements," *J. Lumin.* **51**(6), 297–314 (1992).
23. A. Gaiduk, P. V. Ruijgrok, M. Yorulmaz, and M. Orrit, "Making gold nanoparticles fluorescent for simultaneous absorption and fluorescence detection on the single particle level," *Phys. Chem. Chem. Phys.* **13**(1), 149–153 (2010).
24. J. Miyazaki, H. Tsurui, K. Kawasumi, and T. Kobayashi, "Simultaneous dual-wavelength imaging of nonfluorescent tissues with 3D subdiffraction photothermal microscopy," *Opt. Express* **23**(5), 3647–3656 (2015).

5.1 Gray-level Co-occurrence Matrix Analysis of Several Cell Types in Mouse Brain Using Resolution-enhanced Photothermal Microscopy

1 INTRODUCTION

Optical microscopy has been extensively used for biological imaging for many years but the resolution is limited by diffraction [1]. Recently, in order to visualize a variety of nanoscale cellular components with high specificity and high spatial resolution, several superresolution imaging techniques such as stimulated emission depletion microscopy, structured illumination microscopy, and stochastic optical reconstruction microscopy have been developed reaching sub-100-nm resolution [2, 7]. Such high resolution has revealed fine intracellular structures. Another direction of the development of microscopy is the imaging of nonfluorescent system with high sensitivity and resolution using photothermal (PT) effects with resolution enhancement [8, 12]. In these studies, although the high resolution capability thanks to the pump-probe scheme has revealed fine intracellular structures, intracellular "global" statistical features of the intracellular structure such as uniformity, correlation among the microstructures, and entropic randomness have not been investigated. For example, even though analysis has been performed with the fractal dimensions [13], in this case superresolution was not appreciably utilized and the analysis was only applied to two-dimensional (2-D) images [13]. The accurate detection of a deformation or structural changes at the cellular level depends on the efficiency of the microscope and the skill of the investigator and is prone to human error. A quantitative measure for the classification and detection of some cell types from the pathological

DOI: 10.1201/9781003476030-18

viewpoint at the cellular level has been established using a microscope and texture analysis using the gray-level co-occurrence matrix (GLCM) method. The GLCM method is capable of evaluating fine structural changes in the organelles in cells or in the distribution of cells that are otherwise difficult to properly evaluate using standard visual analysis methods. Thus far, GLCM analysis has been performed using differences in the texture parameters of images obtained through standard microscopic methods. For example, the GLMC method has been used to study the distribution arrangements of cells in two areas of the mouse brain, the cortex and the medulla [14, 15]. However, GLCM-type analysis has not been utilized to study high resolution intracellular images.

In this study, we performed for the first time qualifications of intracellular structure in the resolution-enhanced images of cells in mouse brain obtained by PT imaging, which we developed to perform nondiffraction limited imaging using a compact system without requiring fluorescent capability [16–18]. We examined five types of cells, neurons, and glia cells in basal nuclei, Purkinje cells, and granule cells in cerebellum, and granule cells in dentate gyrus. These cells are distinct in their morphology and the location, which provides reliable identification of them. These cells have been shown to be involved in peculiar diseases: Purkinje cells[1] in Nieman-Pick type[2] C1 ataxia[3] and spinocerebellar[4] degeneration type 6[5], cerebellar granule (CGr)[6] cells in ataxia telangiectasia[7], both neurons and glia cells of basal nuclei in corticobasal[8] degeneration, dentate gyrus[9] granule (DGr) cells in Alzheimer disease.[10]

There have been some reports which correlate GLCM analysis parameters to physiological or pathological meanings: finite nuclear change associated with thymocyte development [14], correlation with cancer prognosis [19] or grading [20], and nuclear change in the early stage of apoptosis [21]. However, those findings are correlations between GLCM parameters and are rather superficial phenomena, and underlying mechanisms are not shown. This is due to the stage at which GLCM analysis has been accomplished and where it still has substantial pathological or medical meanings. In this chapter, we have successfully assigned parameters which are appropriate to characterize individual cells.

As the texture properties obtained by the GLCM method vary depending on distance, parameter d was selected for analysis [14, 15]. Our results exhibit a characteristic variation of the dependence of the five GLCM parameters on d. For analysis, we evaluated the parameters for the five images taken in the order of increasing depth from layer 1 to layer 5 in steps of 0.283 μm in the same lateral region of interest.

This study has been reviewed and approved with License number 280076 by the Institutional Animal Care and Committee of Juntendo University. All the authors have no relevant financial interests in the manuscript and no other potential conflicts of interest to disclose.

2 EXPERIMENT

We performed pump-probe microscopy using an intensity-modulated laser diode (LD) as a pump and a nonmodulated continuous-wave LD as a probe. The schematic of the system is shown in Figure 5.1.1 and the details are described below.

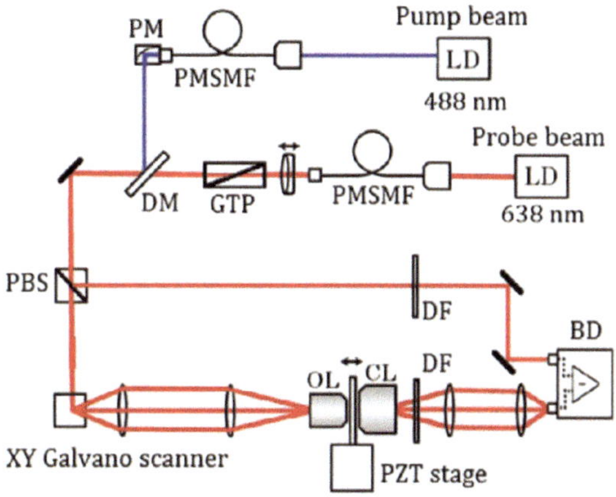

FIGURE 5.1.1 Schematic of the system of a PT microscope equipped with a Galvano scanner and spatially segmented balanced detection. LD, laser diode; PMSMF, polarization maintaining single-mode fiber; PM, parabolic mirror; GTP, Glan-Thompson polarizer; DM, dichroic mirror; PBS, polarizing beam splitter; OL, objective lens; CL, condenser lens; DF, dielectric multilayer filter; and BD, balanced detector. The copyright of this figure is shared between the representative author (Takayoshi Kobayashi) and Nature Portfolio.

The principle of the pump-probe microscopy is based on the PT effect induced by the pump and detected by the probe. The absorbed pump beam energy with Gaussian distribution on the focal plane is converted to heat and changes the refractive index forming a concave thermal lens which deflects the probe beam resulting in the detected probe intensity change. High spatial resolution can be obtained for the pump-probe microscope because the signal intensity is proportional to the product of the pump and probe intensities [10]. The sensitivity of the PT microscope is reported to be a single-molecular level even at room temperature.[22]

The details of the PT microscopy system, which uses simple LDs and a balance detector, are described in our previous papers [16, 17]. Two modes of imaging were used: stage scanning and fast Galvano scanning.[11] These modes are described below.

In the stage-moving mode, each horizontally linear polarized beam was collimated through a spatial filter, and the two beams were combined with a dichroic beam-combining mirror. The sample position was raster-scanned using a three-axis positioning stage driven by piezoactuators (Thorlabs MAX311D). To evaluate the spatial resolution using gold nanoparticles[12] (GNPs), we used a high-accuracy positioning stage (PI P-622.2) with a closed-loop resolution of 0.7 nm. A 1.4-NA oil-immersion condenser (Olympus U-AAC) was used to collect the transmitted probe light. A narrowband filter matching the laser line was placed in front of the detector to filter out both the excitation beam and the fluorescence from the sample, so that only the probe beam was detected. An autobalanced photoreceiver (New Focus, Nirvana) was used to cancel out the noise due to intensity-fluctuation of the probe laser. The probe beam was reflected by a beam splitter placed in front of the focusing lens which reflected the probe beam and then directed it to the reference port of the balanced detector through a multimode fiber. The maximum area size available in the stage scan mode is 20 μm × 20 μm. Image sizes typically described in this chapter are 144 × 144 points, which corresponds to 3.6 μm × 3.6 μm. The lock-in amplifier sensitivity was 5 mV, and the time constant was 500 μs per point.

In the beam-sweeping mode with the Galvano-mirror system, the collimated pump laser and the probe laser through the Glan-Thompson polarizer were overlapped through a dichroic mirror. The probe beam was split with a polarizing beam splitter (PBS) for balance detection and was reflected with the Galvano mirror. One of the split probe beams was used as a reference beam. The center of the 4f-optical system was set at the sample position. The probe transmitting through and reference bypassing the sample were both detected with the split detection section of the balance detector. Three-dimensional (3-D) images of the samples were obtained by scanning in the X–Y plane with the Galvano mirror and by moving the piezo stage[13] in the Z-direction. The probe beam through the sample and the reference beam irradiated the balance detector. Image sizes in this system are 300 × 300 pixels corresponding to 72 μm × 72 μm. In comparison with the 3.6 μm × 3.6 μm of the stage-scan mode, the Galvano system can cover a much wider range. The lock-in amplifier sensitivity was 5 mV, and the time constant was 20 μs per pixels. Dwell time was 50 μs per pixel.

In both modes, two LDs [Nichia NDS4216 (488 nm) and Mitsubishi ML520G55 (638 nm)] were used for the pump and probe, respectively. Their direct output powers from the lasers were ~200 mW, which was reduced to 2.0 mW for the pump and 2.6 mW for the probe at the focal point of the microscope by appropriate filters. The LDs were mounted on temperature-controlled laser-diode mounts (Thorlabs, LDM9T). An objective lens [Olympus, UPLSAPO 40X2, numerical aperture (NA) = 0.95] was used. The beam size was adjusted to fill the back aperture of the objective lens. The lock-in technique was utilized to detect the pump-induced signal modulation with a lock-in amplifier (Signal Recovery 7270) referenced to a frequency of 100 kHz. A modulated signal at the frequency of 100 kHz was generated by the bilinear interaction between the pump field and the probe field in the sample. The output of the balanced detector was connected to the lock-in amplifier through a low-pass filter (Mini-circuit LPF-B0R3). The reference signal for the lock-in amplifier was delivered from the signal generator. GNPs of 20-nm diameter were used for the evaluation of resolution. GNPs were dispersed in water and mixed with an aqueous solution of polyvinyl alcohol. Then the mixed solution is spin coated on a glass substrate.

The details of the tissue sample are described as follows. In this study, sagittal-sliced sections of mouse brain were used. When the sliced section was stained using the hematoxylin-eosin method, the cerebral cortex[14], the hippocampal[15] formation under the corpus callous[16], basal nuclei (BN)[17], and the cerebellar portion[18] can be easily identified by an individual well experienced in the anatomy of the mouse brain from their structures and the well-established analytical atlas of the mouse brain. Cells that were large and stained faintly in the basal nucleus (BN) were considered to be nerve cells (pyramidal cells[19]). Cells in the BN that were small and stained heavily were considered to be glial cells. The hippocampal formation has a very distinctive shape, and cells included within the hippocampus can be easily identified.

We captured images of cells, which were round and not bipolar or pyramidal, and which could be identified as granule cells[20]in the gyrus dentate[21]. In the cerebellar cortex, the border between the Purkinje cell layer and granular cell layer is clear. The granular cells are small and strongly stained, whereas Purkinje cells have large round bodies and thick processes. As such, they were easily identified. The purpose of this study is to demonstrate the capability of the proposed method for extracting the quantitative characteristic features of these cell types in different domains of a brain. This can be useful for the estimation of the degree of pathological process in various cells.

The GLCM analysis was performed using the image data obtained by the Galvano-mode imaging. We evaluated five images taken at five depths increasing from layer 1 to layer 5 with a step of 0.283 μm while maintaining the same lateral region.

It is notable as being one of a select few brain structures which may have significant rates of adult neurogenesis in many species of mammals, from rodents to primates [4]. Other sites of adult neurogenesis may include the subventricular zone, the striatum [5] and the cerebellum [6]. However, whether significant neurogenesis exists in the adult human dentate gyrus has been a matter of debate [7–11].

3 IMAGING RESULTS

First, we studied the spatial resolution of the PT microscope used in the present experiment by imaging GNPs of 20 nm in diameter as shown in Figure 5.1.2. This image was taken in the moving stage mode Fig. 5.1.2(a), one prominent image, three medium intensity images, and two faint images are visible by stage moving measurement. The differences in intensity are due to the difference in the distance in the Z-direction from the central focal plane for these GNPs in the doped samples. Figs. 5.1.2(b) and (c) show 2-D GNP images in the lateral plane and the axial direction by the Galvano-mirror system, respectively. The most intense GNP image was used for the resolution evaluation. The full-width at half-maxima (FWHMs) of the intensity distribution profile was found to be 219.9 nm for the piezo-stage system as shown in Fig. 5.1.2(d). The experimental resolution is defined as the FWHM of Gaussian function obtained by the least squares mean (LSM) fitting to the observed intensity distribution of the GNPs, and values following "±" symbol are the standard deviations obtained by the LSM. If the finite size of the GNP (20 nm) is taken into account, the FWHM is 218.9 nm.

A theoretical evaluation of the resolution of the PT microscopy was conducted for the purpose of comparison with the experimental results.

The "lateral" distribution of the signal intensity depends on the probe and heating beam profiles and also on the dielectric susceptibility profile of the target structured material (in the present case, intracellular structure). However, the "lateral" spatial extension of the susceptibility profile is much smaller than the "lateral" size of the probe beam, and the transverse resolution detected using the phase sensitive detection method as in this study is concluded to be approximated by the product of the two (pump and probe) beam profiles [10]. This procedure was applicable under the experimental conditions in the previous studies [17, 23, 24] as well as in this study.

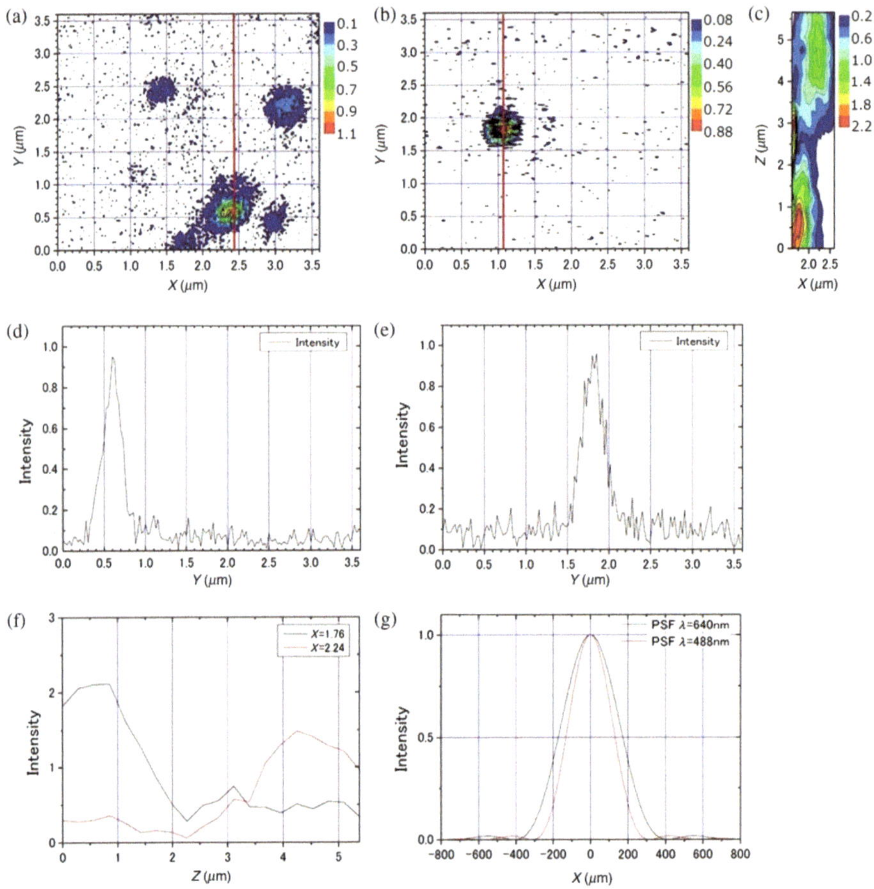

FIGURE 5.1.2 (a) 2-D GNP image obtained by PT microscopy in the piezo-stage moving mode (a) and (d) intensity distribution profile along the red line in (a); (b) 2-D display of a GNP image in the *XY*-plane obtained by PT microscopy in the Galvano-sweeping mode, (b) intensity distribution profile along the red line; (c) 2-D display of a GNP image in *XZ*-plane measured with PT microscopy in the Galvano-sweeping mode; (e) intensity distribution profile at $x = 1.76$ and $x = 2.24$ in (f); (g) calculated PSF at $\lambda = 638$ and 488 nm. The copyright of this figure is shared between the representative author (Takayoshi Kobayashi) and Nature Portfolio.

The FWHMs of the intensity distribution of the pump (488 nm) and probe (638 nm) are calculated to be 264 and 350 nm, respectively, using the Bessel function as the point spread function (PSF) with an NA = 0.95 [see Fig. 5.1.2(g)]. Then, the FWHM of the pump-probe setup obtained by the product of these two Bessel functions corresponding to the pump and probe was calculated to be 210 nm. The resolution obtained by the PT microscopy of the GNP in this study was determined to be 219.9 nm for the piezo-stage system in relatively good agreement with the calculation result of 210 nm. Thus, it is proved that the intensity distribution of the PT microscopy image is approximately given by the product of the PSFs of pump and probe. This means that the resolution is not limited by the spot size of the pump beam (264 nm) but is rather limited by the product of the PSFs of the pump and probe. This indicates that the imaging is beyond the diffraction of both pump and probe beams. Higher resolution is realized by nonlinear PT imaging of the combination of the present resolution enhancement mechanism and nonlinear effect [18]. This resolution improvement is also found in other pump-probe microscopies such as stimulated gain microscopy and stimulated emission-induced fluorescence-quenching microscopy [16].

The FWHM of the intensity distribution profile was found to be 270.8 nm for the Galvano-mirror system as shown in Fig. 5.1.2(e). When including the effect of the GNP size, the FWHM of the intensity distribution profile was 269.8 nm. The degradation of resolution in the Galvano system (269 nm) from the piezo-stage stage system (209 nm) likely occurs because the inertia of a galvanometric scanner leads to time delays when following external commands, which subsequently introduces image distortions that deteriorate as the scan frequency increases.[25]

Thus, the lateral resolution is well explained in terms of the product of the PSFs of pump and probe. On the other hand, the axial resolution is not given by this product [10]. It was found both theoretically and experimentally that the assumption of a product PSF of heating and detection laser in the axial direction is inappropriate for describing the PT image profile. Instead, the two-lobe structure and the size of the two lobes depend sensitively on the axial displacement (Δz_f) of the two laser foci [Figs. 5.1.2(d) and 2(e) in Ref. 10]. It was found that independently of Δz_f, whenever the probe laser focus is in front of the refractive index profile, the detected intensity is decreased as compared to that of the undisturbed probe beam, while a signal increased intensity is measured if the probe laser focus is behind the refractive index profile [Fig. 5.1.1(c) in Ref. 10]. The two lobes are shown in Figs. 5.1.2(c) and (f). There is some peak displacement (in the *X*-direction) along the *Z*-direction in Fig. 5.1.2(c). This is probably due to some small dislocation of the beam axis between pump and probe in the *X*-direction.

4 TEXTURE ANALYSIS

The analysis of textures of the five different cell types described below was performed by using the GLCM method [14, 15]. The texture parameters were calculated by analyzing relevant regions of interest of five 2-D images of the mouse brain of interest captured at increasing depths from layer 1 to layer 5 with a step of 0.283 μm in the axial direction. The PT images of neurons in BN [Figure 5.1.3(a)–(i)] and cerebellum [Figure 5.1.4(a)–(i)] were captured by step-wise shifting along the axial direction with a step of 283 nm. Images (j) in both Figs. 5.1.3 and 5.1.4 show bright-field images of the corresponding region to the PT images.

Figure 5.1.4(k) shows a bright-field image of the cerebellum in a different place from (a) to (j). Purkinje cells are lined side by side next to the granule cell layer. Figure 5.1.5(a)–(i) and (j) shows the PT images and bright-field image, respectively,

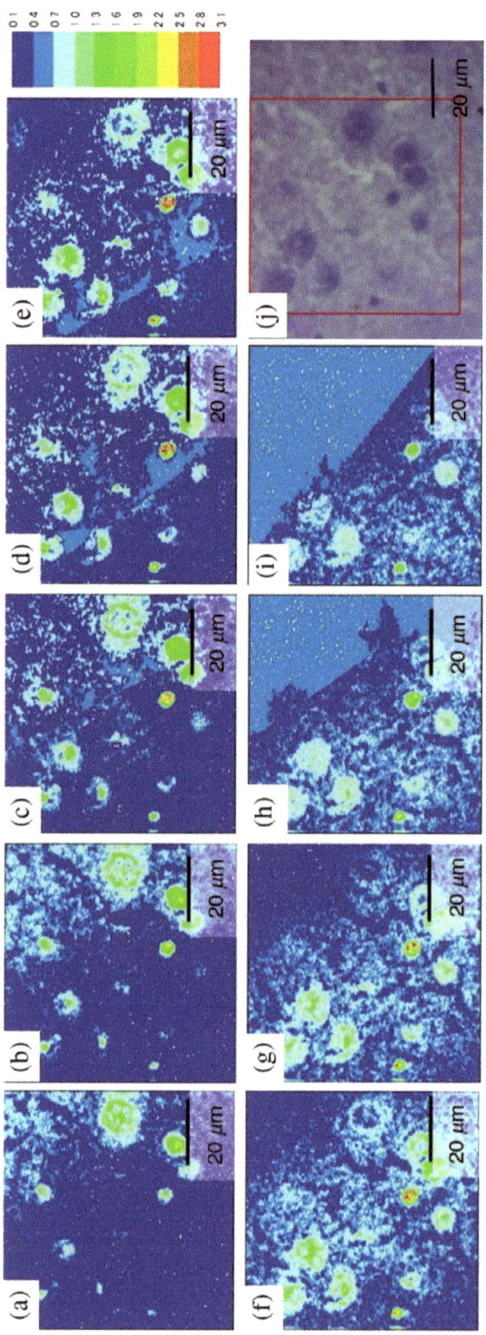

FIGURE 5.1.3 PT images and bright-field image of neurons in BN. Nine images (a)–(i) are taken by shifting along the axial direction by 283 nm at each step. (j) The bright-field image. Red-line marked box in image (j) showing the image area of (a)–(i).

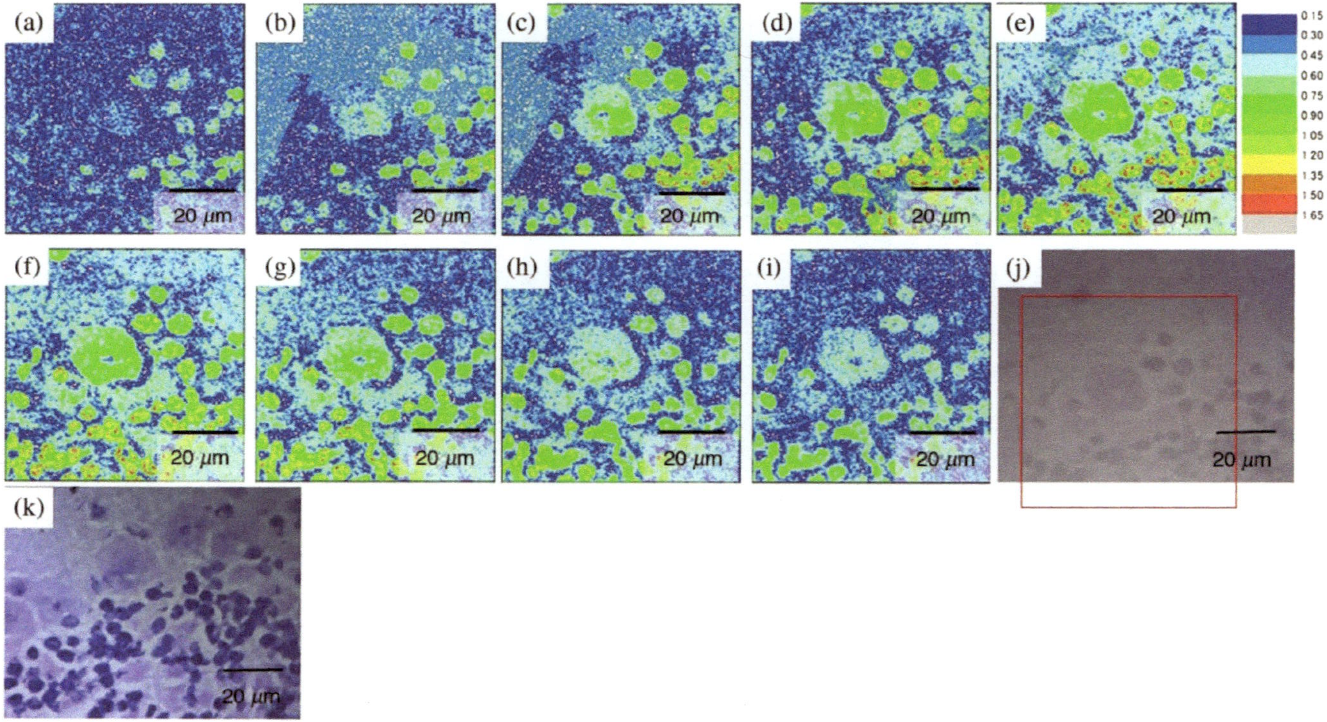

FIGURE 5.1.4 PT images and bright-field image of neurons in cerebellum. Nine images (a)–(i) are taken by shifting along the axial direction by 283 nm at each step. (j) The bright-field image. Red-line marked box in image (j) showing the image area of (a)–(i). (k) A bright-field image of the cerebellum in a different place from (a)–(j). Purkinje cells lined with layers and many granule cells are shown in (k). The copyright of this figure is shared between the representative author (Takayoshi Kobayashi) and Nature Portfolio.

of the hippocampal dentate gyrus. Image (k) in Fig. 5.1.5 shows the PT image of the dentate gyrus in an extended area, which is the same area as in (i) showing the overplayed bright-field image and red-boarded area image in (j) in Fig. 5.1.5.

We studied five different cell types; a neuron (nerve cell), a glial cell, two kinds of granule cells, and a Purkinje cells. The fundamental properties and functions of them are briefly described as follows. The numbers of five cell types studied were 10 for pyramidal neurons in the BN, 10 for glial cells in the basal nucleus (BGl), 5 for cerebellar Purkinje (CP) cells, 8 for CGr cells, and 32 for DGr cells.

In Figure 5.1.6, analysis was performed for (a) nerve cells and (b) glial cells in the BN in cerebrum, (c) Purkinje cells and (d) granule cells in the cerebellum, and (e) granule cells in the dentate gyrus. The inside areas of the encircled parts were used for the analysis.

Fig .5.1.6 shows the images taken by the PT microscope for BN and BGl, CP and CGr, and DGr. The black lines encircling single cells in the domain show the area where the analysis was made. The encircling of the cell was performed by thresholding setting for the digitized 2-D image data of 256 gray levels.

Here, we describe briefly the function of the studied cell types.

Nerve cells are specialized cells found in the nervous system designed to stimulate other cells in the body in order to communicate. Neurons are excitable, which means that they function by using electrical stimulation. Glial cells are nonneuronal cells that maintain homeostasis, form myelin, and provide support and protection for neurons in the central and peripheral (called Schwan cells) nervous systems. Purkinje cells, also called Purkinje neurons, are neurons in vertebrate animals located in the cerebellar cortex of the brain. Purkinje cell bodies are shaped like a flask and have many threadlike extensions called dendrites, which receive impulses from other neurons called cerebellar granule cells. Granule cells of the cerebellum are among the smallest neurons in the brain. (The term granule cell is used for several unrelated types of small neurons in various parts of the brain.) CGr cells are the most numerous neurons in the brain. Granule cells in the molecular layer of the dentate gyrus receive the hippocampal formation's major excitatory input from the cortex. To compare the difference in intracellular structures among the five cell types, we analyzed the PT images by the GLCM method.

The GLCM is a square matrix in which the axes correspond to the number of gray levels in an image. It contains the frequency of neighboring pixel relationships of a 2-D image (of the region of interest) at the position of the corresponding gray level. The value of frequency at which the pixel intensity pair i and j appear together is denoted by $P(i,j)$ (see the detailed definition of this value in Refs. 14 and 15). Several texture properties can be extracted from the GLCM matrix, including angular second moment (ASM), contrast, correlation, inverse difference moment (IDM), and entropy.

ASM is a measure of the homogeneity of an image

$$\mathrm{ASM} = \sum_{i=0}^{G-1}\sum_{j=0}^{G-1}\{P(i,j)\}^2. \quad (1)$$

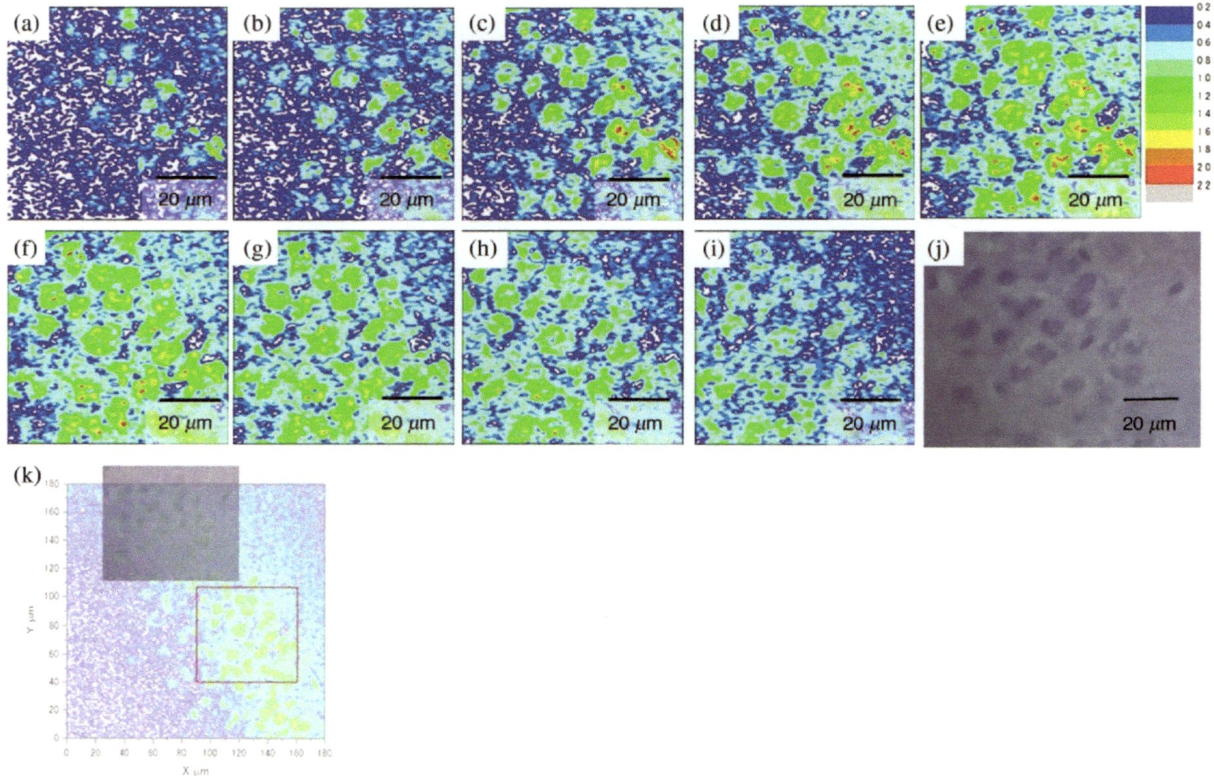

FIGURE 5.1.5 PT image and bright-field image of dentate gyrus. Nine images (a)-(i) are taken by shifting along the axial direction by 283 nm at each step. (j) A bright-field image. (k) shows the extended PT image with an overlaid bright-field image and red-line marked box showing the image area of (a)–(i). The copyright of this figure is shared between the representative author (Takayoshi Kobayashi) and Nature Portfolio.

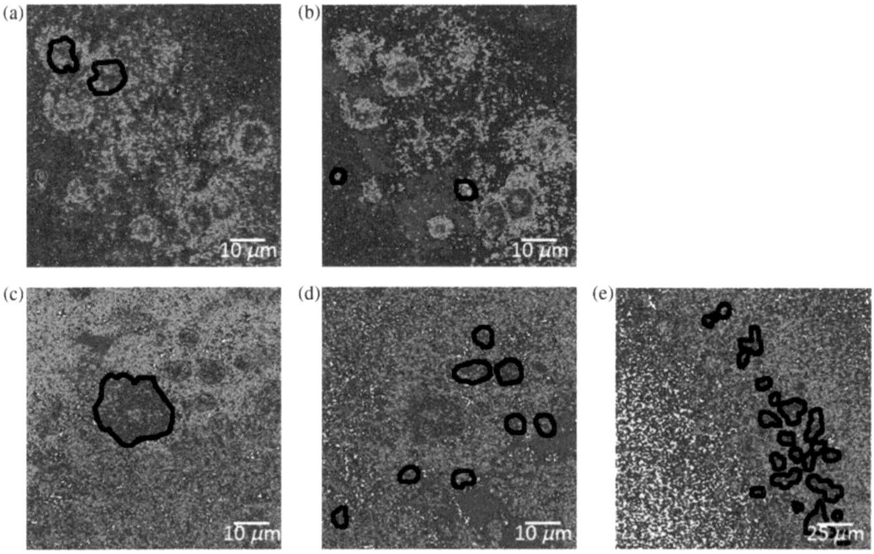

FIGURE 5.1.6 Images taken with Galvano-mode PT microscopy in the domains of (a) and (b) cerebrum, (c) and (d) cerebellum, and (e) hippocampal dentate gyrus. The types of cells indicated by the encircling black lines are (a) nerve cells and (b) glial cells in the BN, (c) Purkinje cells, (d) granule cells in the cerebellum, and (e) granule cells in the dentate gyrus. The inside areas of encircled parts were used for the analysis. The copyright of this figure is shared between the representative author (Takayoshi Kobayashi) and Nature Portfolio.

"ASM" value shows the strength of homogeneity, namely the pair correlation [$P(i, j)$] of the whole area nearly independent pair of i and j. So even if the concentration of chromatin in this case is homogeneous, it becomes smaller if the concentration has slope(s) in any direction showing area-wide uniformity. This can happen in the case of glial cells because the nuclei of glial cell are small and condensed resulting in the homogeneous distribution of chromatin. The average value of ASM measured for pixel distance d = 1 was 0.272 ±0.087 and was the highest for BGI. The value decreased to 0.229 ±0.085 being highest for BGl. The values to 0.229± 0.085 and 0.201 ± 0.083 when the pixel decreased to distance was increased to 2 and 3, respectively. The difference in ASM between the cells was highest for BGl and CP, followed by CGr and BN. CGr and CP had the least difference in magnitude of ASM for pixel lengths (distances) d = 1 and 3, with a mean difference of 0.04. As can be expected from the images in Fig. 5.1.6, Purkinje cells have large and patterned nuclei, and the above-mentioned special properties of its pattern are quantitatively discussed and quite reasonably explained. A homogeneous area is considered to contain only a small number of gray levels, giving a GLCM with only a few but relatively high values of $P(i,j)$ [26–29]. Thus, the sum of squares giving the values of ASM will be high.

Contrast is a measure of the local variations in gray levels present in an image. Images with large neighboring gray level differences are associated with a high contrast

$$\text{contrast} = \sum_{|i-j|=0}^{G-1} |i-j|^2 \left\{ \sum_{i=1}^{G} \sum_{j=1}^{G} P(i,j) \right\} \tag{2}$$

Contrast by definition represents the difference in the concentration in the distal location due to the $|i - j|^2$ term. This means that if some cell has a large value of this parameter, within the cell of this type the concentration is relatively localized in some specific spatial area in the cell. Contrast of CGr is the highest among the five cell types in the case of $d = 6$. It is to be noticed that the value is dependent on the distance parameter d as in the other parameter. For example, the contrast of BGl is larger than the one of BN in the case of d = 1 > contrast (383; BN, d =1). However, contrast (915; BGl, d= 3) <contrast (976; BN, d=3) (Table 5.1.1).

The crossover takes place between d =1 and d =2. This shows that the evaluation of difference of P varies with d. This dependence also provides the important information of the dependence of difference distribution with the unit of d (distance parameter for GLCM analysis = 283 nm) in the GLMC analysis, which is very reasonably used since it is close to the PT microscope resolution (219 nm).

Correlation is a measure of gray level linear dependence between pixels at specified positions relative to each other:

$$\text{correlation} = \sum_{i=0}^{G-1} \sum_{j=0}^{G-1} \left[\frac{\{i \times j\} \times P(i,j) - \{\mu_x \times \mu_y\}}{\sigma_x \times \sigma_y} \right]. \tag{3}$$

Here, μx and μ y are, respectively, the means of $P_x(i) = \sum_{j=0}^{G-1} P(i,j)$ and $P_y(j) = \sum_{i=0}^{G-1} P(i,j)$ and σx and σy are the corresponding standard deviations.

Correlation is different from contrast and IDM in such a way that it is the difference from the mean values of P (i, j), while the other two are the absolute values. Therefore, in the case where the frequency P (i, j) is increased or decreased by the same amount for all of the (i, j) pairs, the values of contrast and IDM proportionally increase or decrease correspondingly, while correlation is maintained.

The correlation of BGl at pixel distance 1 is $(89.9 \pm 3.1) \times 10^{-5}$, which was the highest among all five cell types studied. The value decreased to $(70.3 \pm 5.9) \times 10^{-5}$ when the pixel distance was increased to 2. DGr had an average correlation value of $(88.8 \pm 4.6) \times 10^{-5}$ for distance 1. The average difference is $(85.9 \pm 4.6) \times 10^{-5}$ for distance parameter d=2. The difference in the average correlation value between CGr and BGl was the highest in the case of d = 1. The value of correlation decreased for all cells with an increase in pixel distance value.

The IDM defined by the following equation is also influenced by the homogeneity of the image:

TABLE 5.1.1
Values of Texture Parameters Evaluated for Three Pixel Distances d = 1,2,3. The five cell types studied were pyramidal neurons in the BN, BGI, CP, DGr, CP, CGr, and DGr cells

Cell type	Texture parameter	$d = 1$	$d = 2$	$d = 3$
BN	ASM	0.171 ± 0.022	0.136 ± 0.021	0.115 ± 0.02
	Contrast	383 ± 39	720 ± 69	976 ± 88
	Correlation	$(77.8 \pm 6.9) \times 10^{-5}$	$(62.8 \pm 5.6) \times 10^{-5}$	$(51.4 \pm 4.6) \times 10^{-5}$
	IDM	0.776 ± 0.026	0.668 ± 0.038	0.590 ± 0.046
	Entropy	3.31 ± 0.19	3.37 ± 0.20	3.40 ± 0.20
BGI	ASM	0.272 ± 0.087	0.229 ± 0.085	0.201 ± 0.083
	Contrast	396 ± 64	703 ± 81	915 ± 78
	Correlation	$(89.9 \pm 3.1) \times 10^{-5}$	$(70.3 \pm 5.9) \times 10^{-5}$	$(53.7 \pm 6.5) \times 10^{-5}$
	IDM	0.797 ± 0.034	0.699 ± 0.051	0.629 ± 0.063
	Entropy	2.85 ± 0.39	2.90 ± 0.40	2.93 ± 0.40
CP	ASM	0.088 ± 0.005	0.061 ± 0.003	0.048 ± 0.003
	Contrast	631 ± 96	1131 ± 160	1461 ± 188
	Correlation	$(69.9 \pm 7.6) \times 10^{-5}$	$(46.0 \pm 6.4) \times 10^{-5}$	$(30.3 \pm 5.3) \times 10^{-5}$
	IDM	0.595 ± 0.005	0.447 ± 0.007	0.360 ± 0.008
	Entropy	4.40 ± 0.08	4.44 ± 0.08	4.47 ± 0.08
CGr	ASM	0.091 ± 0.009	0.064 ± 0.008	0.051 ± 0.007
	Contrast	979 ± 132	1713 ± 236	2171 ± 301
	Correlation	$(38.7 \pm 3.8) \times 10^{-5}$	$(26.9 \pm 3.0) \times 10^{-5}$	$(19.5 \pm 2.5) \times 10^{-5}$
	IDM	0.613 ± 0.010	0.472 ± 0.014	0.389 ± 0.018
	Entropy	4.43 ± 0.05	4.47 ± 0.05	4.50 ± 0.05
DGr	ASM	0.179 ± 0.014	0.153 ± 0.019	0.133 ± 0.021
	Contrast	71 ± 29	137 ± 53	202 ± 76
	Correlation	$(88.8 \pm 4.6) \times 10^{-5}$	$(85.9 \pm 5.0) \times 10^{-5}$	$(82.9 \pm 5.6) \times 10^{-5}$
	IDM	0.911 ± 0.034	0.830 ± 0.060	0.758 ± 0.079
	Entropy	1.99 ± 0.10	2.19 ± 0.14	2.34 ± 0.16

$$\text{IDM} = \sum_{i=0}^{G-1} \sum_{j=0}^{G-1} \frac{1}{1+(i-j)^2} P(i,j). \tag{4}$$

Because of the weighting factor $1+(i-j)^2]^{-1}$ IDM will get small contributions from far separate areas (i ≠ $_j$). This makes IDNM lower for "locally" inhomogeneous images and higher for locally homogeneous images. Therefore, contrast is a measure of the appearance frequency for distant pairings while IDM is that of near neighboring pairing.

The mean IDM for DGr was 0.911 ± 0.034 at pixel distance *d* =1. The value for CP cells at a pixel distance d =1 was 0.595 ±0.005. The difference in values was the highest between DGr and CP. The difference in value remained the highest for pixel distance 2 with values of 0.830 ±0.060 for BGl and 0.447 ± 0.007 for CP. The value of IDM decreased for all cell types from pixel distance 1 to pixel distance 3. The difference in the values between each cell type also decreased significantly with increasing pixel length.

Entropy is defined and most frequently used in thermodynamics and it has also been used in information science:

$$\text{entropy} = -\sum_{i=0}^{G-1} \sum_{j=0}^{G-1} P(i,j) \times \log\left[P(i,j)\right]. \tag{5}$$

In the former, it has relevance to the useless (unusable) energy and direction of equilibration, whereas in the latter it is used as a measure of the amount of information and randomness present in a signal. In imaging, it is used in terms of the latter sense and nonhomogeneous images have low first-order entropy, whereas homogeneous scenes have high entropy.

The values of the five texture parameters are defined by the above equations and calculated for the cells shown in Fig. 5.1.6. For the analysis, we evaluated the five images of cells at the same lateral positions with five different depth (axial) positions (layer 1 to layer 5 with a step of 0.283 μm). They are summarized in Tables 5.1.1 and 5.1.2, and Figure 5.1.7. The values for GLCM parameters in Tables 5.1.1 and 5.1.2 are the mean values of statistical analysis of the GLCM parameter values obtained by the calculation of the five PT images of the same cells with the five different depth positions. The parameters are obtained by using ImageJ plugin "texture analysis" and the values after are the standard deviations of the statistical analysis over the five PT images.

The values for d = 4–6 have common features of monotonic increase or decrease as seen in this figure in the same way as for *d* = 1, 2, and 3 in Table 5.1.1.

The values in Tables 5.1.1, 5.1.2 and Fig. 5.1.7, ASM, correlation, and IDM monotonically decrease with increase in *d*, while the value of contrast monotonically increases with increase in *d*. This is reasonable for the following reasons. If d increases in the case of correlation, the above feature is reasonable since the intensity correlation becomes less and less correlated. ASM decreases because the probability P(I,j) decreases with the Eq. (1) decreases even more rapidly. IDM decreases because

TABLE 5.1.2
Values of Texture Parameters Evaluated for Three Pixel Distances d ¼ 4, 5, 6. The five cell types studied were pyramidal neurons in the BN, BGl, DGr cells, CP cells, and CGr cells

Cell type	Texture parameter	*d* = 1	*d* = 2	*d* = 3
BN	ASM	0.171 ± 0.022	0.136 ± 0.021	0.115 ± 0.002
	Contrast	383 ± 39	720 ± 69	976 ± 88
	Correlation	$(77.8 \pm 6.9) \times 10^{-5}$	$(62.8 \pm 5.6) \times 10^{-5}$	$(51.4 \pm 4.6) \times 10^{-5}$
	IDM	0.776 ± 0.026	0.668 ± 0.038	0.590 ± 0.046
	Entropy	3.31 ± 0.19	3.37 ± 0.20	3.40 ± 0.20
BGl	ASM	0.272 ± 0.087	0.229 ± 0.085	0.201 ± 0.083
	Contrast	396 ± 64	703 ± 81	915 ± 78
	Correlation	$(89.9 \pm 3.1) \times 10^{-5}$	$(70.3 \pm 5.9) \times 10^{-5}$	$(53.7 \pm 6.5) \times 10^{-5}$
	IDM	0.797 ± 0.034	0.699 ± 0.051	0.629 ± 0.063
	Entropy	2.85 ± 0.39	2.90 ± 0.40	2.93 ± 0.40
CP	ASM	0.088 ± 0.005	0.061 ± 0.003	0.048 ± 0.003
	Contrast	631 ± 96	1131 ± 160	1461 ± 188
	Correlation	$(69.9 \pm 7.6) \times 10^{-5}$	$(46.0 \pm 6.4) \times 10^{-5}$	$(30.3 \pm 5.3) \times 10^{-5}$
	IDM	0.595 ± 0.005	0.447 ± 0.007	0.360 ± 0.008
	Entropy	4.40 ± 0.08	4.44 ± 0.08	4.47 ± 0.08
CGr	ASM	0.091 ± 0.009	0.064 ± 0.008	0.051 ± 0.007
	Contrast	979 ± 132	1713 ± 236	2171 ± 301
	Correlation	$(387 \pm 3.8) \times 10^{-5}$	$(26.9 \pm 3.0) \times 10^{-5}$	$(19.5 \pm 2.5) \times 10^{-5}$
	IDM	0.613 ± 0.010	0.472 ± 0.014	0.389 ± 0.018
	Entropy	4.43 ± 0.05	4.47 ± 0.05	4.50 ± 0.05
DGr	ASM	0.179 ± 0.014	0.153 ± 0.019	0.133 ± 0.021
	Contrast	71 ± 29	137 ± 53	202 ± 76
	Correlation	$(88.8 \pm 4.6) \times 10^{-5}$	$(85.9 \pm 5.0) \times 10^{-5}$	$(82.9 \pm 5.6) \times 10^{-5}$
	IDM	0.911 ± 0.034	0.830 ± 0.060	0.758 ± 0.079
	Entropy	1.99 ± 0.10	2.19 ± 0.14	2.34 ± 0.16

of $|i - j|^2$ and then the squared value of $P(i, j)$ in the $|i - j|^2$ in the denominator in Eq. (4) even with $P(i, j)$ in the numerator.

The physiological meaning of the parameter values are discussed in the following.

In the present experiment, we used a 488-nm pump laser which is strongly absorbed by hematoxylin, which has positive charge and binds chiefly nucleic acids such as mRNAs and DNAs, hence it is used for examining detailed chromatin structure. Interphase nucleus consists of roughly two portions; a condensed, inactive, and densely stained portion called heterochromatin, and a loosened, actively transcribed, and faintly stained portion called euchromatin. Heterochromatin is further divided into two portions, the constitutive chromatin such as pericentromeric region and telomeric region, which always retains heterochromatin, and the facultative

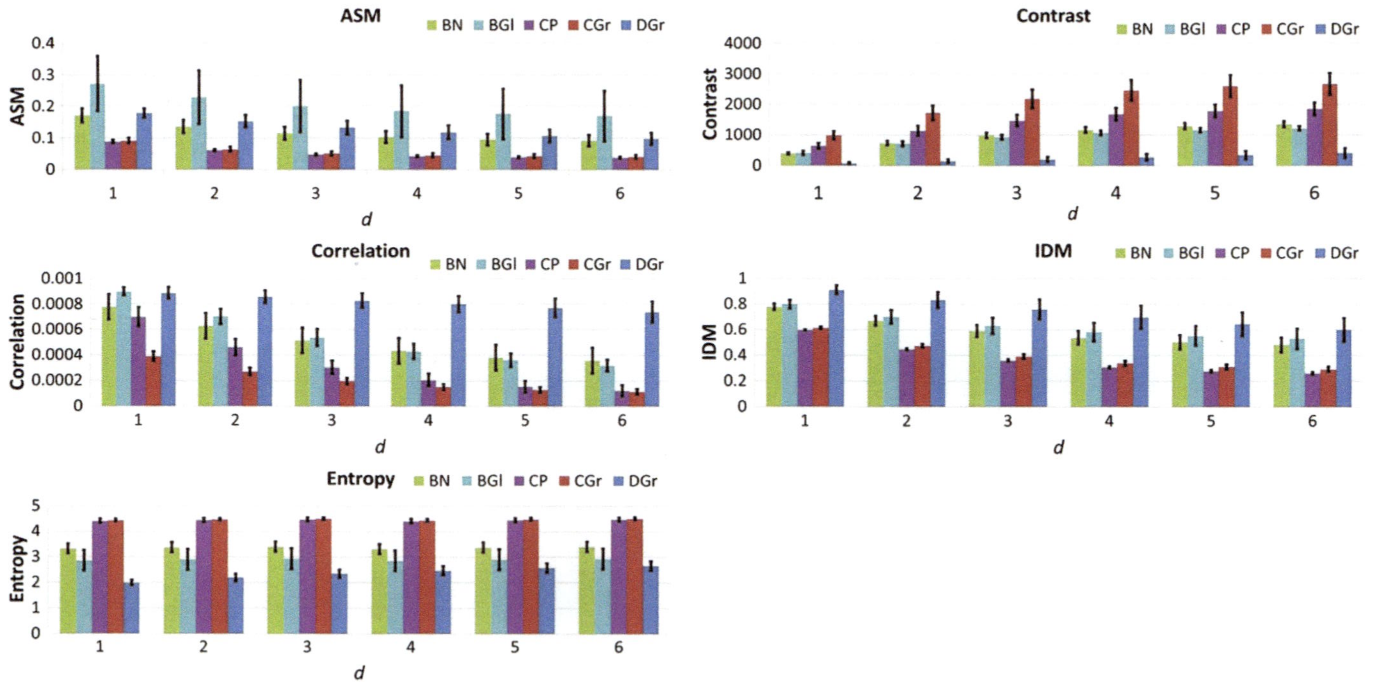

FIGURE 5.1.7 Graphical representation of variations in texture parameters. Among the above graphically represented parameters, correlation, ASM, and contrast showed clear differences among the different cell types. Entropy had the smallest variation among the different cell types. The distribution of the entropy values over the five different cell types had differences smaller than 0.8. The copyright of this figure is shared between the representative author (Takayoshi Kobayashi) and Nature Portfolio.

heterochromatin, which bears heterochromatin state upon the satisfaction of some conditions ceasing the transcription. The texture analysis of GLCM is thought to reflect the transcriptional activity of the cells. Especially in the stages such as initialization of cells occurring during iPS cell induction or early embryogenesis, where many genes are transcribed simultaneously and dynamically, the texture analysis of nuclei seems to provide interesting findings otherwise unavailable.

For the GLCM analysis, there are several important remarks. The spatial resolution is 269 nm, imaging pixel size is 240 nm, and distance parameter d = 240 nm. They are thus nearly equal indicating the analysis of the image using the distance parameter is most appropriate.

The values of ASM have larger dispersion than the others. This is because the parameter includes the squared values of $P(i, j)$ and hence, it is more sensitive to the variation of $P(i, j)$ than the other parameters.

From the detailed investigation of the data depicted in Table 5.1.1 (plus additional columns for d = 4–6), Table 5.1.2 and Fig. 5.1.7, we could find out that correlation, ASM, and contrast showed clear differences among the different cell types. It is also found that they have the following detailed features of the five texture parameters.

1. The texture parameters ASM and IDM have minimum values for CP and CGr among the five different cell types, and entropy has maximum values for the two cell types in all cases of distance. This means that the intracellular structure of these cell types in the cerebellum is more nonhomogeneous than the other cell types, especially for a short distance. This may mean that the distributions of heterochromatin and euchromatin are varying in a short distance indicating more complicated gene-expression in these cell types.
2. The cell types of BN and BGl have the lowest contrast and the highest correlation among the five cell types. This indicates that both neuron cells and glial cells have smaller variations over large distances than the other cell types. These cell types in BN have smaller contrast for d =6 (corresponding to 1440 nm) than the other types for d = 3 (corresponding to 720 nm). This means that the nonhomogeneity is extended even to distances larger than 1.4 μm in these cells.

   The cell type of CGr has the highest contrast and the lowest correlation. This is in contrast to the case of DGr, BN, and BGl. It reflects that CGr has the highest nonhomogeneity among the five cell types. It means that the physiological function of CGr can be maintained even under intracellular nonhomogeneous conditions in the cerebral area. In such a way, we can classify the five cell types into groups. Granule cell in cerebellum CGr has the high contrast and low correlation indicating that the concentration of chromatin is nonhomogeneous. DGr, BN, and BGl have high contrast and low correlation meaning they have an homogeneity of chromatin. CP has medium values.
3. The cell type of CGr has the highest contrast and the lowest correlation. This is in contrast to the case of DGr, BN, and BGl. It reflects that CGr has the highest nonhomogeneity among the five cell types. It means that the

physiological function of CGr can be maintained even under intracellular nonhomogeneous conditions in the cerebral area. In such a way, we can classify the five cell types into groups. Granule cell in cerebellum CGr has the high contrast and low correlation indicating that the concentration of chromatin is nonhomogeneous. DGr, BN, and BGl have high contrast and low correlation meaning they have an homogeneity of chromatin. CP has medium values.

4. The mean entropy was highest for CGr cells, with a value of 4.43 ± 0.05 for pixel length (pixel distance) d = 1. DGr had the lowest value of 1.99 ± 0.10 at pixel length d = 1. The value of entropy increased for all cell types as pixel length increased. The value for DGr and CGr at pixel length 3 was 2.34 ± 0.16 and 4.50 ± 0.05, respectively. The step increase in the entropy decreased with an increase in pixel distance. The value of entropy for cerebellar CP at a pixel length d = 1 was 4.40 ± 0.08. The difference was the lowest between CP and CGr, with a magnitude of 0.04 measured at pixel distance 1 and 0.03 measured at pixel distance 3. From the value analysis, entropy can be classified into two groups; those in group 1 (G1) were in the range of 2.0 to 3.5 for DGr, BN, and BGl in the BN. The others were in the range of 4.40 to 4.43 for CP and CGr. The entropy value of each cell type monotonically increased with d, but the two-grouping feature is maintained for all the d values. From the graphical representation of the results shown in Fig.5.1.7, the relative differences among cell types are increasing with the d value, while the entropy was found to have the smallest variation among the different cell types in comparison with the other four parameters. From Fig. 5.1.7, the grouping into the two is also made for the other parameters. Relatively larger ASM, Correlation, IDM for G1 and relatively smaller Contrast (and Entropy as mentioned above) than those of G2.

The partition function $P(i, j)$ in Eq. (5) defining the entropy is an essential function, which can be defined for a thermodynamically large system in thermal contact with the environment. Entropy is only meaningful for systems with a large number of states to be calculated in the partition function. As the logarithm of $P(i, j)$ was calculated for entropy, the variation in entropy among the different cell types was very small and was thus unfavorable for analysis.

Tables 5.1.1 and 5.1.2 show that all five parameters (ASM, contrast, correlation, IDM, and entropy) for a short distance (d = 1) and a long distance (d = 6) are different among the cell types. For example, in case of ASM, 0.171 0.022, 0.272 0.087, 0.179 0.014, 0.088 0.005, and 0.091 0.009 for BN, BGr, DGr, CP, and CGr, respectively, for d = 1 They are separated from each other even though they have some overlaps of the dispersions. Similar features are also found for the other distance values d = 1 2 – 6. Therefore, we can discuss the features of the different cell types separately using the five parameters.

Different tendencies of the orders in the sizes of the parameters that were observed in the samples of the different cell types are shown in Table 5.1.3.

TABLE 5.1.3
Order of Parameter Size among the Five Different Cell Types for $d = 1-6$. The five cell types are the nerve cells (BN) and the glial cells (BGI) from BN, granule cells from dental gyrus (DGr) and Purkinje (CP) and granule cells (CGr) from cerebellum

Property	ASM	Contrast	Correlation	IDM	Entropy
$d=1$	BGI > DGr > BN > CGr > CP	CGr > CP > BGI > BN > DGr	BGI > DGr > BN > CP > CGr	DGr > BGI > BN > CGr > CP	CGr ~ CP > BN > BGI > DGr
$d=2$	BGI > DGr > BN > CGr > CP	CGr > CP > BN > BGI > DGr	DGr > BGI > BN > CP > CGr	DGr > BGI > BN > CGr > CP	CGr - CP > BN > BGI > DGr
$d=3$	BGI > DGr > BN > CGr > CP	CGr > CP > BN > BGI > DGr	DGr > BGI > BN > CP > CGr	DGr > BGI > BN > CGr > CP	CGr ~ CP > BN > BGI > DGr
$d=4$	BGI > DGr > BN > CGr > CP	CGr > CP > BN > BGI > DGr	DGr > BN > BGI > CP > CGr	DGr > BGI > BN > CGr > CP	CGr ~ CP > BN > BGI > DGr
$d=5$	BGI > DGr > BN > CGr > CP	CGr > CP > BN > BGI > DGr	DGr > BN > BGI > CP > CGr	DGr > BGI > BN > CGr > CP	CGr - CP > BN > BGI > DGr
$d=6$	BGI > DGr > BN > CGr > CP	CGr > CP > BN > BGI > DGr	DGr > BN > BGI > CP > CGr	DGr > BGI > BN > CGr > CP	CGr - CP > BN > BGI > DGr

The above orderings clearly show that the ASM, IDM, and correlation were all correlated with homogeneity. Contrast and entropy were anticorrelated with homogeneity. Therefore, the first three, ASM, IDM, and correlation, were correlated among each other. Contrast and entropy were anticorrelated with these three.

The "relative" standard deviations around the average of the parameters were defined by (standard deviation of the GLCM parameters)/(average of the GLCM parameters for the five different Z). Here the values are taken for angles 0°, 45°, and 90° (concerning the meaning of the angle, see Refs. 14 and 15) for d =1–6. Table 5.1.4 shows the values for d =3, as that value of d was determined as the most suitable for analysis. The relative error obtained was substantially different among the five GLCM parameters for each cell type. From these results, we showed that the most suitable GLCM parameter is IDM for BN and BGl, CP and CGr in the cerebellum, and DGr. ASM was the most suitable parameter for neurons in the basal nucleus (BGl). The same tendencies were observed for different d values.

The difference in parameters and nuances in the microscopic scale analyzed by PT microscopy can be found in Table 5.1.4. The largest distribution of ASM is for BGl and that of contrast is for DGr. Correlation has the largest distribution in the case of CP.

TABLE 5.1.4
Relative Standard Deviations at d ¼ 3 Calculated Using the Values of Averages of Parameters and Their Standard Deviation in Table 1

Parameter	BN	BGl	CP	CGr	DGr
ASM	17.10	41.40	6.50	13.66	15.79
Contrast	8.99	8.60	12.90	13.90	37.62
Correlation	8.91	12.00	17.70	12.91	6.76
IDM	7.84	10.00	2.20	4.51	10.42
Entropy	5.73	13.60	1.70	1.16	6.838

Entropy has the smallest variation in all cell types except BGl. This is because of the logarithmic dependence of the value as discussed before and hence it is not suitable for detecting the change taking place in a cell, while IDM has the second smallest variation for all the cell types except for BGl. Therefore, IDM is the most suitable parameter for the cell state analysis for the detection of changes induced by malignation or other damage taking place in the cell except for BGl. In the case of BGl, contrast is considered to be useful as an indicator to detect cell lesions.

5 CONCLUSION

The qualifications of intracellular structure in the resolution-enhanced images of mouse brain obtained by PT imaging are analyzed to characterize the structural details of several cell types in a mouse brain. The GLCM method has been used to extract the parameters of texture features. The parameters are correlation, contrast, ASM, IDM, and entropy from the images of cells in a mouse brain. By analysis, the cell types can be classified into two groups. One group is G1 (neurons in the BN, BGl, and DGr cells) and the other is G2 (CGr cells and CP cells). In the case of ASM, correlation, and IDM, G1 > G2. For contrast and entropy, G1 < G2 for all cases of d = 1–6 with only one exception. From the values of the parameters of entropy, IDM, and ASM, it can be concluded that the neurons and glial cells have more uniformity in BN than in the cerebellar region. From detailed analysis, it can be concluded that IDM is the most suited parameter for the cell state analysis for the detection of malignation or other damage taking place in the cell except for BGl. In the case of BGl, contrast is considered to be useful as an indicator to detect cell lesions. It is proven that it is possible to discuss the features of the different cell types separately using the five parameters because of large enough differences in the parameters among the different cell types. The texture analysis of GLCM is thought to reflect the transcriptional activity of the cells by utilizing the signal intensity which is expected to provide the distribution and correlation of the heterochromatin or euchromatin [30]. This method may also be helpful for typing and prognostic prediction of cancer [31].

In the present PT imaging system using a Galvano mirror, an image can be taken at the frame rate of 1 frame/5 s. Then, such correlation information can be obtained in real time with this time resolution to study the dynamics of bioenergetics in a cell through the dynamic intracellular distribution change of chromatin. This information cannot or is very difficult to be obtained by the other method. Further simple improvements will enable provision of 3-D information by simply adding a moving stage along the perpendicular direction. By using a pump at 520 nm close to the peak of the Eosin, we can further study the correlation between nucleus and cytoplasm in the early stage of cell division. As a future prospect, it is to be stated that even though we successfully assigned parameters which are appropriate to characterize individual cells, functional or cellular architectural meanings of them are not yet clarified. By applying our method to compare those parameters of model mice or differentiated iPS cells from them with that of wild type specimens, their meanings are expected to be elucidated, hence providing general cell biological tools to infer mechanisms of the diseases. Those are the next steps to pursue.

In summary, the present subsection describes the qualifications of intracellular structure which were performed for the first time using the gray-level co-occurrence matrix (GLCM) method for images of cells obtained by resolution-enhanced photothermal imaging. The GLCM method has been used to extract five parameters of texture features for five different types of cells in mouse brain; pyramidal neurons and glial cells in the basal nucleus (BGl), dentate gyrus granule cells, cerebellar Purkinje cells, and cerebellar granule cells. The parameters are correlation, contrast, angular second moment (ASM), inverse difference moment (IDM), and entropy for the images of cells of interest in a mouse brain. The parameters vary depending on the pixel distance taken in the analysis method. Based on the obtained results, we identified that the most suitable GLCM parameter is IDM for pyramidal neurons and BGI, granule cells in the dentate gyrus, Purkinje cells and granule cells in the cerebellum. It was also found that the ASM is the most appropriate for neurons in the basal nucleus. The research activities described in this subsection are based on the collaborative activity of the following people: T. Kobayashi, D. Sundaram, K. Nakata, and H. Tsurui.[32]

NOTES

1. Purkinje cell: neurons in vertebrate animals located in the cerebellar cortex of the brain. Purkinje cell bodies have flask-like with many threadlike extensions (dendrites). The latter receive impulses from other neurons (granule cells). Each cell also has a single projection called an axon, which transmits impulses to the part of the brain that controls movement, the cerebellum. Purkinje cells are inhibitory secreting neurotransmitters that bind to receptors which inhibit or reduce the firing of other neurons.
2. Nieman-Pick disease: known as acid sphingomyelinase deficiency. It is in a group of rare genetic diseases. These are inherited metabolic disorders in which sphingomyelin accumulates in lysosomes in cells of many organs.
3. ataxia: a neurological sign consisting of lack of voluntary coordination of muscle movements that can include gait abnormality, speech changes, and abnormalities in eye movements indicating partial dysfunction of the nervous system coordinating movement, such as the cerebellum.

4. spinocerebellar: The spinocerebellar tract is a nerve tract originating in the spinal cord and terminating in the same side (ipsilateral) of the cerebellum.
5. degeneration type 6: neurotransmission passes through Rexed (a scientist's name) laminae layers I–VI (1-6) of the posterior grey column (dorsal horn) to form synapses with second order or secondary neurons in layer VII (7) just beneath the dorsal horn.
6. cerebellar granule (CGr) cells: Cerebellar granule cells form the thick granular layer of the cerebellar cortex and are among the smallest neurons in the brain. Cerebellar granule cells are also the most numerous neurons in the brain: in humans, estimates of their total number average around 50 billion corresponding to ~3/4 of the total number of the brain's neurons.
7. ataxia telangiectasia(A–T) is a rare, neurodegenerative, autosomal recessive disease causing severe disability. Ataxia refers to poor coordination and telangiectasia to small, dilated blood vessels, both of which are hallmarks of the disease. A–T affects many parts of the body: It (1) impairs certain areas of the brain including the cerebellum, causing difficulty with movement and coordination, (2) weakens the immune system, causing a predisposition to infection. It prevents repair of broken DNA, increasing the risk of cancer and (3) prevents repair of broken DNA, increasing the risk of cancer.
8. both neurons and glia cells of basal nuclei in corticobasal: Corticobasal degeneration is a rare neurodegenerative disease involving the cerebral cortex and the basal ganglia. It is characterized by marked disorders in movement and cognition. Diagnosis is difficult, as symptoms are often similar to those of other disorders, such as Parkinson's disease.
9. dentate gyrus granule (DGr): The dentate gyrus is part of the hippocampal trisynaptic circuit and is thought to contribute to the formation of new episodic memories, the spontaneous exploration of novel environments and other functions.
10. Alzheimer disease (AD): is a neurodegenerative disease that usually starts slowly and progressively worsens, and is the cause of 60–70% of cases of dementia (general name for a decline in cognitive abilities that impacts a person's ability to perform everyday activities). The most common early symptom is difficulty in remembering recent events. As the disease advances, symptoms can include problems with language, disorientation (including easily getting lost), mood swings, loss of motivation, self-neglect, and behavioral issues. As a person's condition declines, they often withdraw from family and society. Gradually, bodily functions are lost, ultimately leading to death. Although the speed of progression can vary, the typical life expectancy following diagnosis is three to nine years.
11. Galvano scanning: a technique to control a laser beam in any direction and irradiates the laser light at a pinpoint by using a reflecting mirror. The equipment with this scanning function is called a Galvano scanner or Galvano mirror, which is at the core of modern precision processing technology in conjunction with lasers, including laser processing and marking, and high-performance 3D printers. The name comes from the word "galvanometer", which is a machine for detecting and measuring electrical current. It was named in honor of Luigi Galvani, an Italian physician and physicist who discovered in the 18th century that electric sparks in frogs caused their muscles to spasm.
12. Gold nanoparticle: In recent years, nanotechnology has been the topic of extensive research and a great deal of interest among researchers. The manufacturing and utilization of nanoparticles (NPs) has expanded dramatically as a result of the rapid development of nanotechnology. Gold nanoparticles (AuNPs) are one of the most important nanoparticles, and they have been widely used for medical and non-medical applications as ideal material because of their unique distinct features: inert, biocompatible, and especially due to low toxicity.

13. piezo stage (piezoelectric stage) is composed of a stage and a piezo motor (piezoelectric actuator) plus mechanical components which allow the movement of the stage. The piezoelectric effect results from the linear electromechanical interaction between the mechanical and electrical states in crystalline materials with no inversion symmetry. [77] The piezoelectric effect is a reversible process: materials exhibiting the piezoelectric effect also exhibit the reverse piezoelectric effect, the internal generation of a mechanical strain resulting from an applied electric field.
14. cerebral cortex: known as the cerebral mantle, is the outer layer of neural tissue of the cerebrum of the brain in humans and other mammals. The cerebral cortex mostly consists of the six-layered neocortex, with just 10% consisting of the allocortex. It is separated into two cortices, by the longitudinal fissure that divides the cerebrum into the left and right cerebral hemispheres. The two hemispheres are joined beneath the cortex by the corpus callosum. The cerebral cortex is the largest site of neural integration in the central nervous system. It plays a key role in attention, perception, awareness, thought, memory, language, and consciousness. The cerebral cortex is part of the brain responsible for cognition.
15. hippocampal: The hippocampus is a major component of the brain of humans and other vertebrates. Humans and other mammals have two hippocampi, one in each side of the brain. The hippocampus is part of the limbic system, and plays important roles in the consolidation of information from short-term memory to long-term memory, and in spatial memory that enables navigation.
16. corpus callous: is a wide, thick nerve tract, consisting of a flat bundle of commissural fibers, beneath the cerebral cortex in the brain. It spans part of the longitudinal fissure, connecting the left and right cerebral hemispheres, enabling communication between them. It is the largest white matter structure in the human brain, about 10 in (250 mm) in length and consisting of 200–300 million axonal projections.
17. basal nuclei: (also basal ganglia) are associated with a variety of functions, including regulating voluntary motor movements, procedural learning, habit formation, conditional learning, eye movements, cognition, and emotion.
18. cerebellar portion: is a major feature of the hindbrain of all vertebrates. In humans, the cerebellum plays an important role in motor control. It may also be involved in some cognitive functions such as attention and language as well as emotional control such as regulating fear and pleasure responses, but its movement-related functions are the most solidly established.
19. pyramidal cell (pyramidal neuron): a type of multipolar neuron found in areas of the brain including the cerebral cortex, the hippocampus, and the amygdala. The pyramidal cell is the primary excitation unit of the mammalian prefrontal cortex and the corticospinal tract.
20. granule cell: accounts for the majority of neurons in the human brain. These granule cells receive excitatory input from mossy fibers originating from pontine (neural pathway) nuclei and project up through the Purkinje layer into the molecular layer where they branch out into parallel fibers that spread through Purkinje cell (a class of GABAergic inhibitory neurons located in the cerebellum) dendritic arbors. These parallel fibers form thousands of excitatory granule-cell–Purkinje-cell synapses onto the intermediate and distal dendrites of Purkinje cells using glutamate as a neurotransmitter.
21. gyrus dentate: a part of the hippocampal formation in the temporal lobe of the brain, which also includes the hippocampus and the subiculum. The dentate gyrus is a part of the hippocampal trisynaptic circuit and is thought to contribute to the formation of new episodic memories, the spontaneous exploration of novel environments etc.

REFERENCES

1. M. Born and E. Wolf, *Principles of Optics*, Pergamon Press, New York (1965).
2. S. W. Hell and J. Wichmann, "Breaking the diffraction resolution limit by stimulated emission: stimulated emission depletion microscopy," *Opt. Lett.* 19, 780–782 (1994).
3. T. A. Klar et al., "Fluorescence microscopy with diffraction resolution limit broken by stimulated emission," *Proc. Natl. Acad. Sci. U. S. A.* 97, 8206–8210 (2000).
4. L. Meyer et al., "Dual-color STED microscopy at 30-nm focal-plane resolution," *Small* 4, 1095–1100 (2008).
5. P. A. Pellett et al., "Two-color STED microscopy in living cells," *Biomed. Opt. Express* 2(8), 2364–2371 (2011).
6. L. Schermelleh et al., "Subdiffraction multicolor imaging of the nuclear periphery with 3D structured illumination microscopy," *Science* 320, 1332–1336 (2008).
7. M. Bates et al., "Multicolor super-resolution imaging with photoswitchable fluorescent probes," *Science* 317, 1749–1753 (2007).
8. K. Uchiyama et al., "Thermal lens microscope," *Jpn. J. Appl. Phys.* 39, 5316–5322 (2000).
9. D. Boyer et al., "Room-temperature detection of a single molecule's absorption by photothermal contrast," *Science* 330, 353–356 (2010).
10. S. Bercaiaud et al., "Photothermal heterodyne imaging of individual metallic nanoparticles: theory versus experiments," *Phys. Rev. B* 73, 045424 (2006).
11. M. Selmke, M. Braun, and F. Cichos, "Photothermal single-particle microscopy detection of a nanolens," *ACS Nano* 6, 2741–2749 (2012).
12. A. A. van Dijk et al., "Absorption and scattering microscopy of single metal nanoparticles," *Phys. Chem. Chem. Phys.* 8, 3486–3495 (2006).
13. K. Klein et al., "Marker-free phenotyping of tumor cells by fractal analysis of reflection interference contrast microscopy images," *Nano Lett.* 13, 5474–5479 (2013).
14. I. Pantic et al., "Nuclear entropy, angular second moment, variance and texture correlation of thymus cortical and medullar lymphocytes: grey level co-occurrence matrix analysis," *An. Acad. Bras. Ciênc.* 85(3), 1063–1072 (2013).
15. A. Gebejes and R. Huertas, "Texture characterization based on greylevel co-occurrence matrix," in *Proc. of the 2nd Int. Conf. of Informatics and Management Sciences ICTIC*, pp. 375–378 (2013).
16. J. Miyazaki et al., "Sub-diffraction resolution pump–probe microscopy with shot-noise limited sensitivity using laser diodes," *Opt. Exp.* 22(8), 9024–9032 (2014).
17. J. Miyazaki et al., "Simultaneous dual-wavelength imaging of nonfluorescent tissues with 3D subdiffraction photothermal microscopy," *Opt. Exp.* 23(3), 3647–3656 (2015).
18. J.-P. He et al., "Label-free imaging of melanoma with nonlinear photothermal microscopy," *Opt. Lett.* 40(7), 1141–1144 (2016).
19. T. Vujasinovic et al., "Gray-level co-occurrence matrix texture analysis of breast tumor images in prognosis of distant metastasis risk," *Microsc. Microanal.* 21(3), 646–654 (2015).
20. S. Doyle et al., "Automated grading of breast cancer histopathology using spectral clustering with textural and architectural image features," in *Biomedical Imaging: From Nano to Macro*, pp. 496–499 (2008).
21. I. Pantic, S. Pantic, and B. Jovanovic, "Grey level co-occurrence matrix (GLCM) texture analysis of germinal center light zone lymphocyte nuclei: physiology viewpoint with focus on apoptosis," *Microsc. Microanal.* 18, 470–475 (2012).
22. A. Gaiduk et al., "Room-temperature detection of a single molecule's absorption by photothermal contrast," *Science* 330, 353–356 (2010).

23. J. Miyazaki et al., "Optimal detection angle in sub-diffraction resolution photothermal microscopy: application for high sensitivity imaging of biological tissues," *Opt. Exp.* 22(16), 18833–18842 (2014).
24. J. Miyazaki et al., "Fast 3D visualization of endogenous brain signals with high-sensitivity laser scanning photothermal microscopy," *Biomed. Opt. Express* 7(5), 1702–1710 (2016).
25. W. Wang, Z. Wu, and H. Zeng, "Image distortion and its correction in linear galvanometric mirrors-based laser-scanning microscopy," *J. Biomed. Opt.* 20(5), 056001 (2015).
26. M. Benco and R. Hudec, "Novel method for color textures features extraction based on GLCM," *Radio Eng.* 16(4), 64–67 (2007).
27. O. B. Sassi et al., "Improved spatial gray level dependence matrices for texture analysis," *Int. J. Comput. Sci. Inf. Technol.* 4(6), 209–219 (2012).
28. R. C. Gonzalez and R. E. Woods, *Digital Image Processing*, Addison-Wesley Publishing, Boston (1993).
29. R. M. Haralick, K. Sanmugam, and I. Dinstein, "Textural features for image classification," *IEEE Trans. Syst. Man Cybern.* SMC-3, 610–621 (1973).
30. I. Pantic et al., "Nuclear entropy, angular second moment, variance and texture correlation of thymus cortical and medullar lymphocytes: grey level co-occurrence matrix analysis," *Anais Acad. Bras. Ciênc.* 85(3), 1063–1072 (2013).
31. K. Yorgesan et al., "Entropy-based texture analysis of chromatin structure in advanced prostate cancer," *Cytometry* 24(3), 268–276 (1996).
32. T. Kobayashi, D. Sundaram, K. Nakata, and H. Tsurui "Gray-level co-occurrence matrix analysis of several cell types in mouse brain using resolution-enhanced photothermal microscopy," *J. Biomed. Optics* 22(3), 036011(14pages) (2013).

5.2 Label-Free Imaging of Melanoma with Nonlinear Photothermal Microscopy

1 INTRODUCTION OF NONLINEAR PHOTOTHERMAL (NLPT) MICROSCOPY

Melanomas are among the most commonly occurring cancers and are a clinical challenge to diagnose [1]. Presently, the best method for the clinical evaluation of melanoma is visual inspection by dermoscopy; however, this technique is still far from a reliable diagnosis [2]. The optical diagnosis of melanoma is accessible because the two dominant types of melanin (eumelanin and pheomelanin) have a specific and intrinsic molecular contrast when imaged [3, 4]. Several forms of nonfluorescent optical contrast mechanisms can be used to image melanoma without any staining, such as pump–probe [1,3], photoacoustic [5–8], and photothermal [7, 8] microscopy. Photothermal microscopy has been used widely in material and biological research [7–11] because of the high sensitivity and lack of staining requirements. Compared with fluorescence microscopy, photothermal imaging does not suffer from photobleaching and blinking, which makes it attractive for robust single molecule detection [12–14]. Besides the commonly used linear photothermal (LPT) mechanism, nonlinear photothermal microscopy (NLPTM) has been used to image silica windows and gold pads with enhanced imaging contrast [15], in which the pump heating beam is chopped at f and the temperature variation of the sample is detected at $2f$ with a linear temperature sensor. Nonlinear phenomena have also been observed in photothermal optical-deflection imaging experiments on samples of both high-purity aluminum and aluminum alloys [16]. The NLPT response of tungsten [17], thin solid films and coatings [18], carbon/epoxy composite materials [19], and cracks in solid [20] have also been investigated. The resolution of nonlinear thermal wave microscopes has been discussed, and it has been suggested that nonlinear microscopy is superior to linear microscopy [21]. However, to the best of our knowledge, all of the research on the effect of NLPT has been performed on bulk materials, thin

DOI: 10.1201/9781003476030-19

films, and layered-structure samples; there are no experimental results about the NLPT imaging of biological tissues.

2 MECHANISM OF NLPT EFFECTS

There are two possible explanations for the mechanism of the NLPT process. The first attributes the effects to thermal-wave second harmonic generation (SHG) related to the temperature-dependent heat capacity C_T and thermal conductivity κ_T of the materials [22, 23$_{].}$

In the second explanation, a diffraction theory of continuous-wave photothermal deflection spectroscopy with fundamental and harmonic responses is presented, in which the displacement of the probe beam centroid leads to a set of analytical solutions to the fundamental and SHG, and the harmonics are caused by the diffraction of the probe beams in the mirage range [24]. In this chapter, we give a brief explanation of NLPT using the first mechanism. The conventional form of the heat conduction equation in a medium with temperature dependent heat capacity C_T and thermal conductivity κ_T is as follows [25]

$$C\frac{\partial T}{\partial t}-\nabla\cdot\left(k\nabla T\right)=Q\left(\mathbf{r},t\right), \tag{5.3.1}$$

where Q is the external heat source input per unit volume. *C*(*T*) and *k*(*T*) can be expanded with the first Taylor expansion as follows:

$$C\left(T\right)\simeq C\left(T_0\right)+\left[\left.\frac{\partial C}{\partial T}\right|_{T_0}\right]T_1\equiv C_0\left(1+\delta_1 T_1\right), \tag{5.3.2}$$

$$k\left(T\right)\simeq k\left(T_0\right)+\left[\left.\frac{\partial k}{\partial T}\right|_{T_0}\right]T_1\equiv k_0\left(1+\delta_2 T_1\right), \tag{5.3.3}$$

where T_0 is the initial temperature of the material, and T_1 is the temperature change initiated by an external heating pump beam with an intensity modulated to a frequency of *f*. Then, Eq. (1) transforms to

$$\left(\nabla^2-\frac{1}{D_0}\frac{\partial}{\partial t}\right)T_1=-\frac{1}{k_0}Q\left(\mathbf{r},t\right)-\frac{1}{2}\left(\delta_2\nabla^2-\frac{\delta_1}{D_0}\frac{\partial}{\partial t}\right)T_1^2, \tag{5.3.4}$$

where *D*0 = *k*0/ *C*0 Cis the thermal diffusivity. Both the *C* = *C*(*T*) and *k* = *k* (*T*) contribute to dependences.

The quadratic nonlinearity in Eq. (5.3.4) induces the generation of the second harmonic of the thermal wave. With a one-dimensional simplification of the sample, the fundamental and second harmonic of the thermal wave is as follows [22]:

$$T_f\left(z,t\right)=A_f\left(z\right)\cos\left(ft+\phi_f\right), \tag{5.3.5}$$

$$T_{2f}(z,t) = A_{2f}(z)\cos(2ft + \phi_{2f}), \tag{5.3.6}$$

where $Af(z)$ and ϕ_f are the amplitude and phase of the thermal wave with a fundamental frequency, and ϕ2f are those for the second harmonic of the thermal wave. The amplitudes of the fundamental and second harmonic of the thermal wave have relationships with the pump intensity $A_f(z) \propto I_{\text{pump}}$ and $A_{2f}(z) \propto I^2_{\text{pump}}$,[1] respectively. Assuming the response to be instantaneous, the refractive index caused by the temperature profile is $\Delta n = T_1 \partial n / \partial T_1$ and acts as a modulated thermal lens. If T_f+T_{2f} is used to replace $T1$, then we can obtain

$$\Delta n = T_f \frac{\partial n}{\partial T_1} + T_{2f} \frac{\partial n}{\partial T_1} = \Delta n_L + \Delta n_{\text{NL}}, \tag{5.3.7}$$

where Δn_{L} and Δn_{NL} are refractive indices induced by the fundamental and second harmonic thermal wave, respectively. The probe scattered by the thermal lens is proportional to Δn and the probe intensity I_{probe} [26]. The NLPT signal can then be written as

$$S \propto I^2_{\text{pump}} \cdot I_{\text{probe}}. \tag{5.3.8}$$

The relationship is also proved in [16] experimentally. Compared with the LPT signal, which can be written as $S \propto I_{\text{pump}} \cdot I_{\text{probe}}$ [26], the cubic dependence of the NLPT signal on the laser intensity makes the spatial resolution of NLPT microscopy higher than that of the LPT one. If we assume that the pump and probe beams have the same wavelength and intensity distribution, this spatial resolution enhancement can be estimated to be ~18% using scalar diffraction theory. Compared with conventional optical microscopy, the resolution enhancement of NLPT microscopy is 42%.

3 SCHEMATIC OF NLPTM

The schematic of the experimental setup we developed in our group is shown in Figure 5.2.1. Both the pump and probe pulses, with a 10 nm spectral width, are spectrally filtered from a commercial supercontinuum (SC) fiber laser (WL-SC400-4, Fianium, Southampton, UK) using band-pass filters. The pulse duration of a 10 nm spectral width component of the SC is measured as ~100 ps and the repetition frequency is tunable in the range of 0.1–40 MHz. The central wavelengths of the pump and probe beams are 488 and 632 nm, respectively. After collinear combination by a short-pass dichroic mirror (DM) with a cutoff wavelength of 560 nm, the two beams are focused into the specimen by a microscope objective with an NA of 0.9 and amplification of 60×. The forward propagating beam is collected and collimated by a condenser lens (Olympus) with an NA tunable from 0 to 1.4. Then, the probe beam is spectrally filtered out by a band-pass filter with a central wavelength of 632 nm. To eliminate the effects of the laser noise of the probe beam, an auto-balanced detector (BD) (Nirvana 2007, Newport) is used to accumulate the signals. The

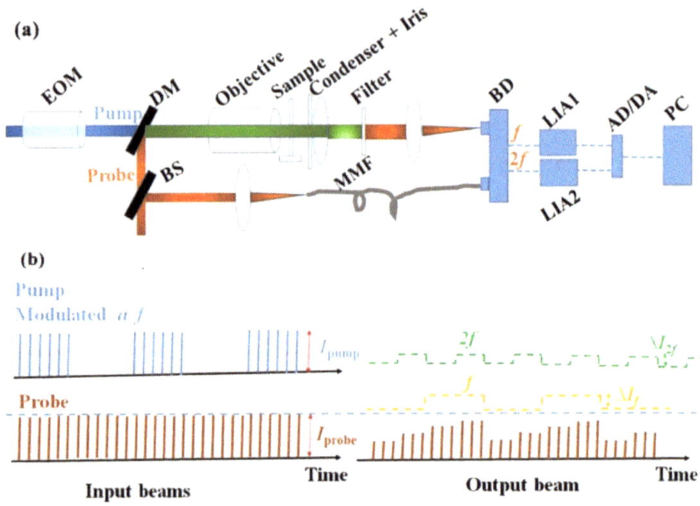

FIGURE 5.2.1 (a) Schematic diagram of the experimental setup. EOM, electro-optic modulator; DM, short-wavelength transmitting dichroic mirror with a switching wavelength of 560 nm; BS, beamsplitter; MMF, multi-mode fiber; Filter, band-pass filter with a central wavelength of 632 nm and a bandwidth of 10 nm; BD, auto-balanced detector with a bandwidth of 125 kHz; LIA1,2, lock-in amplifiers. (b) Temporal intensity modulation behavior of the pump and probe pulse trains before and after interacting with the sample. The intensity of the pump pulse is modulated at f with the EOM and the output probe pulse is modulated at both *f* and *f* with a modulation depth of ΔI_f and ΔI_2f, respectively, because of the generation of the second-harmonic of the thermal wave. The LPT and NLPT signals are demodulated from the output probe beam with two LIAs referring to f and 2f, respectively. The copyright of this figure is partially held between the representative author (Takayoshi Kobayashi) and Optica.

intensity of the pump heating beam is modulated at *f* with an electro-optic modulator (EOM) (LM0202P, Qioptiq Photonics, Germany), creating an LPT signal at f and an NLPT signal at 2f, as depicted in Fig. 5.2.1(b). Two lock-in amplifiers (LIA) (SR844, Stanford Research System, US; Model 7265, Signal Recovery, US) are applied to extract the LPT and NLPT signals from the probe simultaneously, with reference frequencies of *f* and 2*f*, respectively. The sample is mounted on a set of piezo stages (P-622.2CL and P-622.ZCL, Physik Instrumente (PI), Germany), with a resolution of better than 1 nm in all three directions.

The sample of mouse skin melanoma is prepared as follows. B16 melanoma cells RCB-1283 (Riken Bio-Resource Center) are cultured in RPMI-1640[1] supplemented with 10% FBS, 100 units/ml of penicillin[2], and 100μg/ml of streptomycin[3] (Meiji Seika Kaisha Ltd.) at 37°C in a humidified atmosphere of 95% air and 5% CO2. A 50 μl suspension containing 0.5 million melanoma cells for each head is subcutaneously inoculated into female nude mice (six weeks old) as five aliquots on the dorsal sides from the base of the tail to the neck. Tissues containing the inoculated cells were fixed with 4% paraformaldehyde for three days, embedded in melted paraffin, sliced at 15–20 μm, extended on glass slides, and enclosed with a coverslip.

First, we optimize the whole experimental setup and check the linearity of the detection system. The repetition frequency of the pump and probe pulses is set to its maximum value (40 MHz) for a high signal-to-noise ratio (SNR). The LPT and NLPT signals decrease with the increase of the modulation frequency of EOM, as is shown in Figure 5.2.2(a). The optical noises of the pump and probe pulses increase sharply with the frequency decreases to <20 kHz. The modulation frequency of the EOM is set to 30 kHz for the consideration of both the signal and the noise level. The auto-balanced detector is critical for imaging, as it can help to reduce the laser noise of probe pulse by 20–30 dB and obtain images with a higher SNR. The NA of the condenser lens is set within the range of 0.4–0.6 for the maximum LPT and NLPT signals. The time delay between the pump and probe pulses is also adjusted carefully for the highest LPT and NLPT signals (within 1 ns). After optimizing the experimental setup, we check the linearity of the detection system and ensure that the NLPT signal is not due to the saturation and nonlinearity of the detector. As

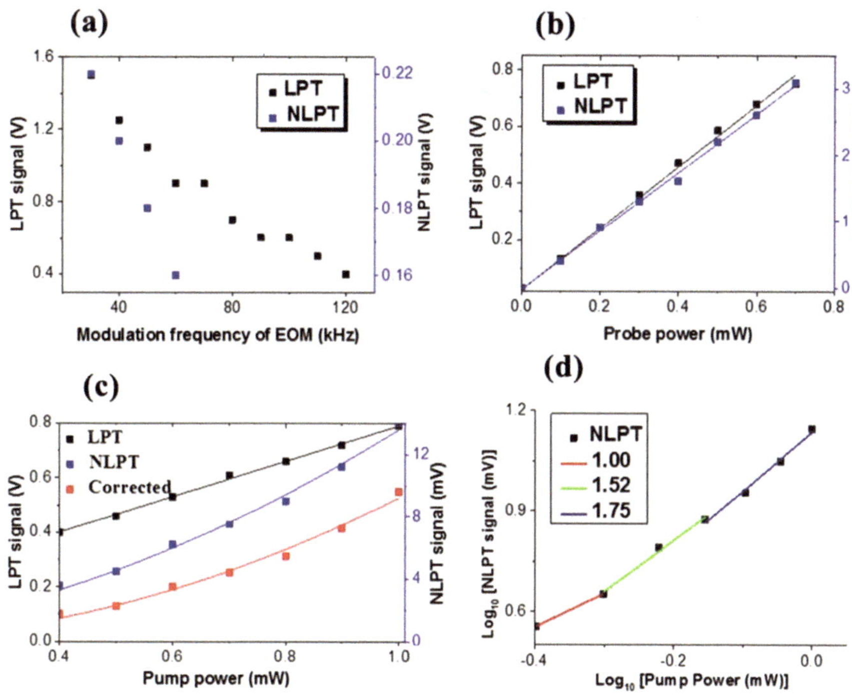

FIGURE 5.2.2 (a) Dependence of the LPT and NLPT signals on the modulation frequency of the EOM. Responses of the LPT and NLPT signals to (b) the probe power and (c) the pump power. The pump power is 1 mW in (b) and the probe power is set to 0.5 mW in (c). In (d), the x(horizontal) and y(vertical) axes are the logarithm to the base 10 of pump power and NLPT signal, respectively. The slopes of the red, green, and blue parts of the fitting curve are 1.00, 1.52, and 1.75, respectively. The copyright of this figure is partially held between the representative author (Takayoshi Kobayashi) and Nature Portfolio.

shown in Fig. 5.2.2(b), both the LPT and NLPT signals increase linearly with the probe power and no obvious saturation is observed. The LPT signal also shows a linear response to the pump power, shown as a black line in Fig. 5.2.2(c). The best fit to the data of NLPT signal is as an equation of the form $y\ ax^2\ bx$, shown as a blue line in Fig. 5.2.2(c). The term bx is because of imperfection of the square wave modulator (EOM in the experiment), which may have generated a small spurious NLPT signal. The corrected curve (without the term bx) for the response of the NLPT signal to the pump power is shown as a red line in Fig. 5.2.2(c). To show the fitting more clearly, Fig. 5.2.2(d) is shown with vertical (x) and longitudinal (y) axes in the logarithmic scale to the base 10 of pump power and NLPT signal, respectively. The slope of the fitting curve increases from 1 to 1.75 with the pump power increased from 0.4 to 1.0 mW, which means that the NLPT signal is mainly the result of the term bx with low pump power, and the effect of the term ax^2 becomes significant at high pump rate.

The LPT and NLPT images of mouse melanoma are obtained simultaneously in the experiment. The pump and probe powers are set to 0.62 and 0.45 mW, respectively, which are in the linear range of the detection system shown in Fig. 5.2.2. The pump power is also far below the damage threshold of the sample. To obtain images with a high SNR, the time constant of the LIA and the dwell time of the imaging are set to 10 and 20 ms, respectively. The LPT and NLPT imaging of the mouse melanoma is then performed in a range of $10\mu m \times 10\mu m$. The melanin in the melanoma is clearly imaged with SNR > 10, as shown in Figures 5.2.3(a) and 5.2.3(b). The NLPT signal is about 100 times smaller than the LPT signal, although the ratio is variable with different pump rate and different sample position. As shown in Fig. 5.3.3(c), the cross section at the same position [the red line in Figs. 5.2.3(a) and (b)] indicates that the resolution of LPT and NLPT can achieve 212 and 174 nm, respectively, with a resolution enhancement of 18% for the NLPT microscopy compared with the linear microscopy. The contrast is also higher for the NLPT imaging, as the obscure structures in Fig. 5.2.3(a) no longer exist in Fig. 5.2.3(b). The contrast enhancement can also be seen in Fig. 5.2.3(c), in which the maximum to minimum value of the NLPT curve is larger than that of the LPT curve. The maximum SNR of Fig. 5.2.3(b) is >150, indicating that a high-quality image is still possible, even if the dwell time is decreased 10 times [27].

The high harmonic demodulation technology is also applied in saturated excitation (SAX) microscopy to obtain sub-diffraction limited resolution [28, 29], although the saturated fluorescence in SAX microscopy may induce photobleaching of the fluorescent dye, which will not happen in the label-free NLPT microscopy. The sensitivity of photothermal microscopy is also higher compared with fluorescence microscopy, as discussed in the introduction.

SUMMARY

In this subsection we demonstrated the label-free imaging of mouse melanoma with both LPT and NLPT microscopy simultaneously [30]. The spatial resolution and contrast of NLPT microscopy are higher than those of LPT microscopy. The pump and probe pulses are filtered from a supercontinuum fiber laser, which makes the

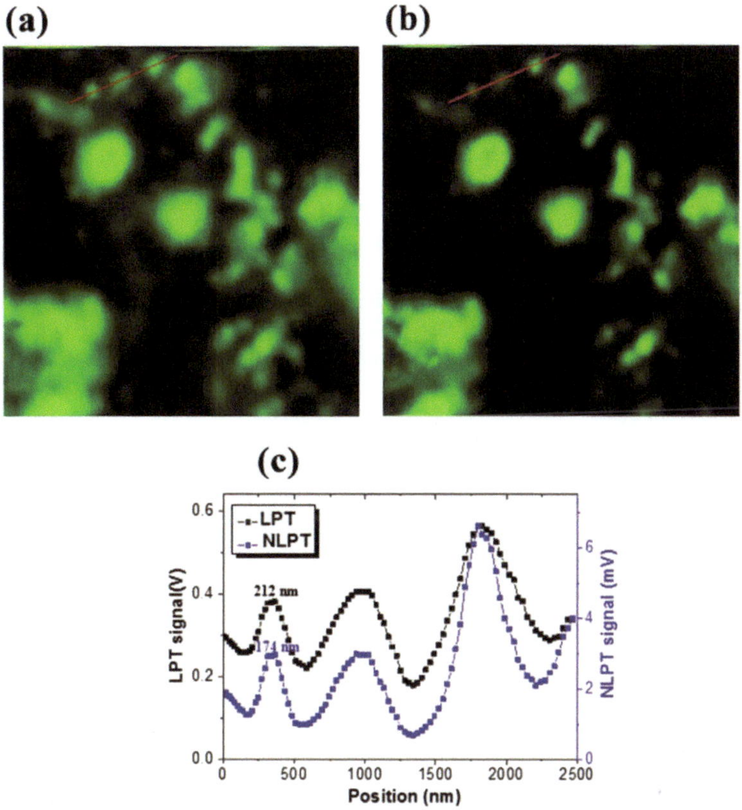

FIGURE 5.2.3 (a) LPT and (b) NLPT imaging of mouse melanoma with the range of 10μm × 10μm and whole pixels of 300 × 300. The powers of the pump and probe pulses are 0.62 and 0.45 mW, respectively. The line profiles in (c) indicated by the red line in (a) and (b) show that the spatial resolutions for the LPT and NLPT imaging are 212 and 174 nm, respectively. The NLPT signal is about 100 times smaller than the LPT signal. The copyright of this figure is partially held between the representative author (Takayoshi Kobayashi) and Nature Portfolio.

experimental setup more compact and cheaper than the Ti:sapphire laser and OPA system. The technique opens further possibilities for the clinical evaluation of stainless melanoma and other bio-tissues with a high resolution and contrast. Nonlinear photothermal microscopy, in which the intensity of the pump heating beam is modulated at f and the photothermal signal is extracted from the probe beam with a lock-in amplifier referred to $2f$, is applied to the imaging of mouse melanoma without any staining. The pump and probe pulses, with central wavelengths of 488 and 632 nm, and a pulse duration of ~100 ps, are filtered from a compact commercial supercontinuum fiber laser source. An auto-balanced detector is applied to accumulate the signal and remove the laser noise of the probe. The spatial resolution of the nonlinear photothermal imaging is enhanced by ~18% in both theoretical calculations and experiments, compared with a linear photothermal mechanism, and the resolution enhancement is theoretically ~42% compared with conventional optical microscopy. This imaging technique shows possibilities for the clinical evaluation of melanoma with a high contrast and spatial resolution [30]. The research output described in this subsection is based on the collaborative work of the following people: J. He, J. Miyazaki, N. Wang, H. Tsurui, and T. Kobayashi [30].

NOTES

1. RPMI-1640: a cell culture medium commonly used to culture mammalian cells. It was originally formulated to support lymphoblastoid cells in suspension cultures, but can also support a wide variety of adherent cells. RPMI 1640 allows the cultivation of many cell types, especially human lymphocytes, HeLa cells, bone marrow cells, hybridomas, and carcinomas.
2. penicillin: Penicillin was among the first medications to be effective against many bacterial infections caused by staphylococci and streptococci. They are still widely used today for different bacterial infections, though many types of bacteria have developed resistance following extensive use. The term "penicillin" was coined by Alexander Fleming in 1929 when he discovered the antibacterial property of Penicillium rubens.
3. streptomycin: Streptomycin is an antibiotic medication used to treat a number of bacterial infections, including tuberculosis, mycobacterium avium complex, endocarditis, brucellosis, plague, tularemia, and rat bite fever. It works by blocking the ability of 30S ribosomal subunits to make proteins, which results in bacterial death.

REFERENCES

1. T. E. Matthews, I. R. Piletic, M. A. Selim, M. J. Simpson, and W. S. Warren, "Pump-probe imaging differentiates melanoma from melanocytic nevi", *Sci. Transl. Med.* 71, 71ra15 (2011).
2. H. Kittler, H. Pehamberger, K. Wolff, and M. Binder, "Diagnostic accuracy of dermoscopy" *Lancet Oncol.* 62, 751 (2010).
3. R. Marchesini, A. Bono, and M. Carrara,"In vivo characterization of melanin in melanocytic lesions: spectroscopic study on 1671 pigmented skin lesions", *J. Biomed. Opt.* 14, 014027 (2009).
4. G. Zonios, A. Dimou, M. Carrara, and R. Marchesini, "In vivo assessment of optical properties of melanocytic skin lesions and differentiation of melanoma from non-malignant lesions by high-definition optical coherence tomography" *Photochem. Photobiol.* 86, 236 (2010).
5. E. I. Galanzha, E. V. Shashkov, P. M. Spring, J. Y. Suen, and V. P. Zharov, "In vivo, noninvasive, label-free detection and eradication of circulating metastatic melanoma cells using two-color photoacoustic flow cytometry with a diode laser",*Cancer Res.* 69, 7926 (2009).
6. M. Mehrmohammadi, S. J. Yoon, D. Yeager, and S. Y. Emelianov, "Image improvement in linear-array photoacoustic imaging using high resolution coherence factor weighting technique", *Curr. Mol. Imaging* 2, 89 (2013).
7. V. P. Zharov, "Ultrasharp nonlinear photothermal and photoacoustic resonances and holes beyond the spectral limit", *Nat. Photonics* 5, 110 (2011).

8. D. A. Nedosekin, M. A. Juratli, M. Sarimollaoglu, C. L. Moore, N. J. Rusch, M. S. Smeltzer, V. P. Zharov, and E. I. Galanzha, "Photoacoustic and photothermal detection of circulating tumor cells, bacteria and nanoparticles in cerebrospinal fluid in vivo and ex vivo", *J. Biophotonics* 6, 523 (2013).
9. D. Lasne, G. A. Blab, F. D. Giorgi, F. Ichas, B. Lounis, and L. Cognet, "Label-free optical imaging of mitochondria in live cells ", *Opt. Express* 15, 14184 (2007).
10. L. J. Inglehart, A. Broniaotowski, D. Fournier, A. C. Boccara and F. Lepoutre, "Photothermal imaging of copper-decorated grain boundary in silicon", *Appl. Phys. Lett.* 56, 1749 (1990).
11. A. H. Sahraoui, C. Kolinsky, S. Delenclos, A. Daoudi, and J. M. Buisine, "Photothermal imaging as a tool to study phase transitions in binary mixtures of liquid crystals *J. Appl. Phys.* 82, 6209 (1997).
12. M. Tokeshi, M. Uchida, A. Hibara, T. Sawada, and T. Kitamori, "Determination of subyoctomole amounts of nonfluorescent molecules using a thermal lens microscope: subsingle-molecule determination", *Anal. Chem.* 73, 2112 (2001).
13. S. Berciaud, L. Cognet, G. A. Blab, and B. Lounis, "Photothermal heterodyne imaging of individual nonfluorescent nanoclusters and nanocrystals". *Phys. Rev. Lett.* 93, 257402 (2004).
14. D. Lasne, G. A. Blab, S. Berciaud, M. Heine, L. Groc, D. Choquet, L. Cognet, and B. Lounis, "Single Nanoparticle Photothermal Tracking (SNaPT) of 5-nm Gold Beads in Live Cells", *Biophys. J.* 91, 4598 (2006).
15. Y. N. Rajakarunanayake and H. K. Wickramasinghe, "Nonlinear photothermal imaging", *Appl. Phys. Lett.* 48, 218 (1986).
16. G. C. Wetsel, Jr. and J. B. Spicer, "Nonlinear effects in photothermal-optical-beam-deflection imaging" *Can. J. Phys.* 64, 1269 (1986).
17. A. Salnick, J. Opsal, A. Rosencwaig, and A. Mandelis, "Nonlinear fundamental photothermal response: experimental results for tungsten", *Solid State Commun.* 114, 133 (2000).
18. V. Gusev, A. Mandelis, and R. Bleiss, "Theory of combined acousto-photo-thermal spectral decomposition in condensed phases: parametric generation of thermal waves by a non-stationary ("breathing") sub-surface defect", *Mat. Sci. Eng. B* 26, 111 (1994).
19. G. Kalogiannakis, D. V. Hemelrijck, and C. Glorieux, "Comparison of intrinsic and acquired nonlinear effects in photothermal detection of delaminations in composites", *NDT &E Int.* 39, 236 (2006).
20. V. Gusev, A. Mandelis, and R. Bleiss, *Int. J. Thermophys.* 14, 321 (1993), *Appl. Phys. A* 57, 229 (1993).
21. O. Dka, A. Mikls, and A. Lrincz, "Resolution of nonlinear thermal wave microscopes ", *Appl. Phys. A* 48, 415 (1989).
22. V. Gusev, A. Mandelis, and R. Bleiss, *Int. J. Thermophys.*"Theory of second harmonic thermal-wave generation: One-dimensional geometry" 14, 321 (1993).
23. A. Mandelis, A. Salnick, J. Opsal, and A. Rosencwaig, "Nonlinear fundamental photothermal response in three-dimensional geometry: Theoretical model", *J. Appl. Phys.* 85, 1811 (1999).
24. J. Zhao, J. Shen, and C. Hu, "Continuous-wave photothermal deflection spectroscopy with fundamental and harmonic responses", *Opt. Lett.* 27, 1755 (2002).
25. H. S. Carslaw and J. C. Jaeger, *Conduction of Heat in Solids* 2nd ed. (Oxford University, 1959).
26. A. Gaiduk, P. V. Ruijgrok, M. Yorulmaz, and M. Orrit, "Detection limits in photothermal microscopy", *Chem. Sci.* 1, 343 (2010).

27. J. F. de Boer, B. Cense, B. H. Park, M. C. Pierce, G. J. Tearney, and B. E. Bouma, "Improved signal-to-noise ratio in spectral-domain compared with time-domain optical coherence tomography", *Opt. Lett.* 28, 2067 (2003).
28. K. Fujita, M. Kobayashi, S. Kawano, M. Yamanaka, and S. Kawata, " High-Resolution Confocal Microscopy by Saturated Excitation of Fluorescence", *Phys. Rev. Lett.* 99, 228105 (2007).
29. S. Kawano, N. I. Smith, M. Yamanaka, S. Kawata, and K. Fujita, "Determination of the Expanded Optical Transfer Function in Saturated Excitation Imaging and High Harmonic Demodulation", *Appl. Phys. Express* 4, 042401 (2011).
30. J. He, J. Miyazaki, N. Wang, H. Tsurui, and T. Kobayashi, "Label-free imaging of melanoma with nonlinear photothermal microscopy", *Opt. Lett.* 40, 1141 (2015).

5.3 Noninvasive, Label-free, Three-dimensional Imaging of Melanoma with Confocal Photothermal Microscopy

Differentiate Malignant Melanoma from Benign Tumor Tissue

1 INTRODUCTION

Skin cancer is one of the most common cancers. Melanoma accounts for less than 2% of skin cancer cases but causes a large majority of skin cancer deaths [1]. The rates of melanoma also have been growing for at least 30 years [2]. The most dangerous characteristic of melanoma is the capability of deep invasion, as it can spread over the body through lymphatic and blood vessels. For this reason, early detection and therapy of melanoma is of crucial importance in saving lives. Presently, the best method for clinical detection of melanoma is dermoscopy [3]. Based on the results of the Consensus Net Meeting on Dermoscopy, the best sensitivity of dermoscopy is 83.7%, and the best specificity is 83.4% at the time of this investigation [4]. In the Report [4] the following are stated:

> There is a need for better standardization of the dermoscopic terminology in assessing pigmented skin lesions. Objective: The virtual Consensus Net Meeting on Dermoscopy was organized to investigate reproducibility and validity of the various features and diagnostic algorithms. Methods: Dermoscopic images of 108 lesions were evaluated via the Internet by 40 experienced dermoscopists using a 2-step diagnostic procedure. The first-step algorithm distinguished melanocytic versus nonmelanocytic lesions. The second step in the diagnostic procedure used 4 algorithms (pattern analysis, ABCD

DOI: 10.1201/9781003476030-20

> rule, Menzies method, and 7-point checklist) to distinguish melanoma versus benign melanocytic lesions. κ values, log odds ratios, sensitivity, specificity, and positive likelihood ratios were estimated for all diagnostic algorithms and dermoscopic features. Results: Interobserver agreement was fair to good for all diagnostic methods, but it was poor for the majority of dermoscopic criteria. Intraobserver agreement was good to excellent for all algorithms and features considered. Pattern analysis allowed the best diagnostic performance (positive likelihood ratio: 5.1), whereas alternative algorithms revealed comparable sensitivity but less specificity. Interobserver agreement on management decisions made by dermoscopy was fairly good (mean κ value: 0.53). Conclusion: The virtual Consensus Net Meeting on Dermoscopy represents a valid tool for better standardization of the dermoscopic terminology and, moreover, opens up a new territory for diagnosing and managing pigmented skin lesions.
>
> **(J. Am. Acad. Dermatol. (2003) 48: 679–93.)**

The reliability of this technique needs further improvement, and melanoma diagnosis is still clinically challenging. Since melanin carries information about the metabolism and location of melanocytes and melanogenesis, the distribution of melanin could act as a marker for melanoma [5, 6]. Two dominant types of melanin, eumelanin and pheomelanin, have high absorption cross section in the visible light range without efficient fluorescence emission [7], which makes it possible to image melanoma with photothermal (PT) microscopy (PTM).

PTM, which relies on the detection of local heating induced by thermalization after a sample's optical absorption due to vibronic relaxation in molecules or electron-phonon coupling, has shown potential in biological imaging and clinical applications. The key advantages of PTM are high sensitivity and no requirement of staining [8–10]. It can image nanometer sized absorbers among scatters with high resolution, high signal-to-noise ratio (SNR) and in real time [11–13]. However, the PT signal in normal PTM (NPTM) has two extrema in axial direction [14], which will introduce distortions and poor axial resolution to three-dimensional (3D) PT imaging. Confocal PTM (CPTM), which has a detection scheme similar to confocal microscopy, can help to remove the drawback and improve the axial resolution [14].

In the present subsection, a using CPTM developed in our group we studied a noninvasive, label-free, 3D imaging system. The performance of the setup is tested with a sample of 20-nm gold nanoparticle. An axial resolution enhancement of ~3 times is achieved compared with NPTM. Then, 3D microscopic distributions of melanin in benign and malignant melanoma tissue are obtained with this setup. The statistical discussions of 16 3D images showed marked differences in the density and shapes of melanin for the benign and malignant tissues. The 3D fractal analysis of all the images is also performed, and the malignant melanoma has a larger fractal dimension. The detection of melanin distributions in melanoma using CPTM can be a new option for melanoma diagnosis.

2 EXPERIMENTAL SETUP

Supplementary Fig. S1 outlines Simple description of the experimental setup utilized in this subsection is described as follows. The pump and probe beams, with central wavelength of 488 and 632.8 nm, respectively, are spectrally filtered from a compact supercontinuum fiber laser source (WL-SC450-2, 20 MHz, Fianium, UK) with bandpass filters (FL488-10, FL632.8-10, Thorlabs). Considering the light absorption coefficient of melanin and other molecules in skin [15] and the optical spectral density of the fiber laser source, the center wavelength of 488 nm seems to be the best choice for the pump beam in our experiment. The powers of pump and probe are 0.43 mW and 0.28 mW, respectively. The intensity of the pump pulse is modulated at 30 kHz with an electro-optic modulator (EOM) (LM202P, Qioptiq, Germany). Two sets of lenses are used to expand the pump and probe beams and adjust the divergence of the two beams. An objective lens (60• /NA 0.9, UPlanFLN, Olympus) is used to focus the two beams into the specimen. The 3D scanning of the samples is performed with a set of piezo stages (PS) (P-622.2CL and P-622.ZCL, Physik Instrumente (PI), Germany).

The detection module can be divided into three parts Detection 1 is for the optimization of the axial overlapping of pump and probe beams. The back scattered pump and probe beams from the sample (silver film) are focused into a fiber with a diameter of 25 μm by an achromatic lens (AC254100-A, Thorlabs) with focal length of 100 mm. The detector in Detection 1 is the CCD camera in a spectrometer (USB 4000, Ocean Optics), which can accumulate both the pump and probe simultaneously together with our home-made data acquisition and processing software. The axial overlap of the pump and probe is optimized by adjusting the beam divergence of the pump and probe. The forward propagating beams are collected and collimated by a condenser lens (100x/NA 1.4, Olympus).

After passing through the condenser lens, the probe beam is spectrally filtered through by a bandpass filter (FL632.810, Thorlabs). Then, the probe beam is split into two by a beam splitter (BSW26R, Thorlabs) for Detections 2 and 3, which are used respectively for NPTM and CPTM. The aperture of the iris in Detection 2 is optimized to obtain the maximum PT signal, while the position of the fiber in Detection 3 is optimized to minimize the side lobe intensity (see in Figure 5.3.1(c)). The focal length of the lens (AC254-100-A, Thorlabs) in Detection 3 is 100 mm. In the section of Test of axial properties of CPTM, the two detection modules are applied simultaneously, while in the section of 3D imaging of benign and malignant melanoma with CPTM, only Detection 3 is used in the experiment. The detectors in Detection 2 and Detection 3 are two auto-balanced detectors (BD) (Nirvana 2007, Newport), which can help to reduce the laser noise of the probe beam by ~20 dB. The PT signals are demodulated by two lock-in amplifiers (LIA1, 2) (SR844, Stanford Research System, US; Model 7265, Signal Recovery, US). An AD/DA converter is applied to transfer the signals from LIAs to computer (PC) and send the command from the PC to the PS.

Three types of samples are used in the research. (1) A sample of silver film is used to calibrate the axial overlapping of pump and probe; (2) a sample of 20-nm gold nanoparticle is used to test the axial properties of CPTM and NPTM; (3) two malignant melanoma samples (MMS1 and MMS2) and one benign melanoma sample (BMS1) are used in the biomedical imaging study. The main object of the study is to differentiate the malignant melanoma from the benign tumor tissue with our home-made CPTM.

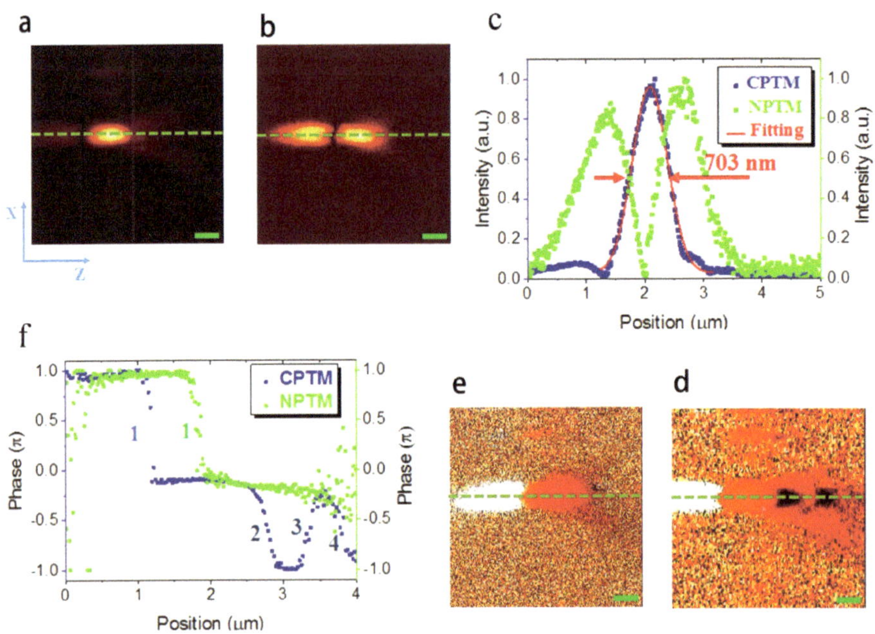

FIGURE 5.3.1 Intensity (a, b) and phase (d, e) distributions in XZ plane for the PT signal obtained with CPTM (a, d) and NPTM (b, e). The sample is a single 20-nm GNP. Scale bar: 500 nm. The cross sections [denoted by the red dashed lines in (a, b, d, e)] of the intensity image (a, b) and phase image (d, e) of the GNP are shown in (c, f) respectively. The axial resolution of CPTM is 703 nm, and that for NPTM is worse than 2 μ m. The copyright of this figure is partially held between the representative author (Takayoshi Kobayashi) and Optica.

The sample of silver film in (1)is grown on a slide glass (Matsunami, Japan) with a CVD method by Dr. Nakata (University of Electro & Communications, Japan). The size of the silver film is 2 mm × 6 mm × 0.01 mm.

Gold nanoparticles (GNP) of 20-nm diameter in (2) stabilized suspension in 0.1 mM PBS with optical density of 1.0 is supplied by Sigma Aldrich. A drop of such suspension is diluted by 100 times with distilled water, and then a drop of the diluted suspension is spread on the slide glasses (Matsunami, Japan). After the suspension become dried, a piece of cover glass (Matsunami, Japan) is applied to cover the sample.

RET-mice that have had an oncogene RET (RFP-RET) introduced under metallothionein-I promoter enhancer spontaneously develop benign melanocytic tumor and melanoma. Macroscopic and microscopic appearances of a benign melanocytic tumor and melanoma were shown in our previous reports[1] [16, 17] (Figure 5.3.3 in Ref. 16 and Fig. 5.3.1 in Ref. 17). Since both benign melanocytic tumors and melanoma in RET-mice are usually developed as a hemispherical shape, there is no sampling direction. The sample in (3) of both benign or malignant melanoma were fixed with 3% buffered formalin, embedded in paraffin, sliced at 10–15 μm thick, spread on slide glasses and sealed with mounting medium. Two malignant melanoma samples (MMS1 and MMS2) and one benign melanoma sample (BMS1) are used in the course of the present study. All the experiments have been reviewed and approved by the Institutional Animal Care and Use Committee, Juntendo University School of Medicine with the approval #270108. The procedures using the animals were conducted according to the institutional guidelines.

RESULTS

Test of the axial properties of CPTM. The axial overlapping of the pump and probe is first optimized by confocal scattering microscope with a sample of silver film. The setup of the confocal scattering microscope is a part of the whole setup as depicted in Fig. 5.2.1 in the previous subsection 5.2, After the optimization, the axial offset of pump and probe is reduced from 140 nm to <20 nm). The excellent axial overlapping of pump and probe will help to enhance the signal intensity [14] and also the lateral resolution [18] for CPTM. Then, the axial distributions of the point spread functions (PSFs) of CPTM and NPTM are tested with a sample of 20-nm GNP. Fig. 5.3.1 depicts an XZ (Z axis is in the axial direction) scan of a single GNP with CPTM and NPTM. The PT signal for NPTM has two extrema in axial direction [see Fig. 5.3.1(b)], resulting in the full width at half maximum (FWHM) of the intensity distribution of ~2.2 μm [see green curve in Fig. 5.3.1(c)]. Compared with the case of NPTM, the PT signal obtained with CPTM has only a single peak in axial direction [Fig. 5.3.1(a)], with an enhanced axial resolution defined by FWHM of ~700 nm [blue and red curves in Fig. 5.3.1(c)] and an improvement of ~3 times. The phases of the PT signals obtained with CPTM and NPTM are shown in Fig. 5.3.1(d, e), respectively. The PT signal phase experiences 4 sharp jumps in the CPTM image, as depicted in Fig. 5.3.1(d) and the blue curve in Fig. 5.3.1(f). The four jumps, separating the two phase states (0 and π/-π), are corresponding to the four peaks [one main peak and three small side lobes, blue curve in Fig. 5.3.1(c)] in the intensity distribution profile, and the phase of the most intense peak is around 0. The phases of the two intensity peaks [green curve in Fig. 5.3.1(c)] for the case of NPTM have a difference of π, and only one phase jump exists in the whole scan range, as shown in Fig. 5.3.1(f) (green curve).

3D imaging of benign and malignant melanoma with CPTM. After the optimization and calibration of the setup, the home-made CPTM is applied to the 3D imaging of mouse melanoma without any staining. To maintain high SNR images, the time constant of the lock-in amplifier (LIA) and the dwell time of the imaging are both set as 3 ms. Under this condition, the output voltages of the maximum PT signals are typically in the range of 0.5–2 V and the noise level of all the images is ~0.05 V with the SNR of higher than 10. To perform the statistical analysis of melanoma, especially to demonstrate the differences between benign and malignant melanoma, we have performed eight 3D scans at different positions of the samples for both the benign and malignant melanoma. The range of the 3D images is 22.5 μm×22.5μm ×11.25μm, with the whole pixel number of 200 × 200 × 50. The 3D images for benign and malignant melanoma are depicted in Figures 5.3.2 and 5.3.3,

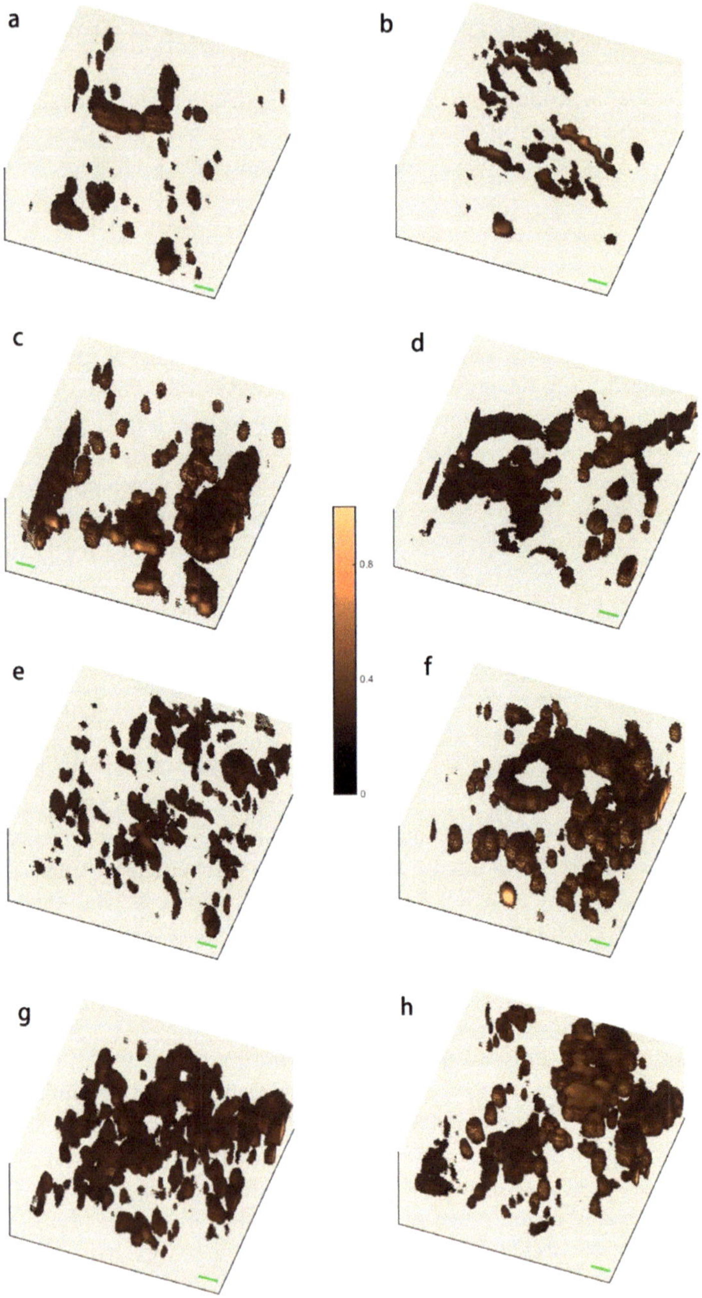

FIGURE 5.3.2 3D distributions of melanin in benign melanoma. (a–f) denote the 3D images of benign melanoma at different positions of the sample, with the images named B1–B8, respectively. Scale bar: 2μm. Pump: 488 nm/0.43 mW/20 MHz/~100 ps. Probe: 632.8 nm/0.28 mW/20 MHz/~100 ps. Dwell time: 3 ms. Copyright of this figure is shared between the representative author (Takayoshi Kobayashi) and Optica.

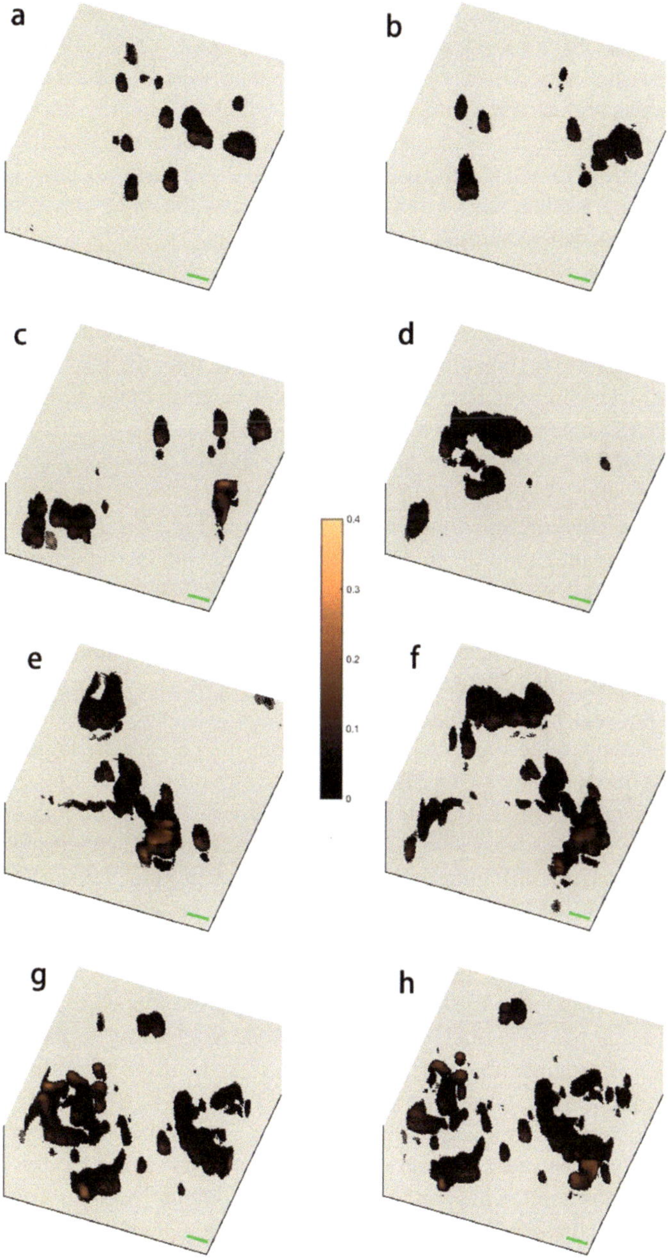

FIGURE 5.3.3 3D distributions of melanin in malignant melanoma. (a–f) denote the 3D images of benign melanoma at different positions of the sample, with the images named M1–M8, respectively. Scale bar: 2 m. Pump: 488 nm/0.43 mW/20 MHz/~100 ps. Probe: 632.8 nm/0.28 mW/20 MHz/~100 ps. Dwell time: 3 ms. Copyright of this figure is shared between the representative author (Takayoshi Kobayashi) and Optica.

respectively. The brown dots or aggregations in these images are melanin, which shows the metabolism and location of melanocytes and melanogenesis. The gray background is the skin tissue, which induces much lower PT signal because of its much lower absorption of pump light compared with melanin [15]. Some other molecules or organelles in the skin, such as hemoglobin, also absorb pump light and induce PT signal. However, because the absorption coefficient of hemoglobin is at least 3 times smaller than that of melanin in the case of the pump wavelength of 488 nm [15], the PT signal generated from hemoglobin also should be smaller than that generated from melanin. The 3D imaging of benign melanoma tissue with very low melanin density was measured, and from the image we can find some structures of organelles, which may be hemoglobin. The PT signal of such structures is just above the noise level, and it is typically 10–20 times smaller than the PT signal generated from melanin. The high contrast of the imaging of melanin in the skin tissue is possible by careful choice of pump wavelength. Utilizing the 16 3D images shown in Figs 5.3.2 and 5.3.3, we discuss melanoma pathology based on classic morphometric methods (size, shape, density...) and fractal analysis.

Pathology of melanoma: melanin density. From Figs. 5.2.2 and 5.2.3, we can find that the melanin density is lower for benign melanoma compared with the malignant one intuitively. However, to demonstrate the difference quantitatively, we first calculated the melanin density in the 16 3D images in Figs. 5.2.2 and 5.2.3. The method for the melanin density calculation is demonstrated in the section Methods. Figure 5.3.4 depicts the melanin density of the 16 3D images (eight for benign melanoma B1–B8, and eight for malignant melanoma M1–M8). We can find that the melanin densities of the eight benign melanoma images are all lower than 10%, with the maximum value of 7.7% (B8) and the minimum value of 1.6% (B1), seeing the black squares in Fig. 5.3.4. The average density of the eight benign melanoma images (B1–B8) is calculated to be 4.6% ± 2.4%. The melanin densities in malignant melanoma tissues are all higher than 10%, with the minimum value of 11.9% (M1) and the maximum value of 34.4% (M8). The average melanin density of the eight malignant melanoma images (M1–M8) is calculated to be 24.3 ± 9.1%, which is ~5 times larger than that of the benign melanoma samples. However, all the 16 3D images are taken in or near the center part (defined by the part with most intense PT signal) of the samples. The melanin density will be much lower in the periphery of the samples, for example, the melanin density can be nearly zero at the edge of the benign melanoma tissue. Based on the results in Fig. 5.3.4, we can use a melanin density of ~10% to differentiate the benign and malignant melanoma. The melanocytes are only concentrated in several localized positions for the benign melanoma tissues. And the melanin, which represents the melanocytes, has spread over and also becomes denser when the melanoma becomes malignant.

Pathology of melanoma: size distribution of melanin particles/aggregations. Besides the difference in the melanin density, the size distribution of melanin particles/aggregations in benign and malignant melanoma also shows some difference, as depicted in Figs. 5.3.2 and 5.3.3. To give a quantitative discussion, we have calculated the melanin size distributions of melanin particles/aggregations in the 16 3D images in Figs. 5.3.2 and 5.3.3. The size information, such as one dimensional (1D)

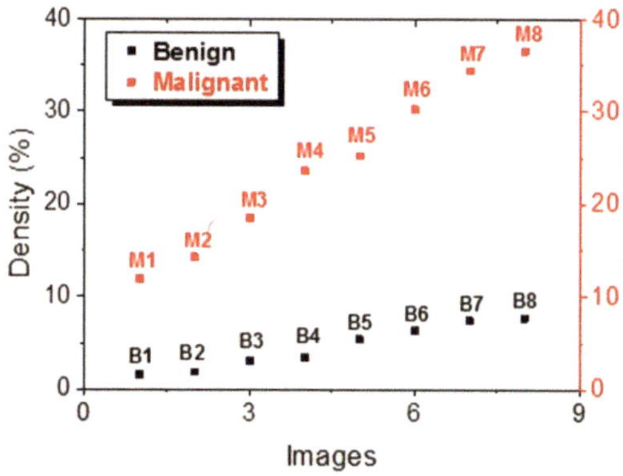

FIGURE 5.3.4 Melanin density of benign and malignant melanoma plotted in the order of the increasing densities of benign and malignant samples. Copyright of this figure is shared between the representative author (Takayoshi Kobayashi) and Optica.

(distance and length), two dimensional (2D) (surface area), and 3D (volume) information, can be calculated with the software ImageJ, and the detailed information is given in the Methods section. We only demonstrate and discuss the surface area and volume in the present work. During the calculation and discussion, units of cube of pixel (pixel^3) and square of pixel (pixel^2) are used for volume and surface area, respectively, with the pixel unit of 112.5 nm. The total numbers of the melanin particles/aggregations in malignant melanoma images (M1–M8) and benign melanoma images (B1–B8) are calculated to be 319 and 105, respectively. In order to depict the statistics of the 2D and 3D size data of all the melanin particles/aggregations graphically, box plot is applied. Conventionally, five values from a set of data are used in box plot: the extremes, the upper and lower quartiles, and the median [19]. Figure 5.3. 5(a, b) show the box plots of volumes and surface areas of the melanin particles/aggregations in the 16 3D images in Figs. 5.3.2 and 5.3.3. As shown by the upper datum (denoted by the word "Max") in Fig. 5.3.5(a, b), the maximum melanin aggregation in benign melanoma, has a volume of 52020 pixel^3 (~74 μm^3) and a surface area of 21561 pixel^2 (273 μm^2), while the volume and surface area of the maximum melanin aggregation in malignant melanoma are calculated to be 116944 pixel^3 (~167 μm^3) and 50424 pixel^2 (638 μm^2), respectively. The largest melanin aggregation in malignant melanoma is ~2 times larger in size than that in the benign melanoma. In order to show the box, which is formed by the first and third quartiles, more clearly, we have reduced the ranges of the box plots, and the plots are depicted in Fig. 5.3.5(c, d). The mean sizes (both volume and surface area) of the melanin particles/aggregations, denoted by the small squares in Fig. 5.3.5(c, d), show a difference of ~10–20% for benign and malignant melanoma. The widths of the boxes in the box plots of malignant melanoma images seem narrower than those in benign images, which means the sizes of the particles/aggregations in malignant melanoma are concentrated in a smaller range. However, more data is needed to justify whether the two parameters are effective criteria for differentiating malignant melanoma from the

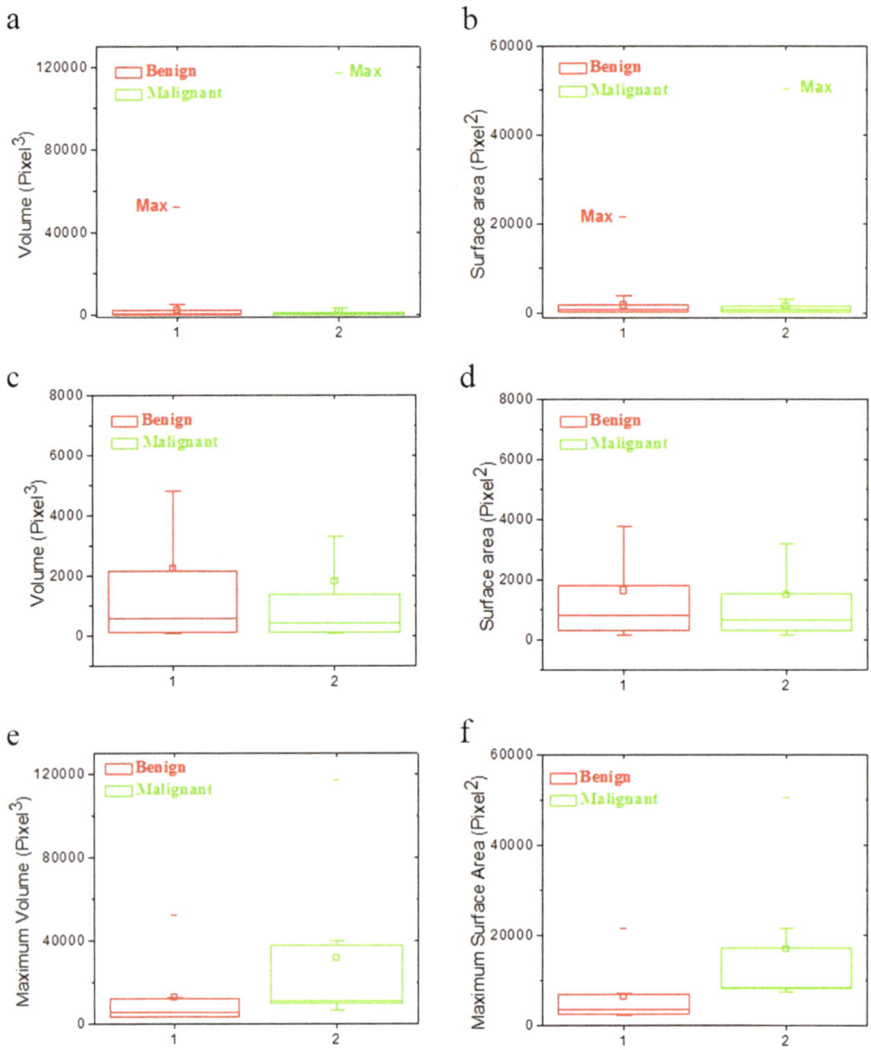

FIGURE 5.3.5 Box plots of the size information of melanin particles/aggregations in benign and malignant melanoma. (a, b) are box plots of volume and surface area, respectively, of the melanin particles/aggregations in the 16 3D images in Figs. 5.3.2 and 5.3.3. (c) The box plot of volume with small plot range compared with (a). (d) The box plot of surface area with small plot range compared with (b). (e, f) are box plots of the volumes and surface areas of the maximum melanin particles/aggregations in each image in Figs. 5.3.2 and 5.3.3. Copyright of this figure is shared between the representative author (Takayoshi Kobayashi) and Optica.

benign one. Besides the box plots of the size data of all the melanin particles/aggregations in the 16 3D images in Figs. 5.3.2 and 5.3.3, the box plots of the volumes and surface areas of the maximum melanin particles/aggregations in each image are also given, and the plots are depicted in Fig. 5.3.5(e, f). From the plots in Fig. 5.3.5(e, f), we find that the maximum values, the first and third quartiles, and the median of the size data for the eight particles/aggregations selected from malignant melanoma are all ~2 times the corresponding values for the eight maximum particles/aggregations selected from the eight benign images, which means that the particles/aggregations grow during the melanoma progression process.

Pathology of melanoma: Fractal dimension. The growth of cancer, including melanoma, is usually a chaotic and irregular process, and it was discussed that a fractal dimension analysis can be well suited to describe the malignancies in a qualitative sense [20–25]. The raw images of cancer for fractal analysis can be obtained with fluorescence microscopy [20], dermoscopy [26] and atomic force microscopy [22], but those images are typically 2D [20, 21, 23–25] or pseudo-3D [22] images at best. In the case of melanoma, several factors (organ or tissue) can be used in the fractal analysis, such as the boundary between cancer and the host [20, 26], and nuclear chromatin [22]. Here, we performed the fractal analysis of benign and malignant melanoma based on 3D distributions of melanin, and the boundary between the melanin (the melanocytes) and the host (skin tissue) plays a key role in the analysis. Unlike topological dimensions, the fractal index can take non-integer values, indicating that a set fills its space qualitatively and quantitatively differently from how an ordinary geometrical set does.

The fractal dimensions of all the 16 3D images are calculated with the software ImageJ, and the details of the calculation are shown in the section of Methods. Figure 5.3.6(a) shows the fractal dimensions of all the 16 3D images in Figs. 5.3.2 and 5.3.3. The fractal dimensions for benign melanoma [black squares in Fig. 5.3.6(a)] are between 1.68 and 2.16, with the mean value of 2.00 and standard deviation of 0.19, whereas those for malignant melanoma [red squares in Fig. 5.3.6(a)]are in the range of 2.10–2.61, with the mean value of 2.37 and standard deviation of 0.16. The malignant melanoma seems to be of larger fractal dimensions compared with the benign one, which means that the melanin distribution becomes more chaotic and poorly regulated during the melanoma cancer growth. Although better understanding of the melanin morphology in melanoma still needs to be revealed by molecular methods [27]. To compare the reliability of 2D and 3D fractal analysis in melanoma diagnosis, we have also performed the fractal analysis of several 2D images for both benign and malignant melanoma. We chose one binary 2D image from B6 (B6-1, depth of 9.9μm) and three binary 2D images from M8 (depth of 3.375μm, 9μm and 10.125μm for M8-1, M8-2 and M8-3, respectively), as depicted in Fig. 5.3.6(b–e). The fractal dimension of B6-1 is calculated to be 1.61, while those of M8-1, M8-2, and M8-3 are 1.56, 1.72, and 1.61, respectively. Since the 2D images from different depths are very different among them, 2D fractal analysis is no longer suitable for differenting benign melanoma from the malignant one.

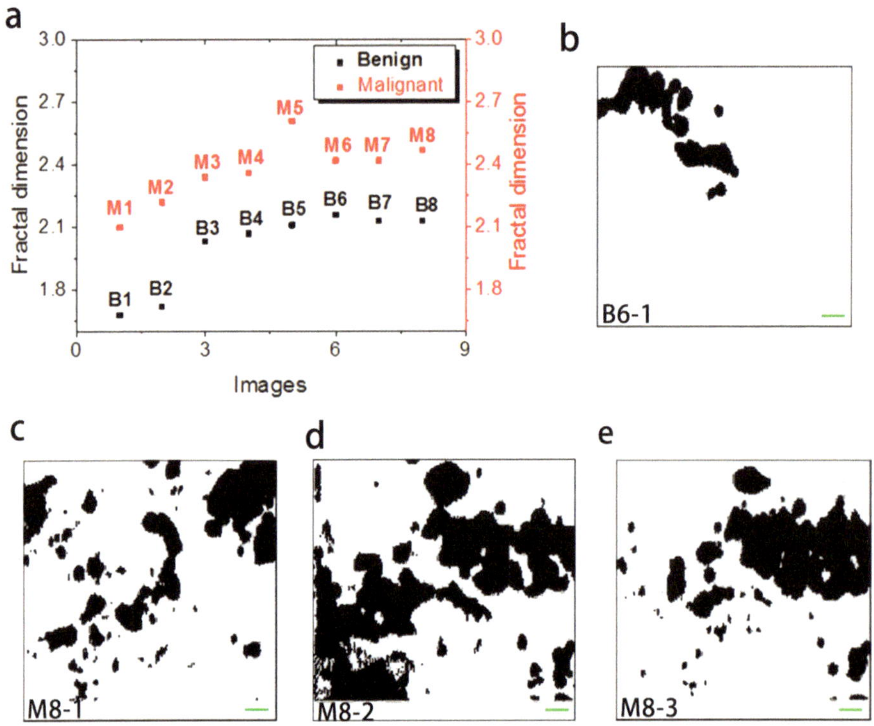

FIGURE 5.3.6 Fractal analysis of benign and malignant melanoma. (a) Fractal dimensions of all the 16 3D images 2D binary images from B6 (b) and M8 (c–e) are used for 2D analysis. The range of the 2D images is 22.5 μm × 22.5 μm. Scale bar: 2μm. The fractal dimensions are 1.61, 1.56, 1.72, and 1.61 for (b–d, f), respectively. Copyright of this figure is shared between the representative author (Takayoshi Kobayashi) and Optica.

DISCUSSION

In the present work, we have obtained high resolution 3D images of benign and malignant melanoma with CPTM. Based on the 16 captured 3D images (eight for benign and the others for malignant melanoma), we have demonstrated an understanding of melanoma pathology using classical morphometric methods (melanin density and size distribution) and fractal analysis. The melanin density in malignant melanoma is as high as 24.3%, and it is ~5 times the value in benign melanoma. The maximum melanin aggregation in malignant melanoma has a volume of ~74 μ m ^{3}, which is ~2 times larger than that in the benign case. Based on the calculated results, we can conclude that the melanin, which represents the melanocytes, becomes denser and more aggregated when the melanoma becomes malignant. The fractal dimensions of benign and malignant melanoma are 2.0 and 2.4, respectively, which indicates that the melanin distributions become more chaotic and poorly organized during the growth of melanoma cancer. In comparison to the 2D fractal analysis, 3D fractal analysis is more useful and reliable in melanoma diagnosis.

Besides the well-established clinical melanoma diagnosis method—dermoscopy [3, 26, 28], several other optical methods also show potentials to be used in melanoma diagnosis, such as confocal scanning laser microscopy (CSLM) [29, 30], optical coherent tomography (OCT) [31], pump-probe microscopy (PPM) [6] and photoacoustic microscopy (PAM) [32, 33]. The main disadvantages of dermoscopy is relatively low sensitivity and lack of deep imaging capability of suspicious pigmented lesions [3, 26, 28]. Based on the summarization by D. S. Rigel [3], CSLM can measure into the papillary dermis with long wavelength pump light, but it gives poor resolution of chromatin patterns, nuclear contours, and nucleoli. The papillary dermis is the uppermost layer of the dermis

Therefore it is important for the change of modification of the structure induced by the carcinogenic process by high-resolution microscope without destructive investigation. OCT can give higher detection depth and resolution, but it is apt to image artifacts, and special materials are needed to reduce the scattering and increase the imaging depth. PPM can differentiate eumelanin and pheomelanin in pigmented lesions, however, the demands of a two femtosecond laser source will lead to high cost and large setup [6]. PAM can detect a deep tumor with relatively high resolution, although it typically needs the staining of metal nanoparticles to enhance the imaging contrast [31, 32]. The method demonstrated in the present work is based on CPTM, which is a pure optical process. It can supply us with a very high imaging sensitivity even reaching a single molecule level [8]. The absorption coefficient of melanin is ten times larger than hemoglobin and other molecules in skin tissues even with pump wavelength in near infrared range [15], which means that the label-free imaging of melanoma with high contrast and great depth is feasible with the present method. Continuous-wave (CW) laser sources, such as diode lasers, can be used as the pump and probe beams for CPTM [18], and it will make the imaging system very cheap and compact. The spatial resolution of CPTM is also better than PAM since CPTM is a PPM process.

The advantages of the present method can be summarized as: (1) label-free; (2) no need of femtosecond lasers; (3) high spatial resolution; (4) 3D capacity; (5) high contrast; (6) high sensitivity. And these advantages make the present method a new option for monitoring the melanoma growth and also the melanoma diagnosis in clinics. The CPTM can also be converted or coupled to a photothermal cancer therapy system [34, 35] by just increasing the pump energy. However, due to two limitations, the experimental setup used to demonstrate the present method is not feasible for *in vivo* imaging of melanoma in the present phase. The first limitation is the imaging speed. Due to the high laser noise level and low average power of the pump beam [36], the imaging dwell time is set as 3 ms to obtain high SNR images. As a result, the imaging time for one 3D image is about 100 min, which is too slow for *in vivo* applications.

To solve this problem, we are going to use a high power stable CW laser source as the pump and probe beams. Typically, the stability of the CW laser source can be better than 1%, and it is >10 times better than the beams spectrally filtered from the SC laser source. It means the imaging dwell time can be reduced by > 10 times without influencing the SNR much. The utilization of a higher power pump and probe can also help to increase the SNR. The second limitation of the current setup is that the home-made sample stage is not ready for live animals currently. We are still proceeding the research to *in vivo* imaging of human melanoma.

METHODS

CPTM. CPTM has been demonstrated both theoretically and experimentally by J. Moreau [14, 37]. The probe beam spot size change on the pinhole (a single mode fiber in the present setup) plane, which can be treated as the photothermal signal, is given by [38]:

$$\frac{\delta\omega}{\omega}(z) = \frac{1}{f_{th}} \frac{(z_0 - x)\left(1 + \frac{xz}{z_{R0}^2}\right)}{1 + \frac{z^2}{z_{R0}^2}} \tag{1}$$

where f_{th} is the focal length of the thermal lens, z_0 is the distance between the probe beam waist and the thermal lens, z_{R0} is the probe beam Rayleigh range in the sample. In the equation, $x = (z_1 + z_2 - z_1 z_2/f)/(1 - z_2/f)$, with f the focal length of the focal lens for the detection, z_1 the distance between the probe beam waist (in the sample) and the focal lens, z_2 the distance between the focal lens and the pinhole. The confocal case corresponds to $x/z_{R0} = \pm 1$, while the standard non-confocal case corresponds to $x \to \infty$.

The size distribution of melanin particles/aggregations in the melanoma tissue sample. The size distribution of melanin particles/aggregations is calculated with the software ImageJ (Fuji Is Just). The 2D 8-bit gray-type image sequence was first imported into the software ImageJ (Fuji Is Just), and then 3D images were reconstructed with the plugin "3D Viewer". After that, we performed the statistics of the particle size (volume and surface area) distribution with "3D Objects Counter" in the "Analyze". The threshold for the calculation is set to be 10. Considering the real spatial resolution of the system and the simplification of the analysis, the size filter is set as $(50\ \text{pixel})^3$ corresponding to $(0.07\ \mu\text{m})^3$. Then, the particle number, the maximum and mean values of the volume and surface area of the melanin in all the 3D images were obtained.

Fractal analysis of benign and malignant melanoma. The fractal dimensions of all the 3D images are calculated with the software ImageJ (Fuji Is Just). The 2D image sequence is first imported into the software ImageJ, and the type of the 2D images is changed to 8-bit gray. Then the fractal dimension is calculated with plugin "Fractal Count". During the 2D fractal analysis, the images are first converted into binary images, and then the plugin "Fractal Count" is used to calculate the fractal dimension. The parameters for the 2D fractal analysis are also same as those in Table S1. In summary in this subsection, the research on noninvasive, label-free, three-dimensional imaging of melanoma with confocal photothermal microscopy is explained and it shows the verification of differentiation capability of malignant melanoma from benign tumor tissue.

The research described in this subsection is based on the collaborative activity of the following group: J. He, N. Wang, H. Tsurui, M. Kato, M. Iida, and T Kobayashi.

REFERENCES

1. American Cancer Society, Melanoma skin cancer overview, http://www.cancer.org/acs/groups/cid/documents/webcontent/003063pdf.pdf, Date of access: 02/09/2015 (2015).
2. Siegel, R. L., Miller, K. D. & Jemal, A. Cancer statistics. *CA Cancer J Clin.* **65**, 5–29 (2015).
3. Regel, D. S., Russak, J. & Friedman, R. The evolution of melanoma diagnosis: 25 years beyond the ABCDs. *CA Cancer J Clin.* **60**, 301–316 (2010).
4. Argenziano, G. *et al.* Dermoscopy of pigmented skin lesions: results of a consensus meeting via the internet. *J. Am. Acad. Dermatol.* **48**, 679–693 (2003).
5. Jimbow, K. *et al.* Characterization of melanogenesis and morphogenesis of melanosomes by physcochemical properties of melanin and melanosomes in malignant melanoma. *Cancer Res.* **44**, 1128–1134 (1984).
6. Matthews, T. E., Piletic, I. R., Selim, M. A., Simpson, M. J. & Warren, W. S. Pump-probe imaging differentiates melanoma from melanocytic Nevi. *Sci. Transl. Med.* **3**, 71ra15 (2011).
7. Zonios, G. *et al.* Melanin absorption spectroscopy: new method for noninvasive skin investigation and melanoma detection. *J. Biomed. Opt.* **13**, 014017 (2008).
8. Tokeshi, M., Uchida, M., Hibara, A., Sawada, T. & Kitamori, T. Determination of subyoctomole amounts of nonfluorescent molecules using a thermal lens microscope: subsingle-molecule determination. *Anal. Chem.* **73**, 2112–2116 (2001).
9. He, J., Miyazaki, J., Wang, N., Tsurui, H. & Kobayashi, T. Label-free imaging of melanoma with nonlinear photothermal microscopy. *Opt. Lett.* **40**, 1141–1144 (2015).
10. Lu, S., Min, W., Chong, S., Holtom, G. R. & Xie, X. S. Label-free imaging of heme proteins with two-photon excited photothermal lens microscopy. *Appl. Phys. Lett.* **96**, 113701 (2010).
11. Boyer, D., Maali, A., Lounis, B. & Orrit, M. Photothermal imaging of nanometer-sized metal particles among scatterers. *Science* **297**, 1160–1163 (2002).
12. Nedosekin, D. A., Galanzha, E. I., Dervishi, E., Biris, A. S. & Zharov, V. P. Super-resolution nonlinear photothermal microscopy. *Small* **10**, 135–142 (2014).
13. Nedosekin, D. *et al.* Photoacoustic and photothermal detection of circulating tumor cells, bacteria and nanoparticles in cerebrospinal fluid *in vivo* and *ex vivo*. *J. Biophotonics* **6**, 523–533 (2013).
14. Moreau, J. & Loriette, V. Confocal thermal-lens microscope. *Opt. Lett.* **29**, 1488–1490 (2004).
15. Meglinski, I. V. & Matcher, S. J. Quantative assessment of skin layers absorption and skin reflectance spectra simulation in the visible and near-infrared spectral regions. *Physiol. Meas.* **23**, 741–753 (2002).
16. Kato, M. *et al.* Transgenic mouse model for skin malignant melanoma. *Oncogene* **17**, 1885–1888 (1998).
17. Kato, M. *et al.* c-Kit-Targeting immunotherapy for hereditary melanoma in a mouse model. *Cancer Res.* **64**, 801–806 (2004).
18. Miyazaki, J., Tsurui, H. & Kobayashi, T. Reduction of distortion in photothermal microscopy and its application to the high resolution three-dimensional imaging of nonfluorescent tissues. *Biomed. Opt. Express* **6**, 3217–3224 (2015).

19. McGill, R., Tukey, J. W. & Larsen, W. A. Variations of box plots. *Am. Stat.* **32**, 12–16 (1978).
20. Ahammer, H., Devaney, T. T. J. & Tritthart, H. A. Fractal dimension of K1735 mouse melanoma clones and spheroid invasion *in vitro*. *Eur. Biophys. J.* **30**, 494–499 (2001).
21. Baish, J. W. & Jain, R. K. Fractals and cancer. *Cancer Res.* **60**, 3683–3688 (2000).
22. Bedin, V., Adam, R. L., Sá, B., Landman, G. & Metze, K. Fractal dimension of chromatin is an independent prognostic factor for survival in melanoma. *Cancer* **10**, 260 (2010).
23. Guz, N. V., Dokukin, M. E., Woodworth, C. D., Cardin, A. & Sokolov, I. Towards early detection of cervical cancer: Fractal dimension of AFM images of human cervical epithelial cells at different stages of progression to cancer. *Nanomed.-Nanothechnol.* **11**, 1667–1675 (2015).
24. Davie, N. A. *et al.* Fractal dimension (df) as a new structure biomarker of clot microstructure in different stages of lung cancer. *Thromb. Haemostasias.* **114**, 1251–1259 (2015).
25. Waliszewski, P., Wagenlehner, F., Gattenlöhner, S. & Weidner. On the relationship between tumor structure and complexity of the spatial distribution of cancer cell nuclei: a fractal geometric model of prostate carcinoma. *The Prostate* **75**, 399–414 (2015).
26. Skvara, H., Teban, L., Fiebiger, M., Binder, M. & Kittler, H. Limitations of dermoscopy in the recognition of melanoma. *Arch. Dermatol.* **141**, 155–160 (2005).
27. Bittner, M. *et al.* Molecular classification of cutaneous malignant melanoma by gene expression profiling, *Nature* **406**, 536–540 (2000).
28. Andreassi, L. *et al.* Digital dermoscopy analysis for the differentiation of a typical Nevi and early melanoma: a new quantitative semiology. *Arch. Dermatol.* **135**, 1459–1465 (1999).
29. Lorber, A. *et al.* Correlation of image analysis features and visual morphology in melanocytic skin tumors using *in vivo* confocal laser scanning microscopy. *Skin Res. Technol.* **15**, 237–241 (2009).
30. Branzan, A. L., Landthaler, M. & Szeimies, R. M. *In vivo* confocal scanning laser microscopy in dermatology. *Lasers Med. Sci.* **22**, 73–82 (2007).
31. Gambichler, T. *et al.* Characterization of benign and malignant melanocytic skin lesions using optical coherence tomography *in vivo*. *J. Am. Acad. Dermatol.* **57**, 629–637 (2007).
32. Mehrmohammadi, M., Yoon, S. J., Yeager, D. & Emelianov, S. Y. Photoacoustic imaging for cancer detection and staging, Curr. *Mol. Imaging* **2**, 89–105 (2013).
33. Galanzha, E. I., Shashkov, E. V., Spring, P. M., Suen, J. Y. & Zharov, V. P. *In vivo*, noninvasive, label-free detection and eradication of circulating metastatic melanoma cells using two-color photoacoustic flow cytometry with a diode laser. *Cancer Res.* **69**, 7926–7934 (2009).
34. Huang, X., El-Sayed, I. H., Qian, W. & El-Sayed, M. A. Cancer cell imaging and photothermal therapy in the near-infrared region by using gold nanorods. *J. Am. Chem. Soc.* **128**, 2115–2120 (2006).
35. Dickerson, E. B. *et al.* Gold nanorod assisted near-infrared plasmonic photothermal therapy (PPTT) of squamous cell carcinoma in mice. *Cancer Lett.* **269**, 57–66 (2008).
36. He, J., Miyazaki, J., Wang, N., Tsurui, H. & Kobayashi, T. Biological imaging with nonlinear photothermal microscopy using a compact supercontinuum fiber laser source. *Opt. Express.* **23**, 9762–9771 (2015).
37. Moreau, J. & Loriette, V. Confocal dual-beam thermal-lens microscope: model and experimental results. *J. J. Appl. Phys.* **45**, 7141–7151 (2006).

38. He, J., Wang, N., Tsurui, H. Kato, Iida, M., & Kobayashi, T. Noninvasive, label-free, three-dimensional imaging of melanoma with confocal photothermal microscopy: Differentiate malignant melanoma from benign tumor tissue. *Sci. Reports*, **6**, 30209 (11 pages) (2006).

Additional Information

Supplementary information accompanies this paper at http://www.nature.com/srep

Competing financial interests: The authors declare no competing financial interests.

How to cite this article: He, J. *et al.* Noninvasive, label-free, three-dimensional imaging of melanoma with confocal photothermal microscopy: Differentiate malignant melanoma from benign tumor tissue. *Sci. Rep.* **6**, 30209; doi: 10.1038/srep30209 (2016).

5.4 Label-Free Imaging of Melanoma with Confocal Photothermal Microscopy

Differentiation between Malignant and Benign Tissue

1 INTRODUCTION

Malignant melanoma (MM) is one of the most common cancers worldwide and also one of most aggressive cancers. It has been increasing at a greater than any other cancer. It has a favorable prognosis only if the affected area is removed at an early stage. MM reportedly causes the large majority of skin cancer deaths despite the fact that it accounts for <2% of skin cancer cases [1]. The incidence of MM has been increasing for >30 years [2] and one of its most ominous characteristics is its high propensity to produce distant metastases, because it can get disseminated throughout the body through lymphatic and hematogenous spread. For this reason, early detection and treatment of MM are crucial life-saving measures [3]. Although dermoscopy is a powerful diagnostic technique [4] and the ABCDE (abbreviation for asymmetrical shape, border, color, diameter, and evolution) rule provides a guide to the identification of involved areas [5], pathological examination is even now the gold standard for MM diagnosis. However, diagnosis remains subjective and highly reliant on the skill level of the pathologist. Interobserver reproducibility of MM diagnosis varies even among experts.

Fractal analysis has been developed as a strategy to improve diagnostic reliability. This method is based on the calculation of the fractal dimension (FD) of the structure of MM cells and their distribution [6–11]. Although this is a very attractive idea, the method has three drawbacks. First, the technique is based on self-similarity, namely, having reduced its image in itself recursively; therefore, if no self-similarity is present in the structure, the arbitrarily defined FD is difficult to connect to the structural features being discussed in the chapter. Second, the FD is based on the

DOI: 10.1201/9781003476030-21

calculation of the line length or area of the covered region, and changes in its value are then visualized by changing the analyzed domain. This procedure limits the ratio between the minimal size of evaluation and the total size, resulting in limited spatial resolution and spatially dependent information. Finally, marker-free phenotyping of tumor cells by fractal analysis of reflection interference contrast microscopy (RICM) images results in a very small difference in FD values between the different types of MM cells [12]. For example, two types of MM cells in one study had FD values of 1.353 ± 0.004 and 1.312 ± 0.005, which corresponds to only a 3.08% difference in the average value of the two groups [12]. Even though these authors claimed that the standard deviations (SDs) were as small as 0.005 and 0.004, the FDs in Figure 5.4.5 of their paper showed nearly 80% overlap, while their RICM images showed very clear visual differences in the apparent structural features between the two types (Figure 5.4.5 (A) of the paper). This means that, even when the image patterns are quite different between two images, similar FDs can be obtained, indicating that the method is difficult to apply to clinical diagnosis. This disadvantage forms the basis of our motivation to find a more reliable and useful diagnostic application of textural structure to skin cancer cells in the present paper.

Melanin carries information about the metabolism and location of melanocytes and melanogenesis; therefore, the melanin distribution could act as a marker for MM [13, 14]. The two dominant types of melanin (eumelanin and pheomelanin) absorb a large cross section of visible light without substantial fluorescence emission [15], resulting in the difficulty in the imaging of MM distribution. However, the non-fluorescent property of melanin enables even more sensitive imaging of MM by employing photothermal (PT) microscopy (PTM), which is the main subject of the present paper.

PTM, which is based on the detection of changes to probe light intensity by thermal lensing due to local heating of the sample by absorption of the laser light, has demonstrated potential for biological imaging and clinical application. The key advantages of PTM are its high sensitivity and lack of a requirement for staining [16–21]. It facilitates the real-time, high-resolution imaging of nanometer-sized absorbers buried among light scatterers with a high signal-to-noise ratio [19, 22, 23]. However, the PT signal intensity in normal PTM has two extremes in the axial direction [24], which introduces distortions resulting in limited axial resolution of three-dimensional PT images. Confocal PTM (CPTM), which has a detection scheme similar to that of confocal microscopy, can help to ameliorate this drawback and improve the axial resolution [24]. Our group has used CPTM to study superresolution microscopic images of neurons in mouse brains [25, 26] and mouse skin MM [27].

In our previous paper [28], we developed a CPTM system and applied it to examine the features of melanin aggregates of both nevus and MM cells in thick (10–15 μm) specimens of a melanoma model mouse. The *ret* gene is a receptor tyrosine kinase type oncogene, and *rfp* is a fusion of 5'-half of ret gene and a finger structure probably capable to bind DNAs. RET-transgenic mice of the 304/B6 (RET-Tg) line develop benign melanocytic tumors and MM in a stepwise manner, and they are widely used for the study of melanoma genesis [29–31]. Obtained 3D images were analyzed to characterize the features of the melanin aggregates such as the density, size, and the surface FD. Even though the differences of those features between the two cell types were clearly demonstrated, we were not able to evaluate very accurately its diagnostic capability because of the limited image quality due to slow stage with which photodegradation during one frame imaging. The photo-damage was inevitable to get images with enough intensity for high S/N but it induced signal intensity reduction in the later stage of the one-frame imaging process.

In this study, we made an improvement upon the CPTM technique and applied it for noninvasive label-free imaging of MM in the mouse model previously used for the purpose of future application to humans. The scanning method was changed from piezo-driven stage to galvanometer, hence the wider swept area from 20×20 μm^2 to 72×72 μm^2 and the shorter scanning period from 3 ms to 20 μs. This enabled solving the problem in the previous study which hampered us in getting enough quality images. The performance of the analytical system was tested with a sample of 20 nm gold nanoparticles. Using the PT imaging data, we then analyzed the structural properties of MM and nevus cells to be compared using the GLCM method [32–38]. We calculated nine different parameters: angular second moment (ASM), Contrast, Correlation, Entropy, inverse difference moment (IDM), Homogeneity, Prominence, Shade, and Variance. These textural parameters were determined by analyzing relevant regions of interest on 12 two-dimensional PT images obtained at two different positions in three tissue sections containing nevus and MM cells. The details of this analysis are described in the following experimental section. Our method provides an objective evaluation independent of the experience, skill, and knowledge of individual medical doctors, and prognostication on each occasion of the pathological diagnosis. Thus, GLCM calculation provides a quantitative indicator that may become a "standard" in the future by the accumulation of cases in real clinical settings from various individuals with a variety of experiences. We first applied difference criterion (DIF) analysis and receiver operating characteristic (ROC) curve analysis. The analyses provided the results that Entropy, Contrast, and Variance were most suited for the discrimination between MM and nevus when 405 nm excitation was used, and Prominence, Variance, and Shade were best at 488 nm excitation. Second, we defined a new index, a "clearness discrimination parameter" (DISC), for the discrimination between nevus and MM cells. This index suggested that Entropy and Homogeneity in the case of the 405 nm pump, and Entropy and Prominence in the case of the 488 nm pump were the two most suited among the nine parameters for discrimination between nevus and MM cells in a mouse. This research received no external funding.

2 MATERIALS AND METHODS

2.1 Mice

Nevus and MM samples, which were diagnosed in advance by the specially trained pathologist, in transgenic mice (RET-Tg[1]) carrying constitutively activated RFP/RET, a hybrid oncogene[2] between RFP[3] and c-RET[4] [31], were prepared and used for analysis. The Animal Care and Use Committee (approval no. 270108 in Juntendo

University) and the Recombination DNA Advisory Committee (approval no. 13–7626-55 in Juntendo University) approved this study.

2.2 Experimental Setup and Image-Taking Procedure

Figure 5.4.1 shows the scheme of the experimental setup. Two sets of pump wavelengths are used in the experiment. A 405 nm laser diode (LD) (NDV4316; Nichia, Tokushima, Japan) and a 488 nm LD (L488P60; Thorlabs, Newton, NJ, USA) were used for the pump laser. A 638 nm LD (ML520G55; Mitsubishi, Tokyo, Japan) was used for the probe. The pump laser was modulated at 100 kHz using signal generators, and one of the two LDs was introduced to a beam collimator. After combination of the pump and probe beams using a dichroic mirror, the probe was split by a polarized beam splitter for balance detection and then directed to a Galvano mirror (VM500PLUS; GSI Group, Bedford, MA, USA). A 4F optical system was located near the sample and objective lens (MPLFNF 40× with numerical aperture of 0.75; Olympus, Tokyo, Japan). Two-dimensional images of the samples were obtained by scanning laser beams with the Galvano mirror in the X–Y plane. The irradiation powers of the pump and probe lasers were 300 μW (405 nm), 1.5 mW (488 nm), and 3 mW, respectively. One of the two pump lasers was used at one time. An auto-balanced detector (New Focus Nirvana; Newport Inc., Irvine, CA, USA) composed of two sets of a photodiode and a lock-in amplifier (7270 Signal Recovery; Ametek,

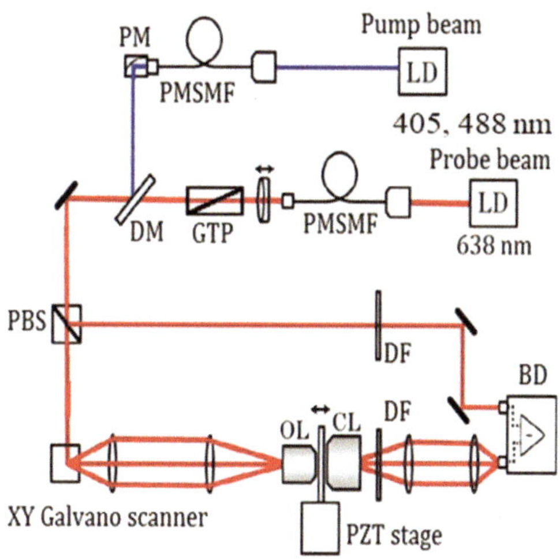

FIGURE 5.4.1 Scheme of the experimental setup. LD: laser diode, PMSMF: polarization maintaining, single-mode fiber, PM: parabolic mirror, GTP: Glan-Thompson polarizer, DM: dichroic mirror, PBS: polarizing beam splitter, OL: objective lens, CL: condenser lens, DF: dielectric multi-layer filter, BD: balanced detector, PZT (Piezoelectric induced alloy system composed of lead (Pb), zircon, and titan (=lead zircon titanate) stage: piezo driven stage. The copyright of this figure is shared between the representative author (Takayoshi Kobayashi) and Optica.

Berwyn, PA, USA) was used to remove the noise from the probe. The frame size of the image was 600 × 600 pixels, which corresponds to a 72 × 72 μm^2 area. The lock-in amplifier sensitivity was 1 mV, and the time constant was 20 μs per point. For the GLCM analysis, a four-section tiled area (18 × 18 μm^2 out of 36 × 36 μm^2) in the X–Y plane was imaged for each 72 × 72 μm^2 image. We have analyzed 48 images of 18 × 18 μm^2 area in total for both nevus and MM samples.

At the image obtaining process, we selected the areas to be investigated based on the following strategic criteria. Because the intensity of the PT signal varies with different areas, we selected those in which the obtained PT signals were strong enough to provide a high signal-to-noise ratio. In practice, we selected areas in which strong pigmentation was recognized.

The aim of the present study is to provide a non-specialist of the pathology of melanoma with a tool to distinguish between MM and nevus. The agreement of the classification of them in the present research was consistent with the classification performed by another specialist.

3 RESULTS

The images of cell samples of about 1 μm on 1 mm thick microscope slides are shown in Figures 5.4.2–5.4.4. Fig. 5.4.2 shows a photograph of cutaneous tissues taken from mice with MM. Microscopic regions of 72 × 72 μm^2 in the PT images were selected from the images for analysis as shown in Figs. 5.4.3 and 5.4.4. The two bright-field images in Fig. 5.4.3 are obtained with charge coupled device (CCD) (DCC1645C, Thorlabs, NJ, USA) for the nevus sample (top left) and the MM (bottom left) with a 40× objective lens. The wavelength and power of excitation of the LD was 405 nm and 0.3 mW, respectively. Top and bottom right are the PT images using a 405 nm pump (600 × 600 pixels, 120 nm/pixel) of nevus and MM cells,

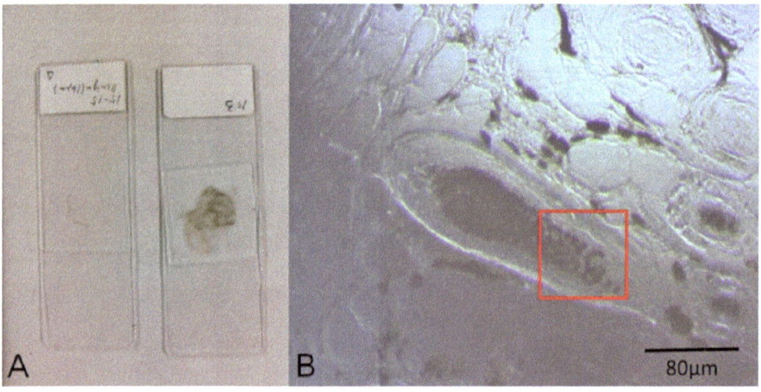

FIGURE 5.4.2 (A) Photo of the prepared specimens. (B) Charge coupled device (CCD) image of a nevus (10× objective lens). The red square area corresponds to the range of the photothermal (PT) image of Figure 5.4.3 B. The copyright of this figure is shared between the representative author (Takayoshi Kobayashi) and Optica.

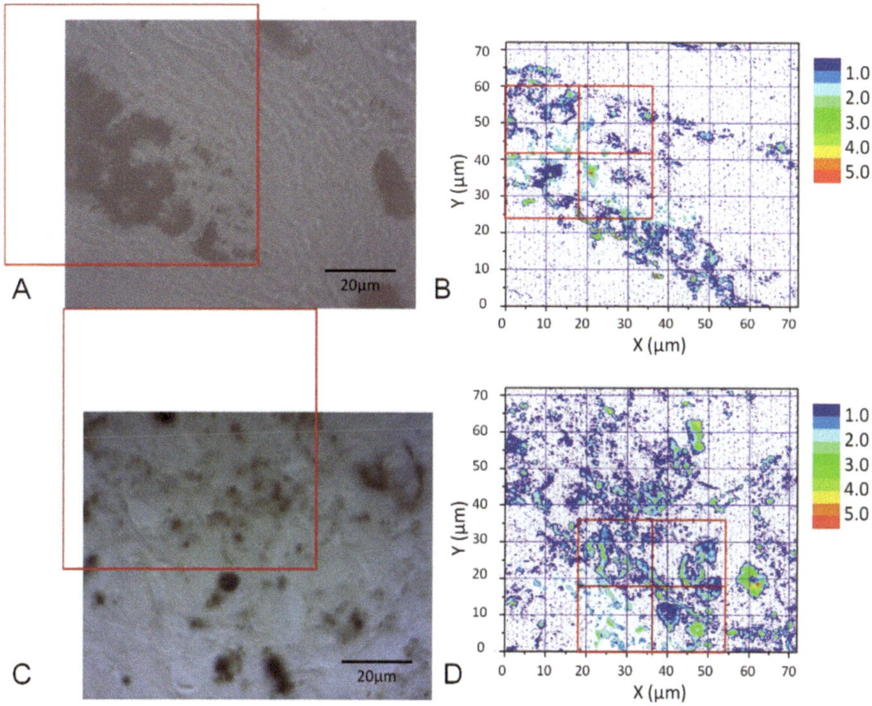

FIGURE 5.4.3 CCD images of a nevus (A) and malignant melanoma (MM) (C) (40× objective lens) and corresponding PT images (B, D) from the red squares obtained with a 405 nm pump. Four equal size (18 × 18 μm²) areas (red squares in B and D) were used for Grey Level Co-occurrence Matrix (GLCM) analysis. The copyright of this figure is shared between the representative author (Takayoshi Kobayashi) and Optica.

respectively. Red squares in the bright field images show the 72 × 72 μm² areas in the PT images. Four equal size (18 × 18 μm²) areas segmented from the red square (36 × 36 μm²) portion in the PT images were used for GLCM analysis. In Fig. 5.4.4, top left and top right are the bright field CCD images of nevus samples and MM samples, respectively, obtained with a 40× objective lens and with 488 nm excitation under the same condition as the ones in Fig. 5.4.3. The top and bottom right images are the corresponding PT images (600 × 600 pixels, 120 nm/pixel) of nevus and MM samples, respectively. Red squares in the bright field images show the 72 × 72 μm² areas of the PT images. Four equal size (18 × 18 μm²) areas were segmented out of the four red square areas in the PT images and used for GLCM analysis. Twelve areas from the PT images were evaluated and 48 calculation regions were selected for both 405 nm and 488 nm excitation.

The textural structure of the images of the mouse skin samples containing both nevus and MM cells taken with the PT imaging method was analyzed by GLCM. The nine parameters were calculated as shown below. The areas of imaging data within the red lines in Fig. 5.4.3 at 405 nm excitation were analyzed by the GLCM

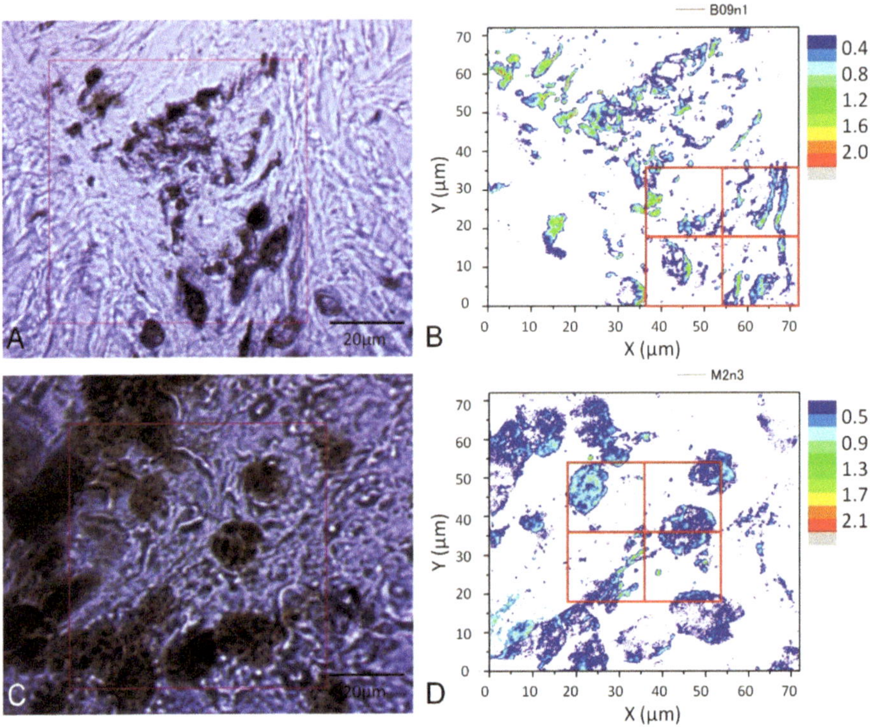

FIGURE 5.4.4 (A): CCD image and PT image of a nevus sample (40× objective lens); (C): MM (40× objective lens) with 488 nm pump. (B, D): four equal size (18 × 18 μm^2) areas segmented from the four red square areas in the PT images are used for GLCM analysis. The copyright of this figure is shared between the representative author (Takayoshi Kobayashi) and Optica.

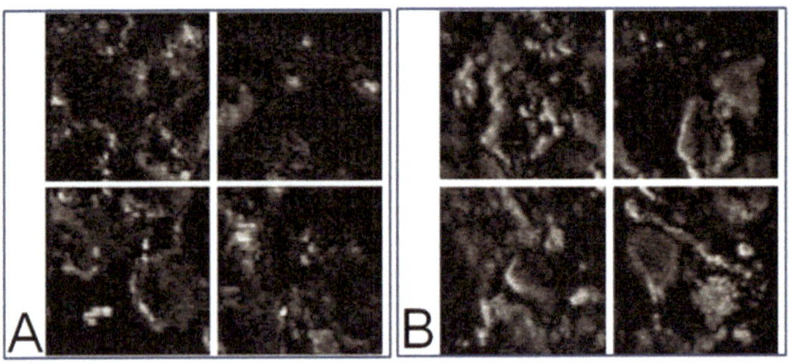

FIGURE 5.4.5 (A): Eight-bit gray-level PT image of a nevus sample (40× objective). The four images correspond to the four segmented areas represented by the four red squares in Figure 5.4.3B; (B): MM sample (40× objective). The four images correspond to segmented areas separated by the four red squares in Figure 3D.

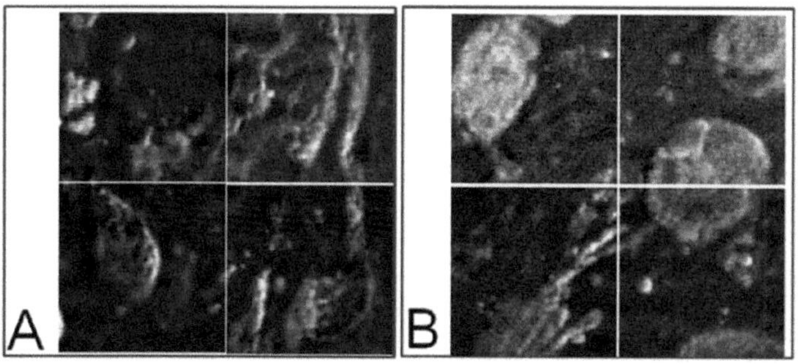

FIGURE 5.4.6 (A): Eight-bit gray level PT image of a nevus sample (40× objective). The four images correspond to the four segmented areas represented by the four red squares in Figure 5.4.4B; (B): MM sample (40× objective). The four images correspond to the four segmented areas represented by the four red squares in Figure 5.4.4D. The copyright of this figure is shared between the representative author (Takayoshi Kobayashi) and Optica.

method, and the 8-bit level gray level intensity distribution of the PT signal is shown in Figure 5.4.5. Twelve images of 72 × 72 μm^2 to multiple samples of excitation at 488 nm for both nevus and MM samples were also obtained. The areas of imaging data within the red lines in Fig. 5.4.4 (left) at 488 nm excitation were analyzed by this method, and the 8-bit level gray level intensity distribution of the PT signal is shown in Figure 5.4.6. Four sets of 18 × 18 μm^2 areas (shown in Figs. 5.4.3 and 5.4.4) with higher intensity out of the 12 images were chosen and analyzed using the GLCM analysis. There were total of 48 images with an area of 18 × 18 μm^2 from both nevus and MM. This provided sufficient data at two sets of pump wavelengths to ensure statistical reliability. As shown in the bright-field images, the sample areas are selected out of various parts of skin. For this technique to become a standard method, many more samples from a larger number of patients are required. However, as discussed below, we discovered that a few of the nine GLCM parameters clearly showed the ability to discriminate between nevus and MM, and can hopefully be used as criteria for pathological diagnosis.

Hereafter, the formalism of GLCM is described [39, 40] by showing nine selected parameters out of the most frequently used indexes.

In all of the following formulas, *P*(*i*, *j*) stands for the (*i*, *j*)th entry or value in a normalized GLCM.

$$ASM = \sum_{i=0}^{G-1}\sum_{j=0}^{G-1}\{P(i,j)\}^2 \tag{1}$$

$$\text{Contrast} = \sum_{i-j=0}^{G-1} |i-j|^2 \left\{\sum_{i=1}^{G}\sum_{j=1}^{G} P(i,j)\right\} \tag{2}$$

$$\text{Correlation} = \sum_{i=0}^{G-1}\sum_{j=0}^{G-1}\left[\frac{\{i \times j\} \times P(i,j) - \{\mu_x \times \mu_y\}}{\sigma_x \times \sigma_y}\right] \tag{3}$$

$$\text{Entropy} = -\sum_{i=0}^{G-1}\sum_{j=0}^{G-1} P(i,j) \times \log\left(P(i,j)\right) \tag{4}$$

$$\text{Homogeneity} = \sum_{i=0}^{G-1}\sum_{j=0}^{G-1} \frac{1}{1+|i-j|} P(i,j) \tag{5}$$

$$IDM = \sum_{i=0}^{G-1}\sum_{j=0}^{G-1} \frac{1}{1+(i-j)^2} P(i,j) \tag{6}$$

$$\text{Prominence} = \sum_{i=0}^{G-1}\sum_{j=0}^{G-1} \left(i+j-\mu_x-\mu_y\right)^4 P(i,j) \tag{7}$$

$$\text{Cluster Shade} = \sum_{i=0}^{G-1}\sum_{j=0}^{G-1} \left(i+j-\mu_x-\mu_y\right)^3 P(i,j) \tag{8}$$

$$\text{Variance} = \sum_{i=0}^{G-1}\sum_{j=0}^{G-1} (i-\mu)^2 P(i,j) \tag{9}$$

$$\mu_x = \sum_{i=0}^{G-1} i \sum_{j=0}^{G-1} P(i,j) \tag{10}$$

$$\mu_y = \sum_{i=0}^{G-1} j \sum_{j=0}^{G-1} P(i,j) \tag{11}$$

where $\mu_x = \mu_y = \mu$ for a symmetric matrix.

Figures 5.4.7 and 5.4.8 depict bar charts of the nine averaged calculated distance parameters d = 1–10 (corresponding to the shift distance 120–1200 nm of the image in GLCM calculations) [39, 40] and their standard deviations (SDs). In the graph, the bar is the SD of corresponding parameters calculated for the four sets of the image data shown in Figs. 5.4.5 and 5.4.6.

The features shown in Figs. 5.4.7 and 5.4.8 are summarized as follows.

The GLCM analysis of the PT images obtained at 405 nm excitation showed that the extent of the difference of nevus–MM distribution depends on the value d, and as it increases from 1 to 10, some parameters (namely Contrast, Entropy, Homogeneity, and IDM) are considerably increased, whereas others (Correlation and Variance) showed very little change. As a whole, d = 10 provided the best separation, and

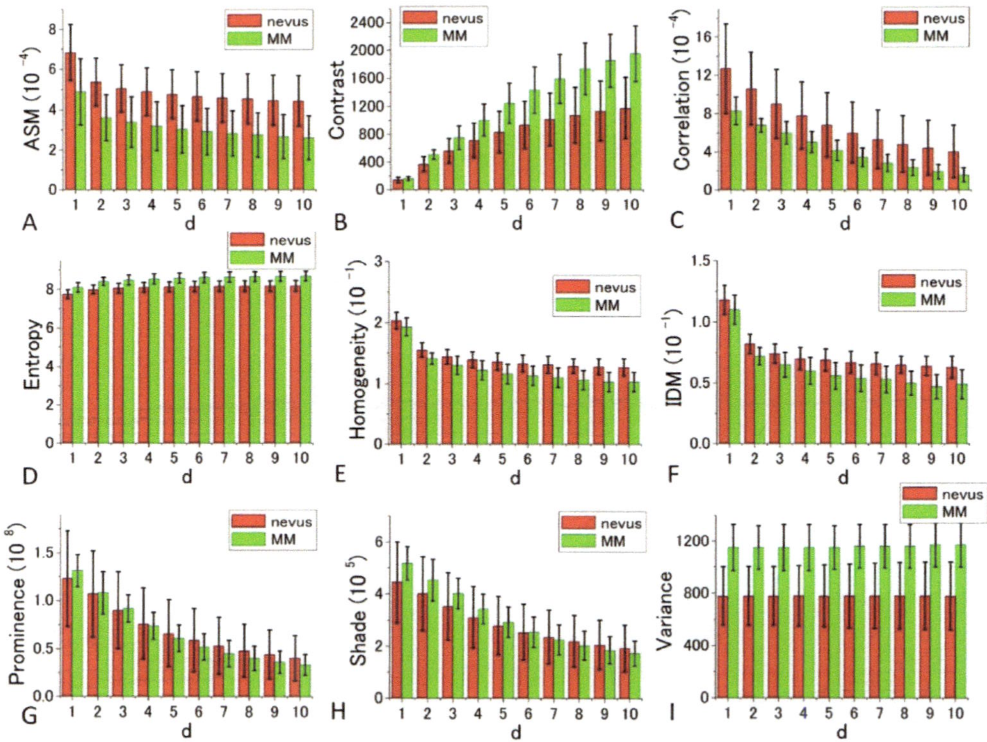

FIGURE 5.4.7 Histograms of nine calculated GLCM parameters (angular second moment (ASM), Contrast, Correlation, Entropy, Homogeneity, inverse difference moment (IDM), Prominence, Shade, Variance) for nevus (red bar), and MM samples (green bar) in 405 nm excitation. *p*-values of the *t*-test for all the parameters at $d = 10$ are, 6.72×10^{-11}, 1.16×10^{-10}, 2.71×10^{-6}, 2.93×10^{-27}, 2.36×10^{-14}, 3.69×10^{-13}, 0.374616, 0.071748, and 2.86×10^{-7} in the order of the panels, respectively. The copyright of this figure is shared between the representative author (Takayoshi Kobayashi) and Optica.

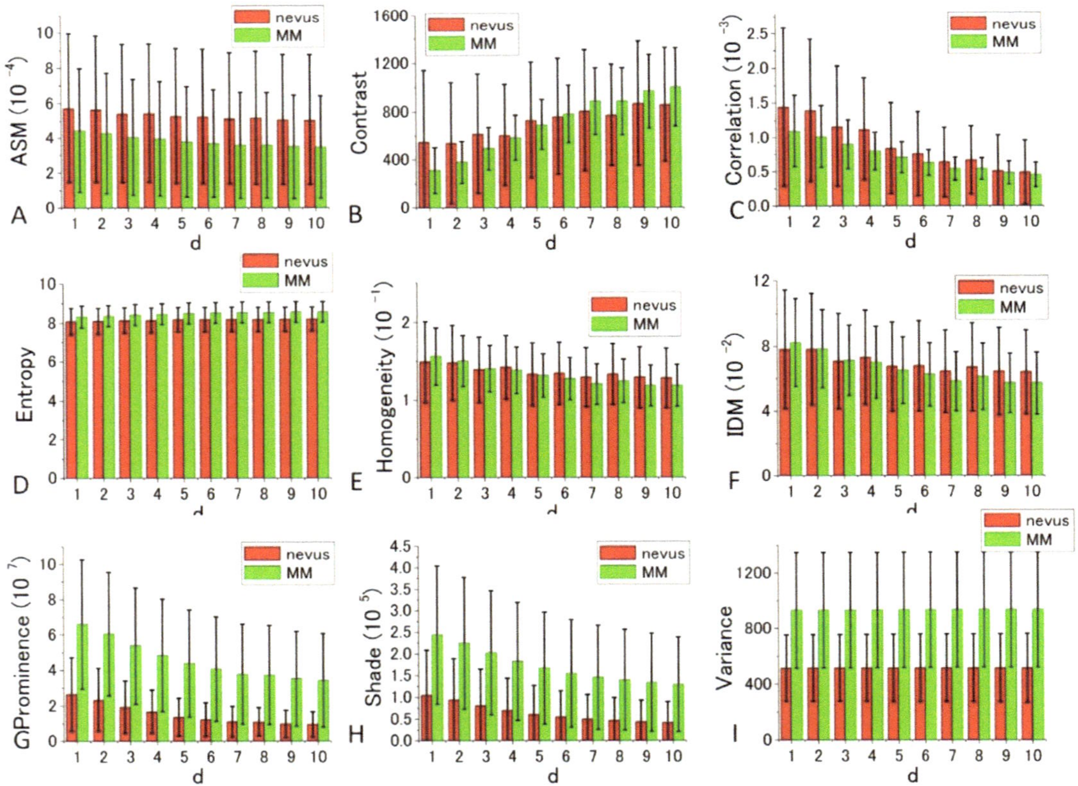

FIGURE 5.4.8 Histograms of nine calculated GLMC parameters (ASM, Contrast, Correlation, Entropy, Homogeneity, IDM, Prominence, Shade, Variance) for nevus (red bar), and MM samples (green bar) in 488 nm excitation. *p*-values of *t*-test for all the parameters at $d = 10$ are, 0.007765, 0.040389, 0.308247, 0.001409, 0.067924, 0.069963, 3.37×10^{-8}, 1.65×10^{-6}, and 2.61×10^{-8} in the order of the panels, respectively. The copyright of this figure is shared between the representative author (Takayoshi Kobayashi) and Optica.

among the nine parameters, Contrast, Entropy, and Variance showed large differences between the means of nevus and MM cells, and they were almost equal to the sum of their SD values (ratios are between 0.9 and 0.95).

GLCM analysis of the PT images obtained by 488 nm excitation was also performed. Among the nine parameters, Prominence, Shade, and Variance were well separated between the nevus and MM cells compared with the other parameters. In these cases, the ratio of the difference between the two cell types to the sum of the values of SDs at $d = 10$ were 0.737, 0.635, and 0.557, respectively. The ratios in the other parameters were smaller than 0.31. To obtain suitable parameters for the identification of MM, we analyzed the data with a commonly used diagnostic method, including the parameters sensitivity, specificity, positive predictive value, and negative predictive value.

For the diagnosis, we used Figs. 5.4.7 and 5.4.8 for each GLCM parameter for $d = 10$. We obtained the Gaussian curves fit to the data with the peaks and the widths being the mean and the SD values respectively in the histograms for $d = 10$ in Figs. 5.4.7 and 5.4.8 for each GLCM parameter; these data are shown in Figures 5.4.9 and 5.4.10.

The differences in the parameters between malignant and benign can be discriminated even though it may not provide good enough discrimination for some parameters. We call this diagnosis the GLMC-DIAG method. The results of the discrimination parameters are listed for sensitivity (Sn), specificity (Sp), accuracy (AC), positive likelihood ratio (LR+), negative likelihood ratio (LR–), positive predictive value (PPV) and negative predictive value (NPV) in Supplementary Tables S1 and S2 using the data shown in Figs. 5.4.9 and 5.4.10 in the cases of 405 nm and 488 nm excitation, respectively. These figures show that the relative positions and widths of the two excitation wavelengths are quite different from each other. This is discussed later in this subsection.

As shown in Supplementary Table S1, Variance, Entropy, and Contrast had large AC and LR+ values and a small LR– value, indicating that these three GLMC parameters are more reliable than the other GLMC parameters at 405 nm excitation. In the case of 488 nm excitation, as tabulated in Supplementary Table S2, Prominence, Shade, and Variance had large AC and LR+ values and a small LR– value, indicating that these three GLMC parameters are more reliable than the other GLMC parameters.

The distribution profiles calculated for the nine GLCM parameters at both 405 nm and 488 nm excitation are shown in Figs. 5.4.9 and 5.4.10.

To obtain suitable parameters for the identification of MM, we defined the difference criterion (DIF) as follows:

$$\mathrm{DIF} = \left|(\mathrm{PM} - \mathrm{PN})\right| / \left[(\mathrm{DM} + \mathrm{DN})/2\right] \tag{12}$$

where PM and PN are the parameter values of the MM and nevus cells, respectively, and DM and DB are the SDs of PM and PN for the four sets of imaging data. The rank orders of the magnitude and absolute values of DIF for the nine parameters

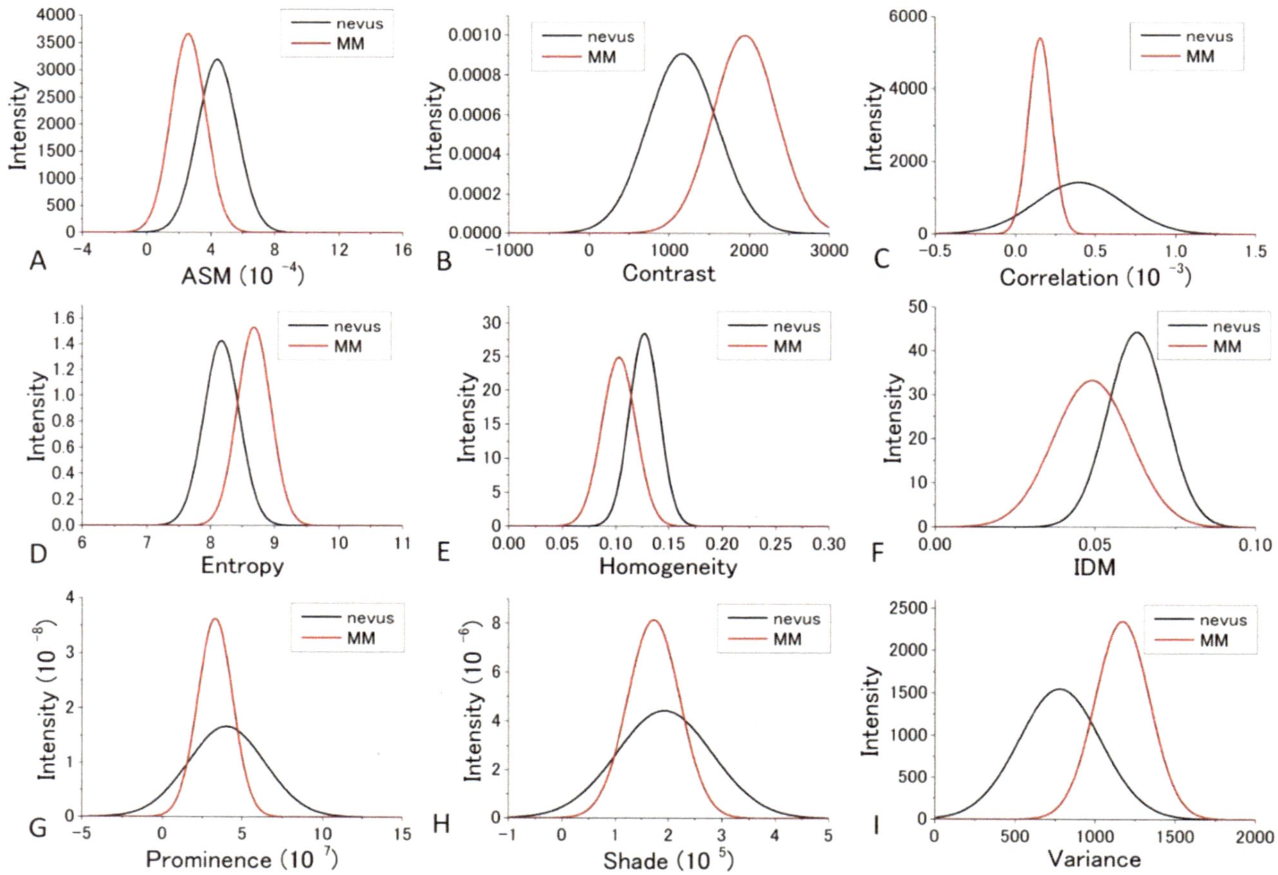

FIGURE 5.4.9 Gaussian distribution of the parameters fitted to the observed values of the parameters at 405 nm excitation. The copyright of this figure is shared between the representative author (Takayoshi Kobayashi) and Optica.

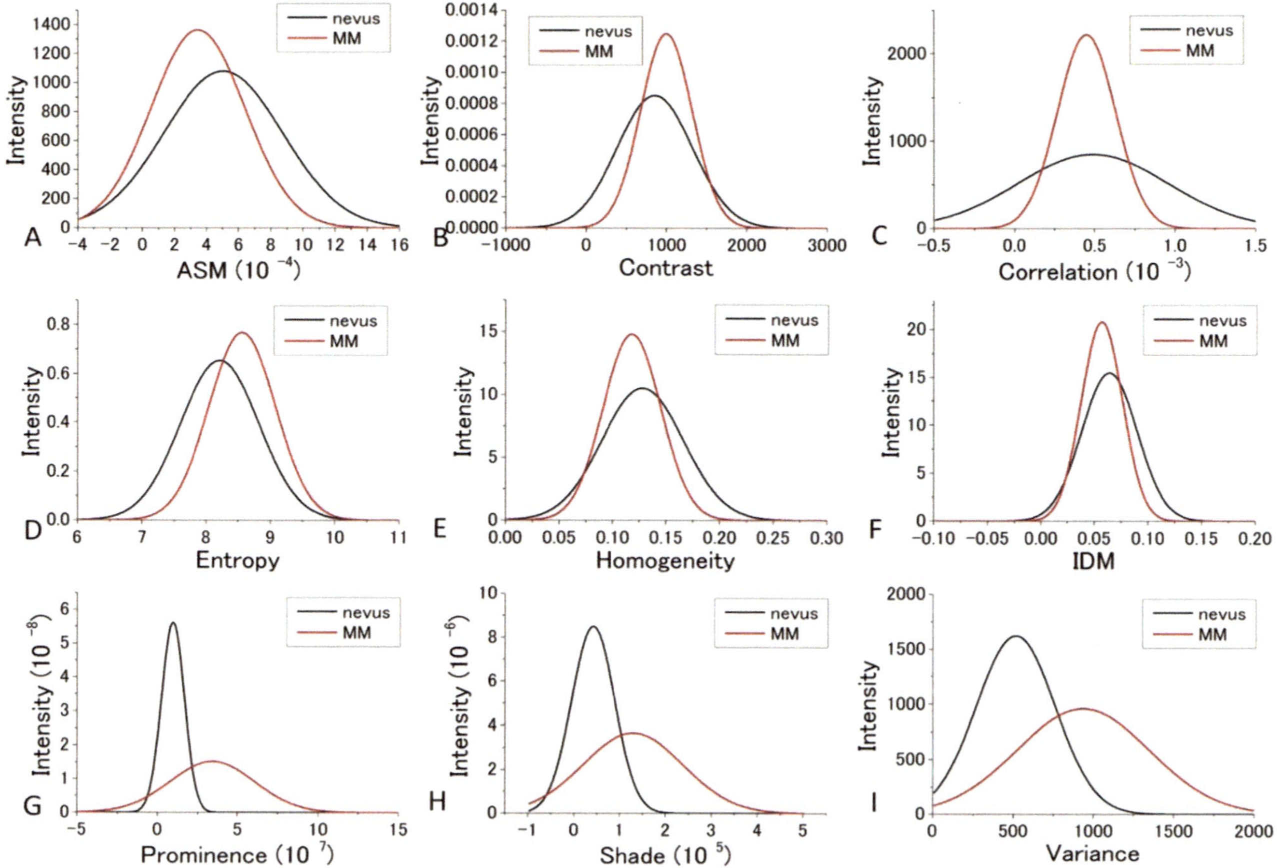

FIGURE 5.4.10 Gaussian distribution of the parameters fitted to the observed values of the parameters at 488 nm excitation. The copyright of this figure is shared between the representative author (Takayoshi Kobayashi) and Optica.

are shown in Supplementary Tables S1–S4, respectively. Entropy, Variance, and Contrast were still ranked within the top three positions when d is near 10 in the 10 different d-value sets. Therefore, these three parameters (Entropy, Variance, and Contrast) are likely to be suited for the discrimination between nevus and MM cells.

From the above orders of the DIF values in Supplementary Tables S3 and S5, it can be concluded that Entropy, Contrast, and Variance are most suited for the discrimination between MM and nevus cells in 405 nm excitation. For Entropy and Contrast, the distance parameter d = 10 gives the highest DIF of 1.879 and 1.847, respectively, and for Variance, d = 2 produced the largest value, although it was not substantially larger than those of other values of d. The probabilities of correct identification of nevus cells were 39.49%, 39.21%, and 39.49% for Entropy (d = 10), Contrast (d = 10), and Variance (d = 10), respectively, while the probabilities of correct identification of MM cells were 39.49%, 39.21%, and 39.49% for Entropy (d = 10), Contrast (d = 10), and Variance (d = 10), respectively. These values are much more accurate than those obtained by fractal analysis [6–11].

From the orders of DIF values in Supplementary Tables S2 and S4, it can be concluded that Prominence, Variance, and Shade are most suited for the discrimination between MM and nevus cells in 488 nm excitation. For Shade and Variance, the distance parameter d = 10 gave the highest DIF values of 1.121 and 1.273, respectively, while for Prominence, d = 9 produced the highest value, although the difference between DIF at other values of d was very small. The probabilities of correct identification of nevus cells are 38.9%, 33.3%, and 33.6% for Prominence (d = 10), Shade (d = 10), and Variance (d = 10), respectively. The probabilities of correct identification of MM cells are 38.9%, 33.3%, and 33.6% for Prominence (d = 10), Shade (d = 10), and Variance (d = 10), respectively. These values are much better than those obtained by fractal analysis [6–11].

These findings indicate that the GLCM parameter method, especially GLCM-DIF analysis, is a simple and useful method for the identification of suitable parameters for differentiation between different stages of cancers and detection of various types of disease that alter cell structure.

We then performed receiver operating characteristic (ROC) curve analysis based on those Gaussian curves. A ROC curve is commonly used to evaluate the diagnostic ability of a test. When a threshold parameter used in the system classifying examinees into two groups, positive and negative for some features, this curve is plotted as the sensitivity against the false positive ratio. As shown in Figs. 5.4.7 and 5.4.8, d = 10 provides the best performance for all nine GLCM parameters in both cases pumped at 405 nm and pumped at 488 nm. We plotted ROC curves for the nine parameters (Figure 5.4.11A, B) based on the Gaussian curves (Figs. 5.4.9 and 5.4.10). The area under the curve (AUC) in both of Figures 5.4.9 and 5.4.10 is an indicator of the diagnostic ability; >0.9, 0.7~0.9, and <0.7 correspond to high accuracy, moderate accuracy and poor accuracy, respectively. As shown in Figure 5.4.11A, Entropy, Contrast, and Variance show high AUCs, namely 0.909, 0.905 and 0.897, respectively. These values indicate that those parameters provide highly accurate methods to distinguish nevus and MM cells. In the case of 488 nm excitation, the AUCs of Prominence, Variance, and Shade are 0.812, 0.808, and 0.768, respectively,

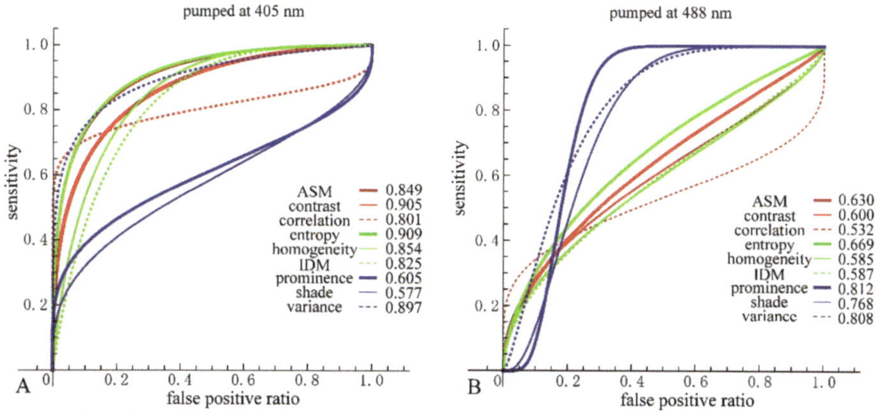

FIGURE 5.4.11 (A) Receiver operating characteristic (ROC) curves and area under the curve (AUC) for the nine parameters at 405 nm excitation. For consistency, directions of threshold-parameters scanning were set from nevus to MM. Nevus cells were considered as positive and MM cells as negative. (B) ROC curves and AUC for the nine parameters at 488 nm excitation. The copyright of this figure is shared between the representative author (Takayoshi Kobayashi) and Optica.

indicating worse performance than 405 nm excitation. Those results, both at 405 nm and 488 nm excitation, agreed with those of the DIF analysis.

4 DISCUSSION

In the previous analysis, DIF was calculated using the average of the parameters obtained from the four images which contain 48 raw images. The calculated values of DIF for the nine GLCM parameters are shown in Table S6. To further utilize the parameters obtained by the GLCM analysis, we assessed them by taking the dispersion of the distribution of the parameters into account. The mean and distribution (= dispersion) were fitted with a Gaussian distribution as shown in Figs. 5.4.9 and 5.4.10 for all nine parameters. These figures show that the values of the parameters are widely distributed and depict that the degree of overlap between the nevus and MM cells is different among the parameters. Using the distribution function, the probability of benign (defined by B = Benign/(Benign + Malignant)) was then calculated and plotted against the parameter values for the distance $d = 10$. The results are shown in Figures 5.4.12 and 5.4.13.

We tried to evaluate the appropriateness of assigning either nevus or MM to cells using another simple discrimination level for each. For this, we adopted a "clearness discrimination parameter" (DISC value), that is defined by the following equation for each parameter using the probability of benign (B) shown in Figs. 5.4.12 and 5.4.13:

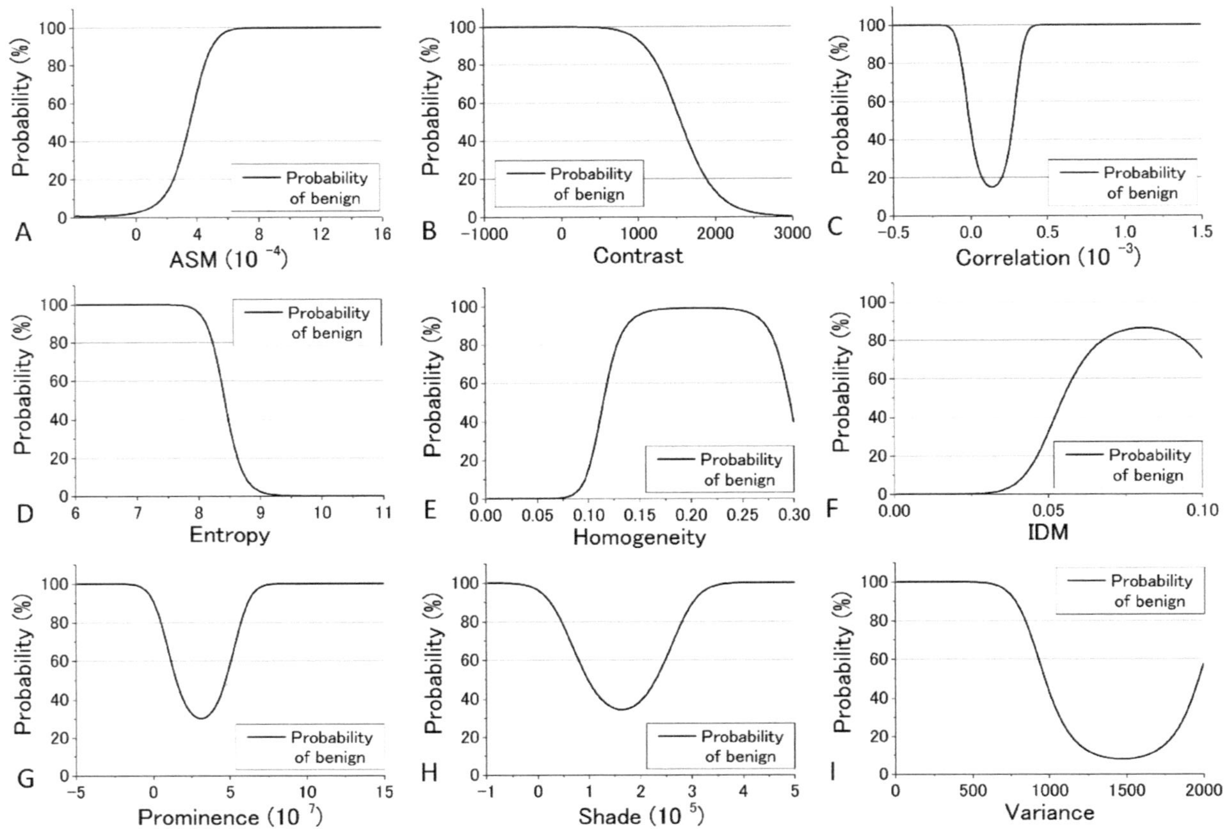

FIGURE 5.4.12 Probability of benign plotted against the parameter values for the distance $d = 10$ at 405 nm excitation.

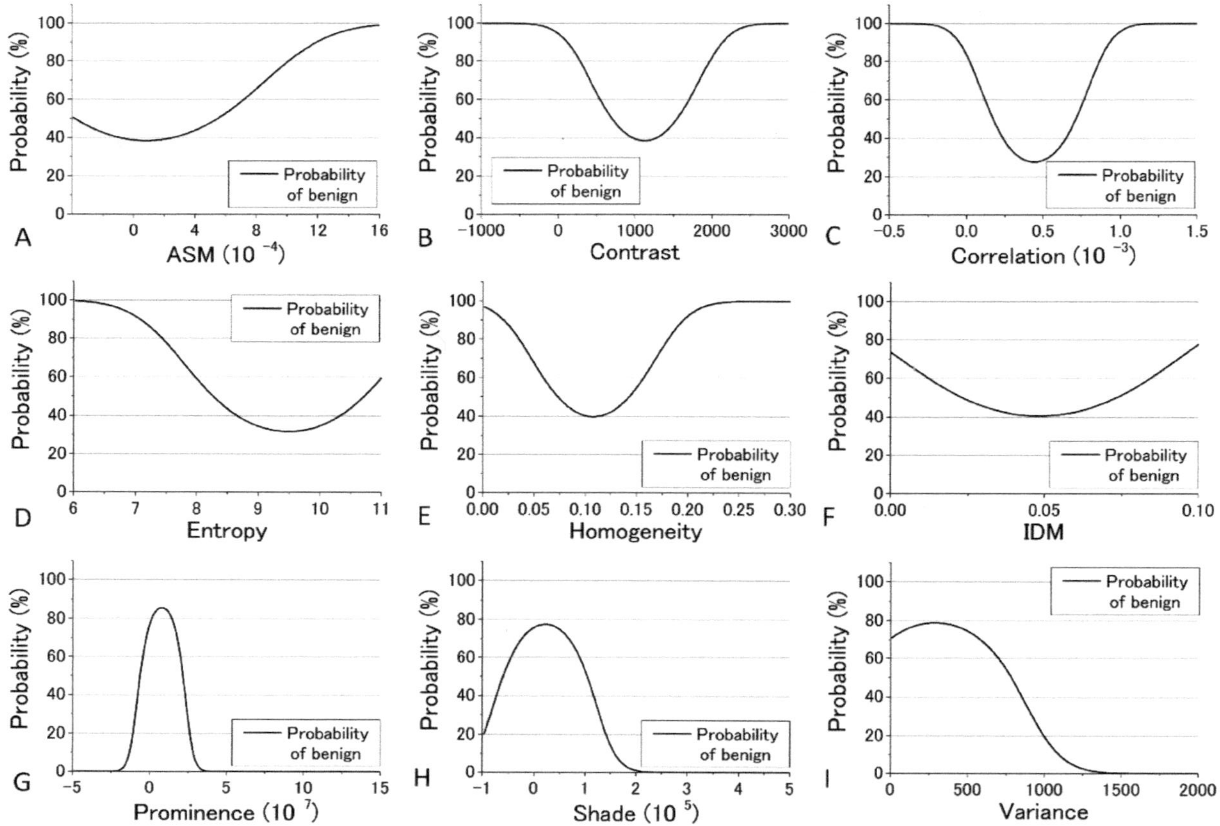

FIGURE 5.4.13 Probability of benign plotted against the parameter values for the distance $d = 10$ at 488 nm excitation. The copyright of this figure is shared between the representative author (Takayoshi Kobayashi) and Optica.

$$\text{DISC} = \{(\text{parameter at } 65\% \text{ of B}) - (\text{parameter at } 35\% \text{ of B})\} / \{(\text{parameter at } 65\% \text{ of B})/2 + (\text{parameter at } 35\% \text{ of B})/2\} \quad (13)$$

Here, "35% of B" means the value of the corresponding parameter at 35% of its maximum intensity. Usually, 90% and 10% are used for the level of discrimination or for steepness evaluation for the distribution curve; for some of the parameters, however, the B value did not reach 10% for any parameter in the present analysis. The smaller the DISC value, the more accurate the discrimination. The calculated results are shown in Tables 5.4.1 and 5.4.2 in ascending order of DISC value at 405 nm and 488 nm excitation, respectively. In 405 nm excitation, the DISCs for nevus and MM, Entropy, Homogeneity, and Variance, were conspicuously small, which

TABLE 5.4.1
The Clearness Discrimination Parameter (DISC) Value for the Nine GLCM parameters (405 nm excitation)

Parameter	65%	35%	DISC
Entropy	8.34	8.52	0.021
Homogeneity	0.121	0.108	0.112
Variance	913	1040	0.13
IDM	0.0598	0.0508	0.162
Contrast	142	1700	0.182
Correlation	3.04	2.52	0.188
ASM	3.98	3.06	0.264
Prominence	5.25×10^8	3.86×10^8	0.306
Shade	256,900	173,000	0.39

TABLE 5.4.2
DISC Value for the Nine GLCM Parameters (488 nm excitation)

Parameter	40%	60%	DISC
Entropy	8.64	7.97	0.080
Prominence	2.34×10^7	1.99×10^7	0.159
Correlation	6.47	7.65	0.166
Variance	856	700	0.200
Shade	117,000	888	0.274
Homogeneity	0.107	0.157	0.378
IDM	0.0480	0.0842	0.547
Contrast	968	544	0.560
ASM	2.60×10^{-4}	7.20×10^{-4}	0.939

was expected to be effective for that discrimination (Table 5.4.1). Additionally, in the GLCM calculation result at 488 nm excitation, the DISCs for nevus and MM, Entropy, Prominence, and Correlation, were the three smallest (Table 5.4.2). Thus, there were differences in effective parameters depending on excitation wavelength. This can be well explained in terms of the sensitivity of the signal intensity to the components in the cells, namely melanin and porphyrin, due to the differences in the absorption cross section between them.

The ROC curve is drawn by plotting the true positive ratio (sensitivity) against the false positive ratio as the threshold parameter goes through the measuring range. We plotted the ROC curves based on Figs. 5.4.9 and 5.4.10, and the curves took on distinct shapes depending on the SDs and the difference of the means of the two cell types. (1) The difference of mean values of both cell types were near the SDs, as seen in the case of ASM, Contrast, Entropy, Homogeneity, IDM, and Variance at 405 nm excitation and Entropy at 488 nm excitation. In this case, the ROC curves rose quite precipitously, gradually decreased their slopes and got almost horizontal at the end, however, always kept convex upward. (2) SDs of the parameters for the nevus cells were large enough that the distribution for them spread to cover the main part of the distribution for MM cells, as seen in the case of Correlation, Prominence, and Shade at 405 nm excitation and Correlation at 488 nm excitation. In this case, the ROC curve rose precipitously, however, after a short time, decreased its slopes to nearly horizontal, and again got steep at the end. This is because at first, only nevus cells were considered as positive and MM cells were not included within the threshold, and hence the false positive was kept very low. As the threshold parameter passed through the main part of MM cells, the cumulating values were mainly due to MM cells, resulting in the horizontal shape. After the threshold-parameter passed the main part of MM cells, nevus cells mainly count for the positives. (3) SDs of the parameters for the MM cells were large enough that the distribution for them spread to cover the main part of the distribution for nevus cells, as seen in the case of Prominence, Shade, and Variance at 488 nm excitation. In this case, the ROC curves first went horizontally, however, after a short time, rose up very steeply and got to nearly horizontal at the end. The reason for this behavior is the reverse of (2).

As mentioned previously, analyses of ROC and DIF agreed quite well at both 405 nm and 488 nm. However, there were some differences between the analyses results of ROC or DIF and DISC, especially at 488 nm excitation. The correlation diagrams between AUC and DISC at 405 nm and at 488 nm excitation are shown in Figure 5.4.14. The correlation coefficient at 405 nm excitation was −0.844, indicating a strong correlation, and at 488 nm excitation, it was −0.3735, showing a weak correlation.

Analyses based on DIF, ROC, and DISC for evaluating the diagnostic ability of GLCM-parameters showed little difference between 405 nm and 488 nm excitation. The difference between the calculation results for experiments performed at 405 nm and 488 nm excitation can be explained in more detail as follows.

The molar extinction coefficients of melanin at 405 nm and 488 nm in the literature are approximately 2500 and 1500/mol/cm, respectively [41]. In contrast, the

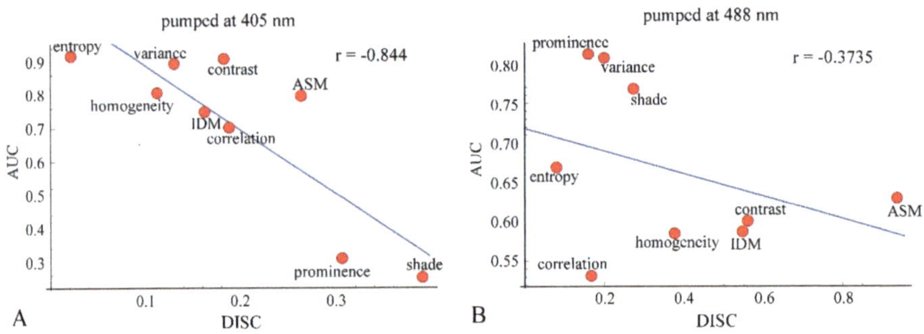

FIGURE 5.4.14 Correlation diagrams between AUC and DISC at 405 nm excitation (A) and at 488 nm excitation (B). The copyright of this figure is shared between the representative author (Takayoshi Kobayashi) and Optica.

molar extinction coefficient of hemoglobin at 405 nm and 488 nm is about 275,000 and 16,000/mol/cm, respectively. At both wavelengths, the molar extinction coefficient of hemoglobin is high, but the absolute values are quite different. The extinction coefficient of hemoglobin at 488 nm is 10 times larger than that of melanin, while the coefficient of hemoglobin at 405 nm is 100 times larger than that of melanin. The amount of melanin contained in the cell slice is 10 times greater than that of hemoglobin. Thus, it might be possible to assess the state or degree of transformation of the skin tissue by monitoring with GLCM analysis of 488 nm pump images. It may be possible that at 488 nm excitation, Prominence and Entropy detected the changes of melanin distribution induced by the transformation of cells.

At 405 nm excitation, Homogeneity and Entropy might detect the changes in hemoglobin distribution induced by cellular transformation.

We defined DISC using 65% and 35% DIF values instead of 10% and 90% values, which are commonly used. This might diminish the performance of DISC analysis and might provide some discordance with ROC analysis (Fig. 5.4.14 A, B). We devised the DISC analysis as a simple and versatile method to evaluate the diagnostic abilities of the GLCM parameters. Applying this to a wider range of data will hopefully improve the performance of DISC analysis.

In view of the purpose of this study, it can be concluded that melanin observation by 488 nm excitation is more suitable for the determination of cancerous tumors, so the conclusion is that Entropy and Prominence at 488 nm excitation are suitable for benign-malignancy determination. By utilizing 405 nm excitation, we may be able to study the effect of cancerous tumors on the hemoglobin-containing tissues, such as muscle attached to the sample slices. In this case, Homogeneity and Entropy can be used for benign-malignancy determination. This may correspond to the morphological change in hemoglobin distribution induced by cellular tumorigenesis. This means that the spatial distribution of cancerous tumors can be investigated through hemoglobin.

5 CONCLUSIONS

Label-free confocal photothermal (CPT) microscopy combined with a texture analysis method was utilized for the first time to investigate malignancy in mouse skin cells in terms of benign-malignancy determination. A Grey Level Co-occurrence Matrix (GLCM) method for texture analysis was applied to the CPT images of malignant melanoma (MM) to study the differences in intracellular super-resolved structural properties between MMs and nevus. We first applied DIF (difference) analysis and obtained the results that Entropy, Contrast, and Variance were most suited for the discrimination between MM and nevus when 405 nm excitation was used, and Prominence, Variance, and Shade were most suited at 488 nm excitation. Analysis based on ROC gave the same results. DISC, which we introduced as a new discrimination parameter, suggested Entropy, Homogeneity, and Variance were best at 405 nm excitation, and Entropy, Prominence, and Correlation were best at 488 nm excitation. Observation at 405 nm excitation might detect the morphological change in hemoglobin distribution induced by cellular transformation, whereas observation at 488 nm excitation caught the change of melanin distribution. The differences in effective parameters due to differences in excitation wavelength can be well explained in terms of the sensitivity of the signal intensity to the components in the cells, namely melanoma and porphyrin (hemoglobin), due to the difference in the absorption cross section between them. The contents described in this subsection are the product of the collaborative activity of the following people: T. Kobayashi, K. Nakata, I. Yajima, M. Kato, and H. Tsurui.

Supplementary Materials: The following are available online at www.mdpi.com/xxx/s1, Table S1: Results of calculated GLCM diagnosis method (405 nm excitation), Table S2: Results of calculated GLCM diagnosis method (488 nm excitation), Table S3: Order of the DIF values among the nine parameters (405 nm excitation), Table S4: Order of the DIF values among the nine parameters (488 nm excitation), Table S5: Calculated values of DIF for the nine GLCM parameters (average of 48 data from 48 images), Table S6: Calculated values of DIF for the nine GLCM parameters (average of data from 48 images).

Table S1. Results of calculated GLCM diagnosis method (405 nm excitation).

Parameter	SN	SP	AC	LR+	LR−	PPV	NPV
ASM	0.8017	0.7630	0.7823	3.382	0.2600	0.7717	0.7938
IDM	0.6972	0.8224	0.7599	3.927	0.3681	0.7970	0.7310
Contrast	0.8350	0.8100	0.8225	4.394	0.2038	0.8146	Accuracy

Table S2. Results of calculated GLCM diagnosis method (488 nm excitation).

Parameter	SN	SP	AC	LR+	LR−	PPV	NPV
ASM	0.722	0.461	0.592	1.34	0.603	0.573	0.624
Contrast	0.826	0.370	0.598	1.31	0.471	0.567	0.680
Correlation	--	--	--	--	--	--	--
Entropy	0.706	0.547	0.627	1.56	0.537	0.609	0.651
Homogeneity	0.813	0.356	0.585	1.26	0.526	0.558	0.656
IDM	0.811	0.350	0.580	1.25	0.540	0.555	0.649
Prominence	0.683	0.954	0.819	14.8	0.332	0.937	0.751
Shade	0.592	0.908	0.750	6.42	0.450	0.865	0.690
Variance	0.647	0.858	0.753	4.57	0.411	0.820	0.709

Table S3. Order of the DIF values among the nine parameters (405 nm excitation).

d	ASM	Contrast	Correlation	Entropy	Homogeneity	IDM	Prominence	Shade	Variance
1	4	8	3	2	6	7	9	5	1
2	5	4	3	2	6	7	9	8	1
3	3	6	4	2	5	7	9	8	1
4	3	4	5	2	6	7	9	8	1
5	3	4	7	2	5	6	8	9	1
6	4	3	7	2	5	6	8	9	1
7	4	3	7	2	5	6	8	9	1
8	6	3	7	1	5	4	8	9	2
9	6	3	7	1	5	4	8	9	2
10	4	2	7	1	5	6	8	9	3

Table S4. Order of the DIF values among the nine parameters (488 nm excitation).

d	ASM	Contrast	Correlation	Entropy	Homogeneity	IDM	Prominence	Shade	Variance
1	7	4	5	6	8	9	1	3	2
2	7	5	4	6	8	9	1	3	2
3	6	7	5	4	8	9	1	3	2
4	6	9	4	5	8	7	1	3	2
5	5	7	6	4	9	8	1	3	2
6	5	9	6	4	8	7	1	3	2
7	5	9	8	4	7	6	1	3	2
8	5	7	6	4	8	9	1	3	2
9	5	8	9	4	7	6	1	3	2
10	5	6	9	4	7	8	1	3	2

Table S5. Calculated values of DIF for the nine GLCM parameters (average of 48 data from 48 images).

d	ASM	Contrast	Correlation	Entropy	Homogeneity	IDM	Prominence	Shade	Variance
1	1.274	0.589	1.423	1.504	0.623	0.602	0.2413	0.6622	1.832
2	1.525	1.581	1.659	1.799	1.408	1.369	0.0372	0.4588	1.860
3	1.376	1.059	1.256	1.641	1.065	1.000	0.0783	0.5325	1.844
4	1.419	1.225	1.205	1.685	1.169	1.066	0.0692	0.3591	1.835
5	1.452	1.365	1.195	1.708	1.291	1.206	0.1987	0.1744	1.827
6	1.465	1.486	1.212	1.734	1.332	1.240	0.2988	0.0120	1.814
7	1.498	1.604	1.256	1.797	1.430	1.329	0.3670	0.1179	1.802
8	1.524	1.710	1.308	1.820	1.658	1.665	0.4063	0.2081	1.802
9	1.543	1.792	1.347	1.854	1.750	1.789	0.4205	0.2625	1.816
10	1.555	1.847	1.350	1.879	1.512	1.386	0.4054	0.2837	1.836

Table S6. Calculated values of DIF for the nine GLCM parameters (average of data from 48 images).

d	ASM	Contrast	Correlation	Entropy	Homogeneity	IDM	Prominence	Shade	Variance
1	0.334	0.603	0.415	0.372	0.175	0.127	1.374	1.052	1.268
2	0.365	0.474	0.507	0.429	0.049	0.005	1.412	1.063	1.268
3	0.376	0.367	0.428	0.473	0.042	0.015	1.451	1.074	1.269
4	0.407	0.074	0.614	0.528	0.112	0.123	1.452	1.08	1.27
5	0.421	0.11	0.291	0.549	0.082	0.094	1.477	1.084	1.271
6	0.449	0.061	0.336	0.582	0.217	0.228	1.465	1.088	1.272
7	0.456	0.206	0.279	0.606	0.282	0.285	1.453	1.094	1.273
8	0.462	0.327	0.399	0.61	0.26	0.248	1.459	1.102	1.273
9	0.472	0.248	0.084	0.627	0.317	0.325	1.479	1.108	1.273
10	0.473	0.367	0.137	0.628	0.315	0.312	1.472	1.121	1.273

NOTES

1. transgenic mice (RET-Tg): RET is the abbreviation of "rearranged during transfection". Transgenic mouse (singular form of mice) is genetically modified mouse or genetically engineered mouse model (GEMM), which is a mouse that has had its genome altered through the use of genetic engineering techniques. Genetically modified mice are commonly used for research or as animal models of human diseases and are also used for research on genes.
2. hybrid oncogene: An oncogene is a gene that has the potential to cause cancer. In tumor cells, these genes are often mutated, or expressed at high levels.
3. RFP: RFP is the abbreviation of "Ret finger protein (RFP)". Its basic helix–loop–helix (bHLH) transcription factors play a pivotal role in the regulation of tumorigenesis, and also in a wide range of other developmental processes in diverse species from yeast to humans.
4. c-RET:RFP-RET is a hybrid oncogene (a gene that has the potential to cause cancer. In tumor cells, these genes are often mutated, or expressed at high levels) between c-RET and RFP.

REFERENCES

1. American Cancer Society. Melanoma Skin Cancer Overview. 2015. Available online: https://www.cancer.org/cancer/melanoma-skin-cancer.html (accessed on 14 August **2018**.
2. Siegel, R.L.; Miller, K.D.; Jemal, A. Cancer statistics. *CA Cancer J. Clin.* **2015**, *65*, 5–29.
3. Rigel, D.S.; Carucci, J.A. Malignant melanoma: Prevention, early detection, and treatment in the 21st century. *CA Cancer J. Clin.* **2000**, *50*, 215–236.

4. Soyer, H.P.; Smolle, J.; Kerl, H.; Stetner, H. Early diagnosis of malignant melanoma by surface microscopy. *Lancet* **1987**, *2*, 803.
5. Nachbar, F.; Stolz, W.; Merkle, T.; Cognetta, A.B.; Vogt, T.; Landtthaler, M.; Bilek, P.; BraunFalco, O.; Plewig, G. The ABCD rule of dermatscopy. *J. Am. Acad. Dermatol.* **1994**, *30*, 551–559.
6. Ahammer, H.; DeVaney, T.T.J.; Tritthart, H.A. Fractal dimension of K1735 mouse melanoma clones and spheroid invasion in vitro. *Eur. Biophys. J.* **2001**, *30*, 494–499.
7. Baish, J.W.; Jain, R.K. Fractals and cancer. *Cancer Res.* **2000**, *60*, 3683–3688.
8. Bedin, V.; Adam, R.L.; de Sa, B.C.S.; Landman, G.; Metze, K. Fractal dimension of chromatin is an independent prognostic factor for survival in melanoma. *Cancer* **2010**, *10*, 260, doi:10.1186/1471-2407-10-260.
9. Guz, N.V.; Dokukin, M.E.; Woodworth, C.D.; Cardin, A.; Sokolov, I. Towards early detection of cervical cancer: Fractal dimension of AFM images of human cervical epithelial cells at different stages of progression to cancer. *Nanomed. Nanothechnol.* **2015**, *11*, 1667–1675.
10. Davies, N.A.; Harrison, N.E.; Morris, R.H.K.; Noble, S.; Lawrence, M.J.; D'Silva, L.A.; Broome, L.; Brown, M.R.; Hawkins, K.M.; Williams, P.R.; et al. Fractal dimension (df) as a new structure biomarker of clot microstructure in different stages of lung cancer. *Thromb. Haemost.* **2015**, *114*, 1251–1259.
11. Waliszewski, P.; Wagenlehner, F.; Gattenlöhner, S.; Weidner, W. On the relationship between tumor structure and complexity of the spatial distribution of cancer cell nuclei: A fractal geometric model of prostate carcinoma. *Prostate* **2015**, *75*, 399–414.
12. Klein, K.; Maier, T.; Warneken, V.C.H.; Spatz, J.P. Marker-free phenotyping of tumor cells by fractal analysis of reflection interference contrast microscopy images. *Nano Lett.* **2013**, *13*, 5474–5479.
13. Jimbow, K.; Miyake, Y.; Homma, K.; Yasuda, K.; Izumi, Y.; Tsutsumi, A.; Ito, S. Characterization of melanogenesis and morphogenesis of melanosomes by physicochemical properties of melanin and melanosomes in malignant melanoma. *Cancer Res.* **1984**, *44*, 1128–1134.
14. Matthews, T.E.; Piletic, I.R.; Selim, M.A.; Simpson, M.J.; Warren, W.S. Pump-probe imaging differentiates melanoma from melanocytic Nevi. *Sci. Transl. Med.* **2011**, *3*, 71ra15, doi:10.1126/scitranslmed.3001604.
15. Zonios, G.; Dimou, A.; Bassukas, I.; Galaris, D.; Tsolakidis, A.; Kaxiras, E. Melanin absorption spectroscopy: New method for noninvasive skin investigation and melanoma detection. *J. Biomed. Opt.* **2008**, *13*, 014017, doi:10.1117/1.2844710.
16. Tokeshi, M.; Uchida, M.; Hibara, A.; Sawada, T.; Kitamori, T. Determination of subyoctomole amounts of nonfluorescent molecules using a thermal lens microscope: Subsingle-molecule determination. *Anal. Chem.* **2001**, *73*, 2112–2116.
17. Selmke, M.; Braun, M.; Cichos, F. Photothermal single-particle microscopy: Detection of a nanolens. *ACS Nano* **2012**, *6*, 2741–2749.
18. Selmke, M.; Cichos, F. Photothermal single particle Rutherford scattering microscopy. *Phys. Rev. Lett.* **2013**, *110*, 103901, doi:10.1103/physrevlett.110.103901.
19. Boyer, D.; Tamarat, P.; Maali, A.; Lounis, B.; Orrit, M. Photothermal imaging of nanometer-sized metal particles among scatterers. *Science* **2002**, *297*, 1160–1163.
20. He, J.; Miyazaki, J.; Wang, N.; Tsurui, H.; Kobayashi, T. Label-free imaging of melanoma with nonlinear photothermal microscopy. *Opt. Lett.* **2015**, *40*, 1141–1144.
21. Lu, S.; Min, W.; Chong, S.; Holtom, G.R.; Xie, X.S. Label-free imaging of heme proteins with two-photon excited photothermal lens microscopy. *Appl. Phys. Lett.* **2010**, *96*, 113701, doi:10.1063/1.3308485.
22. Nedosekin, D.A.; Galanzha, E.I.; Dervishi, E.; Biris, A.S.; Zharov, V.P. Super-resolution nonlinear photothermal microscopy. *Small* **2014**, *10*, 135–142.

23. Nedosekin, D.A.; Juratli, M.A.; Sarimollaoglu, M.; Moore, C.L.; Rusch, N.J.; Smeltzer, M.S.; Zharov, V.P.; Galanzha, E.I. Photoacoustic and photothermal detection of circulating tumor cells, bacteria and nanoparticles in cerebrospinal fluid in vivo and ex vivo. *J. Biophotonics* **2013**, *6*, 523–533.
24. Moreau, J.; Loriette, V. Confocal thermal-lens microscope. *Opt. Lett.* **2004**, *29*, 1488–1490.
25. Miyazaki, J.; Tsurui, H.; Hayashi-Takagi, A.; Kasai, H.; Kobayashi, T. Sub-diffraction resolution pump-probe microscopy with shot-noise limited sensitivity using laser diodes. *Opt. Exp.* **2014**, *22*, 9024–9032.
26. Miyazaki, J.; Iida, T.; Tanaka, S.; Hayashi-Takagi, A.; Kasai, H.; Okabe, S.; Kobayashi, T. Fast 3D visualization of endogenous brain signals with high-sensitivity laser scanning photothermal microscopy. Biomed. *Opt. Exp.* **2016**, *7*, 1702, doi:10.1364/boe.7.001702.
27. Miyazaki, J.; Tsurui, H.; Kawasumi, K.; Kobayashi, T. Sensitivity enhancement of photothermal microscopy with radially segmented balanced detection. *Opt. Lett.* **2015**, *40*, 479–482.
28. He, J.; Wang, N.; Tsurui, H.; Kato, M.; Iida, M.; Kobayashi, T. Noninvasive, label-free, three-dimensional imaging of melanoma with confocal photothermal microscopy: Differentiate malignant melanoma from benign tumor tissue. *Sci. Rep.* **2016**, *6*, 30209, doi:10.1038/srep30209.
29. Kato, M.; Takeda, K.; Kawamoto, Y.; Tsuzuki, T.; Hossain, K.; Tamakoshi, A.; Kunisada, T.; Kambayashi, Y.; Ogino, K.; Suzuki, H.; et al. c-Kit-targeting immunotherapy for hereditary melanoma in a mouse model. *Cancer Res.* **2004**, *64*, 801–806.
30. Kumasaka, M.Y.; Yajima, I.; Hossain, K.; Iida, M.; Tsuzuki, T.; Ohno, T.; Takahashi, M.; Yanagisawa, M.; Kato, M. A novel mouse model for de novo Melanoma. *Cancer Res.* **2010**, *70*, 24–29.
31. Ohshima, Y.; Yajima, I.; Takeda, K.; Iida, M.; Kumasaka, M.; Mastumoto, M.; Kato, M. c-RET molecule in malignant melanoma from oncogenic RET-carrying transgenic mice and human cell lines. *PLoS One.* **2010**, *5*, e10279.
32. Pantic, I.; Pantic, S.; Paunovic, J.; Perovic, M. Nuclear entropy, angular second moment, variance and texture correlation of thymus cortical and medullar lymphocytes: Grey level co-occurrence matrix analysis. *An. Acad. Bras. Ciênc.* **2013**, *85*, 1063–1072.
33. Gebejes, A.; Huertas, R. Texture characterization based on grey-level co-occurrence matrix. In *Proceedings of the 2nd International Conference of Informatics and Management Sciences (ICTIC 2013)*, Zilina, Slovakia, 25–29 March **2013**; pp. 375–378.
34. Losa, G.A.; Castelli, C. Nuclear patterns of human breast cancer cells during apoptosis: Characterisation by fractal dimension and co-occurrence matrix statistics. *Cell Tissue Res.* **2005**, *322*, 257–267.
35. Honeycutt, C.E.; Plotnick, R. Image analysis techniques and grey-level co-occurrence matrices (GLCM) for calculating bioturbation indices and characterizing biogenic sedimentary structure. *Comput. Geosci.* **2008**, *34*, 1461–1472.
36. Xian, G. An identification method of malignant and benign liver tumors from ultrasonography based on GLCM texture features and fuzzy SVM. *Expert Syst. Appl.* **2010**, *37*, 6737–6741.
37. Kim, T.Y.; Cho, N.H.; Jeong, G.B.; Bengtsson, F.; Choi, H.K. 3D texture analysis in renal cell carcinoma tissue image grading. *Comput. Math. Methods Med.* **2014**, *12*, 536217, doi:10.1155/2014/536217.
38. Pantic, I.; Pantic, S.; Jovanovic, B. Microsc. Grey level co-occurrence matrix (GLCM) texture analysis of germinal center light zone lymphocyte nuclei: Physiology viewpoint with focus on apoptosis. *Microsc. Microanal.* **2012**, *18*, 470–475.
39. Parker, J.R. *Algorithms for Image Processing and Computer Vision*; John Wiley and Sons: New York, NY, USA, **1997**; pp. 150–175.

40. Walker, R.F.; Jackway, P.; Longstaff, I.D. Improving co-occurrence matrix feature discrimination. In *Proceedings of the 3rd Conference on Digital Image Computing: Techniques and Application (DICTA'95)*, Brisbane, Australia, 6–8 December **1995**; pp. 643–648.
41. Zijlstra, W.G.; Buursma, A. Spectrophotometry of hemoglobin: Absorption spectra of bovine oxyhemoglobin, deoxyhemoglobin, carboxyhemoglobin, and methemoglobin. *Comp. Biochem. Physiol. B Biochem. Mol. Biol.* **1997**, *118*, 743–749.
42. Kobayashi, T. Nakata, K, Yajima. I, Kato, M. Tsurui. H. Label-free imaging of melanoma with confocal photothermal microscopy: Differentiation between malignant and benign tissue. *Bioengineering.* **2018,** 67–83 (2018).

6.1 Development of a Multiplex Stimulated Raman Microscope for Spectral Imaging through Multi-channel Lock-in Detection

1 INTRODUCTION

Various chemical imaging methods have been developed to study the spatial dependency of the characteristics and functionalities of materials such as living cells from the distribution of their chemical structures. Raman imaging [1–3], in which an image is constructed using molecular vibrations, has several advantages. It enables contactless, non-destructive, non-labeled, and minimally invasive analysis, and is favorable for biological samples because of its insusceptibility to water compared with infrared (IR) microscopy. Spatial resolution is in the sub-micrometer order when visible probe light is used in Raman microscopy, whereas that of IR microscopy is in the 10-μm order.

The signal intensity of spontaneous Raman scattering, however, is significantly low, which results in a long acquisition time for the image. Imaging by coherent anti-Stokes Raman scattering (CARS) has been used to overcome this problem [4–8]. This method produces signal intensity of about three orders of magnitude higher than the spontaneous Raman microscopy, and has even achieved video-rate imaging [6–8].[8] The CARS spectra, however, are distorted by non-resonant backgrounds (NRB) that complicate the assignment of signals and cause image artifacts. Optical and mathematical methods have been devised to overcome this problem.[4,5]

Recently, stimulated Raman microscopy has been extensively used in high-speed chemical imaging, circumventing the NRB problem [9–19]. Stimulated Raman spectra conserve high fidelity to spontaneous Raman spectra as they are NRB-free and the signal intensity is proportional to the Raman scattering cross-section, which ensures the facile assignment of observed signals. The signal intensities are also proportional to the chemical concentration, which enables straightforward quantitative

DOI: 10.1201/9781003476030-22

analysis [16]. In the experimental setup of the stimulated Raman observation, precisely synchronized pulse lasers of at least two wavelengths are spatially overlapped on the sample. The higher energy laser in Figure 6.1.1 is labeled the pump beam, and its counterpart is labeled the Stokes beam. When the difference between the photon energies is set at the vibrational energy of the molecule, one photon of the pump beam is annihilated and one photon of the Stokes beam is generated. This annihilation and generation is known as Raman loss and gain, and the stimulated Raman signals are observed as either loss or gain processes.

Stimulated Raman spectra can be observed with white light and monochromatic beams using a spectrograph [20] (upper right panel of Fig. 6.1.1). This method is applied for stimulated Raman microscopy for spectral imaging with a charge-coupled device (CCD) [9]. A stimulated Raman spectrum is obtained by the quotient between the white light spectra with or without the monochromatic beam. The dynamic range of the CCD, however, is not very wide (~60–70 dB) and it is prone

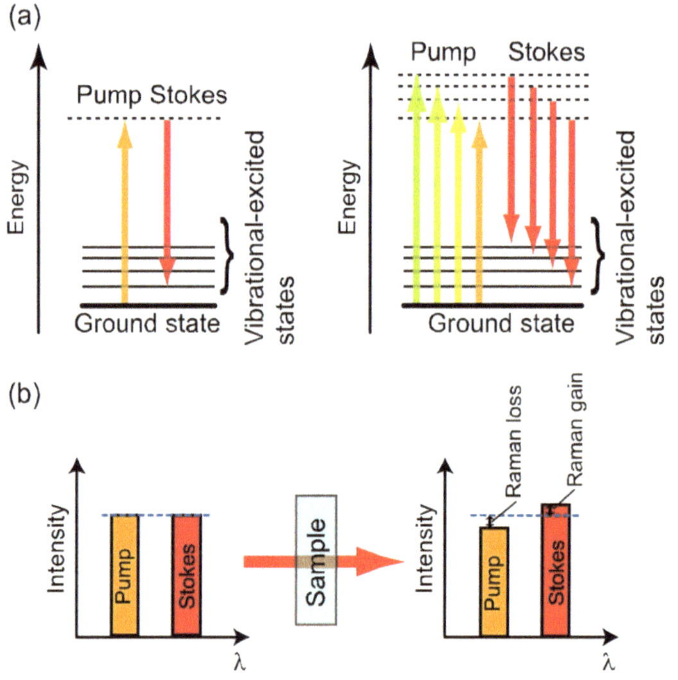

FIGURE 6.1.1 Schematics of interaction between the pump beam and Stokes beam through a molecule with a vibrational energy level (a) and intensity modulation of the beams induced by Raman loss and gain (b). The left panel of (a) illustrates the detection of a single vibration of the molecule with the monochromatic pump and Stokes beams in conjunction with the molecular vibrational levels. The right panel of (a) illustrates our multiplex detection of the molecular vibrations with the broad-band pump beam and monochromatic Stokes beam. The copyright of this figure is shared between the representative author (Takayoshi Kobayashi) and John Wiley.

to saturation, which limits the signal-to-noise ratio (SNR) even when the pump laser is at an elevated intensity. To achieve a wider dynamic range, the lock-in detection of the pump or Stokes beam is used with the intensity modulation of the counterpart beam [12, 21–23]. The lock-in detection is essentially advantageous for weak signals as in stimulated Raman microscopy to reduce the light-source noise over other methods like just signal accumulation with the CCD. The CCD accumulates all frequency components of the noise, whereas the lock-in detection rejects noise different from the lock-in frequency. Therefore, wider dynamic range and higher SNR are readily achieved with the lock-in detection. Single lock-in detection, however, is limited to a single detector. When the wavelengths of the two beams are fixed, only a single frequency Raman image is obtained on a single Raman band.

Several methods have been developed to obtain multicolor stimulated Raman images with the lock-in detection. One method involves scanning the difference in the wavelength, and is achieved by filtering the broad-band femtosecond pulse laser [18] or controlling the temporal difference of the chirped femtosecond pulses (also known as spectral focusing) [15]. However, this method creates system complexities because it requires optical length maintenance during scanning or precise chirp control. Furthermore, the temporal synchronization among wavelengths is limited by the scan speed and the time-constant nature of the lock-in detection, which can cause spectral distortion and in turn render difficult the assignment of spectra and the quantitative analysis of chemical species. Another method involves the use of a few pump beams at various wavelengths that are modulated at different low frequencies and the lock-detection of the pump beams at a high modulation frequency of the Stokes beam, followed by Fourier transformation to distinguish the signals of the respective wavelengths [16]. This method, however, limits the spectra to a few discrete wavelengths, thus limiting the spectral information, although the signals are observed with a high degree of temporal synchronization.

A single light source is desirable when preparing the pump and Stokes beams [15]. Plural light sources may cause inter-source jitter, and precise synchronization [12, 24] is required to achieve a high SNR, thus resulting in a highly complex and costly system.

We have developed a multiplex stimulated Raman microscope for spectral imaging that uses parallel lock-in detection at respective wavelengths with a 128-channel lock-in amplifier [25] and a white pump beam. Parallel lock-in detection is the most straightforward method of observing a spectral image using lock-in detection, and ensures a high degree of temporal synchronization among the respective Raman shifts. Furthermore, the light sources of the pump and Stokes beams come from a single mode-locked picosecond titanium-sapphire oscillator, which guarantees a jitter-free pump and Stokes beam pair. The white light pump beam is generated from a piece of photonic crystal fiber (PCF) [4, 15, 26–28]. The cross-sectional structure of the PCF reveals a core surrounded by regularly arranged air voids. These voids significantly reduce the effective refractive index of the core's surrounding area, which functions as the clad. The light in the PCF is tightly confined in the core owing to the large degree of difference in the refractive indices between the core and clad area. This tight confinement produces a large effective nonlinear coefficient that enables white light (supercontinuum) generation at just one order of kilowatt peak power of the pulse laser from the oscillator.

2 SETUP

A. Optical Setup

The experimental setup is shown in Figure 6.1.2. The light source is a mode-locked titanium-sapphire oscillator (Mira 900P, COHERENT) pumped by a 532-nm diode-pumped solid-state laser (Verdi V5, COHERENT). The duration time, reputation rate, and average power are 2.5 ps, 76.3 MHz, and 450 mW, respectively. The back-reflected light from the optics is rejected by a Faraday isolator (I-7090C-2H, ISOWAVE) to protect the laser from instability induced by the interference of the reflected beam in the oscillator. The pulse laser beam is split into two by a beam splitter (BS1) with 60% transmissivity (T) and 40% reflectivity (R), corresponding to the pump line and Stokes line, respectively.

The 60% transmitted beam is introduced into a 2× telescope to fill the entrance pupil of the microscope objective [40× magnification, numerical aperture (NA) = 0.65] and coupled with a 30-cm-long piece of PCF. The PCF (PM-NL750, NKT) is of the polarization-maintaining type. Hence, the effective nonlinear coefficient is twice as large as that of the non-polarization-maintaining type when polarized light is coupled. The angle of the polarizing plane of the input light is adjusted by a half-wave plate to maximize the efficiency of the conversion into white light. The output light is collimated by another microscope objective lens (40×, NA = 0.65). Components of more than 780 nm in the white light with unconverted 800 nm light are rejected by an inclined 800-nm cut-off interference dichroic mirror (FES0800, THOLABS).

The 40% reflected Stokes beam is chopped at a frequency of 1130 Hz with a mechanical chopper (Model 3501, New Focus). The beam is guided to an inverse telescope with a $f = 170$-mm convex lens and $f = -150$-mm concave lens.

The functions of the inverse telescope are to adjust the beam diameter to fill the entrance pupil of the microscope objective to focus on the sample and to compensate for the axial chromatic aberration of the objective between the pump and Stokes beams. The collimated white pump beam and 800-nm Stokes beam are not focused at the same point because of the objective's chromatic aberration. Therefore, it is necessary to offset the focal point using the defocused inverse telescope, which allows the two beams to overlap at the same point. Compensation is performed by adjusting the inverse telescope's focus. The Stokes beam is sent to a delay stage to adjust the timing of its arrival in the sample to ensure that it overlaps temporally with the pump beam. The beam's power is adjusted with a variable neutral density filter (VND). The pump and Stokes beams are overlapped on the dichroic mirror (DM) and are aligned collinearly.

The beam diameters are expanded using a 1.5× Galilean telescope with a $f = -100$-mm concave lens and $f = 150$-mm convex lens to fill the entrance pupil of the objective. When measuring a solution sample, a 10×, NA = 0.25 objective is used, whereas

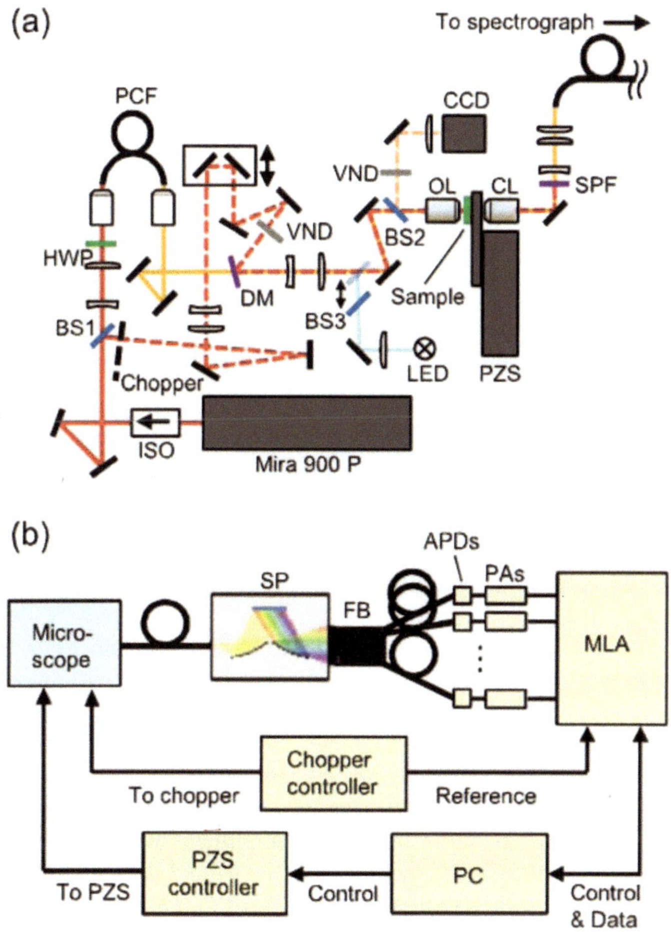

FIGURE 6.1.2 Schematics of optical setup (a) and measurement system for spectral imaging by the multi-channel lock-in amplifier (b). (a) The red line indicates the Stokes beam at 800 nm, and the yellow line is the white pump beam. ISO: isolator that rejects backing scattered or reflected light; BS1: beam splitter for the Stokes and pump line; VND: variable neutral density filter to adjust power; HWP: half-wave plate that optimizes the polarizing plane; PCF: photonic crystal fiber; DM: dichroic mirror; OL: objective lens; CL: collector lens; PZS: piezo scanning stage; SPF: short-pass filter to reject the Stokes beam; BS2: beam splitter used to observe a sample and beam spots; BS3: beam splitter for epi-Köhler illumination; LED: light-emitting diode for epi-Köhler illumination; and CCD: charge-coupled device used to confirm the overlap of the beams and take photos of the samples. (b) SP: spectrograph; FB: fiber bundle; APDs: 128 avalanche photodiodes biased at 140 V; PAs: 128 10^7 gain transimpedance pre-amplifiers; MLA: 128-channel multi-channel lock-in amplifier; and PC: personal computer. The copyright of this figure is shared between the representative author (Takayoshi Kobayashi) and John Wiley.

a 40×, NA = 0.85 objective is used in the measurement of solid plates or films. The beams are tightly focused on the sample by the objective, and the sample is placed on a computer-controlled piezo-scanning stage (MAX311D/M, THOLABS). The light reflected from the sample is collected with the objective and reflected by a pellicle beam splitter (BS2) with 92% T and 8% R to a CCD camera to confirm the spatial overlap of the beams. A sample image is also observed by the CCD camera using epiKöhler illumination with a white light-emitting diode (LED). The beams transmitted through the sample are collimated by an objective (20×, NA = 0.4). The collimated pump beam is sent to a 3× Galilean telescope with a $f = -50$-mm concave lens and $f = 150$-mm convex lens through a 750-nm cut-off interference short-pass filter (FES0750, THOLABS) to reject the Stokes beam. The 3× telescope optimizes the coupling efficiency to a subsequent optical fiber and reduces the beam's swing effect.

B. Multi-channel lock-in detection

The pump beam is coupled with a multi-mode optical fiber (core diameter: 600 μm; clad diameter: 660 μm; NA = 0.22) by a convex lens ($f = 100$ mm) and sent to a 300-mm focal-length spectrograph (SpectraPro-300i, ACTON) equipped with a 1200 g/mm grating. The specified resolution power is 0.1 nm, and the dispersion 2.7 nm/mm.

The spectrum is imaged onto the facet of a 16 × 128 quartz fiber bundle array in the spectrograph [Fig. 6.1.2(b)]. The core and clad diameters are 100 and 110 μm, respectively, and the entrance size is 1.8 mm × 16.5 mm. The spectrum is divided into 128 wavelengths, and each of the 16 fibers of the same wavelength component is bundled. These 128 fibers are then coupled to 140-V biased avalanche photodiodes (APDs) (S5343, HAMAMATSU). The current signals from the APDs are converted to voltage signals by 10^7 gain transimpedance amplifiers (Model 7210/90, SIGNAL RECOVERY) and sent to four synchronized 32-channel digital signal-processing lock-in amplifiers (Model 7210, SIGNAL RECOVERY). The reference signal is generated by the chopper controller. Because the Raman loss signal is observed as the amplitude modulation depth (described later in the chapter), it is preferable that the power of the white pump beam be kept as weak as possible to avoid sample degradation. Furthermore, the respective power levels of the channels are significantly weak, which requires the APDs and high gain preamplifiers.

3 EXPERIMENT

A Wavelength Calibration, Wavelength Range, Spectrum of White Light, and Noise

The wavelengths are calibrated with the bright line spectra of a neon tube, and the sensitivities of the channels are calibrated with a standard light source (color temperature = 2995 K ± 5%) for a fluorescence spectrometer (F-4500, HITACH). The center channels at the peak wavelengths of the bright lines are estimated by the peak position of the Gauss function fitted to these lines and measured with the 250-μm entrance-slit width of the spectrograph. The calibration curve is prepared as a linear function of the channel, and the detection range is calculated by extrapolating this curve. The difference in wavenumber between neighboring channels (wavenumber step) is evaluated by dividing the detection range by the number of channels (128). The spectrum of the white pump beam generated by the PCF is measured with a spectrometer (USB2000+, Ocean Optics) through the DM. The noise spectrum of the white pump beam is observed with a photodiode (DET36A/M, THOLABS) and an oscilloscope (TDS2022B, TEKTRONIX). The spectrum is obtained through fast Fourier transformation of the signal.

The root mean square (RMS) electrical noise (*EN*) of the entire output is evaluated by the lock-in detected noises on the 128 channels at 1130 Hz of various time constants:

$$EN = \sqrt{\frac{1}{128}\sum_{i=1}^{128}\{e(i)\}^2}, \tag{1}$$

where $e(i)$ is the electrical noise observed on the ith channel. The electrical noise normalized to the white light signal (EN/I_P) is calculated as the RMS of the noise signals of the 128 channels divided by the lock-in detected signals of the chopped white light at a time constant of 10 s, with the 100-μm slit width of the spectrograph:

$$\frac{EN}{I_P} = \sqrt{\frac{1}{128}\sum_{i=1}^{128}\left\{\frac{e(i)}{I_P(i)}\right\}^2}, \tag{2}$$

where $I_P(i)$ is the signal of the chopped white light on the ith channel. The noise of the white pump beam (I_PN/I_P) is evaluated as the RMS of the lock-in detected signals of the 128 channels divided by the lock-in detected signals of the chopped white light at a time constant of 10 s:

$$\frac{I_PN}{I_P} = \sqrt{\frac{1}{128}\sum_{i=1}^{128}\left\{\frac{I_PN(i)}{I_P(i)}\right\}^2}, \tag{3}$$

where $I_PN(i)$ is the noise signal of the white light observed on the ith channel.

B Measurement of Raman Loss Spectra

To confirm the proposed system's ability to measure the Raman loss spectra, we measured cyclohexane, polystyrene (PS), and polymethylmethacrylate (PMMA). Cyclohexane was prepared in a 5-mm cell-length quartz cell, and PS and PMMA into 1.7-mm- and 1-mm-thick plates, respectively. The observed signals were normalized by the white light spectrum which was observed with the system to be corrected into Raman loss signals (modulation depths). The spontaneous Raman spectra of the substances were observed on NRS-3100 (JASCO) with 532-nm excitation and a resolution power of 1 cm^{-1}. The linearity of the Raman loss signals obtained by our system was evaluated by the power dependence of the Stokes beam with cyclohexane in the quartz cell. The quantitative capability of the chemical concentration was evaluated by measuring cyclohexane and carbon tetrachloride mixtures at various molar ratios in the quartz cell.

C Raman Loss Imaging

The piezo stage and lock-in amplifiers are computer controlled (Lab View 8.5 software), and the spectra are observed point-by-point with the scanning of the sample. In this study, the samples were prepared as follows. A dispersion of 4-μm PS beads (Invitrogen) was dropped onto a glass slide and left to dry. A mixed PS and PMMA film was prepared by drop-casting a 1:1 (by weight) mixed toluene solution of PS and PMMA on a glass slide plate.

4 RESULTS AND DISCUSSION

A Wavelength Calibration, Wavelength Range, Spectrum of White Light, and Noise

The bright line spectrum of the neon tube with the 250-μm slit-width is shown in Figure 6.1.3[1]. The peak channels of the bright lines, which were estimated by fitting the Gauss function, are listed in Table 6.1.1. The calibration curve was estimated as a linear function [Fig. 6.1.3[2]]:

$$\lambda = 0.28666 \times \text{ch.} + 629.37(\text{nm}), \tag{4}$$

where λ is the wavelength and ch. is the channel. The determination coefficient (R^2) is 0.99999, and the relationship between the wavelength and channel is sufficiently linear. The 1st and 128th channels are extrapolated to 629.65 and 666.06 nm, and these wavelengths correspond to 3382 and 2514 cm^{-1} Raman shifts, respectively. The detection range is 868 cm^{-1}, and the wavenumber step between neighboring channels is 6.78 cm^{-1}. The dispersion of the spectrograph is 2.7 nm/mm, and the resolution power with the 100-μm entrance-slit width is 6.8 cm^{-1} at 630 nm. Thus, the Raman loss spectra shown hereafter were observed with the 100 μm slit width. The spectrum of the white pump beam through the DM is shown in Figure 6.1.4(a). The wavelength range is 575–780 nm, corresponding to 320–4890 cm^{-1} of the Raman shift range. The spectral range is sufficient to cover the frequency range of the molecular vibration.

The system's electrical noise and the intensity noise of the white light are reported in Table 6.1.2. The RMS electrical noise of the system is less than a few microvolts when the time constant is 1 s. The RMS electrical noise, normalized by the signal intensity of the white light with the 100-μm spectrograph slit width, is less than 10^{-5} when the time constant is 1 s. The electrical noise level is acceptable because the signal intensity of the white light is typically around 0.3 V at a spectrograph

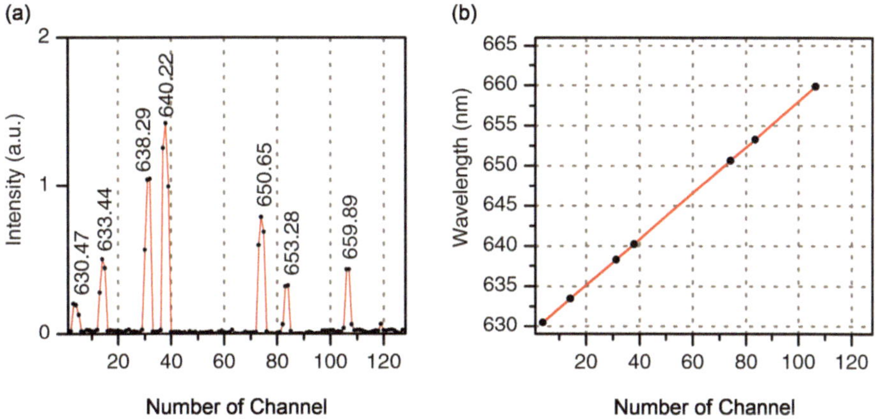

FIGURE 6.1.3 The bright line spectrum from the neon tube with the 250-μm entrance-slit width on the spectrograph (a) and the correction curve (b). (a) The peak wavelengths of the neon bright lines are labeled on the corresponding bands, and (b) the dots represent the relationship between the peak channels estimated by fitting the Gauss function and the peak wavelengths. The red line is the calibration curve. The copyright of this figure is shared between the representative author (Takayoshi Kobayashi) and John Wiley.

TABLE 6.1.1
Wavelengths of Bright Lines from the Neon Tube and Estimated Channels at the Peak Lines

λ[a] (nm)	ch.[b]
630.47	3.7703
633.44	14.213
638.29	31.287
640.22	37.884
650.65	74.071
653.28	83.481
659.89	106.50

a λ: Peak wavelength of the bright lines from the neon tube.
b ch.: Estimated channel at the peak of the bright line by fitting the Gauss function to the bright line observed with a 250-μm entrance-slit width.

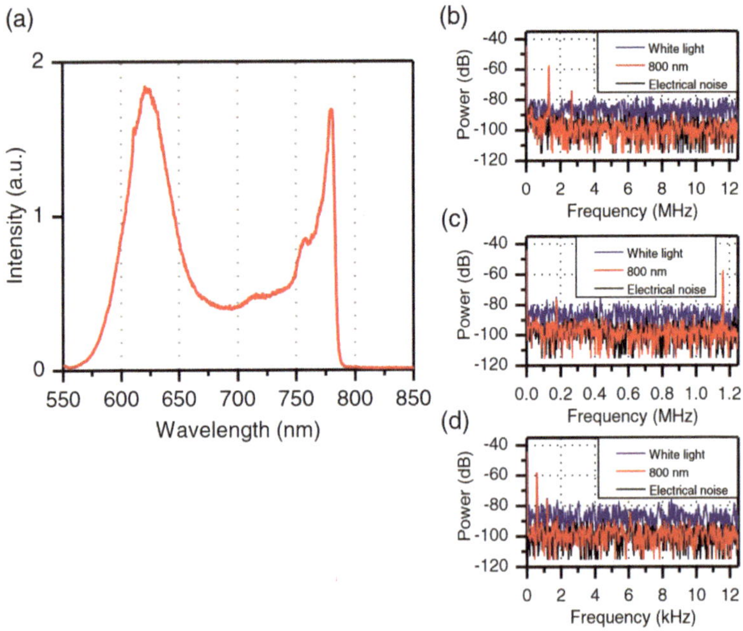

FIGURE 6.1.4 Spectrum of the white pump beam through the dichroic mirror (a); light noise spectra in 0–12.5 MHz (b); in 0–1.25 MHz (c); and in 0–12.5 kHz (d). The copyright of this figure is shared between the representative author (Takayoshi Kobayashi) and John Wiley.

TABLE 6.1.2
RMS Electrical Noise and RMS Intensity Noise of White Pump Beam

TC^a (ms)	EN^b (μV)	$EN/I_P{}^c$ (10^{-4})	$I_PN/I_P{}^d$ (10^{-4})
3	19	1.4	28
10	11	0.81	25
30	5.9	0.43	12
100	3.1	0.23	5.9
300	2.1	0.16	3.7
1000	1.4	0.099	2.1
3000	0.77	0.060	1.2
10000	0.47	0.033	0.71

[a]*TC*: Time constant of the lock-in detection. [b]*EN*: RMS electrical noise of the system output.

[c]EN/I_P: RMS electrical noise relative to the white pump beam signal at the 100 μm entrance-slit width on the spectrograph. [d]I_PN/I_P: RMS intensity noise of the white pump beam. The copyright of this table is shared between the representative author (Takayoshi Kobayashi) and John Wiley.

entrance-slit width of 100 μm, and the modulation depths of the Raman loss signals are typically 10^{-3}–10^{-5}. The white light generated from the PCF, however, is significantly noisy, which requires a long time constant for lock-in detection, although the system is simple. The RMS intensity noise is around 2×10^{-4} for a time constant of 1 s, which limits the SNR of the Raman loss signal.

The noise spectra in the kilo-hertz region and mega-hertz region are shown in Figs. 6.1.4(b)–(d). The noise is white from the kilo-hertz to 10 mega-hertz. Although mega-hertz order high-frequency lock-in detection is often used to reduce observed noise [10, 12, 13, 16–19, 29], in the present case, it cannot be reduced with mega-hertz lock-in detection. This frequency-independent nature of the white light noise is also mentioned by another research group [30].

Thus, noise cancelation by some form of balanced detection [31, 32] is required for high-speed imaging. Because the noise varies widely by wavelength, and transmission through a sample varies during imaging, the application of balanced detection to our Raman-loss detection system requires the signal and reference beams to have exactly the same optical path from the microscope through the spectrograph to the APDs. Collinear balanced detection (CBD) [32] is a balanced-detection method in which the two have the same optical path, but it requires megahertz-order high frequency lock-in detection for realistic optics, and its performance depends on the equal splitting of the pump beam. In our system, the equal splitting of the pump beam is required in a broad wavelength range, which is not realistic with practical optics. We plan to develop a new balanced-detection method in a future study.

B Measurement of Raman Loss Spectra

Figures 6.1.5(a) and (b) show the observed spectra of cyclohexane, PS, and PMMA on the proposed system and their spontaneous Raman spectra, respectively. The average power of the white pump beam and 800-nm Stokes beam are 3.0 and 53.4 mW, respectively, and the time constant of the lock-in detection is 10 s. The spectra on our system conform to the spontaneous Raman spectra. They disappear when the white pump beam and/or 800-nm Stokes beam are blocked, or the focal spots overlap and/or the time delay is displaced. The Raman loss signal is proportional to the Raman scattering cross-section given in the following equation [29]:

$$\frac{\Delta I_P}{I_P} \propto N \times \sigma_{\text{Raman}} \times I_S \tag{5}$$

where I_P, ΔI_P, I_S, N, and σ_{Raman} are the pump intensity, reduction in pump intensity, Stokes intensity, concentration, and Raman scattering cross-section, respectively. The relation ensures the high fidelity of the stimulated Raman loss spectra to the spontaneous Raman spectra. Thus, the results confirm that the stimulated Raman loss spectra are observed on our system. They also demonstrate the strong capability of qualitative analysis to identify the C–H symmetric stretching (~2850 cm^{-1}) and asymmetric stretching (~2930 cm^{-1}) modes of the methylene groups (cyclohexane), CH_3 symmetric stretching (~2870 cm^{-1}), and asymmetric stretching (~2960 cm^{-1})

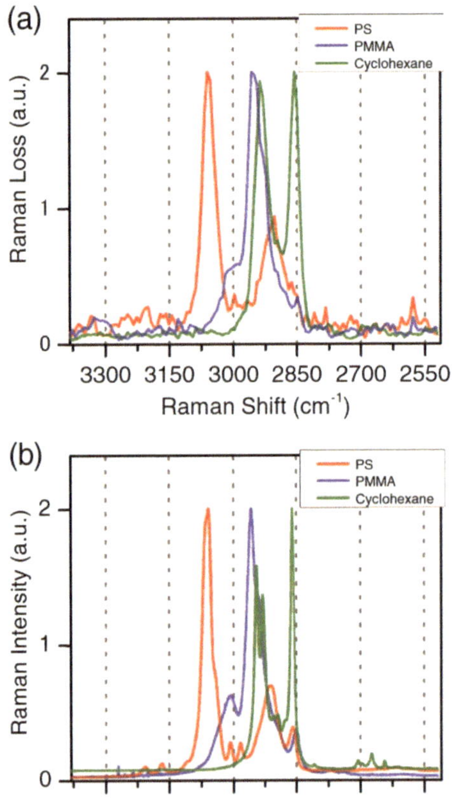

FIGURE 6.1.5 Stimulated Raman loss spectra of polystyrene (PS), polymethylmethacrylate (PMMA), and cyclohexane observed with the system constructed in this study (a) and spontaneous Raman spectra of the same samples (b). The copyright of this figure is shared between the representative author (Takayoshi Kobayashi) and John Wiley.

modes of the methyl groups (PMMA), and aromatic C–H stretching mode (~3030 cm^{-1}, PS).

Figure 6.1.6(a) shows the Raman loss spectra of cyclohexane with various powers of the Stokes beam in the sample position. The average power of the white-light pump beam is 1.1 mW. The Raman loss intensity at the 65th channel, which is the strongest peak in the spectra and corresponds to a white-light spectrum wavelength of 648.00 nm and 2932 cm^{-1} Raman shift, is linearly dependent on the power [Fig. 6.1.6(b)].

The determination coefficient (R^2) is 0.998. The Raman loss signal is linearly dependent on the Stokes power, as in Eq. (5), thus confirming the system's linearity. Fig. 6.1.6(c) shows the Raman loss spectra of cyclohexane diluted with carbon tetrachloride to various concentrations. The average power of the white pump beam and the power of the 800-nm Stokes beam are 1.1 and 83.3 mW, respectively. No signal is observed in the spectra of carbon tetrachloride (indicated as 0 mol/l) in

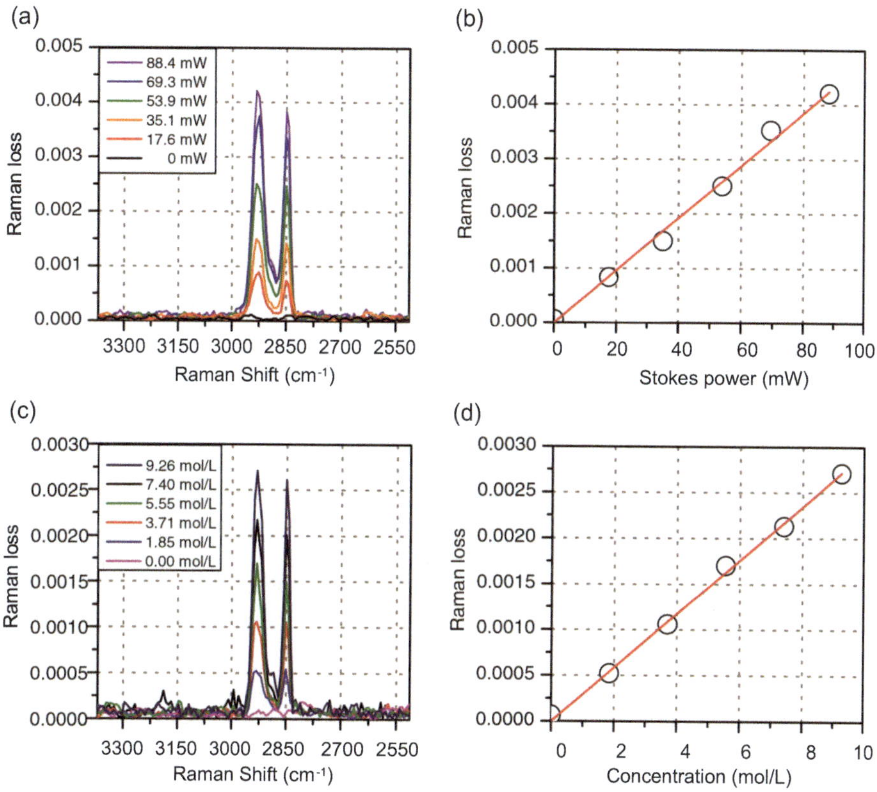

FIGURE 6.1.6 The Raman loss spectra of cyclohexane at various Stokes beam powers (a); the relationship between the Raman loss signal at the 65th channel (the strongest peak of cyclohexane) and the power (b); the Raman loss spectra of cyclohexane diluted by carbon tetrachloride at various concentrations (c); and the relationship between the Raman loss signal at the 65th channel and the concentration. The copyright of this figure is shared between the representative author (Takayoshi Kobayashi) and John Wiley.

the observed range of Raman shifts. The Raman loss intensity at the 65th channel exhibits excellent linear dependence on the concentration (Fig. 6.1.6(d)). The determination coefficient (R^2) of the linearity is 0.999. These results indicate the system's high degree of quantitative capability, as predicted by Eq. (5).

C Raman Loss Imaging

Figure 6.1.7 presents a CCD image of the 4-μm PS beads by white LED illumination and an image of the 4-μm PS beads in the same area based on the summation of the signals on the 40–50th channels (3103–3033 cm^{-1} Raman shifts). The image and pixel sizes are 10 μm × 10 μm and 0.25 μm × 0.25 μm, respectively. The average

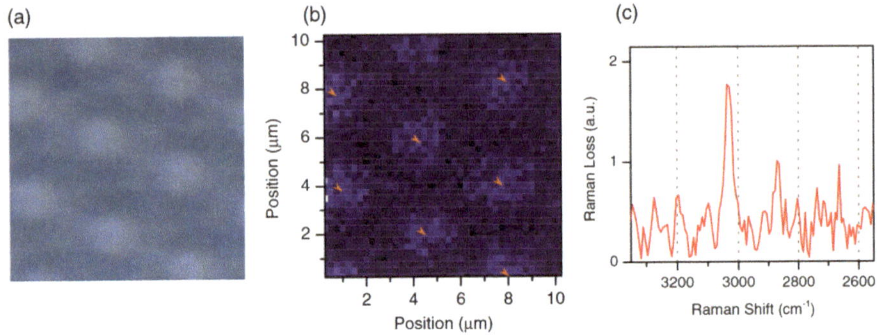

FIGURE 6.1.7 CCD image of the 4-μm PS beads (10 × 10 μm^2 area) (a); image of the same area constructed on the summation of the signals of the 40–50th channels (b); and the averaged spectrum of the spectra at the centers of the beads [indicated by the orange marks in (b)],. The averaged Raman loss spectrum at the centers of the beads (indicated by the orange marks in (b)) (c)copyright of this figure is shared between the representative author (Takayoshi Kobayashi) and John Wiley.

power of the white pump beam and the power of the 800-nm Stokes beam are 0.42 and 67.6 mW, respectively, and the time constant and dwelling times are 300 ms and 600 ms, respectively. The spatial distribution of the lock-in signal strengths conforms to the PS beads observed by the CCD. The averaged spectrum at the centers of the PS beads [Fig. 6.1.7(c)] conforms to the spontaneous Raman spectrum of PS [Fig. 6.1.5(b)], which confirms that the image strength is proportional to the Raman loss signal intensity of PS. Thus, the 4-μm PS beads are successfully imaged.

The CCD image of the mixed PS and PMMA film and the Raman loss images of the same area are shown in Figure 6.1.8(a)–(c). The black lines in the CCD image are cracks. The image and pixel sizes are 20 μm × 20 μm and 0.5 μm × 0.5 μm, respectively. The average power of the white pump beam and the power of the 800-nm Stokes beam are 3.6 and 54.5 mW, respectively, and the time constant and dwelling times are shown in [Fig. 6.1.8(e)]. Although the polymers appear to be mixed homogeneously using the CCD, the differences in contrast of the images in Figs. 6.1.8(b) and (c) suggest a different distribution and some aggregate material structures. The inhomogeneous distribution of mixed polymer films has also been observed by another research group:11000 ms and 2000 ms, respectively. These images are obtained through single imaging, and image (b) is constructed on the summation of the signals from the 40th–50th channels (3103–3033 cm^{-1} Raman shift), whereas (c) is constructed on the signals from the 55th–65th channels (2998– 2930 cm^{-1}). The averaged Raman loss spectrum of the area in which intense signals in the range of the 40th–50th channels are observed [surrounded by a green square in Figs. 6.1.8(b) and 8(c)] is shown in Fig. 6.1.8(d), and that of the 55th–65th channels (surrounded by a red square) is shown in Fig. 6.1.8(e). The PS Raman bands are dominant in Fig. 6.1.8(d), and the PMMA band (which appears around 2960 cm^{-1}) is more intense in the counterpart.

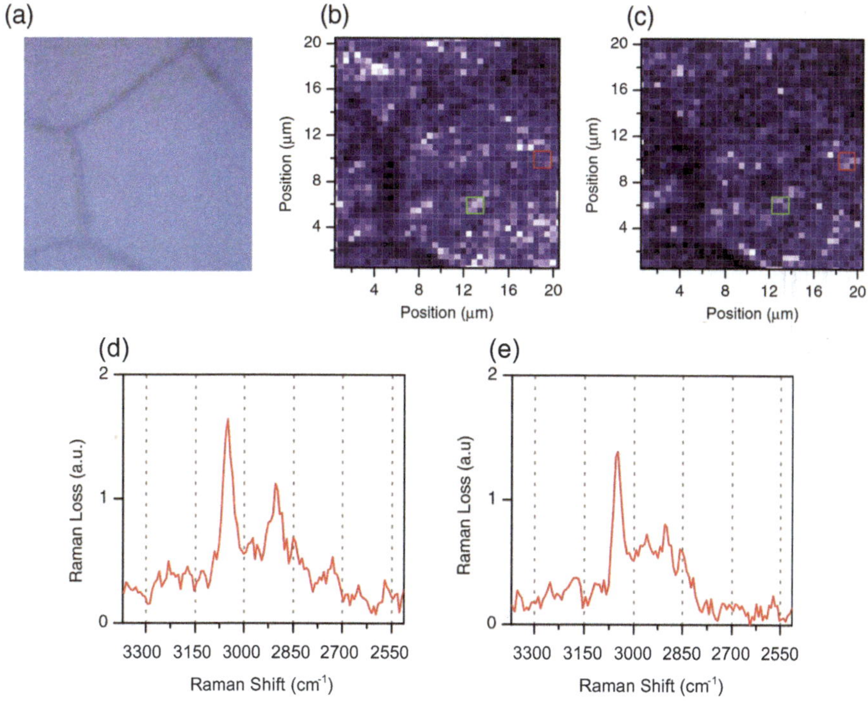

FIGURE 6.1.8 CCD image of the mixed polymer film of PS and PMMA (20 µm × 20 µm area) (a); image of the same area based on the summation of the signals of the 40th–50th channels (3103–3033 cm^{-1} Raman shift) (b); image based on the summation of the 55th–65th channels (2998–2930 cm^{-1}) (c); averaged spectrum in the area of the strong signals of the 40th–50th channels (surrounded by the red square) (d); and the averaged spectrum in the area of the strong signals of the 55th–65th channels (surrounded by the red square) (e). The copyright of this figure is partially held between the representative author (Takayoshi Kobayashi) and John Wiley.

Note that the multi-channel detection of the spectra ensures a high degree of temporal synchronization among the Raman bands, which minimizes the spectral distortion caused by sample degradation during measurement. Such distortion can create difficulties in assigning the Raman bands and/or in quantitative analysis. Temporal synchronization is also important in the high-speed multi-color imaging of moving samples such as living cells and fluids.

The proposed method is highly suited for measurements that require high precision of Raman shifts such as imaging by Raman shifts and/or bandwidth variations since this method does not require wavenumber scanning and is robust in Raman shifts. This imaging is applicable to, for example, the detection of surface defect and spatial non-uniformity of ceramics, inorganics, and doped-polymer film systems used in automotive, plastic displays, and others which are important in industry.

The chromatic aberration of the microscope objective causes the displacement of the focus points between the pump and Stokes beams. This problem not only causes a significant reduction in the intensity of the Raman loss signals, but also creates an artifact signal via the thermal-lens effect [33, 34]. When the artifact generated by the thermal effect appears, the baseline of the spectrum shifts, and the contrast between the Raman signal and background is reduced. This aberration can be corrected by offsetting the focal point of the Stokes beam with the lens pair before the objective. However, the correction is possible only in a narrow wavelength detection range. Although the reflection objective can be used, the NA and throughput of this type of objective are limited. Thus, a more sophisticated achromatic objective correction with a high NA would be desirable. In our system, the Raman loss signals are easily distinguishable from this type of artifact, which is not the case with the conventional single wavelength detection method.

5 CONCLUSION

We have developed a multiplex stimulated Raman microscope for spectral imaging with multi-channel lock-in detection and a white pump beam with a single light source [35]. Our system requires no synchronization of plural light sources or wavelength scanning. In addition, it suffers no jitter or optical-length mismatch problems arising from the multiple sources with different wavelengths used in conventional stimulated Raman microscopy. The ADPs and high-gain preamplifiers eliminate the need for optical amplification of the white pump beam, thereby reducing the possibility of sample degradation and simplifying the system. The 128-channel lock-in detection of the white pump beam enables observation of pseudo-continuous Raman loss spectra and demonstrates strong qualitative capability in identifying various kinds of C–H stretching modes. The multi-channel lock-in detection improves imaging not only by increasing the speed being multiplied by the channel number, but also by alleviating problems such as laser fluctuations and sample damage by simultaneously measuring the full Raman spectral range of interest. The noise of the white light generated by the PCF limits high-speed imaging, but we expect to overcome this problem with a new balanced detection method in a future study. The contents of this subsection are the product of collaborative work among the following people: K. Seto, Y. Okuda, E. Tokunaga, and T. Kobayashi [35].

NOTES

1. λ: Peak wavelength of the bright lines from the neon tube.
2. ch.: Estimated channel at the peak of the bright line by fitting the Gauss function to the bright line observed with a 250-μm entrance-slit width.

REFERENCES

1. M. D. Schaeberle, C. G. Karakatsanis, C. J. Lau, and P. J. Treado, *Anal. Chem.* **67**, 4316–4321 (1995). Raman Chemical Imaging: Noninvasive Visualization of Polymer Blend Architecture
2. M. Okada, N. I. Smith, A. F. Palonpon, H. Endo, S. Kawata, M. Sodeoka, and K. Fujita, *Proc. Natl. Acad. Sci. U.S.A.* **109**, 28–32 (2012). Proc. Natl. Acad. Sci. U.S.A. 109, 28–32 (2012). Label-free Raman observation of cytochrome c dynamics during apoptosis Study on solid structure of pentacene thin films using Raman imaging
3. K. Seto and Y. Furukawa, *J. Raman Spectrosc.* **43**, 2015–2019 (2012).Study on solid structure of pentacene thin films using Raman imaging,
4. M. Okuno, H. Kano, P. Leproux, V. Couderc, J. P. R. Day, M. Bonn, and H.-o. Hamaguchi, *Angew. Chem., Int. Ed.* **49**, 6773–6777 (2010). Quantitative CARS Molecular Fingerprinting of Single Living Cells with the Use of the Maximum Entropy Method†
5. T. Suzuki and K. Misawa, *Opt. Express* **19**, 11463–11470 (2011). Efficient heterodyne CARS measurement by combining spectral phase modulation with temporal delay technique
6. C. L. Evans, E. O. Potma, M. Puoris'haag, D. Côté, C. P. Lin, and X. S. Xie, *Proc. Natl. Acad. Sci. U.S.A.* **102**, 16807–16812 (2005).Chemical imaging of tissue in vivo with video-rate coherent anti-Stokes Raman scattering microscopy
7. M. D. Duncan, J. Reintjes, and T. J. Manuccia, *Opt. Lett.* **7**, 350–352 (1982). Scanning coherent anti-Stokes Raman microscope
8. T. Meyer, N. Bergner, A. Medyukhina, B. Dietzek, C. Krafft, B. F. M. Romeike, R. Reichart, R. Kalff, and J. Popp, *J. Biophoton.* **5**, 729–733 (2012). Interpreting CARS images of tissue within the C-H-stretching region
9. E. Ploetz, S. Laimgruber, S. Berner, W. Zinth, and P. Gilch, *Appl. Phys. B* **87**, 389–393 (2007). Femtosecond stimulated Raman microscopy
10. P. Nandakumar, A. Kovalev, and A. Volkmer, *New J. Phys.* **11**, 033026 (2009). Vibrational imaging based on stimulated Raman scattering microscopy
11. E. Ploetz, B. Marx, T. Klein, R. Huber, and P. Gilch, *Opt. Express* **17**, 18612–18620 (2009). A 75 MHz light source for femtosecond stimulated Raman microscopy
12. Y. Ozeki, Y. Kitagawa, K. Sumimura, N. Nishizawa, W. Umemura, S. i. Kajiyama, K. Fukui, and K. Itoh, *Opt. Express* **18**, 13708–13719 (2010). High-speed molecular spectral imaging of tissue with stimulated Raman scattering
13. B. G. Saar, C. W. Freudiger, J. Reichman, C. M. Stanley, G. R. Holtom, and X. S. Xie, *Science* **330**, 1368–1370 (2010). Video-Rate Molecular Imaging in Vivo with Stimulated Raman Scattering
14. E. R. Andresen, P. Berto, and H. Rigneault, *Opt. Lett.* **36**, 2387–2389(2011). Stimulated Raman scattering microscopy by spectral focusing and fiber-generated soliton as Stokes pulse
15. H. T. Beier, G. D. Noojin, and B. A. Rockwell, *Opt. Express* **19**, 18885– 18892 (2011). Stimulated Raman scattering using a single femtosecond oscillator with flexibility for imaging and spectral applications
16. D. Fu, F.-K. Lu, X. Zhang, C. Freudiger, D. R. Pernik, G. Holtom, and X. S. Xie, *J. Am. Chem. Soc.* **134**, 3623–3626 (2012). Quantitative chemical imaging with multiplex stimulated Raman scattering microscopy
17. C. W. Freudiger, W. Min, G. R. Holtom, B. Xu, M. Dantus, and X. S. Xie, *Nature Photon.* **5**, 103–109 (2011). Highly specific label-free molecular imaging with spectrally tailored excitation-stimulated Raman scattering (STE-SRS) microscopy

18. Y. Ozeki, W. Umemura, K. Sumimura, N. Nishizawa, K. Fukui, and K. Itoh, *Opt. Lett.* **37**, 431–433 (2012). Stimulated Raman hyperspectral imaging based on spectral filtering of broadband fiber laser pulses
19. X. Zhang, M. B. J. Roeffaers, S. Basu, J. R. Daniele, D. Fu, C. W. Freudiger, G. R. Holtom, and X. S. Xie, *Chem. Phys. Chem.* **13**, 1054–1059 (2012). Label-Free Live-Cell Imaging of Nucleic Acids Using Stimulated Raman Scattering Microscopy
20. W. J. Jones and B. P. Stoicheff, *Phys. Rev. Lett.* **13**, 657–659 (1964).Inverse Raman Spectra: Induced Absorption at Optical Frequencies
21. Y. Ozeki, F. Dake, S. I. Kajiyama, K. Fukui, and K. Itoh, *Opt. Express* **17**, 3651–3658 (2009). Analysis and experimental assessment of the sensitivity of stimulated Raman scattering microscopy
22. A. Owyoung and E. D. Jones, *Opt. Lett.* **1**, 152–154 (1977). Stimulated Raman spectroscopy using low-power cw lasers
23. A. Owyoung, *Opt. Commun.* **22**, 323–328 (1977). Sensitivity limitations for cw stimulated Raman spectroscopy
24. D. J. Jones, E. O. Potma, J.-X. Cheng, B. Burfeindt, Y. Pang, J. Ye, and X. S. Xie, *Rev. Sci. Instrum.* **73**, 2843–2848 (2002). Synchronization of two passively mode-locked, picosecond lasers within 20 fs for coherent anti-Stokes Raman scattering microscopy
25. T. Kobayashi, W. Zhuan, and I. Izumi, *New J. Phys.* **10**, 065009 (2008). The relation between the symmetry of vibrational modes and the potential curve displacement associated with electronic transition studied by using real-time vibration spectroscopy
26. A. V. Husakou and J. Herrmann, *Phys. Rev. Lett.* **87**, 203901 (2001). Supercontinuum Generation of Higher-Order Solitons by Fission in Photonic Crystal Fibers
27. J. Herrmann, U. Griebner, N. Zhavoronkov, A. Husakou, D. Nickel, J. C. Knight, W. J. Wadsworth, P. S. J. Russell, and G. Korn, *Phys. Rev. Lett.* **88**, 173901 (2002). Experimental Evidence for Supercontinuum Generation by Fission of Higher-Order Solitons in Photonic Fibers
28. H. Kano and H.-O. Hamaguchi, *Opt. Lett.* **28**, 2360–2362 (2003). Characterization of a supercontinuum generated from a photonic crystal fiber and its application to coherent Raman spectroscopy
29. C. W. Freudiger, W. Min, B. G. Saar, S. Lu, G. R. Holtom, C. He, J. C. Tsai, J. X. Kang, and X. S. Xie, *Science* **322**, 1857–1860 (2008). Label-free biomedical imaging with high sensitivity by stimulated Raman scattering microscopy
30. K. L. Corwin, N. R. Newbury, J. M. Dudley, S. Coen, S. A. Diddams, K. Weber, and R. S. Windeler, *Phys. Rev. Lett.* **90**, 113904 (2003).Fundamental Noise Limitations to Supercontinuum Generation in Microstructure Fiber
31. A. Gambetta, V. Kumar, G. Grancini, D. Polli, R. Ramponi, G. Cerullo, and M. Marangoni, *Opt. Lett.* **35**, 226–228 (2010). Fiber-format stimulated-Raman-scattering microscopy from a single laser oscillator
32. K. Nose, Y. Ozeki, T. Kishi, K. Sumimura, N. Nishizawa, K. Fukui, Y. Kanematsu, and K. Itoh, *Opt. Express* **20**, 13958–13965 (2012). Sensitivity enhancement of fiber-laser-based stimulated Raman scattering microscopy by collinear balanced detection technique
33. M. Tokeshi, M. Uchida, A. Hibara, T. Sawada, and T. Kitamori, *Anal. Chem.* **73**, 2112–2116 (2001). Determination of Subyoctomole Amounts of Nonfluorescent Molecules Using a Thermal Lens Microscope: Subsingle-Molecule Determination
34. E. Tamaki, K. Sato, M. Tokeshi, K. Sato, M. Aihara, and T. Kitamori, *Anal. Chem.* **74**, 1560–1564 (2002). Single-Cell Analysis by a Scanning Thermal Lens Microscope with a Microchip: Direct Monitoring of Cytochrome c Distribution during Apoptosis Process

35. K. Seto, Y. Okuda, E. Tokunaga, and T. Kobayashi, *Rev. Sci. Instrum.* **84**, 083705 (2013). Citation: Rev. Sci. Instrum. **84**, 083705 (2013); doi: 10.1063/1.4818670. Development of a multiplex stimulated Raman microscope for spectral imaging through multi-channel lock-in detection

6.2 Multiplex Stimulated Raman Imaging with White Probe-light from a Photonic-crystal Fiber and with Multi-wavelength Balanced Detection

1 INTRODUCTION

Stimulated Raman scattering (SRS) microscopy has been studied as one of the promising tools for high speed no-labeling chemical imaging of a biological sample, based on molecular vibration, in many studies [1–5]. As the high speed chemical imaging method, coherent anti-Stokes Raman scattering (CARS) microscopy has also been extensively studied [6–12]. However, a CARS signal contains a non-resonant background (NBR) component, and this NBR generates an artificial image and spectral distortion. Additionally, the signal intensity is proportional to the square of a molecular concentration, and results in the degradation of its identification ability and quantitative capacity. Various methods have been developed to solve these problems [11, 12]. SRS microscopy is advantageous over CARS microscopy, since an SRS signal does not contain an NBR signal, and the signal intensity is proportional to the concentration. With these advantages, the imaging based on a Raman band in the fingerprint region [5] and quantitative analysis without data processing has been demonstrated [13].

In SRS microscopy, at least two pulsed beams with different wavenumbers are irradiated on the sample. When the difference between these wavenumbers is equal to a molecular vibration wavenumber, a photon of the pump beam is annihilated and that of the Stokes beam is created. Thus, through adjusting the difference to the target molecular vibration and observing the intensity of the pump or Stokes beam, the chemical image on this vibration is obtained by scanning the beams or sample. However, since this intensity modulation depth is typically of the order of 10^{-4}, the

DOI: 10.1201/9781003476030-23

intensity noise of the light source is a serious problem. Usually for SRS microscopy, to reject light intensity noise and enhance the signal-to-noise ratio (SNR), the intensity of the pump or Stokes beam is modulated and the counterpart is lock-in detected at the modulation frequency [1, 14, 15].

In commonly used lock-in detection, a single wavelength probe beam is detected. One of the recent trends of study is to then develop a method to observe a distribution of more than one substance at one shot imaging and a method for spectral imaging. Although various methods for spectral imaging and multi-element imaging have been developed [3, 13, 16–21], they require complicated optics and adjustment, or only a few spectral components are observed. Previously, we developed a simple method with a white probe beam from a photonic crystal fiber (PCF) [22–24] and multi-wavelength lock-in detection with a multi-channel lock-in amplifier (MLA) through a spectrometer [25]. In this method, a common laser oscillator is used to prepare both the white probe and the monochromatic pump beams, and hence the probe and pump are automatically synchronized, resulting in a simple and robust system. However, the intensity noise of the white light is so prominent in a broad frequency range [26–28] that it is difficult to achieve high speed imaging with high SNR, even using the lock-in detection.

The intensity noise of the probe beam can be canceled by a balanced detection method [29–33]. In this method, the light source is split into probe and reference beams, and after measurement of a sample with the probe, the signal is obtained as a difference between the probe and reference signals. Here, the light intensity noise of the probe beam is the product of the intensity fluctuation of the white light source and the fluctuation of the splitting ratio into the probe and reference beams due to the nature of the particle of the photon. The fluctuation of the white light source causes common intensity fluctuations between the probe and reference beams, and hence this fluctuation is canceled through the balanced detection. The fluctuation of the splitting ratio results in independent intensity fluctuations between the probe and reference beams, and thus, the noise intensity caused by this effect cannot be canceled by the balanced detection. Hereafter, we define the intensity fluctuation originated in the light source as "light intensity noise", whereas we define the fluctuation originated in the splitting-ratio fluctuation due to the particle nature of the photon as "shot noise". The SNR limit of this detection is determined by the sum of the shot noises of the probe and reference beams. To apply balanced detection to the multi-wavelength detection for spectral imaging, at least the following three requirements must be fulfilled. Firstly, the probe and reference beams are spectrally dispersed and detected in exactly the same bandwidth, since intensity fluctuations are completely different for different wavelengths. Secondly, signal gains for the beams are adjusted independently for each corresponding wavelength, since the splitting ratio is different for each wavelength due to wavelength dependencies of the practical optics. Without this adjustment, when the optical intensity ratio is balanced at a particular wavelength, those at the other wavelengths are unbalanced. Finally, the unbalance caused by reflection, scattering, and absorption by the sample is to be corrected during an imaging process. The first requirement can be fulfilled by using a common photodetector for the probe and reference beams [30, 32]. This makes it possible to use a common spectrometer for the beams, and the spectral condition is automatically unified. The method to fulfil the second requirement has been developed in our previous study [32]. The detail of this method is mentioned later, and in this study, automatic adjustment (auto-balance) has been developed for a practical application of this method. The final requirement is assumed to be fulfilled with the irradiation of the beams at the same position of the sample, and detection by the common photodetector [30, 32], called collinear balanced detection (CBD). Through the same optical path, the intensity modulations by the sample are assumed to be common. However, we have found that in a practical application, the balance is displaced even with the CBD. In this study, we point out this problem and report the developed auto-balance scheme that is also helpful to solve this problem. Although the idea of auto-balance is well known [31, 33], usually the beams are detected by different photodetectors, and hence it is difficult to fulfil the first requirement. Based on our method, we demonstrate simultaneous two-wavelength SRS imaging on simple optics with the PCF, in preparation for practical high speed spectral SRS imaging.

The target of the present subsection is as follows. In the previous papers we described a straightforward method for spectral stimulated Raman scattering (SRS) microscopy which can measure the scanned gain/loss spectrum of a white probe light from a photonic crystal fiber (PCF) [25, 32]. However, the intensity of the white light noise was a serious problem for SRS microscopy. In this subsection, we have demonstrated simultaneous two-wavelength SRS microscopy with PCF through balanced detection suitable for spectroscopy with a modification of an auto-balance scheme. The developed auto-balance detection system suppresses the degradation of noise cancelation performance caused by a sample, and is suitable for spectral SRS imaging with simple and robust optics.

METHODS

2.1 Stimulated Raman Microscope

Figure 6.2.1 illustrates the schematic of the experimental setup. A titanium–sapphire laser oscillator (Mira 900P, COHERENT) generates 2.5 ps duration pulsed light of 76.3MHz repetition cycle at 800 nm. The average output power is 600 mW. The pulse is split into two; one is used for white-light generation, and the other is for a pump beam.

The former pulse of the two is coupled into a 1m-length PCF of the zero-dispersion wavelength of 790 nm (NL-2.3-790-2, THORLABS) by a microscope objective [× 40, numerical aperture (NA) = 0.65] through a half-wave plate, and is then converted to the white-light continuum. This half-wave plate is for adjustment of the polarization of the input pulse, to maximize the efficiency of the continuum generation; the efficiency is found to be dependent on the polarization angle due to a small fraction of birefringence of this fiber. The white light output is collimated by a microscope objective (× 40, NA = 0.65), and its polarization angle is adjusted by a Glan–Thompson prism polarizer; the splitting ratio into the probe and reference beams is adjusted by this polarization angle. The generated continuum pulse

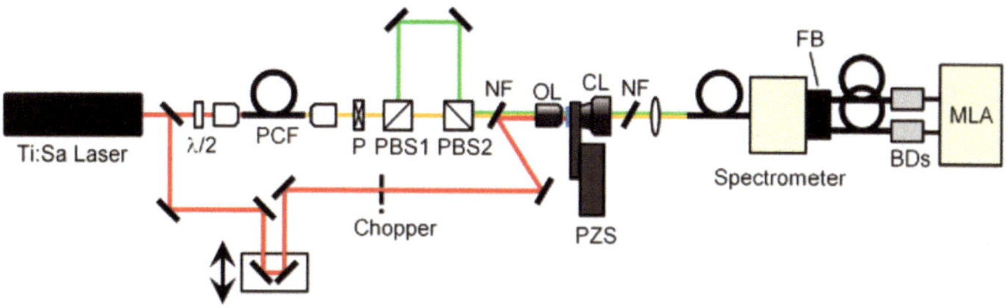

FIGURE 6.2.1 Schematic of the SRS microscope with multi-wavelength balanced detection. The red, orange, and green lines are the 800 nm pulse beam, white probe, and reference pulses, respectively. λ/2: half-wave plate; PCF: photonic crystal fiber; P: Glan–Thompson polarizer; PBS1 and PBS2: polarizing beam splitters; NF: notch filter; OL: microscope objective; CL: condenser lens; PZS: piezo stage; FB: 16 × 128 fiber bundle; BD: the developed balanced detector; MLA: multi-channel lock-in amplifier. The copyright of this figure is partially held between the representative author (Takayoshi Kobayashi) and IOP (Institute of Physics).

beam is split into the probe and reference beams (shown as the orange and green lines, respectively) by a broad band polarizing beam splitter (PBS) (CM1-PBS252, THORLABS). The reference beam is delayed by a quarter cycle of the pulse repetition by an additional 98-cm optical path, and spatially overlapped on the probe beam in a PBS. The alignment for the spatial overlap is checked with an alignment scope (TEL300-38, TRIOPPTICS) with a 3.5 arcsec accuracy to confirm that the probe and reference beams pass the same position of a sample.

Spectral components around 800 nm in the probe and reference beams are rejected by a notch filter (NF808-34, THORLABS).

The latter 800-nm pulse for pump beam is sent to a delay stage to be temporally overlapped with the probe pulse. Then the beam is chopped at 4.48 kHz, and spatially overlapped and collinearly aligned with the probe beam on the notch filter.

The three beams are tightly focused on the sample by a microscope objective (× 40, NA = 0.85). The sample is scanned by a piezo actuator on a stage (NanoMax300, THORLABS), and an SRS image is captured. The light transmitted through the sample is collected by a condenser lens (NA = 1.4). After the pump beam (component of 800 nm) is rejected by a notch filter (NF808-34, THORLABS), the probe and reference beams are guided into a spectrometer (SpectraPro-300i, ACTON) by a multi-mode optical fiber. Thus, the probe and reference beams are spectrally dispersed by the common spectrometer. The focal length and *F* number of the spectrometer are 30 cm and 4, respectively. The groove number (*g*), blaze wavelength, and line dispersion of the grating are 300*g*/mm, 500 nm, and 10.8 nm/mm, respectively. The spectrally dispersed beams are imaged on the entrance facet of a 16 × 128 fiber (core/clad: 100/110 μm) bundle. The spectrum on the facet is split into 128 wavelengths, and each of the sixteen fibers are bundled. In this study, two components of the wavelength are simultaneously observed by two balanced detectors. The outputs are simultaneously measured with a multi-channel lock-in amplifier (Model 7210, SIGNAL RECOVERY).

2.2 Balanced Detector

The developed balanced detector is based on the scheme described in our previous study [32]. Figure 6.2.2 shows the block diagram of the balanced detector.

A common spectral component of the probe and reference beams from the common spectrometer is guided by the sixteen bundled fibers into a common avalanche photodiode (APD) (S5343, HAMAMATSU). The optical current signal is converted to a voltage signal by a resonator of the resonant frequency at 76.3 MHz. This resonator is composed of a coil and a capacitor, and hence free of thermal noise. The signals of the probe and reference beams are expressed as Aa (1 +m) cos(ωt) and Absin(ωt), respectively, with the phase relation due to the quarter cycle optical delay of the reference beam. Here, A, a, b, m, and ω are the intensity of the white pulse, splitting ratios ($a + b = 1$), modulation by the SRS, and the repetition frequency of the pulse source, respectively. Thus, the voltage signal from the resonator is the sum of the signals expressed by

$$A\{a\ (1 + \mathrm{m})\cos(\omega t) + b \sin(\omega t)\}. \tag{6.2.1}$$

This signal is amplified by a pre-amplifier (PA) and transferred to a multiplier (ML).

Also, a synchronous signal of the light source is prepared by observing the optical pulse with a photodetector. The fundamental component of the synchronous pulse is extracted thorough a band-pass filter (BPF). The phase (δ) of this signal is adjusted by a phase sifter, and the amplitude (D) is fixed by a limiter. This signal is expressed as $D\cos(\omega t + \delta)$. Then, the voltage signal of the probe and reference beams is multiplied by this synchronous signal in the ML. The multiplied result is proportional to

$$\begin{aligned} AD[a(1+m)\{\cos(2\omega t+\delta)+\cos\delta\} \\ +b\{\sin(2\omega t+\delta)-\sin\delta\}]. \end{aligned} \tag{6.2.2}$$

The high frequency [radio frequency (RF)] components in equation 6..2.2 are removed by a low-pass filter (LPF1), and the output signal is proportional to

$$A\{a\ (1 + m)\cos\delta - b \sin \delta\}. \tag{6.2.3}$$

Therefore, the output is the difference between the probe and reference signals. Note that these signal gains are adjusted by the δ. By setting the δ as $\tan^{-1}(a/b) + 2n\pi$, the carrier signal of $Aa\cos\delta$ is subtracted and the only modulation signal is output as $(Aa\cos\delta)$m. A phase error due to displacement of the optical delay of the reference beam effectively results in an alternation of the splitting ratio, and thus the complete cancelation of the carrier is still possible (appendix A). When the balanced detection is not applied, the output signal amplitude is proportional to A+Am. Thus, a small modulation (m) (typically 10^{-4} order) on the large carrier signal (A) should be detected. This is problematic, especially when the amplitude of the carrier is strongly fluctuating as in the case of the white-continuum light.

When the δ is not adjusted, as the carrier signal is completely subtracted and/or the ratio of a and b are disturbed during imaging, the direct-current (dc) component appears in the output. As in this diagram, the value of δ is controlled by a feedback mechanism based on the integrated signal of the dc component so that the dc component is zero (the integral control). The disturbances of a and b are thus corrected by this feedback control. The response time of this feedback should be set shorter than the dwelling time at one pixel in the imaging process, and appropriately longer than the modulation cycle of the Raman signal. If the response time is longer than the dwelling time, the feedback control cannot respond to the disturbance during the imaging. Also, if the response time is shorter than the modulation cycle, the modulation signal is also fed back so that the signal vanishes.

By insertion of the LPF2 in the feedback loop, the response speed can be increased through the adequate rejection of the modulation signal. However, the phase retardation of the loop at the cutoff frequency must be carefully checked. The phase retardation of an integrator is 90° at any frequency, and hence feedback is always stable without LPF2. The LPF2 adds the retardation, and if the total retardation exceeds 180° at the cutoff frequency, the feedback results in catastrophe due to the positive feedback of the dc component. Additionally, the phase shifter and ML effectively change the gain of the control. Therefore, in setting the integration time, the gain and phase retardation of the LPF2, the gain in the ML (dependent on the δ), and the gain of the δ in the phase shifter against the control signal must be counted. The detailed scheme is described in appendix B, and concrete responses of the components depend on the electrical circuit. The concrete responses are described with the circuit diagram in the online supplemental information (stacks.iop.org/JPhysD/47/345401/mmedia).

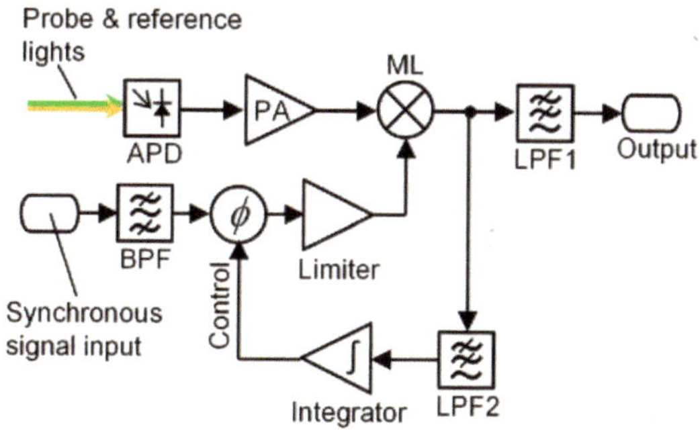

FIGURE 6.2.2 The block diagram of a balanced detector for spectral imaging with a feedback loop for auto-balanced detection. APD: avalanche photodiode; PA: pre-amplifier; ML: multiplier; BPF: band-pass filter; LPF1: low-pass filter to reject RF signal; LPF2: low-pass filter to reject RF and SRS signals. The copyright of this figure is partially held between the representative author (Takayoshi Kobayashi) and IOP (Institute of Physics).

EXPERIMENT

3.1 Noise Cancelation Performance

The noise cancelation performance was initially measured in the case without a sample. The light intensity was adjusted by a neutral-density (ND) filter so as not to saturate the detector. The signal of the light intensity noise was measured 140 times with the lock-in detection of the time constant of 10 ms. The dwelling time was set at 30 ms. The modulation equivalent noise density was evaluated as a tenth of the root-mean-square of these 140 data, normalized by the probe signal intensity. The value of δ was adjusted and fixed by an extra phase shifter so that output signals of the chopped probe and reference were equalized with the same optical input power. In measurements of the signal intensity of the probe (reference) beam, the auto-balance was turned off and the light was chopped with blocking the reference (probe) beam.

The probe beam noise was measured with turning off the feedback, without chopping the light, and with blocking of the reference beam. The fixed balance detected noise was evaluated with inputting the reference light, and then the auto-balance detected noise was measured with turning on the feedback. These evaluations were repeated in various intensity balances between the probe and reference by adjusting the polarizer. The probe light intensities were measured at the corresponding optical balances for the normalization factor of the modulation equivalent noise density, except for the auto-balanced case. For the auto-balanced case, the probe light intensity of the optically balanced case ($a = b$) was used for the normalization factor, since the gain for the probe beam was automatically controlled to the optically balanced case.

3.2 Two-Color SRS Imaging

A sample made up of 4 μm polystyrene (PS) beads embedded in a polyvinyl alcohol (PVA) film on a glass slide was prepared. The observed Raman shifts were selected as the peaks of Raman bands of PS (3054 cm^{-1}) and PVA (2914 cm^{-1}). The optical balance was set at 3054 cm^{-1} by the polarizer, and the δ *s* were adjusted so that the probe and reference signals were equated in respective of wavelength. The images were captured in three cases; without the reference beam and without the auto-balance, with the reference beam and without the auto-balance, and with the reference and with the auto-balance. Also, to investigate the disturbance of the optical balance by the sample, both the probe and reference beams were directly chopped and irradiated on the sample without the pump beam.

RESULTS AND DISCUSSION

4.1 Noise Cancelation Performance

Figure 6.2.3(a) shows the spectra of the probe and reference beams after the microscope. The white continuum contains 600–850 nm components, and covers whole Raman shifts of molecular vibrations. The dip around 800 nm is due to the spectral rejection by the notch filters. Figure 6.2.3(b) shows spectral unbalance of (probe–reference)/probe, and it is found that the unbalance fluctuated within 10% depending

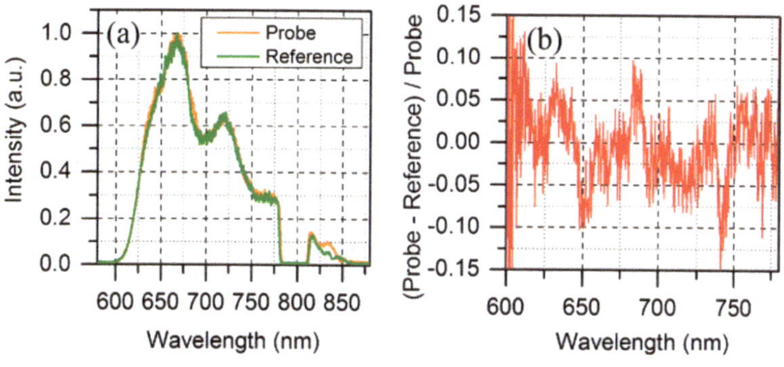

FIGURE 6.2.3 Spectra of probe and reference beams (a), intensity unbalance at the microscope (b). The copyright of this figure is partially held between the representative author (Takayoshi Kobayashi) and IOP (Institute of Physics).

on the wavelength. This is due to the unbalance of optics for the probe and reference beams, and should be corrected for each wavelength for the multi-wavelength balanced detection. The balanced detector in this paper can correct this unbalance.

The modulation equivalent noise against various optical unbalance, (probe–reference)/probe is depicted in Figure 6.2.4. The optical input power of the probe and reference beams at the APD is 6 μW. The noise density of the probe light (the red square plots) is ~1.5×10^{-4} $Hz^{-1/2}$. In the fixed-balanced case (the green round plots), the noise is 5.6×10^{-6} $Hz^{-1/2}$ in the optically best balanced case, and thus 28 dB cancelation is achieved. However, when the optical unbalance is increased in both directions of probe > reference, or probe < reference, the noise proportionally increases due to imperfect subtraction. This is caused by the above-mentioned unbalance of optics, long term instabilities of the optics, and/or the inhomogeneous nature of a sample (see the later discussion). In the auto-balanced case (blue triangle plots), the values of the noise are independent of the optical unbalance and they are almost the same as the value of the optically balanced case in the fixed balanced case. This is due to the automatic correction of the unbalance, and thus the residual noise caused by the optical unbalance is completely canceled out. It is found that the above-mentioned feedback successfully functions for noise cancelation.

However, SNR is affected by the splitting ratio since the gain for the SRS signal is dependent on the ratio, whereas the noise is independent of it. In equation 6.2.3, the signal intensity is proportional to Aam cosδ, where cosδ is set as $b/(a^2 + b^2)^{1/2}$ for the noise cancelation. SNR is dependent on the splitting ratio, as follows,

$$\text{SNR} \propto \frac{ab}{\left(a^2 + b^2\right)^{1/2}} \tag{6.2.4}$$

From equation 6.2.4, it is found that SNR is maximized when $a = b$. Therefore, the condition of as equal splitting as possible is desirable for imaging, although the noise is canceled at maximum efficiency at any splitting ratio.

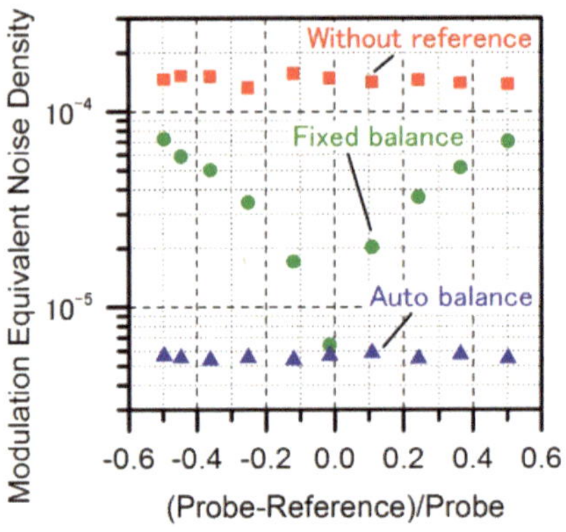

FIGURE 6.2.4 Noise cancelation performance and effect of the auto-balanced detection against the displacement of the optical unbalance between the probe and reference lights. The copyright of this figure is partially held between the representative author (Takayoshi Kobayashi) and IOP (Institute of Physics).

4.2 Two-color SRS Imaging

Spontaneous Raman spectra of PS and PVA are shown in Figure 6.2.5(a). A Raman band of PS assigned to CH stretching mode of the benzene ring and that of PVA assigned to CH stretching mode of the methylene groups are observed at 3054 and 2914cm^{-1} Raman shifts, respectively. The SRS images of the sample are shown in Figures 6.2.5(b)–(f). These images were captured with the time constant of 30 ms, and dwelling time of 60 ms. The pixel size is 0.25 × 0.25 μm, and image size is 41 × 41pixels. The images are illustrated in a unified rainbow color scale. The reddish color represents the stronger signal, and the purplish color represents the weaker signal. The white color represents a strong signal out-of-range of the scale.

Figures 6.2.5(b) and (c) are the SRS images captured on the 3054 and 2914cm^{-1} Raman bands, respectively, with the fixed balance detection. In image (b), ~4 μm discoid distribution of strong signal is captured, corresponding to the PS bead. In (c), the strong signal distribution is opposite to (b), due to the homogeneous distribution of PVA with the discoid exclusion by the PS bead. Image (d) is captured on 3054 cm^{-1} Raman shifts with blocking reference light. In this image, the SRS signal is buried in the light intensity noise of the probe light, and no structure is observed. The situation is the same as the image captured on 2914 cm^{-1} Raman shifts (not shown). Therefore, two-color SRS imaging with the multi-wavelength balanced detection through a spectrometer is attained. Through increasing the number of channels, spectral SRS imaging of the simultaneous detection of the Raman shifts is readily realized.

The images (e) and (f) are captured on 3054 and 2914cm^{-1} Raman bands, respectively, with the auto-balanced scheme. The contrasts of them are correspondingly similar to (b) and (c). These results indicate that light intensity noise is canceled, while the SRS modulation signals are successfully obtained without being vanished

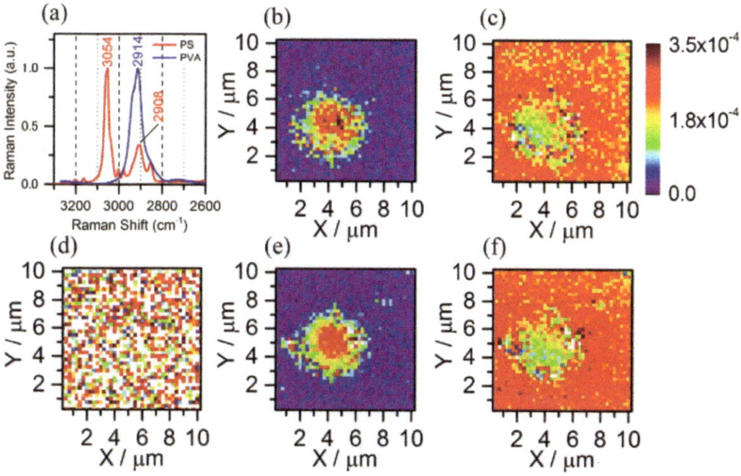

FIGURE 6.2.5 Spontaneous Raman spectra of PS and PVA (a), and SRS images of a 4 μm PS bead in PVA film (b)–(f). (b) and (c) SRS images captured with 3054 and 2914cm^{-1} Raman bands through fixed-balanced detection, respectively. (d) SRS image with 3054cm^{-1} without reference beam. (e) and (f) SRS images with 3054 and 2914cm^{-1} Raman shifts through auto-balanced detection, respectively. The copyright of this figure is partially held between the representative author (Takayoshi Kobayashi) and IOP (Institute of Physics).

during the imaging. The functionality of the scheme to design the parameters of the LPF2 and integrator in the feedback loop is verified. Note that the SNR is better especially around the PS bead than that without the auto-balance.

Figures 6.2.6(a) and (b) are the cross-sections along the X = 4.25 μm lines on the images of the fixed- and auto-balanced cases, respectively. In the fixed-balanced case, the signal of the 3054 cm^{-1} Raman band is strong, and that of the 2914cm^{-1} band is dipped on the PS bead region. However, the signals are strongly fluctuated on this region despite the sphere shape of the bead. In the auto-balanced case, this fluctuation is suppressed. The cause of this fluctuation is explained by the disturbance of the optical balance between the probe and reference beams during imaging, as follows. Figure 6.2.6(c) is captured by directly chopped probe and reference beams through the fixed-balanced detection, and is normalized by the probe signal intensity. If the optical balance is maintained during the imaging, as assumed in the CBD, the value is zero or at least constant at any positions. However, for this sample, the value is upraised to 0.12, especially around the bead. This result clearly indicates the disturbance, and as a result, the light noise of the probe is insufficiently or overly subtracted. In the auto-balanced case, the unbalance caused by this disturbance is corrected, and the light noise is well canceled.

This disturbance is caused by at least three possible factors. One is imperfect alignment, although it is confirmed by the alignment telescope; the same optical path requires significantly severe alignment, and it is susceptible to mechanical displacement, especially in microscopy. Another one is the difference in the transmittance

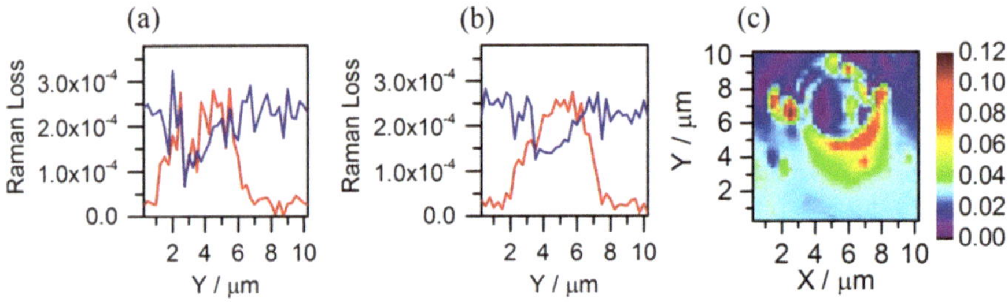

FIGURE 6.2.6 Cross-sectional intensity distributions of SRS images and optical unbalance caused by a sample. (a) Cross-sectional intensity distributions of the fixed-balanced case (Figures 5(b) and (c)) on the X= 4.25 μm lines. The blue and red lines represent the signal intensities of 3054 and 2914cm^{-1} Raman bands, respectively. (b) Those of the auto-balanced case (Figures 5(e) and (f)). (c) The image captured on the chopped probe and reference beams through fixed-balanced detection, normalized by the probe signal intensity. The copyright of this figure is partially held between the representative author (Takayoshi Kobayashi) and IOP (Institute of Physics).

between the probe and reference against the inclined interface of different refractive indices. The probe and reference beams are prepared with the PBSs to maintain high throughput of the optics, and thus the polarization is perpendicular to each other. Thus, the reflectivity on the inclined interface differs in the beams. The last one is birefringence and the dichroic nature of a sample. In our sample, unbalance is about 10%, and it reduces the cancelation to ~20dB. Additionally, this disturbance depends on the sample and position; this distribution may cause an artificial image. These introduce serious errors, especially in applications using a light source of large intensity noise such as the white light continuum. The auto-balanced scheme is required not only to make the system robust and to apply to spectral imaging, but also to eliminate the disturbance induced by the samples.

The contrast of the image of the 2914 cm^{-1} Raman band is lower than that of the 3054 cm^{-1} band. This is due to the overlap of the 2908 cm^{-1} band of PS on the 2914 cm^{-1} band of PVA (Fig. 6.2.5(a)). The ratio between the band areas of the 3054 and 2908 cm^{-1} bands in the spontaneous Raman spectrum is 1:0.68 with fitting of the Voigt function. Equally, the average signal intensity in the range of Y = 4–5 μm of Fig. 6.2.6(b) of the 3054 and 2914 cm^{-1} bands is 2.34×10^{-4} and 1.45×10^{-4}, respectively. The signal intensity ratio is 1:0.61, and is close to the band-area ratio. Therefore, the dominant origin of the signal on the PS bead in Figs. 6.2.5(c) and (f) is estimated as not the 2914cm^{-1} band of PVA film but the 2904cm^{-1} band of the PS bead.

In this study, and also for the auto-balanced detection, both the probe and the reference beams are transmitted through the same position of the sample to compare it with that of the fixed-balanced scheme. However, this configuration is redundant for practical auto-balance detection; the reference beam is not required to pass through the sample, and can bypass the microscope and be overlapped on the probe beam before the spectrometer. By this bypassed configuration, the risk of sample degradation by the reference beam is reduced.

CONCLUSION

Simultaneous two-color SRS imaging with the picosecond continuum probe light is realized through the balanced detector for a multi-wavelength measurement. The probe and reference beams are detected by a common photodetector, and thus this balanced detection is readily applicable to spectroscopy. Additionally, an auto-balance scheme is developed for this balanced detection for a practical multi-wavelength observation and an imaging application. In this scheme, the phase of the synchronous signal, which determines the gains for the probe and reference signal, is successfully feedback controlled. There is limited effectiveness in the case of the method of setting the same optical path for the probe and reference beams through a sample to maintain optical balance during imaging. This limitation is especially serious for an application with the noisy white-continuum probe beam. This problem is well overcome by the developed auto-balance scheme with appropriate setting of the feedback response.

In this SRS microscopy, a single light source is used through preparing a white-continuum probe beam by a PCF. The system does not require synchronization of plural light sources or scanning wavelength of the light, and thus the optics are simple and robust. Although the number of observed wavelengths is two in the present study, spectral imaging by multi-wavelength observation is readily realized through increasing the number of observation wavelengths. Based on this study, high speed spectral SRS imaging is readily realized with high temporal coincidence in respective Raman shifts. The contents in the present subsection are based on the collaborative research of the following members: K Seto, Y. Okuda, E Tokunaga, and T. Kobayashi [34]. Copyright of the contents of this subsection is partially held between the representative author (Takayoshi Kobayashi) and IOP (Institute of Physics).

APPENDIX A

A phase error (δ_e) of the reference beam from the quarter cycle of the pulse repetition effectively causes an alternation of the splitting ratio, as follows. The signal from the resonator is [equation (1) containing δ_e],

$$A\{a\,(1 + m)\cos(\omega t) + b \sin(\omega t + \delta_e)\}. \tag{A1}$$

This is rewritten as

$$A\left\{a'\left(1+m'\right)\cos\left(\omega t\right)+b'\sin\left(\omega t\right)\right\}. \tag{A2}$$

Here,

$$a' = a + b\sin\delta_e, \tag{A3}$$

$$b' = b\cos\delta_e \tag{A4}$$

$$m' = \frac{a}{a + b\sin\delta_e} m \tag{A5}$$

Equation (A2) is effectively the same as equation (A1). Therefore, the light intensity noise can still be completely canceled through the adjustment of the phase of the synchronous signal as $\tan^{-1}(a'/b') + 2n\pi \cdot \Phi$.

APPENDIX B

The system diagram of the feedback loop is illustrated in Figure 6.2.B1. In this case, a controlled object P and detector D are the ML and LPF2, respectively, and controller C is composed of the integrator and phase shifter.

The d and r indicate the dc component in the output and the target value of the dc component. In this study, the value of r is set to be zero. The value of d for the close loop response is given by

$$d = \frac{\text{DPC}}{1 - \text{DPC}} r \tag{B1}$$

Here, the product of the response functions of D, P, and C is the open-loop response function in which D and P are the responses of the LPF2 and ML, respectively, and C is the product of the responses of the integrator and phase shifter. Note, the values of the responses are different in dimensions. The output of the phase shifter is phase, δ, and input is the voltage from the integrator, V_i. Thus, the C including the integrator is expressed as

$$\text{C} = \frac{\text{d}\delta}{\text{d}V_i} \frac{1}{T_i \text{i}\omega}. \tag{B2}$$

The latter term is the response of the integrator. The response of P against δ is

$$\text{P} = \frac{\text{d}(y(\text{i}\omega))}{\text{d}\delta} \tag{B3}$$

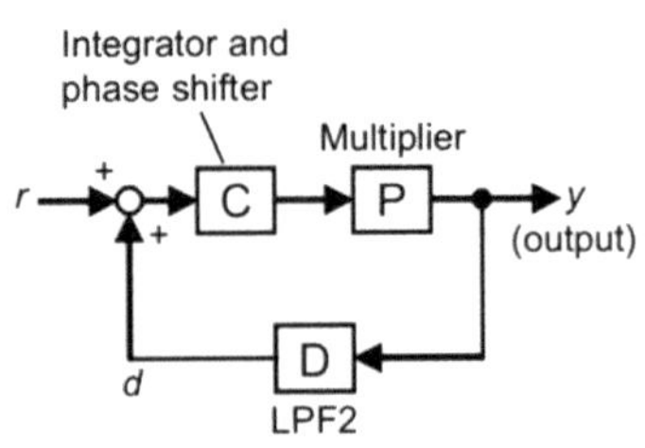

FIGURE 6.2.B1 System diagram for feedback loop. C: controller (integrator and phase shifter in this study); P: controlled object (multiplier); D: detector (LPF2). The copyright of this figure is partially held between the representative author (Takayoshi Kobayashi) and IOP (Institute of Physics).

Here, $y(i\omega)$ is the output of the balanced detector. Therefore, the open-loop response is expressed as follows

$$\mathrm{DPC} = \mathrm{D}\frac{\mathrm{d}y}{\mathrm{d}\delta}\frac{\mathrm{d}\delta}{\mathrm{d}V_i}\frac{1}{T_i i\omega} \tag{B4}$$

From equation (B1) and (B4), the sign of the DPC must be negative, and the phase retardation at the cutoff frequency (when |DPC| = 1) must be less than 180° to make the feedback loop stable. To avoid ringing, it is desirable that the retardation is less than 120° (or phase margin is more than 60°). The concrete response depends on the circuit, and is described in online supplementary data information (stacks.iop.org/JPhysD/47/345401/mmedia).

REFERENCES

1. Freudiger C W, Min W, Saar B G, Lu S, Holtom G R, He C, Tsai J C, Kang J X and Xie X S. 2008. Label-free biomedical imaging with high sensitivity by stimulated Raman scattering microscopy. *Science* **322** 1857–60.
2. Saar B G, Freudiger C W, Reichman J, Stanley C M, Holtom G R and Xie X S. 2010. Video-rate molecular imaging *in vivo* with stimulated Raman scattering. *Science* **330** 1368–70.
3. Ozeki Y, Umemura W, Otsuka Y, Satoh S, Hashimoto H, Sumimura K, Nishizawa N, Fukui K and Itoh K. 2012. High-speed molecular spectral imaging of tissue with stimulated Raman scattering. *Nature Photon.* **6** 845–51.
4. Ozeki Y and Itoh K. 2010. Stimulated Raman scattering microscopy for live-cell imaging with high contrast and high sensitivity. *Laser Phys.* **20** 1114–8.
5. Zhang X, Roeffaers M B J, Basu S, Daniele J R, Fu D, Freudiger C W, Holtom G R and Xie X S. 2012. Label-free live-cell imaging of nucleic acids using stimulated Raman scattering microscopy. *Chem. Phys. Chem.* **13** 1054–9.
6. Duncan M D, Reintjes J and Manuccia T J. 1982. Scanning coherent anti-Stokes Raman microscope. *Opt. Lett.* **7** 350–2.
7. Zumbusch A, Holtom G R and Xie X S. 1999. Three-dimensional vibrational imaging by coherent anti-stokes Raman scattering. *Phys. Rev. Lett.* **82** 4142–5.
8. Kee T W and Cicerone M T. 2004. Simple approach to one-laser, broadband coherent anti-Stokes Raman scattering microscopy. *Opt. Lett.* **29** 2701–3.
9. Evans C L, Potma E O, Puoris'haag M, Côté D, Lin C P and Xie X S. 2005. Chemical imaging of tissue *in vivo* with video-rate coherent anti-Stokes Raman scattering microscopy. *Proc. Natl Acad. Sci. USA* **102** 16807–12.
10. Lutgens M, Chatzipapadopoulos S and Lochbrunner S. 2012. Coherent anti-Stokes Raman scattering with broadband excitation and narrowband probe. *Opt. Express* **20** 6478–87.
11. Okuno M, Kano H, Leproux P, Couderc V, Day J P R, Bonn M and Hamaguchi H. 2010. Quantitative CARS molecular fingerprinting of single living cells with the use of the maximum entropy method. *Angew. Chem. Int. Ed.* **49** 6773–7.
12. Fu D, Lu F-K, Zhang X, Freudiger C, Pernik D R, Holtom G and Xie X S. 2012. Quantitative chemical imaging with multiplex stimulated Raman scattering microscopy. *J. Am. Chem. Soc.* **134** 3623–6.

13. Ozeki Y, Kitagawa Y, Sumimura K, Nishizawa N, Umemura W, Kajiyama S i, Fukui K and Itoh K. 2010. Stimulated Raman scattering microscope with shot noise limited sensitivity using subharmonically synchronized laser pulses. *Opt. Express* **18** 13708–19.
14. Nandakumar P, Kovalev A and Volkmer A. 2009. Vibrational imaging based on stimulated Raman scattering microscopy. *New J. Phys.* **11** 1–9.
15. Lu F-K, Ji M, Fu D, Ni X, Freudiger C W, Holtom G and Xie X S. 2012. Multicolor stimulated Raman scattering microscopy. *Mol. Phys.* **110** 1927–32.
16. Fu D, Holtom G, Freudiger C, Zhang X and Xie X S. 2012. Hyperspectral imaging with stimulated Raman scattering by chirped femtosecond lasers. *J. Phys. Chem. B* **117** 4634–40.
17. Freudiger C W, Min W, Holtom G R, Xu B, Dantus M and Xie X S. 2011. Highly specific label-free molecular imaging with spectrally tailored excitation-stimulated Raman scattering (STE-SRS) microscopy. *Nature Photon.* **5** 103–9.
19. Ozeki Y, Umemura W, Sumimura K, Nishizawa N, Fukui K and Itoh K. 2012. Stimulated Raman hyperspectral imaging based on spectral filtering of broadband fiber laser pulses. *Opt. Lett.* **37** 431–3.
20. Andresen E R, Berto P and Rigneault H. 2011. Stimulated Raman scattering microscopy by spectral focusing and fiber-generated soliton as Stokes pulse. *Opt. Lett.* **36** 2387–9.
21. Beier H T, Noojin G D and Rockwell B A. 2011. Stimulated Raman scattering using a single femtosecond oscillator with flexibility for imaging and spectral applications. *Opt. Express* **19** 18885–92.
22. Dudley J M and Coen S. 2006. Supercontinuum generation in photonic crystal fiber. *Rev. Mod. Phys.* **78** 1135–84.
23. Husakou A V and Herrmann J. 2001. Supercontinuum generation of higher-order solitons by fission in photonic crystal fibers. *Phys. Rev. Lett.* **87** 203901.
24. Herrmann J, Griebner U, Zhavoronkov N, Husakou A, Nickel D, Knight J C, Wadsworth W J, Russell P S J and Korn G. 2002. Experimental evidence for supercontinuum generation by fission of higher-order solitons in photonic fibers. *Phys. Rev. Lett.* **88** 173901.
25. Seto K, Okuda Y, Tokunaga E and Kobayashi T. 2013. Development of a multiplex stimulated Raman microscope for spectral imaging through multi-channel lock-in detection. *Rev. Sci. Instrum.* **84** 083705.
26. Corwin K L, Newbury N R, Dudley J M, Coen S, Diddams S A, Weber K and Windeler R S. 2003. Fundamental noise limitations to supercontinuum generation in microstructure fiber. *Phys. Rev. Lett.* **90** 113904.
27. Brauckmann N, Kues M, Groß P and Fallnich C. 2011. Noise reduction of supercontinua via optical feedback. *Opt. Express* **19** 14763–78.
28. Møller U, Sørensen S T, Jakobsen C, Johansen J, Moselund P M, Thomsen C L and Bang O. 2012. Power dependence of supercontinuum noise in uniform and tapered PCFs. *Opt. Express* **20** 2851–7.
29. Gambetta A, Kumar V, Grancini G, Polli D, Ramponi R, Cerullo G and Marangoni M. 2010. Fiber-format stimulated-Raman-scattering microscopy from a single laser oscillator. *Opt. Lett.* **35** 226–8.
30. Nose K, Ozeki Y, Kishi T, Sumimura K, Nishizawa N, Fukui K, Kanematsu Y and Itoh K. 2012. Sensitivity enhancement of fiber-laser-based stimulated Raman scattering microscopy by collinear balanced detection technique. *Opt. Express* **20** 13958–65.
31. Hobbs P C D. 1997. Ultrasensitive laser measurements without tears. *Appl. Opt.* **36** 903–20.
32. Seto K, Tsukada T, Okuda Y, Tokunaga E and Kobayashi T. 2014. Development of a balanced detector with biased synchronous detection and application to near shot noise limited noise cancelling of supercontinuum pulse light. *Rev. Sci. Instrum.* **85** 023702.

33. Freudiger C W, Yang W, Holtom G R, Peyghambarian N, Xie X S and Kieu K Q. 2014. Stimulated Raman scattering microscopy with a robust fibre laser source. *Nature Photon.* **8** 153–9.
34. Seto K, Okuda Y, Tokunaga, E and Kobayashi T. 2014. Multiplex stimulated Raman imaging with white probe-light from a photonic-crystal fibre and with multi-wavelength balanced detection. *J. Phys. D Appl. Phys.* **47** 345401-9.
35. Suzuki T and Misawa K. 2011. Efficient heterodyne CARS measurement by combining spectral phase modulation with temporal delay technique *Opt. Express* **19** 11463–70

Online supplementary data available from tacks.iop.org/JPhysD/47/345401/mmedia

J. Phys. D: Appl. Phys. 47 (2014) 345401 (9pp) doi:10.1088/0022-3727/47/34/345401

6.3 Theoretical Description for Nonlinear Dynamic Light Scattering Based on Stimulated Raman Effect (NLDLS-SRS)

1 INTRODUCTION

Dynamic light scattering (DLS) is a well-established method for studying the structure of colloidal systems such as particulate dispersions, polymer solutions, and gels. DLS evaluates density fluctuations and diffusion constants of colloidal systems through spatiotemporal fluctuations of the refractive index [1, 2]. For example, DLS elucidates the hydrodynamic radius distribution of the constituent particles in a particulate dispersion based on the translational diffusion coefficient and viscosity of the medium [3]. DLS also measures drift velocity. In combination with electrophoresis [4], DLS is applied to the ζ-potential measurement, which is related to the stability of particulate dispersing solutions. It is also employed to analyze molecular motion in polymer solutions [5, 6]. Developments of the cooperative diffusion theory and the concept of non-ergodicity have led to its application in gels.

That is, DLS analyzes the typical length of the cooperative diffusion (gel mesh structure size), gelation point [7], and structural change in phase transition [8]. However, DLS lacks chemical identifiability since it detects refractive index fluctuations. Providing the chemical structure sensitivity to DLS allows chemical composition analysis of the colloid system. Furthermore, it is expected to clarify molecular mechanisms imparting properties of the target colloidal system. That is, the chemical origin of the surface forces [e.g., van der Waals forces, electrostatic repulsion (known as the DLVO theory) and steric repulsion, and other interactions], which are the dominant factor. Indeed, the chemical structure in the phase transition of microgels was studied employing Raman spectroscopy as well as DLS [9].

Fluorescence correlation spectroscopy (FCS) allows a chemical composition-specific investigation of the fluctuations by selective fluorescence detection. The idea of

 DOI: 10.1201/9781003476030-24

FCS was proposed in the 1970s, and the theory is now well established [10–12]. The development of equipment in the 1990s has allowed for a wide range of applications [13]. For example, FCS detects fluctuation in fluorescence intensity within a tightly focused illuminated volume, and the autocorrelation of the fluctuation is evaluated. Based on this autocorrelation, FCS measures the particle number in the illumination volume, translational and/or rotational diffusion coefficients of the fluorescent molecule, and more. However, the target system is limited to fluorescent molecules. Although fluorescence labeling can identify large non-fluorescent molecules such as proteins [14, 15], it may perturb the fluctuation characteristics of small non-fluorescent molecules.

Raman correlation spectroscopy (RCS) was proposed [16, 17] to investigate the chemically specific fluctuation without labeling. RCS measures the autocorrelation of the fluctuation of spontaneous Raman scattering induced by molecular vibrations instead of that of fluorescence in the illumination volume. RCS evaluated chemically specific diffusion constant from the autocorrelation. However, spontaneous Raman scattering is inherently weak and has a low signal-to-noise ratio (S/N), which has hindered wide application. Coherent anti-Stokes Raman scattering (CARS) spectroscopy is one of the coherent Raman spectroscopies, which enables molecular vibrational spectroscopy with an S/N (in the power ratio) three orders of magnitude higher than that of spontaneous Raman spectroscopy [18] under the typical measurement conditions. Taking this advantage, correlation spectroscopy by CARS (CARS-CS) has been developed [19–22]. However, CARS produces a non-resonant background (NRB) that is not specific to molecular vibrations. This NRB can lead to an artifact of non-chemically-specific fluctuations.

The principle to detect the fluctuation is similar in FCS, RCS, and CARS-CS; that is, these methods measure the spatiotemporal fluctuations of the probed particles within the tightly focused illumination volume. This has the advantage of detecting fluctuations in the local area of focus. It has the disadvantage, however, that when the number of particles in the illumination volume is too large, the average signal hinders the fluctuation signal. This feature may obstruct the application to bulk samples.

This chapter provides an idea and discusses the theory of correlation spectroscopy of stimulated Raman scattering (SRS), which has the same principle as in DLS. This is, namely, nonlinear DLS based on the SRS effect (NLDLS-SRS). SRS spectroscopy is another type of coherent Raman spectroscopy and has been extensively studied in the field of non-labeled imaging [23]. The shot-noise-limited S/N of SRS is similar to that of CARS spectroscopy [18, 24]. Furthermore, SRS is free from NRB associated with CARS, which allows higher chemical selectivity. Stimulated Raman scattering detects the imaginary part of the third-order nonlinear susceptibility that resonates with molecular vibrations [25, 26], while the scattered electric field in DLS is proportional to the spatial Fourier component of the spatiotemporal fluctuations of the linear susceptibility [1]. NLDLS-SRS provided in this paper detects the spatial Fourier component of the spatiotemporal fluctuations of the third-order nonlinear susceptibility, which is the origin of the SRS. In other words, NLDLS-SRS provides molecular vibration-selective DLS. In addition, since the principle of spatiotemporal fluctuation detection is the same as that of DLS, the resources of analysis methods developed for DLS can be employed without modification.

2 FLUCTUATION OF MESOSCALE STRUCTURE AND OPTICAL ELECTRIC FIELD IN DLS

Hereafter, electromagnetism is described in the SI unit system. Although the unit of wavenumber (ν) is "m[1]," it can be converted into the unit of "cm[1]" by being divided by 100 without changing the physical content. A typical optical configuration [1] for DLS is depicted in Figure 6.3.1A. Here, we assume the sample is nonmagnetic, nonconductive, and non-absorptive. The multiple scattering is also negligible. A monochromatic beam of the wavevector k_i and the polarization direction of n_i is incident on a sample. Then, the scattered light of the wavevector k_f and the polarization direction of n_f is detected, where the difference between k_i and k_f is defined as the scattering vector q:

$$\mathbf{q} = \mathbf{k}_i - \mathbf{k}_f, \tag{1}$$

Suppose the dielectric constant of the sample is given by

$$\varepsilon \mathbf{I} + \delta\varepsilon(\mathbf{r}, t) \tag{2}$$

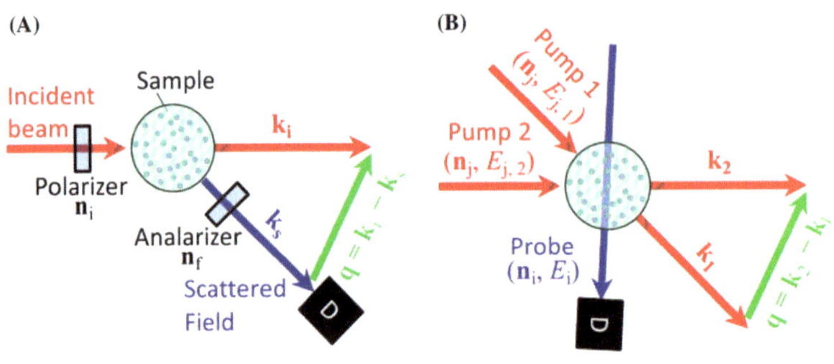

FIGURE 6.3.1 Optical configurations for the conventional dynamic light scattering (DLS) (A) and present nonlinear DLS on stimulated Raman effect (B). Here, D depicts the detector. Note the similarity in the methods. DLS analyzes the spatial Fourier component of the scattering vector (q) of linear susceptibility. q is regarded as the "wavevector" of the standing wave generated by the interference of the incident and scattered fields. The present method analyzes the spatial Fourier component of q of the imaginary part of third-order nonlinear susceptibility. q is the wavevector of the standing wave generated by the interference of two pump beams. The copyright of this figure is shared between the representative author (Takayoshi Kobayashi) and John Wiley & Sons.

where $\delta\varepsilon(\mathbf{r},t)$ is the spatiotemporal deviation from the average permittivity ε, which is generally a second-order tensor quantity, and **I** is the unit tensor. The scattered field (E_s) detected at the distance of R from the sample center is given by

$$E_s(R,t) = \frac{E_0 k_f^2}{4\pi\varepsilon R}\exp\left[i\left(k_f R - 2\pi c\tilde{\nu}_i t\right)\right]\delta\varepsilon_{if}(\mathbf{q},t) \tag{3}$$

Where $E_0, \tilde{\nu}_i$ and c are the amplitude and wavenumber of the incident field, and speed of light in vacuum, respectively. The term of $\delta\varepsilon_{if}(\mathbf{r},t)$ is the spatial Fourier component of q in:

$$\delta\varepsilon_{if}(\mathbf{q},t) = \int_V d^3r\left[\exp\left[i\mathbf{q}\cdot\mathbf{r}\right]\delta\varepsilon_{if}(\mathbf{r},t)\right] \tag{4}$$

and $\delta\varepsilon_{if}(\mathbf{r},t)$ is given by

$$\delta\varepsilon_{if}(\mathbf{r},t) = \mathbf{n}_f\cdot\left(\delta\boldsymbol{\mu}(\mathbf{r},t)\right)\cdot\mathbf{n}_i \tag{5}$$

The V is the scattering volume, which is the spatial overlap of the illuminated region by the incident beam and the detected region. Therefore, DLS detects the electric field proportional to the spatial Fourier component of **q** of the linear dielectric fluctuations or linear susceptibility fluctuations.

Equation (3) indicates that dielectric fluctuations appear as sidebands in the spectrum of the scattered field. However, sidebands typically appear on the order of 10^7 cm^1 (kilohertz in terms of frequency), and measurement with an optical spectrometer is challenging. Therefore, heterodyne detection is employed, where the sidebands are down converted into the vicinity of DC by mixing with the local oscillator. Another method is intensity measurement of the scattered field. This intensity measurement is homodyne detection, where the scattered field itself functions as the local oscillator. The sideband due to the fluctuation is analyzed with the power spectral density (PSD) or autocorrelation function of the detected signal. Note that the photodetector acts as a squared detector. The autocorrelation function for the heterodyne detection is given by

$$\begin{aligned}&\left|E_L(t)+E_s(t)\right|^2\left|E_L(t+\tau)+E_s(t+\tau)\right|^2\\&\simeq I_L^2 + 2I_L\mathrm{Re}\left[E_s^*(t)E_s(t+\tau)\right],\end{aligned} \tag{6}$$

Here, the second term reflects the fluctuation of the system:

$$2I_L\mathrm{Re}\left[E_s^*(t)E_s(t+\tau)\right] \propto \mathrm{Re}\left[\delta\varepsilon_{if}^*(\mathbf{q},t)\delta\varepsilon_{if}(\mathbf{q},t+\tau)\right]. \tag{7}$$

For the homodyne detection, the autocorrelation function is

$$\left\langle\left|E_s(t)\right|^2\left|E_s(t+\tau)\right|^2\right\rangle \propto \left\langle\left|\delta\varepsilon_{if}(\mathbf{q},t)\right|^2\left|\delta\varepsilon_{if}(\mathbf{q},t+\tau)\right|^2\right\rangle \tag{8}$$

Here, the ensemble average of the physical quantity of A is depicted by $\langle A \rangle$. E_L and I_L are the local field and its intensity, respectively. Here, analyses with the PSD or autocorrelation function are equivalent since they are a Fourier transform pair (the Winner–Hinchin theorem [27]).

3 MESOSCALE STRUCTURAL ANALYSIS BASED ON STIMULATED RAMAN EFFECT

In a typical setup for SRS spectroscopy, the difference between the higher photon energy of the pump light and the lower photon energy of the Stokes light is scanned. When the photon energy difference resonates with molecular vibration energy, the radiation field by the imaginary part of the third-order nonlinear polarization interferes destructively with the pump field and constructively with the Stokes field, respectively (Figure 6.3.2A) [28]. That is, the SRS effect causes intensity modulation of the pump and Stokes beams (Fig. 6.3.2B). In order to observe the SRS effect, therefore, the intensity of either the pump or Stokes beams is modulated, and the counterpart is lock-in detected. This setup is commonly called the pump/probe method, where the intensity-modulated light is regarded as the pump light to stimulate the sample, and the other is the probe light to observe the stimulation. Hereafter, let us comply with the terminology for the pump/probe measurement; that is, the light to observe the intensity change is called "probe", and the other light of intensity modulated is called "pump".

As mentioned later, the third-order nonlinear polarization associated with the SRS process is proportional to the probe field amplitude and pump intensity (not the field amplitude). Here, by causing a standing wave of the pump light to modulate its intensity spatially, we can access a spatial Fourier component of third-order nonlinear polarization (Fig. 6.3.1B). In this way, fluctuation analysis based on molecular

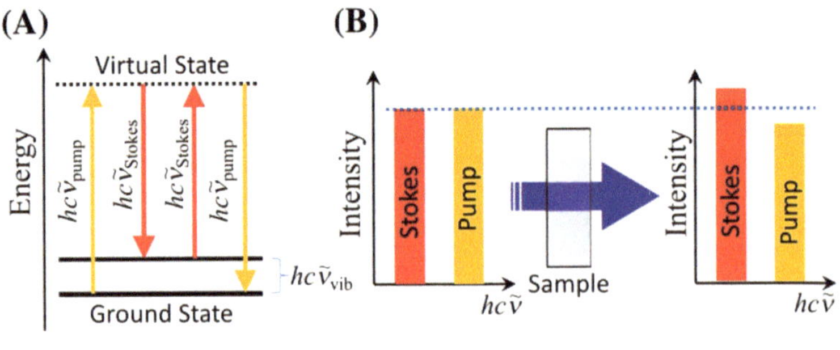

FIGURE 6.3.2 Energy diagram (A) and intensity change of the pump and Stokes beams (B) in the stimulated Raman process. Here, $hc\nu_{vib}$ is molecular vibrational energy. (A) The stimulated Raman process occurs when the difference of the photon energies of the pump and Stokes beams matches the vibrational energy. (B) The intensity of the Stokes beam of lower photon energy is increased in the stimulated Raman process, while that of the pump beam of higher photon energy is decreased. The copyright of this figure is shared between the representative author (Takayoshi Kobayashi) and John Wiley & Sons.

vibrations sensitive to the chemical structure is realized along the well-established DLS methodology.

Maxwell's equations explain the relationship between the spatial evolution of the probe light field and induced polarization [25]. We assume the medium is nonmagnetic, nonconductive, and non-absorbing. From Maxwell's equations, the wave equation for the photoelectric field (**E**) is given by

$$\nabla^2 \mathbf{E} - \frac{1}{c^2}\frac{\partial^2}{\partial t^2}\mathbf{E} = \mu_0 \frac{\partial^2}{\partial t^2}\mathbf{P} \tag{9}$$

where μ_0 and P are the permeability of vacuum and macroscopic polarization, respectively. Although the pump and probe beams are usually focused on the sample to observe the SRS effect, we restrict ourselves to consider a loosely focusing condition [e.g., a numerical aperture (N_A) is about 0.1]. Under this condition, the plane wave approximation is reasonable, especially near the focus, where the SRS effect is pronounced. Then, both the electric fields of the pump and probe beams are approximated to the plane waves. We assume that the probe field (E_{pr}) is linearly polarized to the i axis, whose instantaneous amplitude (E_i) changes slowly. The real electric field is given by

$$E_{\text{pr}}(t,\mathbf{r}) = \frac{1}{2}\left[E_i\left(\tilde{\nu}_{\text{pr}}\mathbf{r}\right)\exp\left[i\left(\mathbf{k}_{\text{pr}}\cdot\mathbf{r} - 2\pi c\tilde{\nu}_{\text{pr}}t\right)\right] + \text{c.c.}\right] \tag{10}$$

where c.c. depicts complex conjugate. Then, substituting Equation (10) into the first term on the left-hand side of Equation (9), we obtain Equation (11).

When the amplitude of the probe beam (E_i (ν_{pr})) is nearly constant in the order of wavelength, Equation (11) is approximated to

$$\begin{aligned}\nabla^2 E_{\text{pr}}(t,\mathbf{r}) = \frac{1}{2}&\exp\left(-2\pi i c\tilde{\nu}_{\text{pr}}t\right)\\ &\left[\exp\left(i\mathbf{k}_{\text{pr}}\cdot\mathbf{r}\right)\nabla^2 E_i\left(\tilde{\nu}_{\text{pr}}\mathbf{r}\right)\right.\\ &+2i\exp\left(i\mathbf{k}_{\text{pr}}\cdot\mathbf{r}\right)\mathbf{k}_{\text{pr}}\cdot\nabla E_i\left(\tilde{\nu}_{\text{pr}}\mathbf{r}\right)\\ &\left.-k_{\text{pr}}^2\exp\left(i\mathbf{k}_{\text{pr}}\cdot\mathbf{r}\right)E_i\left(\tilde{\nu}_{\text{pr}}\mathbf{r}\right)\right] + \text{c.c.}\end{aligned} \tag{12}$$

where ν_{pr} is the linear refractive index for the probe beam. In an isotropic medium with no linear absorption, there is no second-order nonlinearity. Thus, the polarization density is written in terms of the first- and third-order contributions as follows:

$$\begin{aligned}\mathbf{P}(t,\mathbf{r}) &= \mathbf{P}^{(1)}(t,\mathbf{r}) + \mathbf{P}^{(3)}\left(\tilde{\nu}_{\text{pr}},t,\mathbf{r}\right)\\ &= \varepsilon_0\chi^{(1)}\left(\tilde{\nu}_{\text{pr}}\right)\mathbf{E}_{\text{pr}}(t,\mathbf{r}) + \mathbf{P}^{(3)}\left(\tilde{\nu}_{\text{pr}},t,\mathbf{r}\right)\end{aligned} \tag{13}$$

where ε_0 and $\chi^{(1)}$ are the permittivity of vacuum and linear susceptibility, respectively. Now, substituting Equations (10), (12), and (13) into Equation (9) with attention to

$$n_{\mathrm{pr}}^2 = \frac{\varepsilon\left(\tilde{v}_{\mathrm{pr}}\right)}{\varepsilon_0} = 1 + \chi^{(1)}\left(\tilde{v}_{\mathrm{pr}}\right), \tag{14}$$

we obtain

$$\begin{aligned} &i\exp\left[i\left(\mathbf{k}_{\mathrm{pr}} \cdot \mathbf{r} - 2\pi c\tilde{v}_{\mathrm{pr}}t\right)\right]\mathbf{k}_{pr} \cdot \nabla E_i\left(\tilde{v}_{\mathrm{pr}}, \mathbf{r}\right) + \text{c.c.} \\ &= \mu_0\left[\frac{\partial^2}{\partial t^2}P_i^{(3)}\left(\tilde{v}_{\mathrm{pr}}, t, \mathbf{r}\right)\right] \end{aligned} \tag{15}$$

where $\varepsilon(\nu_{\mathrm{pr}})$ is the linear dielectric constant for the probe field.

Let us consider the time dependence of the third-order nonlinear polarization, although it also depends on the spatial location. Since the response speed of the system is finite, the projected component of the third-order nonlinear polarization onto the i-axis is given as a multivariate convolution. The projected nonlinear polarization is as follows, using the j-, k-, and l-axis polarization components of the three real electric fields and the real response function $R^{(3)}_{\mathrm{ijkl}}$ for the electric fields:

$$P_i^{(3)}(t) = \int_0^\infty\int_0^\infty\int_0^\infty R_{ijkl}^{(3)}\left(\tau_1\tau_2\tau_3\right)E_j\left(t-\tau_1\right)E_k\left(t-\tau_2\right)E_l\left(t-\tau_3\right)\mathrm{d}\tau_1\,\mathrm{d}\tau_2\,\mathrm{d}\tau_3, \tag{16}$$

where we take the sum for the *j*-, *k*-, and *l*-axes (the Einstein rule for tensor products). Here, the causality rule

$$R_{ijkl}^{(3)}\left(\tau_1\tau_2\tau_3\right) \equiv 0 \text{ for } \tau_1 < 0, \tau_2 < 0, \tau_3 < 0 \tag{17}$$

is taken into account. The Fourier transform pair of the incident electric field is defined as follows:

$$\begin{aligned} E_\alpha\left(\tilde{v}_\beta\right) &= c\int_{-\infty}^{\infty} E_\alpha(t)\exp\left(-2\pi ic\tilde{v}_\beta t\right)\mathrm{d}t \\ E_\alpha(t) &= \int_{-\infty}^{\infty} E_\alpha\left(\tilde{v}_\beta\right)\exp\left(2\pi ic\tilde{v}_\beta t\right)\mathrm{d}\tilde{v}_\beta. \end{aligned} \tag{18}$$

Substituting Eq. (18) into Eq. (16) yields

$$\begin{aligned} P_i^{(3)}(t) = \varepsilon_0 \int_{-\infty}^{\infty}\mathrm{d}\tilde{v}_1 \int_{-\infty}^{\infty}\mathrm{d}\tilde{v}_2 \int_{-\infty}^{\infty}\mathrm{d}\tilde{v}_3 &\left[E_j\left(\tilde{v}_1\right)E_k\left(\tilde{v}_2\right)E_l\left(\tilde{v}_3\right)\right. \\ &\left.\cdot\chi_{ijkl}^{(3)}\left(\tilde{v}_1\tilde{v}_2\tilde{v}_3\right)\exp\left\{2\pi ic\left(\tilde{v}_1+\tilde{v}_2+\tilde{v}_3\right)t\right\}\right] \end{aligned} \tag{19}$$

where the multivariate Fourier transform of the third-order nonlinear response function (susceptibility) is defined below:

$$\chi_{ijkl}^{(3)}\left(\tilde{\nu}_1\tilde{\nu}_2\tilde{\nu}_3\right) := \frac{1}{\varepsilon_0}\int_0^{\infty}\mathrm{d}\tau_1\int_0^{\infty}\mathrm{d}\tau_2\int_0^{\infty}\mathrm{d}\tau_3\Big[R_{ijkl}^{(3)}\left(\tau_1\tau_2\tau_3\right) \cdot\exp\left\{-2\pi ic\left(\tilde{\nu}_1\tau_1+\tilde{\nu}_2\tau_2+\tilde{\nu}_3\tau_3\right)\right\}\Big]. \tag{20}$$

Generally, this amount is a complex function. When considering the spatial dependence of Eq. (19), it further becomes a spatial convolution of the local electric field and the local nonlinear susceptibility. However, since the spatial variation of the incident field is negligible in the molecular size range that determines the microscopic polarization, the spatial convolution is approximately a product of the field and susceptibility (the long-wave approximation) [26]. That is, the spatial response function for the nonlinear susceptibility of the molecule is regarded as the spatial Dirac delta function. Then, the polarization at location **r** is

$$P_i^{(3)}\left(t,\mathbf{r}\right) = \varepsilon_0\int_{-\infty}^{\infty}\mathrm{d}\tilde{\nu}_1\int_{-\infty}^{\infty}\mathrm{d}\tilde{\nu}_2\int_{-\infty}^{\infty}\mathrm{d}\tilde{\nu}_3\Big[E_j\left(\tilde{\nu}_1,\mathbf{r}\right)E_k\left(\tilde{\nu}_2,\mathbf{r}\right)E_l\left(\tilde{\nu}_3,\mathbf{r}\right) \cdot\chi_{ijkl}^{(3)}\left(\tilde{\nu}_1\tilde{\nu}_2\tilde{\nu}_3\right)\exp\left\{2\pi ic\left(\tilde{\nu}_1+\tilde{\nu}_2+\tilde{\nu}_3\right)t\right\}\Big]. \tag{21}$$

In this study, two pump beams are provided, both of which are assumed to be linearly polarized in the j-axis (Fig. 6.3.1B). The respective pump fields are

$$E_1\left(t,r\right) = \frac{1}{2}\Big[E_{j,1}\left(\tilde{\nu}_{\mathrm{pu}}\right)\exp\left[i\left(\mathbf{k}_1\cdot\mathbf{r}-2\pi c\tilde{\nu}_{\mathrm{pu}}t\right)\right]+\mathrm{c.c.}\Big] \tag{22}$$

and

$$E_2\left(t,r\right) = \frac{1}{2}\Big[E_{j,2}\left(\tilde{\nu}_{\mathrm{pu}}\right)\exp\left[i\left(\mathbf{k}_2\cdot\mathbf{r}-2\pi c\tilde{\nu}_{\mathrm{pu}}t\right)\right]+\mathrm{c.c.}\Big], \tag{23}$$

respectively. In the SRS process, the intensity of the pump beams also changes in the interaction volume. Still, since the intensity change is typically 10^4, the pump field amplitude is approximated to be constant. As a result of the interference of these fields, the pump field becomes

$$\begin{aligned} E_{\mathrm{pu}}\left(t,\mathbf{r}\right) &= E_1\left(t,\mathbf{r}\right)+E_2\left(r,\mathbf{r}\right) \\ &= \frac{1}{2}E_j\left(\tilde{\nu}_{\mathrm{pu}}\mathbf{k}_1\mathbf{k}_2\mathbf{r}\right)\exp\left[-2\pi ic\tilde{\nu}_{\mathrm{pu}}t\right]+\text{ c.c.} \end{aligned}, \tag{24}$$

where

$$E_j\left(\tilde{\nu}_{\mathrm{pu}}\mathbf{k}_1\mathbf{k}_2\mathbf{r}\right) = exp\left[\mathbf{k}_1\cdot\mathbf{r}\right]\Big[E_{j,1}\left(\tilde{\nu}_{\mathrm{pu}}\right)+E_{j,2}\left(\tilde{\nu}_{\mathrm{pu}}\right)\exp\left[i\left(\mathbf{k}_2-\mathbf{k}_1\right)\cdot\mathbf{r}\right]\Big] \tag{25}$$

Substituting the Fourier transform of Eq. (10) (the probe field) and Eq. (24) (the pump field) into Eq. (21), 64 terms appear. Among them, the real third-order nonlinear polarization related to the SRS process, that is, interfering with the probe field, is as follows:

$$P_i^{(3)}\left(\tilde{v}_{\mathrm{pr}},t,\mathbf{r}\right)=\frac{6}{4}\varepsilon_0\left|E_j\left(\tilde{v}_{\mathrm{pu}}\mathbf{k}_1\mathbf{k}_2\mathbf{r}\right)\right|^2 E_i\left(\tilde{v}_{\mathrm{pr}}\mathbf{r}\right)\exp\left[i\mathbf{k}_{\mathrm{pr}}\cdot\mathbf{r}\right] \\ \cdot\chi_{ijj}^{(3)}\left(\tilde{v}_{\mathrm{pr}}-\tilde{v}_{\mathrm{pr}}-\tilde{v}_{\mathrm{pu}}\tilde{v}_{\mathrm{pu}}\right)\exp\left(-2\pi ic\tilde{v}_{\mathrm{pr}}t\right)+\text{c.c.} \tag{26}$$

Here, due to the symmetry of the susceptibility, the polarization is the same amount for substitutions among i, j, and k, and hence, the factor 3! = 6 is multiplied. Furthermore, noting the symmetry of the susceptibility and

Eq. (20), the relation of

$$\chi_{ijkl}^{(3)}\left(\tilde{v}_1\tilde{v}_2-\tilde{v}_3\right)=\chi_{ijkl}^{(3)^*}\left(\tilde{v}_1\tilde{v}_2\tilde{v}_3\right)=\chi_{ijkl}^{(3)}\left(\tilde{v}_1-\tilde{v}_2\tilde{v}_3\right) \tag{27}$$

is taken into account. Note that

$$\left|E_j\left(\tilde{v}_{\mathrm{pu}}\mathbf{k}_1\mathbf{k}_2\mathbf{r}\right)\right|^2=\left|E_{j,1}\left(\tilde{v}_{\mathrm{pu}}\right)\right|^2+\left|E_{j,2}\left(\tilde{v}_{\mathrm{pu}}\right)\right|^2 \\ +E_{j,1}\left(\tilde{v}_{\mathrm{pu}}\right)E_{j,2}^*\left(\tilde{v}_{\mathrm{pu}}\right)\exp\left[-i\left(\mathbf{k}_2-\mathbf{k}_1\right)\cdot\mathbf{r}\right] \\ +E_{j,1}^*\left(\tilde{v}_{\mathrm{pu}}\right)E_{j,2}\left(\tilde{v}_{\mathrm{pu}}\right)\exp\left[i\left(\mathbf{k}_2-\mathbf{k}_1\right)\cdot\mathbf{r}\right]. \tag{28}$$

This represents a standing wave whose amplitude depends on r.

Hereafter, it is assumed that the third-order nonlinear polarization density fluctuates due to the translational and/or rotational motion of the molecule. Then, the non-linear polarization density is a spatiotemporal function. The timescale of this fluctuation is assumed to be very slow compared to that of the photoelectric fields and their difference (related to molecular vibrations in the SRS process). That is,

$$\chi_{ijj}^{(3)}\left(\tilde{v}_{\mathrm{pr}},\tilde{v}_{\mathrm{pr}},-\tilde{v}_{\mathrm{pu}},\tilde{v}_{\mathrm{pu}};t,\mathbf{r}\right)=:\chi_{iijj}^{(3)}\left(t,r\right) \tag{29}$$

and

$$\left|\frac{\partial}{\partial t}\left[\chi_{iijj}^{(3)}\left(t,\mathbf{r}\right)\right]\right|<<\left|2\pi c\tilde{v}_{\mathrm{pr}}\chi_{ijj}^{(3)}\left(t,\mathbf{r}\right)\right|. \tag{30}$$

Substituting Eq. (26) into Eq. (15), we obtain the following equation for the probe amplitude:

$$\mathbf{k}_{pr}\cdot\nabla E_i\left(\tilde{v}_{\mathrm{pr}}\mathbf{r}\right)=6\pi^2\tilde{v}_{\mathrm{pr}}^2\left|E_j\left(\tilde{v}_{\mathrm{pu}}\mathbf{k}_1\mathbf{k}_2\mathbf{r}\right)\right|^2\chi_{ijj}^{(3)}\left(t,r\right)E_i\left(\tilde{v}_{\mathrm{pr}}\mathbf{r}\right). \tag{31}$$

Actually, the probe field is scattered due to the fluctuations of the linear and nonlinear susceptibilities in the direction perpendicular to k_{pr}, and there are some field components in a different direction from the incident wavevector. However, when the concentration of the scatterer is dilute enough, most of the probe light passing through the sample keeps the direction of the incident wavevector. Or, under the condition that only the probe light in the direction of $\mathbf{k}_{pr}$ is detected, the solution of Eq. (31) is approximated to

$$\begin{aligned} E_i\left(\tilde{v}_{pr} z_0\right) \simeq \\ E_i\left(\tilde{v}_{pr} - z_0\right)\exp\left[iK\Phi\left(\mathbf{k}_1\mathbf{k}_2 t\right)\right] \\ \exp\left[-K\int_V \left|E_j\left(\tilde{v}_{pu}\mathbf{k}_1\mathbf{k}_2\mathbf{r}\right)\right|^2 \operatorname{Im}\left[\chi_{iij}^{(3)}(t,\mathbf{r})\right]\mathrm{d}^3 r\right], \end{aligned} \tag{32}$$

Where

$$\begin{aligned} K := 3\pi\tilde{v}_{pr}\sqrt{\frac{\varepsilon_0}{\varepsilon(\tilde{v}_{pr})}}, \\ \Phi\left(\mathbf{k}_1\mathbf{k}_2 t\right) := \int_V \left|E_j\left(\tilde{v}_{pu}\mathbf{k}_1\mathbf{k}_2\mathbf{r}\right)\right|^2 \operatorname{Re}\left[\chi_{iijj}^{(3)} t,\mathbf{r})\right]\mathrm{d}^3 r, \end{aligned} \tag{33}$$

and V and z_0 in (32) and (33) are the interaction volume and length, respectively. The probe field intensity at the detector is
proportional to

$$\left|E_i\left(\tilde{v}_{pr}\mathbf{r}_0\right)\right|^2 = \left|E_i\left(\tilde{v}_{pr} - \mathbf{r}_0\right)\right|^2 \exp\left[-2K\int_V \left|E_j\left(\tilde{\tau}_{pu}\mathbf{k}_1\mathbf{k}_2\mathbf{r}\right)\right|^2 \operatorname{Im}\left[\chi_{ijj}^{(3)}(t,\mathbf{r})\right]\mathrm{d}^3 r\right]$$

$$\simeq \left|E_l\left(\tilde{v}_{pr} - \mathbf{r}_0\right)\right|^2 \left[1 - 2K\int_V \left|E_j\left(\tilde{v}_{pu}\mathbf{k}_1\mathbf{k}_2\mathbf{r}\right)\right|^2 \operatorname{Im}\left[\chi_{ijj}^{(3)}(t,\mathbf{r})\right]\mathrm{d}^3 r\right], \tag{34}$$

where a slight change in probe field intensity is assumed in the latter approximation. The lock-in detection readily extracts the second term in the parenthesis with intensity modulated pump beam. The detected signal [s(t); the modulation depth of the square of the probe field] with substitution of Eq. (28) is given by

$$\frac{s(t)}{KI_{pr}} = I_s X(t,0) + 2\operatorname{Re}\left[I_{12} X(t,\mathbf{q})\right], \tag{35}$$

where

$$I_{pr} = \left|E_i\left(\tilde{v}_{pr} - \mathbf{r}_0\right)\right|^2, \tag{36}$$

$$I_s := \left|E_{j,1}\left(\tilde{\nu}_{pu}\right)\right|^2 + \left|E_{j,2}\left(\tilde{\nu}_{pu}\right)\right|^2, I_{12} := E_{j,1}\left(\tilde{\nu}_{pu}\right)E^*_{j,2}\left(\tilde{\nu}_{pu}\right), \tag{37}$$

$$q := \mathbf{k}_2 - \mathbf{k}_1, \tag{38}$$

$$X(t,q) := \int_V Im\left[\chi^{(3)}_{iijj}(t,\mathbf{r})\right] exp\left[-iq\cdot r\right]\mathrm{d}^3 r. \tag{39}$$

Here, the second term in Eq. (34) is halved in the lock-in detection. This is the fundamental result of the SRS signal obtained with the interfered pump field. Note that the conventional SRS output is obtained by setting $\mathbf{q} = \mathbf{0}$. According to Eq. (39), the received signal contains the scattering vector of the q component in the spatial Fourier transform of the susceptibility, as in DLS [Eq. (4)]. Therefore, we can expect to detect density fluctuations from the autocorrelation as in DLS. Furthermore, the molecular vibration is selectively detected in this component since it is the imaginary part of the third-order nonlinear susceptibility.

As in the conventional DLS, let us take the autocorrelation of Eq. (35). When the system is stationary, the autocorrelation does not depend on absolute time but the lag of the autocorrelation, and thus,

$$\begin{aligned}
\frac{r(\tau)}{K^2 I_{pr}^2} &= \frac{\langle s(0)s(\tau)\rangle}{K^2 I_{pr}^2} \\
&= I_s^2\langle X(0,0)X(\tau 0)\rangle + 2\left|I_{12}\right|^2 \mathrm{Re}\left[X(0,\mathbf{q})X^*(\tau\mathbf{q})\right] \\
&+2I_s\mathrm{Re}\left[I_{12}\langle X(0,0)X(\tau\mathbf{q})\rangle\right] + 2I_s\mathrm{Re}\left[I_{12}\langle X(\tau 0)X(0,\mathbf{q})\rangle\right] \\
&+2\,\mathrm{Re}\left[I_{12}^2\langle X(0,\mathbf{q})X(\tau\mathbf{q})\rangle\right]
\end{aligned} \tag{40}$$

Here, when the system is macroscopically homogeneous (e.g., not close to the macroscopic interface),

$$\frac{r(\tau)}{K^2 I_{pr}^2} = I_s^2\langle X(0,0)X(\tau 0)\rangle + 2\left|I_{12}\right|^2 Re\left[X(0,\mathbf{q})X^*(\tau\mathbf{q})\right]. \tag{41}$$

The first term is independent of the coherence between the pump intensity distribution and spatial structure. It is similar to FCS and RCS, reflecting the particle number in the illumination volume, fluctuation of the particle number, and so on. The second term is derived from the coherent output and isomorphic to the heterodyne DLS representation. Here, the selective detection of the second term is essential because the fluctuation component can be overwhelmed by the mean component and because the overlap between the first and second terms can complicate the analysis. As discussed in the next section, consideration of a specific system readily demonstrates the situation.

4 SUSPENSION OF MONODISPERSE PARTICLES

4.1 INFLUENCE OF Q = 0 COMPONENT

For an intuitive understanding of the above result [Eq. (41)], we apply it to a system composed of dispersed identical spherical particles of uniform composition as a standard system [1]. We discuss the available information with the fundamental application. Suppose N particles are diffusing in the medium without correlating (Figure 6.3.3). Here, we assume that the difference in refractive index between the medium and the material comprising the particles is negligible. This assumption assures that the phase of the incident beams does not change after particle transmission (the Rayleigh-Gans approximation). Furthermore, we assume no interaction between the particles. For a suspension of a few tens of nanometers to micrometers in size of the particle with a volume fraction of a few percent or less, the experimental result is well explained [29], even if particle-particle interactions are neglected.

The phases of the irradiated pump beams depend on the location of the particle. We then divide this particle into n identical segments, assuming that the size of one segment is sufficiently smaller than the wavelength and that the phases within the segments are identical (Fig. 6.3.3). When the imaginary part of the susceptibility of the segment is depicted as Imh i$\chi^{\text{ð Þ}}{}_{iijj}{}^{3}$, and the position of the v-th segment in particle k in the laboratory coordinate system is $r^{k}{}_{v}$ ð Þt , the imaginary part of the susceptibility of the particle system is as follows:

$$\mathrm{Im}\left[\chi_{ijj}^{(3)}(t,\mathbf{r})\right] = \sum_{k=1}^{N}\sum_{v=1}^{n}\mathrm{Im}\left[\chi_{iijj}^{(3)}\right]\delta\left(\mathbf{r}-\mathbf{r}_{v}^{k}(t)\right), \tag{42}$$

where δ(r) is Dirac's delta. The signal occurs when the particles are present in the interaction volume. To explicitly express the fact, we introduce b^k, which takes the value 1 or 0 when the particle is in or out of the interaction volume, respectively. The number of particles in the interaction volume (N(t)) is

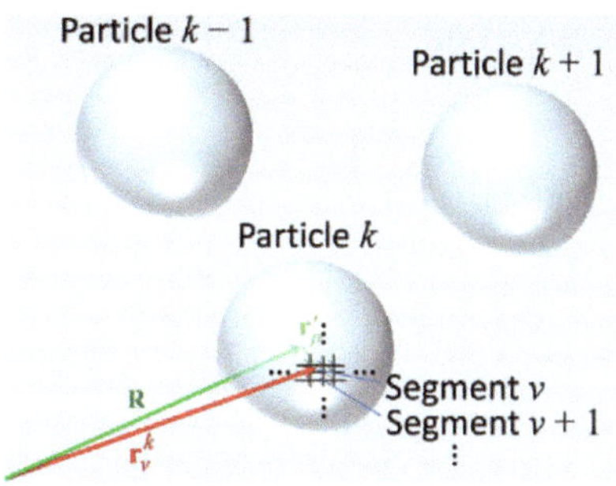

FIGURE 6.3.3 Dispersion system of identical spheric particles and schematic for segmentation. Coordinate systems and suffixes for the segments are indicated. The copyright of this figure is shared between the representative author (Takayoshi Kobayashi) and John Wiley & Sons.

$$N(t)=\sum_{k=1}^{N} b^k(t). \tag{43}$$

Then, Eq. (39) is given by

$$X(t,\mathbf{q})=\mathrm{Im}\left[\chi_{iijj}^{(3)}\right]\sum_{k=1}^{N}\sum_{v=1}^{n} b^k(t)\exp\left[-i\mathbf{q}\cdot\mathbf{r}_v^k(t)\right]. \tag{44}$$

The first term in Eq. (41) is proportional to

$$\langle X(0,0)X(\tau 0)\rangle = n^2 Im\left[\chi_{iijj}^{(3)}\right]^2\langle N(0)N(\tau)\rangle. \tag{45}$$

Here, when N(t) is expressed by the sum of the average number of <N> and the fluctuation of $\delta N(t)$,

$$\langle N(0)N(\tau)\rangle = \langle N\rangle^2+\langle\delta N(0)\delta N(\tau)\rangle \tag{46}$$

When the intensity distribution of the field in the interaction volume has a Gaussian shape given by

$$exp\left[-\frac{x^2+y^2}{2\sigma_1^2}\right]exp\left[-\frac{z^2}{2\sigma_2^2}\right], \tag{47}$$

and the motion of the particle is only diffusion, the second term in Equation (46) is calculated to [1, 10]

$$\langle\delta N(0)\delta N(\tau)\rangle = \langle N\rangle\left(1+\frac{D\tau}{\sigma_1^2}\right)^{-1}\left(1+\frac{D\tau}{\sigma_2^2}\right)^{-\frac{1}{2}}, \tag{48}$$

where D is the diffusion coefficient.

$$X(0,\mathbf{q})X^*(\tau\mathbf{q})$$

$$=\mathrm{Im}\left[\chi_{iij)}^{(3)}\right]^2\left\langle\sum_{k=1}^{N}\sum_{p=1}^{n}\sum_{q=1}^{n} b^k(0)b^k(\tau)\exp\left[-i\mathbf{q}\cdot\left\{\mathbf{r}_p(0)-\mathbf{r}_q(\tau)\right\}\right]\right\rangle. \tag{49}$$

Concerning the second term in Eq. (41), since the particles are identical and their positions are statistically independent, the timescale (τ_b) of $b^k(0)b^k(t)$ is equivalent to the time required for the particle k to move a characteristic size (L) of the interaction volume. Regarding the diffusive motion, $\langle L\rangle = (Dt)^{1/2}$. Hence, $\tau_b = L^2/D$. On the contrary, the timescale for the factor of displacement $|-i\mathbf{q}\cdot\{\mathbf{r}p(0)-\mathbf{r}q(t)|$ is equivalent to the time for displacement $|\mathbf{r}p(0)-\mathbf{r}q(t)|$ to change by q^{-1} ; that is, $\tau_q = \left(q^2 D\right)^{-1}$. Therefore, the ratio of the timescale is $\tau_b / \tau_q = q^2 L^2$. In a typical SRS measurement, visible to near-infrared focused beams are employed, and thus, q ~10^7 m^{-1} and L~ 5 μm (λ= 800 nm, N_A= 0.1). In this case, $\tau_b / \tau_q \sim 2{,}500$, and the term of $b^\kappa(0) b^\kappa(t)$ is approximately regarded as a time-independent constant, and $b^k(0) b^k(t) \simeq b^k(0) b^k(0) = b^k(0)$. Then, Equation (49) isreorganized to

$$\left\langle X(0,\mathbf{q}) X^*(\tau\mathbf{q}) \right\rangle = n^2 \mathrm{Im}\left[\chi_{ijij}^{(3)}\right]^2 \langle N \rangle \frac{1}{n^2} \left\langle \Sigma_{p=1}^{n} \Sigma_{q=1}^{n} \exp\left[-i\mathbf{q} \cdot \left\{\mathbf{r}_p(0) - \mathbf{r}_q(\tau)\right\}\right] \right\rangle \tag{50}$$

To investigate the effect of the translational motion of the particles, the position of the segment in the experimental coordinate system is divided into the center of gravity (R(t)) and relative coordinates (r'_p (t)):

$$\mathbf{r}_p(t) = \mathbf{r}'_p(t) + \mathbf{R}(t). \tag{51}$$

Equation (50) is rewritten with this coordinate transformation as follows:

$$\begin{aligned} &\left\langle X(0,\mathbf{q}) X^*(\tau\mathbf{q}) \right\rangle \\ &= n^2 \mathrm{Im}\left[\chi_{ijj}^{(3)}\right]^2 \langle N \rangle \frac{1}{n^2} \left\langle \exp\left[i\mathbf{q} \cdot \Delta\mathbf{R}(\tau)\right] \sum_{p=1}^{n} \sum_{q=1}^{n} \exp\left[-i\mathbf{q} \cdot \left(\mathbf{r}'_p(0) - \mathbf{r}'_q(\tau)\right)\right] \right\rangle, \end{aligned} \tag{52}$$

$$\Delta\mathbf{R}(t) = \mathbf{R}(t) - \mathbf{R}(0). \tag{53}$$

For the spherical particles, the relative motion of the segments is rotation. Since the sphere is invariant by rotation, the summation for the segments is independent of time and constant. Thus,

$$X(0,\mathbf{q}) X^*(\tau\mathbf{q}) = \chi_{\mathrm{p}}^2 \langle N \rangle P(q) \left\langle exp\left[iq \cdot \Delta R(\tau)\right] \right\rangle, \tag{54}$$

where

$$\chi_{\mathrm{p}} := nIm\left[\chi_{iij}^{(3)}\right], \tag{55}$$

and

$$\begin{aligned} P(\mathbf{q}) &= \frac{1}{n^2} \left\langle \sum_{p=1}^{n} \sum_{q=1}^{n} \exp\left[-i\mathbf{q} \cdot \left(\mathbf{r}'_p(0) - \mathbf{r}'_q(0)\right)\right] \right\rangle \\ &= \left(\frac{3}{(qa)^3} [\sin(qa) - qa\cos(qa)] \right)^2. \end{aligned} \tag{56}$$

Eq. (56) is the particle structure factor for the sphere of the radius a. Here, the factor of

$$F_{\mathrm{s}}(\mathbf{q}\tau) := \left\langle \exp\left[i\mathbf{q}\cdot\Delta\mathbf{R}(\tau)\right]\right\rangle \tag{57}$$

is called the self-intermediate-scattering function. This function is the Fourier transform of a probability distribution for the particle to displace by ΔR during time τ:

$$G_{\mathrm{s}}(\mathbf{R}\tau) = \left\langle \delta\left(\mathbf{R} - \Delta\mathbf{R}(\tau)\right)\right\rangle. \tag{58}$$

This function is known as the self-spatiotemporal correlation function, and the following diffusion equation gives the equation of motion:

$$\frac{\partial}{\partial\tau} G_{\mathrm{s}}(\mathbf{R}\tau) = D\nabla^2 G_{\mathrm{s}}(\mathbf{R}\tau), \tag{59}$$

where $D = k_B T/\zeta$ and k_B, T, and ζ are the Boltzmann constant, absolute temperature, and friction coefficient on the particle, respectively. For the spherical particle in the liquid of the viscosity of η, $\zeta = 6\pi\ \eta a$, and

$$D = \frac{k_{\mathrm{B}}T}{6\pi\eta a}, \tag{60}$$

which is the Stokes-Einstein relation. Thus,

$$F_{\mathrm{s}}(\mathbf{q}\tau) = \exp\left(-q^2 D\tau\right) \quad . \tag{61}$$

Here, the diffusion process is described by the Langevin equation, in which the velocity of the particle is determined by random force and mobility (friction) on the particle. When the interaction of the particle is considerable, or the interaction is objective to be analyzed, it can be treated as a driving force applied on one particle from another and mobility is regarded to be determined by the relative positions of the particles [30]. The Langevin equation, therefore, incorporates the potential energy between particles, which determines the force between particles. Additionally, the mobility is given by the hydrodynamic interaction between particles. This hydrodynamic interaction results from the correlation between the random forces on one particle and another.

From the above, Eq. (41) for the diffusion of identical spherical homogeneous particles is summarized to

$$\begin{aligned}\frac{r(\tau)}{K^2 I_{\mathrm{pr}}^2} &= I_{\mathrm{s}}^2 \langle N\rangle^2 \chi_{\mathrm{p}}^2 + I_{\mathrm{s}}^2 \langle N\rangle \chi_{\mathrm{p}}^2 \left(1 + \frac{D\tau}{\sigma_1^2}\right)^{-1} \left(1 + \frac{D\tau}{\sigma_2^2}\right)^{-\frac{1}{2}} \\ &+ 2\left|I_{12}\right|^2 \langle N\rangle \chi_{\mathrm{p}}^2 P(\mathbf{q}) \exp\left(-q^2 D\tau\right)\end{aligned} \tag{62}$$

The first and second terms are isomorphic to FCS and RCS, while the third is the term that appears in the heterodyne DLS. Here, the first term is proportional to the square of the average number of particles, and the second and third terms are linearly proportional to the average number. Therefore, when many particles are present in the interaction volume, the information on the fluctuations will be hindered, as in FCS.

4.2 Selective Detection of Spatial Fourier Component with Phase Modulation

Note that according to Eq. (37), when the phase of one of the pump beams is shifted by π, the first term in Eq. (35) is left intact, while the sign of the second term is inverted. This phase shift corresponds to swapping the node and antinode positions of the standing wave of the two pump beams. In the case that the phase of the pump beam 2 [Eq. (23)] is sinusoidally modulated by $m_{pm} \sin(\omega_{pm} t)$, the phase-modulated pump field 2 is given by

$$E_2(t,\mathbf{r}) = \frac{1}{2}\Sigma_{n=-\infty}^{\infty} J_n(m_{pm})\left[\exp(in\omega_{pm}t)E_{j,2}\exp\left[i\left(\mathbf{k}_2\cdot\mathbf{r} - 2\pi\, c_{pu}\, t\right)\right] + \text{c.c.}\right], \quad (63)$$

where $J_n(x)$ is the n-th order Bessel function of the first kind. Accordingly, the second term in Eq. (35) is converted into

$$\begin{aligned}
&2KI_{pr}\int_V \left|E_j\left(\tilde{v}_{pu},\mathbf{k}_1,\mathbf{k}_2,\mathbf{r}\right)\right|^2 \text{Im}\left[\chi_{iijj}^{(3)}(t,\mathbf{r})\right]d^3r \\
&= 2KI_{pr}I_sX(t,0) \\
&+2KI_{pr}E_{j,1}E_{j,2}^{*}X(t,\mathbf{q})\sum_{n=-\infty}^{\infty} J_n(m_{pm})\exp(-in\omega_{pm}t) \\
&+2KI_{pr}E_{j,1}E_{j,2}^{*}X(t,\mathbf{q})\sum_{n=-\infty}^{\infty} J_n(m_{pm})\exp(in\omega_{pm}t) \\
&= 2KI_{pr}I_sX(t,0) + 4KI_{pr}\text{Re}\left[I_{12}X(t,\mathbf{q})\right]J_0(m_{pm}) \\
&+8KI_{pr}\text{Re}\left[I_{12}X(t,\mathbf{q})\right]\sum_{n=1}^{\infty} J_{2n}(m_{pm})\cos(2n\omega_{pm}t) \\
&+8KI_{pr}\text{Im}\left[I_{12}X(t,\mathbf{q})\right]\sum_{n=1}^{\infty} J_{2n-1}(m_{pm})\sin\{(2n-1)\omega_{pm}t\}
\end{aligned} \quad (64)$$

Such phase modulation is readily operated with an electro-optic modulator (EOM), photoelastic modulator (PEM), or oscillating mirrors driven by voice coils or piezoelectric elements. In any case, the second term in Eq. (35) is selectively detected by the lock-in detection with the phase modulation of the pump beam. Here, ω_{pm} must be set at one more than double the reciprocal of the shortest relaxation time of the system, and the time constant of the lock-in detection must be set shorter than the relaxation time.

Then, when the third or fourth term in Eq. (64) is detected,

$$s(t) = \frac{J_{2n-1}(m_{pm})}{2} \cdot 8KI_{pr}\mathrm{Im}\left[I_{12}X(t,\mathbf{q})\right] \text{ or } \frac{J_{2n}(m_{pm})}{2} \cdot 8KI_{pr}\mathrm{Re}\left[I_{12}X(t,\mathbf{q})\right], \tag{65}$$

respectively, by the lock-in detection at the frequency of (2n – 1)ω_{pm} or 2nω_{pm}. The factor of 1/2 is due to the lock-in detection. Accordingly, Eq. (41) becomes

$$r(\tau) = 8K^2 J^2_{2n-1 \text{ or } 2n}(m_{pm}) I^2_{pr} |I_{12}|^2 \mathrm{Re}\left[X(0,\mathbf{q})X^*(\tau\mathbf{q})\right], \tag{66}$$

for both cases, and we find that the autocorrelation function of the spatial Fourier component is extracted. For the dispersion of identical particles, the autocorrelation is given by a single exponential function [cf. Equations (54), (57), and (61)]:

$$r(\tau) = 8K^2 J^2_{2n-1 \text{ or } 2n}(m_{pm}) I^2_{pr} |I_{12}|^2 \chi^2_p \langle N \rangle P(\mathbf{q}) \exp\left(-q^2 D\tau\right) \tag{67}$$

This is a similar result obtained with the heterodyne DLS. From the decay constant, the translational diffusion constant is obtained. Further, the particle size (the hydrodynamic radius) is evaluated with the Stokes-Einstein relation. Note that a combination of Eq. (34) for the probe beam intensity before the lock-in detection and Equation (64) dictates that the same signal is received solely by phase modulating and lock-in detection without intensity modulation with the pump beam.

Another method is the measurement of the PSD with a spectrum analyzer or FFT analyzer. The PSD without the phase modulation is given by the Fourier transform of the autocorrelation of Eq. (34). The detected probe beam is

$$s(t) = I_{pr} - 2KI_{pr}I_s X(t,0) + 4KI_{pr}\mathrm{Re}\left[I_{12}X(t,\mathbf{q})\right]. \tag{68}$$

The autocorrelation function is

$$r(\tau) = I^2_{pr} + 4K^2 I^2_{pr} I^2_s \langle X(0,0)X(\tau 0)\rangle + 8K^2 I^2_{pr} |I_{12}|^2 \mathrm{Re}X(0,\mathbf{q})X^*(\tau\mathbf{q}). \tag{69}$$

Then, the PSD is given by

$$\frac{1}{2\pi}\int_{-\infty}^{\infty} r(\tau)\exp\left[-i\omega\tau\right]\mathrm{d}\tau = I^2_{pr}\delta(\omega) + 4K^2 I^2_{pr} I^2_s PSD_{X,0}(\omega) + 8K^2 I^2_{pr} |I_{12}|^2 PSD_{X,\mathbf{q}}(\omega) \tag{70}$$

where the PSDs for $X(0,0)$ and $X(0,\mathbf{q})$ are defined as

$$\frac{1}{2\pi}\int_{-\infty}^{\infty} X(0,0)X^*(\tau 0)\exp(-i\omega\tau)d\tau = \mathrm{PSD}_{X,0}(\omega) \tag{71}$$

and

$$\frac{1}{2\pi}\int_{-\infty}^{\infty}\mathrm{Re}\left[X(0,\mathbf{q})X^*(\tau\mathbf{q})\right]\exp(-i\omega\tau)d\tau = \mathrm{PSD}_{X,\mathbf{q}}(\omega), \tag{72}$$

respectively. Especially for monodisperse spherical particle dispersions,

$$\begin{aligned}\mathrm{PSD}_{X,0}(\omega) &= \langle N\rangle^2 \chi_{\mathrm{p}}^2\delta(\omega)\\ &+\langle N\rangle\chi_{\mathrm{p}}^2\mathrm{FT}\left[\left(1+\frac{D|\tau|}{\sigma_1^2}\right)^{-1}\left(1+\frac{D|\tau|}{\sigma_2^2}\right)^{-\frac{1}{2}}\right]\end{aligned} \tag{73}$$

and

$$\mathrm{PSD}_{X,\mathbf{q}}(\omega) = \frac{1}{2\pi}\langle N\rangle\chi_{\mathrm{p}}^2 P(\mathbf{q})\frac{2q^2D}{\left(q^2D\right)^2+\omega^2}, \tag{74}$$

respectively, where FT[f(t)] represents the Fourier transform of $f(t)$. Note

$$X(0,\mathbf{q})X^*(-\tau\mathbf{q}) = X^*(0,\mathbf{q})X(\tau\mathbf{q}) \tag{75}$$

since the system is stationary. The second term in Eq. (73) cannot be expressed with the elementary functions. The PSD comprises sidebands of the second term in Equations (73) and (74) around the DC peak. Although both the sidebands reflect the fluctuation of the system, the overlap of the sidebands complicates the analysis.

The phase modulation is also effective in simplifying the PSD. The phase modulation converts the sideband due to Eq. (72) around the harmonics of the phase modulation frequency. The phase-modulated PSD is calculated as follows. The phase-modulated signal given by Eq. (64) is not stationary; that is, its autocorrelation function depends on both time and lag. In this case, the PSD is given by the Fourier transform of the time-averaged autocorrelation function [27]. The autocorrelation function is

$$\begin{aligned}r(t,\tau) &= \langle s(t)s(t+\tau)\rangle\\ &= I_{\mathrm{pr}}^2 + 4K^2I_{\mathrm{pr}}^2I_{\mathrm{s}}^2\langle X(0,0)X(\tau,0)\rangle\\ &+8K^2I_{\mathrm{pr}}^2|I_{12}|^2\,\mathrm{Re}\Big[X(0,\mathbf{q})X^*(\tau,\mathbf{q})\\ &\sum_{n,m=-\infty}^{\infty}\mathrm{J}_n(m_{\mathrm{pm}})\mathrm{J}_m(m_{\mathrm{pm}})\exp(-i(n-m)\omega_{\mathrm{pm}}t)\exp(in\omega_{\mathrm{pm}}\tau)\Big]\end{aligned} \tag{76}$$

where the stationarity of $X(t, \mathrm{q})$ is considered. Then, the PSD is

$$\begin{aligned}\mathrm{PSD}(\omega)=&\frac{1}{2\pi}\int_{-\infty}^{\infty}\overline{r(t,\tau)}\exp(-i\omega\tau)\mathrm{d}\tau\\&=I_{\mathrm{pr}}^2\delta(\omega)+4K^2I_{\mathrm{pr}}^2I_{\mathrm{s}}^2PSD_{X,0}(\omega)\\&+8K^2I_{\mathrm{pr}}^2\left|I_{12}\right|^2\ \mathrm{J}_0^2(m_{\mathrm{pm}})PSD_{X,\mathbf{q}}(\omega)\\&+8K^2I_{\mathrm{pr}}^2\left|I_{12}\right|^2\ \Sigma_{n=1}^{\infty}\mathrm{J}_n^2(m_{\mathrm{pm}})\Big[PSD_{X,\mathbf{q}}(\omega+n\omega_{\mathrm{pm}})\\&+PSD_{X,\mathbf{q}}(\omega-n\omega_{\mathrm{pm}}\Big),\end{aligned} \tag{77}$$

where $\overline{r(t\tau)}$ is the time average of $r(t,\tau)$. As indicated by

the third term, $PSD_{X,\mathbf{q}}(\omega)$, reflecting the fluctuations, is shifted by the harmonics of the phase modulation frequency apart from the incoherent sideband near DC. Note the PSD reflects the Doppler shift of the Raman band due to the fluctuations in the system. The PSD, expressed in wavenumbers, is as follows:

$$\begin{aligned}\mathrm{PSD}(\tilde{\nu})=&2\pi cI_{\mathrm{pr}}^2\delta(\tilde{\nu})+8\pi cK^2I_{\mathrm{pr}}^2I_{\mathrm{s}}^2PSD_{X,0}(\tilde{\nu})\\&+16\pi cK^2I_{\mathrm{pr}}^2\left|I_{12}\right|^2\ \mathrm{J}_0^2(m_{\mathrm{pm}})PSD_{X,\mathbf{q}}(\tilde{\nu})\\&+16\pi cK^2I_{\mathrm{pr}}^2\left|I_{12}\right|^2\ \Sigma_{n=1}^{\infty}\Big[\mathrm{J}_n^2(m_{\mathrm{pm}})\\&\Big[PSD_{X,\mathbf{q}}(\tilde{\nu}+\tilde{\nu}_{\mathrm{pm}})+\mathrm{PPD}_{X,\mathbf{q}}(\tilde{\nu}-\tilde{\nu}_{\mathrm{pm}})\Big]\Big],\end{aligned} \tag{78}$$

where the phase modulation angular frequency is $2\pi c\tilde{\nu}_{\mathrm{pm}}$. The objective component of Eq. (74) is rewritten in

$$\mathrm{PSD}_{X,\mathbf{q}}\left(\nu\right)=c\langle N\rangle\chi_{\mathrm{p}}^2P(\mathbf{q})\frac{2q^2D}{\left(q^2D\right)^2+(2\pi c\nu)^2}, \tag{79}$$

5 FEASIBILITY

5.1 Possible Implementations

We discuss some optical systems to implement NLDLS-SRS. In the experimental setups, q should be variable without changing the pulse timing of the two pump beams. In a possible configuration, the pump beam is split, and the two beams are opposed in a system similar to the Sagnac interferometer (Figure 6.3.4A). The probe pulse timing is matched to that of the pump 1 pulse. The pump 2 pulse is phase modulated by the EOM, PEM, vibrated mirror (e.g., DM in Figure 6.3.4A), and so forth. A lens pair is inserted into the opposing beams, and the focuses of the lenses are adjusted so that the focal points overlap in the sample. The pulse timing of the pump beams is adjusted by translating the lens pair and sample along the optical

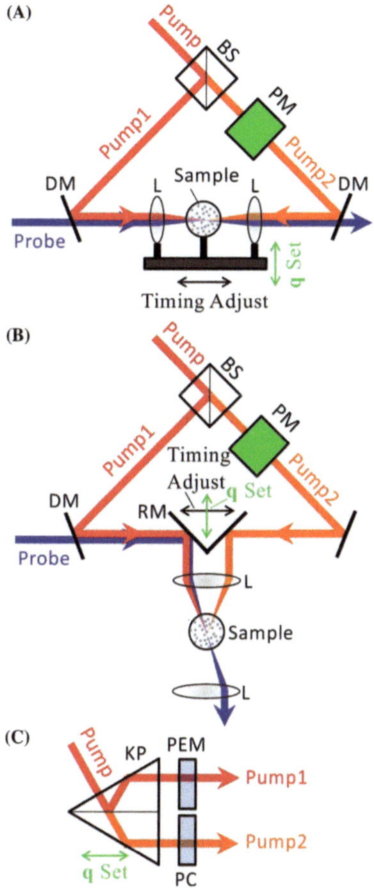

FIGURE 6.3.4 Possible optics for NLDLS-SRS. (A) is a system for a large q, and (B) is for a small q. (C) depicts other optics for the displaced parallel pump beams in (B). Here, BS is a beam splitter, PM is a phase modulator, DM is a dichroic mirror, L is a focusing lens, RM is a right-angle mirror, KP is a Koester prism, PEM is a photoelastic modulator, and PC is a phase compensator. The copyright of this figure is shared between the representative author (Takayoshi Kobayashi) and John Wiley & Sons.

axis. The q is set by moving them perpendicular to the optical axis. This optical system is appropriate for a large q.

For a small q, a right-angled mirror (RM) is inserted in the optical axis of the opposed pump beams to obtain parallel and displaced beams. The parallel beams are focused in the sample by a common lens. The pulse timing and q are adjusted by translating the RM. The displaced parallel beams are also prepared with the Koester prism (KP) instead of the beam splitter and RM (Fig. 6.3.4C). The phase of one beam is modulated by the PEM, while the pulse timing of the other beam is compensated by a phase compensator (e.g., a glass plate) for the PEM. The beam displacement is adjusted by translating the Koester prism to set the desired q.

To vary q significantly in a single optical system, one can transmit the split pump beam with a polarization maintaining optical fiber. To set the incident angle for a desired q, the condenser system is installed on a goniometer or similar mechanics. The undesirable nonlinear effects in the optical fiber are suppressed by employing a large mode-field photonic crystal fiber, which was demonstrated for a CARS and SRS fiber optic probe (endoscope).[31]

5.2 Signal Intensity Estimation

The measurement with both intensity and phase modulations for the pump beam is possible in the present methodology. Specifically, the signal after the lock-in detection at the intensity modulation frequency is subsequently lock-in detected at the phase modulation frequency. In this case, both quantitative and fluctuation analysis is possible. Note that the quantitative analysis is complicated in DLS; thus, this configuration is advantageous. However, the dual modulation reduces the signal intensity. Since the subject of this work is the DLS-like measurement to measure fluctuations, let us discuss the signal intensity solely with the phase modulation instead of the dual modulation.

5.2.1 Time Domain Measurement

The autocorrelation function through the locked-in detection of the square of the absolute probe field is given in Eq. (66). The experimental autocorrelation of the photocurrent is given by further multiplication of the square of the instrument function. The instrument function involves the pulse width, pulse repetition frequency, N_A of the focusing system, and responsibility of the photodetector. The value of the Bessel function is maximal when the order 2n - 1 = 1, and the variable is m_{pm} ' 1.84. The value is $J_1(1.84)$ ' 0.582. Hereafter, we assume the case of lock-in detection at the fundamental frequency of the phase modulation with a phase modulation index of 1.84. To estimate the magnitude of the correlation function, we identify the property of the particulate dispersion, the pump beam, and the probe beam. In addition, we require the relationship between the average incident power and I_{pr} and jI_{12}j, as well as V, <N>, χ_p, D, and P(q).

Let us assume the sample is polystyrene particles dispersed in water. The pulse width (Δt) and pulse repetition frequency (f_{rep}) of the laser are considered 2.5 ps and 76 MHz, respectively. The wavelengths of the pump and probe beams are set at λ_{pu} = 800 nm and λ_{pr} = 643 nm, respectively, to observe the Raman loss at 3,054 cm^1 of polystyrene. For simplicity, the two pump beams and the probe beam are Gaussian beams, and the two pump beams are opposite each other in an optical system focused on N_A = 0.1 (Fig. 6.3.4A). The probe beam is coaxial with one of the pump beams. The q is calculated to be 2.08 10^7 m^1 since the refractive index of water is 1.33. The radius (a) of the particle is set at a similar value to 100 nm, which is the minimum detectable one in the literature [17]. The signal intensity is proportional to the

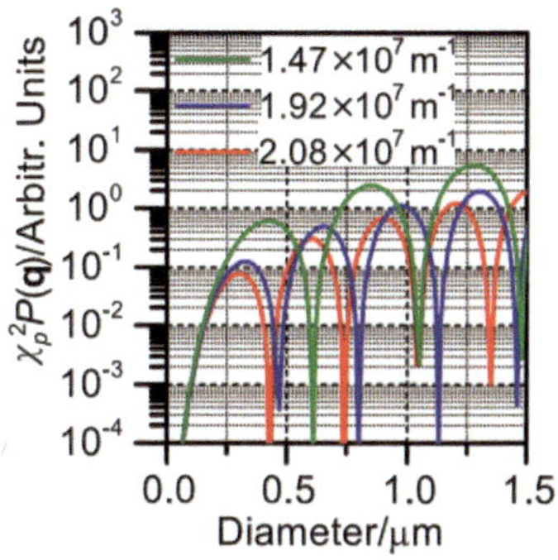

FIGURE 6.3.5 Product of $P(\mathbf{q})$ and χ^2_p for the spheric particle dispersed in water at 20°C. The values in the legend with units of m^{-1} are for $q=|\mathbf{q}|$. The green, blue, and red colored lines correspond to the cases that the 800-nm pump beams are crossed at 90°, 135°, and 180°, respectively. Note the dips actually reach zero, which cannot be illustrated in the logarithmic scale. The copyright of this figure is shared between the representative author (Takayoshi Kobayashi) and John Wiley & Sons.

product of P(q) and χ^2_p ($/a^6$). This product oscillatory increases with (a), taking null and maxima points (Figure 6.3.5). Thus, we set the (a) to 150 nm at the first maxima point. As in the literature [17], the volume fraction (V_{fr}) is assumed to be 2.5 vol%.

The probe intensity modulation by the SRS effect is induced in the region given by the product of the pump field intensity and the probe field intensity. Therefore, V is provided by the spatial distribution of the product of the pump and probe field intensities. Thus, V is estimated as a spheroid body around the optical axis. The minor axis diameter is defined as the beam waist radius of the product of the intensities, and the primary axis diameter is the corresponding Rayleigh length. The diameter $w_{0,pu,}$ pr at the beam waist, which are products of the pump and probe filed intensities, is given by

$$\frac{1}{w^2_{0,\mathrm{pu,pr}}} = \frac{1}{w^2_{0,\mathrm{pu}}} + \frac{1}{w^2_{0,\mathrm{pru}}}, \tag{80}$$

where the beam waist of the pump and probe beams are

$$w_{0,\mathrm{pu}} = \frac{\lambda_{\mathrm{pu}}}{\pi N_{\mathrm{A}}}, w_{0,\mathrm{pr}} = \frac{\lambda_{\mathrm{pr}}}{\pi N_{\mathrm{A}}}. \tag{81}$$

The Rayleigh lengths are given by

$$z_{\mathrm{R,pu}} = \frac{n_{\mathrm{pu}} \pi w^2_{0,\mathrm{pu}}}{\lambda_{\mathrm{pu}}}, z_{\mathrm{R,pr}} = \frac{n_{\mathrm{pr}} \pi w^2_{0,\mathrm{pr}}}{\lambda_{\mathrm{pr}}} \tag{82}$$

respectively. The primary axis diameter is estimated based on the equation,

$$\sqrt{2} = \left(\sqrt{1+\left(\frac{z_{\mathrm{R,pu,pr}}}{z_{\mathrm{R,pr}}}\right)^2}\right)\left(\sqrt{1+\left(\frac{z_{\mathrm{R,pu,pr}}}{z_{\mathrm{R,pu}}}\right)^2}\right). \tag{83}$$

Then V is

$$\begin{aligned} V &= \frac{4\pi}{3}\frac{w_{0,\mathrm{pu,pr}}^2}{2} z_{\mathrm{R,pu,pr}}, \\ &= 1.03 \times 10^{-16}\ \mathrm{m}^3 \end{aligned} \tag{84}$$

and

$$\begin{aligned} \langle N \rangle &= \frac{V}{\frac{4\pi}{3}a^3} V_{\mathrm{fr}}. \\ &= 1.82 \times 10^2 \end{aligned} \tag{85}$$

Then, we assess $I_{\mathrm{pr}} = |E_{\mathrm{pr}}|^2$ and $|I_{12}|^2 = |E_{\mathrm{pu1}}|^2 |E_{\mathrm{pu2}}|^2$. In general,

$$\bar{A} = \frac{1}{2}\varepsilon_0 cn\,|\,E\,|^2, \tag{86}$$

where $\bar{A}$ is the cycle-averaged intensity in a medium of refractive index n. For the given Δt and f_{rep},

$$\bar{P}_{\mathrm{pu}} = \pi \frac{w_{0,\mathrm{pu}}^2}{2} \bar{A}\, f_{\mathrm{rep}} \Delta t, \tag{87}$$

where $\overline{P_{\mathrm{pu}}}$ is the average input pump beam power. Therefore,

$$I_{\mathrm{pr}} = Z_{\mathrm{pr}}\, \bar{P}_{\mathrm{pr}}, \tag{88}$$

and

$$|I_{12}|^2 = Z_{\mathrm{pu}}^2\, \bar{P}_{\mathrm{pu1}}\, \bar{P}_{\mathrm{pu2}}, \tag{89}$$

defining the constants

$$Z_{\mathrm{pu}} := \frac{4}{\pi\varepsilon_0 c n_{\mathrm{pu}} w_{0,\mathrm{pu}}^2 f_{\mathrm{rep}} \Delta t}, Z_{\mathrm{pr}} := \frac{4}{\pi\varepsilon_0 c n_{\mathrm{pr}} w_{0,\mathrm{pr}}^2 f_{\mathrm{rep}} \Delta t}. \tag{90}$$

The χ_p is given by the product of the nonlinear polarizability density of the material and the volume of the particle. The polarizability density is estimated regarding the results (modulation depth of 2.3×10^{-4} at Raman shift[32] of 3,054 cm^{-1}) of our previous study [32]. The employed N_A was 0.85, and the average pump beam power was 50 mW, which was referred to in another previous study [33] of a similar optics system. Under these conditions and the assumption of the Gaussian beam profile, V and $|E_{pu}|^2$ are calculated. The nonlinear polarizability is 10^{-10} m^{-5} V^{-2} based on Eq. (34), with a refractive index of 1.58 for polystyrene. Therefore,

$$\chi_p = 5.9 \times 10^{-10} \frac{4\pi}{3} a^3 \left(m^{-2}\ V^{-2} \right). \tag{91}$$

Finally, D is given by Eq. (60), where η of water is $1.00\text{x}\ 10^{-3}$ Pas at 20°C.

Using the above values, the autocorrelation of the photocurrent [$i_{sig}(t)$] component of the sample is given by conversion of Eq. (66) into photocurrent:

$$\begin{aligned} i_{sig}(0) i_{sig}(\tau) &= S_{pr}^2 \frac{r(\tau)}{Z_{pr}^2} \\ &= 8 \left(S_{pr}\, \bar{P}_{pr}\, K\, J_1(1.84) Z_{pu} \right)^2 \bar{P}_{pu1}\, \bar{P}_{pu2}\, \chi_p^2 \langle N \rangle P(\mathbf{q}) \exp\left(-q^2 D\tau \right) \end{aligned}, \tag{92}$$

where S_{pr} is the responsibility of the detector for the probe beam. The typical value for 650 nm is 0.35

A/W. Under the condition of $\bar{P}_{pu1} = \bar{P}_{pu2} = 25$ mW, $\bar{P}_{pr} = 1$ mW,

$$i_{sig}(0) i_{sig}(\tau) = 2.59 \times 10^{-18} \exp\left(-6.24 \times 10^2 \tau \right) \left(A^2 \right). \tag{93}$$

Then, we discuss the S/N. We define the noise as the error from the autocorrelation function obtained from an infinite number of measured values. In the shot-noise-limited measurement, the noise is composed of the shot noise of the probe light and the fluctuations of the sample. The variance of the photocurrent due to the shot noise is given by $2eS_{pr}\bar{P}_{pr}B$. When the lock-in detection phase is adjusted to the phase modulation phase, it is $eS_{pr}P_{pr}B$. Hereafter, we assume this lock-in phase adjustment. Here, e is the elementary charge, and B is the equivalent noise bandwidth. B is inversely proportional to the integration time and is determined by the time constant (τ_{lock}) and roll-off characteristic of the low-pass filter in the lock-in amplifier. For example, $B= (8\tau_{lock})^{-1}$ for a typical 12 dB/oct roll-off. A longer constant improves the S/N while slowing down the response speed of the lock-in detection; a too-long time constant makes it unable to detect fluctuations. According to the autocorrelation function of the sample, the time constant of the fluctuation is $(q^2D)^{-1}$. For the 12 dB/oct roll-off, τ_{lock} should be set shorter than $(38.1q^2D)^{-1}$ to obtain data with 90% accuracy in 1/10 of $(q^2D)^{-1}$ since the settling time at 90% is $3.81\tau_{lock}$.

The measurement time interval should be larger than τ_{lock}, so that memory between the data points due to the low-pass filter is negligible. The autocorrelation of the shot noise is zero or $eS_{pr}P_{pr}B$ for $\tau \neq 0$ or $\tau = 0$, respectively. For an ergodic sample such as the particulate dispersion, the experiment to determine the autocorrelation involves measuring and recording the photocurrent [$i_{pr;}$ lock(t)] for a finite period. Then, the recorded data are multiplied by themselves at the displaced time of lag (τ), and product terms are averaged to estimate the autocorrelation. This estimate is subject to error since it is obtained from a finite amount of data. Furthermore, the estimator at $\tau = 0$ is not unbiased because the autocorrelation of the shot noise is not zero at $\tau = 0$. Thus, we should eliminate data at $\tau = 0$ to avoid the bias error.

The $i_{pr;\,lock}(t)$ is given by

$$i_{\text{pr;lock}}(t) = i_{\text{sig;lock}}(t) + \delta i_{\text{sh;lock}}(t), \tag{94}$$

where $i_{sig;\,lock}(t)$ and $\delta i_{sh;\,lock}$ $\delta\, i_{sh}$(t) are the signal of the sample and shot noise current, respectively. Here, $i_{sig;}$ lock(t) is given by

$$i_{\text{sig:lock}}(t) = 4\,\mathrm{J}_1(1.84)KS_{\text{pr}}\,\bar{P}_{\text{pr}}\,\mathrm{Im}\left[I_{12}X(t,\mathbf{q})\right], \tag{95}$$

according to Eq. (65). The nature of the shot noise is

$$\left\langle \delta i_{\text{sh;lock}}(t)\right\rangle = 0, \tag{96}$$

and

$$\left\langle \delta i_{\text{sh;lock}}(t)\,\delta i_{\text{sh;lock}}(t+\tau)\right\rangle = \begin{cases} eS_{\text{pr}}\,\bar{P}_{\text{pr}}\,B; \tau = 0 \\ 0 \qquad ;\tau \neq 0 \end{cases}. \tag{97}$$

When $\tau \neq 0$, the variance of the estimator for the autocorrelation of the photocurrent is given by

$$\begin{aligned} &\left\langle \left[i_{\text{pr;lock}}(t)\, i_{\text{pr;lock}}(t+\tau) - i_{\text{pr;lock}}(t)\, i_{\text{pr;lock}}(t+\tau)\right]^2 \right\rangle \\ &= \left\langle \left\{ i_{\text{sig;lock}}(0)\, i_{\text{sig;lock}}(\tau)\right\}^2 \right\rangle - \left\langle i_{\text{sig;lock}}(0)\, i_{\text{sig;lock}}(\tau)^2 \right\rangle \\ &+ \left(eS_{\text{pr}}\,\bar{P}_{\text{pr}}\,B \right)^2 \\ &+ 2\left(eS_{\text{pr}}\,\bar{P}_{\text{pr}}\,B \right)\left\langle i^2_{\text{sig;lock}}(0)\right\rangle \end{aligned} \tag{98}$$

Here, the time-invariant nature of the sample and shot noise is accounted for. The combination of the first and second terms is the variance of the autocorrelation estimates, and the third and fourth terms are the influence of the shot noise. The overall noise is given by

$$\left\langle\left[i_{\mathrm{pr,\,lock}}(t)i_{\mathrm{pr,\,lock}}(t+\tau)-i_{\mathrm{pr;\,lock}}(t)i_{\mathrm{pr,\,lock}}(t+\tau)\right]^2\right\rangle$$
$$=64\left(S_{\mathrm{pr}}\,\bar{P}_{\mathrm{pr}}\,K\,\mathrm{J}_1(1.84)\right)^4\left[|I_{12}|^2\,\chi_{\mathrm{p}}^2P(\mathbf{q})\right]^2$$
$$\cdot\left[\langle N\rangle\langle N+5\rangle+\langle N\rangle\langle N+1\rangle\exp\left[-2q^2D\tau\right]\right], \quad (99)$$
$$+\left(eS_{\mathrm{pr}}\,\bar{P}_{\mathrm{pr}}\,B\right)^2$$
$$+16\left(eS_{\mathrm{pr}}\,\bar{P}_{\mathrm{pr}}\,B\right)\left(S_{\mathrm{pr}}\,\bar{P}_{\mathrm{pr}}\,K\,\mathrm{J}_1(1.84)\right)^2|I_{12}|^2\,\chi_{\mathrm{p}}^2\langle N\rangle P(\mathbf{q})$$

accounting for Equations (44), (51), (55)–(57), (61), and

$$i_{\mathrm{sig;lock}}^2(0)=8\left(S_{\mathrm{pr}}\,\bar{P}_{\mathrm{pr}}\,K\,\mathrm{J}_1(1.84)\right)^2|I_{12}|^2\,\chi_{\mathrm{p}}^2\langle N\rangle P(\mathbf{q}). \quad (100)$$

Then, the S/N in energy ratio is calculated to

$$(S/N)_{\mathrm{eng}}=\frac{\left\langle i_{\mathrm{pr,lock}}(t)i_{\mathrm{pr,lock}}(t+\tau)\right\rangle^2}{\left\langle\left[i_{\mathrm{pr,lock}}(t)i_{\mathrm{pr,lock}}(t+\tau)-i_{\mathrm{pr,lock}}(t)i_{\mathrm{pr,lock}}(t+\tau)\right]\right\rangle^2}$$
$$=\exp\left(-2q^2D\tau\right)\left[\begin{array}{l}\left(1+\dfrac{1}{\langle N\rangle}\right)\exp\left(-2q^2D\tau\right)+\left(1+\dfrac{5}{\langle N\rangle}\right)\\+\dfrac{\left(eS_{\mathrm{pr}}\,\bar{P}_{\mathrm{pr}}\,B\right)^2}{64\left(S_{\mathrm{pr}}\,\bar{P}_{\mathrm{pr}}\,K\,\mathrm{J}_1(1.84)\right)^4\left[Z_{\mathrm{pu}}^2\,\bar{P}_{\mathrm{pu1}}\,\bar{P}_{\mathrm{pu2}}\,\chi_{\mathrm{p}}^2P(\mathbf{q})\right]^2\langle N\rangle^2}\\+\dfrac{eB}{4S_{\mathrm{pr}}\,\bar{P}_{\mathrm{pr}}\left(K\,\mathrm{J}_1(1.84)\right)^2\left[Z_{\mathrm{pu}}^2\,\bar{P}_{\mathrm{pu1}}\,\bar{P}_{\mathrm{pu2}}\,\chi_{\mathrm{p}}^2P(\mathbf{q})\right]\langle N\rangle}\end{array}\right]^{-1}$$
$$(101)$$

per measurement data point. When the autocorrelation is averaged over N_{m} data points, the $(S/N)_{\mathrm{eng}}$ is enhanced by N_{m}. When the S/N is defined as the relative standard deviation (RSD), it is the square root of the $(S/N)_{\text{eng.}}$. As a rule of thumb, the required RSD for the limit of quantification is 10. This condition is written in

$$(S/N)_{\text{eng}}(\tau \to 0)\cdot N_{\text{m}} = N_M \left[2+\frac{6}{\langle N\rangle} + \frac{\left(eS_{\text{pr}}\,\bar{P}_{\text{pr}}\,B\right)^2}{64\left(S_{\text{pr}}\,\bar{P}_{\text{pr}}\,K\,\mathrm{J}_1(1.84)\right)^4\left[Z_{\text{pu}}^2\,\bar{P}_{\text{pu1}}\,\bar{P}_{\text{pu2}}\,\chi_{\text{p}}^2 P(\mathbf{q})\right]^2\langle N\rangle^2} + \frac{eB}{4S_{\text{pr}}\,\bar{P}_{\text{pr}}\left(K\,\mathrm{J}_1(1.84)\right)^2\left[Z_{\text{pu}}^2\,\bar{P}_{\text{pu1}}\,\bar{P}_{\text{pu2}}\,\chi_{\text{p}}^2 P(\mathbf{q})\right]\langle N\rangle}\right]^{-1} = 100 \tag{102}$$

The first and second terms originate in the fluctuation of the estimator, the third term is the shot noise, and the fourth term is the cross effect of the autocorrelation and shot noise. The third and fourth terms are proportional to

$$\left(\bar{P}_{\text{pr}}\,\bar{P}_{\text{pu1}}\,\bar{P}_{\text{pu2}}\right)^{-2} \text{ and } \left(\bar{P}_{\text{pr}}\,\bar{P}_{\text{pu1}}\,\bar{P}_{\text{pu2}}\right)^{-1}, \text{ respectively.}$$

According to Eq. (102), we can estimate the required data points for the averaging, and the measurement time requires $3.81\tau_{\text{lock}} \times N_{\text{m}}$ for the 12 dB/oct filter.

Under the assumed experimental condition, $\tau_{\text{lock}} = (38.1q^2D)^1 = 42$ μs, $B = (8\tau_{\text{lock}})^1 = 2.97$ kHz, and hence $N_{\text{m}} = 2.16\ 10^2$, where the fluctuation of the estimator of (2 + 6/<N>) is dominant. The minimum required averaging time is about 35 ms.

5.2.2 Frequency Domain Measurement

The PSD of the square of the absolute probe field is given by Equation (77), where the term similar to the DLS is the fourth one. For the identical spherical particles, $\text{PSD}_{X,q}$ is given by Eq. (74). This fourth term depicts the spectrum in which comb-like bands of the Lorenz function appear around the harmonics of the phase modulation frequency. The respective peak heights are the Bessel function values of $J_n(m_{\text{pm}})$. The maximum height is $J_1(m_{\text{pm}}) = 0.582$, where the phase modulation index of $m_{\text{pm}} = 1.84$. With the spectrum analyzer, the spectrum in the negative frequency region is folded back into positive frequency and added to the positive frequency component. Thus, the maximum band around ω_{pm} is given by the following equation:

$$8K^2 I_{\mathrm{pr}}^2 \left|I_{12}\right|^2\ \mathrm{J}_1^2(1.84)\cdot 2PSS_{X,\mathbf{q}}\left(\omega-\omega_{\mathrm{pm}}\right) = \frac{8}{\pi}K^2 I_{\mathrm{pr}}^2\left|I_{12}\right|^2\ \mathrm{J}_1^2(1.84)\langle N\rangle \chi_{\mathrm{p}}^2 P(\mathbf{q})\frac{2q^2D}{\left(q^2D\right)^2+\left(\omega-\omega_{\mathrm{pm}}\right)^2} \tag{103}$$

Under the assumed condition, the PSD of the photocurrent is

$$\begin{aligned}\frac{8}{\pi}&\left(S_{\mathrm{pr}}\,\bar{P}_{\mathrm{pr}}\,K\,\mathrm{J}_1(1.84)Z_{\mathrm{pu}}\right)^2\bar{P}_{\mathrm{pu1}}\,\bar{P}_{\mathrm{pu2}}\,\chi_{\mathrm{p}}^2\langle N\rangle P(\mathbf{q})\frac{2q^2D}{\left(q^2D\right)^2+\left(\omega-\omega_{\mathrm{pm}}\right)^2}\\ &=2.64\times10^{-21}\cdot\left[1+\left(\frac{\omega-\omega_{\mathrm{pm}}}{6.24\times10^2}\right)^2\right]^{-1}\left(\mathrm{A}^2\cdot\mathrm{s}\cdot\mathrm{rad}^{-1}\right)\\ &=1.66\times10^{-20}\cdot\left[1+\left(\frac{f-f_{\mathrm{pm}}}{9.93\times10^1}\right)^2\right]^{-1}\left(\mathrm{A}^2\cdot\mathrm{Hz}^{-1}\right)\end{aligned} \tag{104}$$

The spectrum analyzer measurement sets the detection bandwidth, called the resolution bandwidth (RBW), and the respective value is the power, which is the integral of the PSD over the RBW. Therefore, the RBW should be set sufficiently narrow compared to the bandwidth of $\mathrm{PSD}_{X,q}$. With a narrow enough RBW at B Hz, the measured spectrum is approximated to

$$1.66\times10^{-20}\cdot\left[1+\left(\frac{f-f_{\mathrm{pm}}}{9.93\times10^1}\right)^2\right]^{-1}B\left(\mathrm{A}^2\right) \tag{105}$$

Then, we discuss the S/N. The photocurrent from the photodetector $\left(i_{\mathrm{pr}}(t)\right)$ is the sum of the signal of the sample $\left(i_{\mathrm{sig}}(t)\right)$ and shot noise $\left(\delta i_{\mathrm{sh}}(t)\right)$:

$$i_{\mathrm{pr}}(t)=i_{\mathrm{sig}}(t)+\delta i_{\mathrm{sh}}(t) \tag{106}$$

where

$$\left\langle\delta i_{\mathrm{sh}}(t)\right\rangle=0, \tag{107}$$

and

$$\left\langle\delta i_{\mathrm{sh}}(t)\,\delta i_{\mathrm{sh}}(t+\tau)\right\rangle=2eS_{\mathrm{pr}}\,\bar{P}_{\mathrm{pr}}\,\delta(\tau). \tag{108}$$

The $\delta(t)$ is the Dirac delta function. The PSD, including the shot noise, is given by

$$\begin{aligned}\mathrm{PSD}(\omega) &= \frac{1}{2\pi}\int_{-\infty}^{\infty}\overline{i_{\mathrm{pr}}(t)\,i_{\mathrm{pr}}(t+\tau)}\exp(-i\omega\tau)\,\mathrm{d}\tau \\ &= PSD_{\mathrm{sig}}(\omega) + \frac{1}{2\pi}2eS_{\mathrm{pr}}\,\bar{P}_{\mathrm{pr}}\end{aligned}, \tag{109}$$

where $\mathrm{PSD}_{\mathrm{sig}}(\omega)$is Equation (104). The second term is the noise floor due to the shot noise. If the RBW is set to be inappropriately broad (this corresponds to a too-short accumulation time), the estimated power spectrum fluctuates, and the objective bands of the sample fluctuations are blurred. Therefore, we consider the S/N with a sufficiently narrow RBW; that is, we define the noise as the shot noise floor. The S/N is the ratio of Eq. (104) and the shot noise power of $2eS_{\mathrm{pr}}\,\bar{P}_{\mathrm{pr}}\,B$:

$$\frac{8\,S_{\mathrm{pr}}\,\bar{P}_{\mathrm{pr}}}{e}\left(K\,\mathrm{J}_1(1.84)Z_{\mathrm{pu}}\right)^2\bar{P}_{\mathrm{pu1}}\,\bar{P}_{\mathrm{pu2}}\,\chi_{\mathrm{p}}^2\langle N\rangle P(\mathbf{q})\frac{2q^2D}{\left(q^2D\right)^2+\left(\omega-\omega_{\mathrm{pm}}\right)^2}\propto \bar{P}_{\mathrm{pr}}\,\bar{P}_{\mathrm{pu1}}\,\bar{P}_{\mathrm{pu2}}, \tag{110}$$

and we find the S/N is independent of the RBW. To attain the quantitation limit, the S/N defined by this power ratio should be greater than 100. Under the assumed experimental conditions, the S/N is 148, beyond the quantitation limit.

To further improve the S/N, reducing the RBW, which is equivalent to signal accumulation, is ineffective. Instead, we can only increase the power of the probe and pump beams. This contrasts time-domain measurements, where the S/N improves by increasing the number of data points. This difference is because the autocorrelation of the shot noise becomes the delta function of τ. The shot noise causes the bias errors in autocorrelation estimate at $\tau = 0$, whereas it raises the noise floor in the spectral domain. Since measurements in the spectral domain cannot exclude the autocorrelation estimate at $\tau = 0$, the S/N is independent of the RBW.

6 COMPARISON AMONG FCS, DLS, AND NLDLS-SRS

Every method typically evaluates the diffusion constant and drift velocity in the particle dispersion system.[1] Thus, let us summarize the characteristics.

The probes to detect the fluctuation differ in all methods, which is an essential characteristic. The probes in FCS, DLS, and NLDLS-SRS are fluorescent particles, everything that scatters from particles of different refractive indexes from the medium, and molecular vibration, respectively. As a result, FCS has chemical sensitivity identified by fluorescence. However, for non-fluorescent particles, it requires fluorescence labeling. DLS does not require fluorescence labeling but instead has no chemical sensitivity. NLDLS-SRS has chemical sensitivity without labeling.

Concerning the sample fluctuation, the relationship between the particle motion and the detected signal differs. This leads to differences in the applicable average number of particles (<N>) in the detection volume.

In FCS, the diffusion constant and drift velocity are estimated from the fluctuation of the particle number in a tightly focused illumination volume. Therefore, if <N> is large, the fluctuations will be buried in the average signal. For example, assuming the particles in the illumination volume are statistically independent, the particle number follows a Poisson distribution; that is, the standard deviation is $\sqrt{N}$. If <N> = 100, a 1% fluorescence fluctuation must be detected accurately. Therefore, <N> typically should be less than a few.

DLS detects the spatial Fourier component of the refractive index fluctuation. When $q \neq 0$, the autocorrelation is proportional to <N> and not affected by the average contribution of <N>^{2} since there is no contribution from the average refractive index proportional to <N>. Therefore, the signal intensity enhances along with <N>. However, the conventional DLS assumes single scattering; in other words, if the difference in the refractive indices of the particle and medium is significant, and <N> is large enough to induce multiple scattering, the theory for DLS cannot be applied as it is.

The signal as it is in NDLS-SRS contains both the particle number fluctuations within the illumination volume [the second term in Eq. (62)], as in FCS, and the spatial Fourier component (the third term), as in DLS. The conventional SRS method can also detect the former quantity of the particle number fluctuation in the illumination volume. Here, the latter spatial Fourier component, which is the objective of this chapter, is extracted by the phase modulation of the pump beam. As in DLS, the signal intensity of this component improves as <N> increases. Here, employing the confocal optical system, the probe beam unaffected by the multiple scattering processes is dominantly detected. Additionally, the phase modulation allows the detection of the SRS effect due to the standing wave of the interfering pump beams without much effect of the scattering. Therefore, the influence of multiple scattering may be mitigated compared to that in DLS.

7 CONCLUSIONS

To realize DLS that has chemical-structure informativity, we discussed the feasibility of the "nonlinear light scattering" method based on the SRS effect and the theory of its detection method. In DLS, the spatial Fourier component of the linear susceptibility is detected. Noting that the SRS effect is proportional to the intensity of the pump field, the present methodology employs the standing wave induced by the interference of the two pump beams. The sinusoidal spatial modulation of the intensity of the standing wave allows access to the spatial Fourier component of the imaginary part of the nonlinear susceptibility of the SRS effect [Eq. (35)]. Here, the difference between the wavevectors of the two pump beams corresponds to the scattering vector in DLS.

Furthermore, a monodisperse spherical particle dispersion was examined as a specific and basic application system. The signal includes the SRS signals due to the average concentration and concentration fluctuations in the interaction volume, as well. The autocorrelation derived from the concentration fluctuation and the spatial Fourier component is overwhelmed in the average concentration signal at a high concentration [Eq. (62)]. Also, in the PSD, the component of q= 0 overlaps that of desired q [Eq. (70)], which may complicate analysis. However, as depicted in Equations (64) and (77), the spatial Fourier component can be selectively extracted by phase modulation of the pump beam to modulate the position of the nodes and antinode of the standing wave.

In this method, the resources of analytical methods that have been accumulated in DLS can be referred to for mesostructure analysis. An example of a study of the particle interaction is found in the literature [34]. In this literature, the particle–particle (protein) interaction is considered a two-body problem, where the excluded volume effect, Coulomb forces, and dispersion forces are considered (known as the DLVO theory). The literature presents the analytical calculation of the ensemble averaging of the theoretical result [30] for a bulk suspension of identical particles. The two-body interparticle potential is determined based on the theory and dependences on the volume fraction, ionic strength, and temperature. Another example is ionic particles [1] and ionic liquids [35], which interact with each other due to Coulomb forces. The observed time constant of the autocorrelation function is dependent on q. The q dependence reflects that the motion is observed in the mean force field when the q is small, and the spatial region is large. NLDLS-SRS can distinguish between particles by vibrational bands, and exciting results are expected.

Not only the particulate dispersion systems discussed in this chapter but also colloidal systems such as polymer solutions and gels can be composition-specifically examined like DLS. Furthermore, since Raman spectroscopy allows analyzing molecular orientation [36–38], conformation [39–41], and hydrogen bonding [42, 43], it is expected to reveal the molecular orientation, molecular structure, and hydrogen bonding in connection with the mesostructure in the colloidal systems. Vibrational spectroscopy provides the analysis of the local structure around molecules, while the present methodology extends the range of spatial information that vibrational spectroscopy detects, in other words.

REFERENCES

1. B. J. Berne, R. Pecora, *Dynamic Light Scattering: With Applications to Chemistry, Biology, and Physics*, 2nd ed., Dover Publications, New York, 2000.
2. B. Chu, *Laser Light Scattering: Basic Principles and Practice*, 2nd ed., Dover Publications, New York, 1991.
3. P. M. Carvalho, M. R. Felício, N. C. Santos, S. Gonçalves, M. M. Domingues, *Front. Chem.* 2018, 6, 6.
4. B. R. Ware, W. H. Flygare, The simultaneous measurement of the electrophoretic mobility and diffusion coefficient in bovine serum albumin solutions by light scattering. *Chem. Phys. Lett.* 1971, 12, 81.
5. Y. Tsunashima, N. Nemoto, M. Kurata, Dynamic light scattering studies of polymer solutions. 1. Histogram analysis of internal motions. *Macromolecules* 1983, 16, 584.
6. Y. Tsunashima, N. Nemoto, M. Kurata, Dynamic light scattering studies of polymer solutions. 2. Translational diffusion and intramolecular motions of polystyrene in dilute solutions at the θ temperature. *Macromolecules* 1983, 16, 1184.

7. T. Matsunaga, M. Shibayama, Gel point determination of gelatin hydrogels by dynamic light scattering and rheological measurements. *Phys. Rev. E* 2007, 76, 030401.
8. T. Kureha, K. Hayashi, M. Ohira, X. Li, M. Shibayama, Dynamic Fluctuations of Thermoresponsive Poly(oligo-ethylene glycol methyl ether methacrylate)-Based Hydrogels Investigated by Dynamic Light Scattering. *Macromolecules* 2018, 51, 8932.
9. V. Nigro, F. Ripanti, R. Angelini, A. Sarra, M. Bertoldo, E. Buratti, P. Postorino, B. Ruzicka, Molecular mechanisms driving the microgels behaviour: A Raman spectroscopy and dynamic light scattering study. *J. Mol. Liq.* 2019, 284, 718.
10. E. L. Elson, D. Magde, Fluorescence correlation spectroscopy. I. Conceptual basis and theory. *Biopolymers* 1974, 13, 1.
11. M. Ehrenberg, R. Rigler, Rotational brownian motion and fluorescence intensify fluctuations. *Chem. Phys.* 1974, 4, 390.
12. S. R. Aragon, R. Pecora, Fluorescence correlation spectroscopy as a probe of molecular dynamics. *J. Chem. Phys.* 1976, 64, 1791.
13. L. Yu, Y. Lei, Y. Ma, M. Liu, J. Zheng, D. Dan, P. Gao, *Front. Phys.* 2021, 9, 9.
14. M. Oura, J. Yamamoto, H. Ishikawa, S. Mikuni, R. Fukushima, M. Kinjo, Polarization-dependent fluorescence correlation spectroscopy for studying structural properties of proteins in living cell. *Sci. Rep.* 2016, 6, 31091.
15. R. F. Wissner, A. Steinauer, S. L. Knox, A. D. Thompson, A. Schepartz, *ACS Cent. Sci.* 2018, 4, 1379.
16. W. Schrof, J. F. Klingler, S. Rozouvan, D. Horn, Raman correlation spectroscopy: A method for studying chemical composition and dynamics of disperse systems. *Phys. Rev. E* 1998, 57, R2523.
17. A. Barbara, F. Dubois, P. Quémerais, Non-resonant and non-enhanced Raman Correlation Spectroscopy. *L. Eng, Opt. Express* 2013, 21, 15418.
18. Y. Ozeki, K. Itoh, *Laser Phys.* 2010, 20, 1114. © 2023 The Authors. Journal of Raman Spectroscopy published by John Wiley & Sons Ltd.
19. T. Hellerer, A. Schiller, G. Jung, A. Zumbusch, Coherent Anti-Stokes Raman Scattering (CARS) Correlation Spectroscopy. *Chem. Phys. Chem.* 2002, 3, 630.
20. J.-X. Cheng, E. O. Potma, S. X. Xie, Coherent Anti-Stokes Raman Scattering Correlation Spectroscopy: Probing Dynamical Processes with Chemical Selectivity. *J. Phys. Chem. A* 2002, 106, 8561.
21. K. A. Bailey, Z. D. Schultz, Tracking Bulk and Interfacial Diffusion Using Multiplex Coherent Anti-Stokes Raman Scattering Correlation Spectroscopy. *J. Phys. Chem. B* 2016, 120, 6819.
22. M. D. Rabasovic, E. Sisamakis, S. Wennmalm, J. Widengren, Label-Free Fluctuation Spectroscopy Based on Coherent Anti-Stokes Raman Scattering from Bulk Water Molecules. *Chem. Phys. Chem.* 2016, 17, 1025.
23. Y. Li, B. Shen, S. Li, Y. Zhao, J. Qu, L. Liu, Review of Stimulated Raman Scattering Microscopy Techniques and Applications in the Biosciences. *Adv. Biol.* 2021, 5, 2000184.
24. Y. Ozeki, F. Dake, S. Kajiyama, K. Fukui, K. Itoh, Analysis and experimental assessment of the sensitivity of stimulated Raman scattering microscopy. *Opt. Express* 2009, 17, 3651.
25. G. L. Eesley, *Coherent Raman Spectroscopy*, Pergamon Press, Oxford, 1981.
26. D. L. Mills, *Nonlinear Optics: Basic Concepts*, Springer, Berlin Heidelberg 1998.
27. S. L. Miller, D. Childers, *Probability and Random Processes*, 2nd ed., Academic Press, Boston 2012.
28. H. L. Offerhaus, E. T. Garbacik, A. C. W. Rhijn, A. L. Fussell, J. L. Herek, *Rev. Anal. Chem.* 2012, 31, 1.
29. F. Ross Hallett, *Food Res. Int.* 1994, 27, 195.

30. G. D. J. Phillies, Dynamics of Brownian Probes in the Presence of Mobile or Static Obstacles. *J. Phys. Chem.* 1995, 99, 4265.
31. I. Latka, S. Dochow, C. Krafft, B. Dietzek, J. Popp, *Laser Photon. Rev.* 2013, 7, 698.
32. K. Seto, Y. Okuda, E. Tokunaga, T. Kobayashi, Multiplex stimulated Raman imaging with white probe-light from a photonic-crystal fiber and with multi-wavelength balanced detection. *J. Phys. D: Appl. Phys.* 2014, 47, 345401.
33. K. Seto, Y. Okuda, E. Tokunaga, T. Kobayashi, Development of a multiplex stimulated Raman microscope for spectral imaging through multi-channel lock-in detection. *Rev. Sci. Instrum.* 2013, 84, 083705.
34. S. Li, D. Xing, J. Li, Dynamic light scattering application to study protein interactions in electrolyte solutions. *J. Biol. Phys.* 2004, 30, 313.
35. M. Piszko, C. J. Kankanamge, M. H. Rausch, T. Klein, A. P. Fröba, Dynamic Light Scattering for Studying Mutual Diffusion Coefficients in Electrolyte Systems Comprised Entirely of Ions. *J. Electrochem. Soc.* 2020, 167, 133502.
36. K. Seto, Y. Furukawa, Study on solid structure of pentacene thin films using Raman imaging. *J. Raman Spectrosc.* 2012, 43, 2015.
37. M. Tanaka, R. J. Young, Polarised Raman spectroscopy for the study of molecular orientation distributions in polymers. *J. Mater. Sci.* 2006, 41, 963.
38. T. V. Basova, B. A. Kolesov, A. G. Gürek, V. Ahsen, Raman polarization study of the film orientation of liquid crystalline NiPc. *Thin Solid Films* 2001, 385, 246.
39. O. Hirofumi, O. Masataka, K. Teizo, M. Tatsuo, *Bull. Chem. Soc. Jap.* 1974, 47, 1075.
40. K. Seto, Y. Hosoi, Y. Furukawa, Raman spectra of Langmuir–Blodgett and Langmuir–Schaefer films of polydiacetylene prepared from 10,12-pentacosadiynoic acid. *Chem. Phys. Lett.* 2007, 444, 328.
41. H. Abe, H. Kishimura, M. Uruichi, A phase variety of fluorinated ionic liquids: Molecular conformational and crystal polymorph. *Spectrochim. Acta A: Mol. Biomol. Spectrosc.* 2023, 286, 121948.
42. A. Singh, D. Gangopadhyay, R. Nandi, P. Sharma, R. K. Singh, Raman signatures of strong and weak hydrogen bonds in binary mixtures of phenol with acetonitrile, benzene and orthodichlorobenzene. *J. Raman Spectrosc.* 2016, 47, 712.
43. Y. Ling, W. C. Xie, G. K. Liu, R. W. Yan, D. Y. Wu, *J. Tang, Sci. Rep.* 2016, 6, 31981.
44. K. Seto, T. Kobayashi, E. Tokunaga, *J. Raman. Spectrosc.* 2023, 54(7), 719. https://doi.org/10.1002/jrs.653410974555, 2023, 7, Downloaded from https://analyticalsciencejournals.onlinelibrary.wiley.com/doi/10.1002/jr

Index

Note: The page numbers with letter n denotes note numbers.

D

Q

R

S

W

X

Y

Z